Enzymatic and Model Carboxylation and Reduction Reactions for Carbon Dioxide Utilization

NATO ASI Series

Advanced Science Institutes Series

A Series presenting the results of activities sponsored by the NATO Science Committee, which aims at the dissemination of advanced scientific and technological knowledge, with a view to strengthening links between scientific communities.

The Series is published by an international board of publishers in conjunction with the NATO Scientific Affairs Division

A Life Sciences	Plenum Publishing Corporation
B Physics	London and New York
C Mathematical and Physical Sciences	Kluwer Academic Publishers Dordrecht, Boston and London
D Behavioural and Social Sciences	
E Applied Sciences	
F Computer and Systems Sciences	Springer-Verlag
G Ecological Sciences	Berlin, Heidelberg, New York, London,
H Cell Biology	Paris and Tokyo

Series C: Mathematical and Physical Sciences - Vol. 314

Enzymatic and Model Carboxylation and Reduction Reactions for Carbon Dioxide Utilization

edited by

M. Aresta

Department of Chemistry,
University of Bari,
Bari, Italy

and

J. V. Schloss

Central Research & Development Department,
E.I. du Pont de Nemours & Co.,
Wilmington, Delaware, U.S.A.

Kluwer Academic Publishers

Dordrecht / Boston / London

Published in cooperation with NATO Scientific Affairs Division

Proceedings of the NATO Advanced Study Institute on
Enzymatic and Model Carboxylation and Reduction Reactions for Carbon Dioxide
Utilization
Riva dei Tessali, Italy
June 17–28, 1989

Library of Congress Cataloging in Publication Data

```
Enzymatic and model carboxylation and reduction reactions for carbon
   dioxide utilization / edited by M. Aresta and J.V. Schloss.
      p.    cm. -- (NATO ASI series. Series C, Mathematical and
   physical sciences ; vol. 314)
      papers from a NATO Advanced Study Institute summer school held
   June 17-28, 1989 in Ginosa , Italy.
      fIncludes bibliographical references and index.
      ISBN-13: 978-94-010-6783-6        e-ISBN-13:978-94-009-0663-1
      DOI: 10.1007/978-94-009-0663-1

      1. Carbonic anhydrase--congresses.   2. Carbonic dioxide-
   -Congresses.   I. Aresta, M. (Michele), 1940-   . II. Schloss, John
   V.  III. NATO Advanced Study Institute.   IV. Series: NATO ASI
   series.  Series C, Mathematical and physical sciences ; no. 314.
   QP613.C37E58   1990
   665.8'9--dc20                                              90-4968
```

ISBN-13: 978-94-010-6783-6

Published by Kluwer Academic Publishers,
P.O. Box 17, 3300 AA Dordrecht, The Netherlands.

Kluwer Academic Publishers incorporates the publishing programmes of
D. Reidel, Martinus Nijhoff, Dr W. Junk and MTP Press.

Sold and distributed in the U.S.A. and Canada
by Kluwer Academic Publishers,
101 Philip Drive, Norwell, MA 02061, U.S.A.

In all other countries, sold and distributed
by Kluwer Academic Publishers Group,
P.O. Box 322, 3300 AH Dordrecht, The Netherlands.

Printed on acid-free paper

To our Families

TABLE OF CONTENTS

x

Acknowledgements

We wish to thank NATO for having granted this ASI Summer School on

"Enzymatic and Model Carboxylation and Reduction Reactions for Carbon Dioxide Utilization".

which was the second one supported by NATO on the topic

"Carbon Dioxide Utilization".

The first one was organized in 1986 and held in Pugnochiuso (Vieste - Italy).
(Proceeding: Vol. 206 Series C, NATO ASI Series, D. Reidel Publishing).

The present School was held in *Riva dei Tessali* (Ginosa - Italy) at the *Golf Hotel* in June (17 - 28) 1989.

It represented a great opportunity for Chemists and Biochemists to discuss in detail reaction mechanisms related to carbon dioxide utilization and to suggest the "work we have ahead of us". All participants, Lecturers and Students, took a great advantage of its interdisciplinary approach. They also enjoyed the Golf Hotel hospitality.

We equally wish to express our gratitude to Prof. Carlo Fragale, Dr. Eugenio Quaranta, Dr. Ada Tommasi, Dr. Lisiena Simone and to Brunella Aresta for their precious assistance in the preparation of the scientific sessions and for having made ready the copious scientific material participants have used.

We wish to thank also:
the University of Bari and Prof. Attilio Alto, Rector; the Department of Chemistry - University of Bari; the Italian Chemical Society - Inorganic Chemistry Division; the National Research Council (CNR) - Rome and Prof. Romano Cipollini, President of the Chemistry Committee; ENEA; ENEL; the Regione Puglia; the Ente Provinciale del Turismo of Matera; the Ente Provinciale del Turismo of Lecce; Pergine Industries - Firenze; Perkin Elmer Italiana; the Azienda Vinicola Leone De Castris - Salice Salentino and Dr. Salvatore Leone De Castris; Dr. Angelo Dell'Atti - Campi Salentina for the interest demonstrated and for helping in the organization of the School.

Utilization of Carbon Dioxide as a Source of Carbon : a Challenge of Great Intellectual Significance for Scientists.

Michele   Aresta,
Dipartimento di Chimica,Campus Universitario,
Università,70126 Bari, Italy.

Abstract.
Carbon dioxide is one of the gases responsible for the Greenhouse Effect (GHE)
Its emission into the atmosphere is estimated today to range around 5 Gt per year
as carbon. The forecast is that the rate at which it is accumulating in the
atmosphere will increase with time. Although an exact prediction of the
concentration is not possible, it is a general belief that the amount of carbon
dioxide discarged into the atmosphere by the year 2010 will range around 6.8 Gt
per year as carbon. A severe cut down of carbon dioxide discharge has recently
been agreed upon. Several strategies have been proposed for putting this plan
into action. A very stimulating one,from the intellectual point of view, is the
"recovery and utilization of carbon dioxide". This requires the collaboration of
those scientists working in the area of carbon dioxide reactions, considering
both natural and man made systems.
Chemists,biochemists and plant physiologists can really make a considerable
contribution to solving the problem of controlling the level of carbon dioxide in
the atmosphere.

1. Introduction

The emission into the atmosphere of those gases causing the
"Greenhouse Effect" (GHE) can produce a change in the average
seasonal temperature of our planet.[1]
The following activities can be considered as the main sources of GHE
gases :

Energy production (contributes 57% of the total amount of GHE
gases).
CFC use (17 %)
Agriculture (14%)
Land use (10%)
Others (2%).

M. Aresta and J. V. Schloss (eds.),
Enzymatic and Model Carboxylation and Reduction Reactions for Carbon Dioxide Utilization, 1–18.
© 1990 *Kluwer Academic Publishers.*

Carbon dioxide,which is generated in the following activities:
    Electric power generation (36%)
    Transportation (30 %)
    Industry (24%)
    Residences (10 %),
is one of the main gases responsible for the GHE and in fact for sure gives the highest relative contribution to global warming (Table 1). However the concentration of other gases in the atmosphere,namely the CFCs,has shown the greatest rate of growth in recent years. (Table 2) These gases also have the longest residence time. (Table 1).

TABLE 1
Percentage Contribution to the Global Warming.

Gas	Percent	Main Source	Time of residence (years)
Carbon dioxide	50	Coal (10) Gas (6) Oil (14) Cement and other (5) Land use and Deforestation (15)	7
Methane	18	Agriculture,livestock, biomass,natural gas	10
Dinitrogen oxide,N_2O	10	Biomass,transportation, fuel burning,fertilizers, natural emission.	180
Chlorofluorocarbons	13	Sprays,air conditioners, refrigerators,electronic cleaners,plastics,insulators.	Several centuries?
Ozone	15	Transportation means.	0.3
Others	5		

TABLE 2
Rate of Increase of the Concentration of GHE Gases

Gas	Rate of increase (%/year)
Dinitrogen oxide	0.2
Carbon dioxide	0.4
Methane	1.0
Ozone	1.5
Chlorofluorocarbons	5.5

Table 3 gives the estimated present and preindustrial era concentration of GHE gases in the atmosphere. It shows that some species that could be considered as absent in the atmosphere in the last century are now accumulating at an ever increasing rate and constitute a real problem nowadays. Obviously, GHE gases show a different efficacy towards the warming effect. It has been estimated that if carbon dioxide is taken as the reference molecule, the warming effect of methane is *ca*. 25 times higher while that of CFCs is 20 000 times higher.

TABLE 3
Variation Trend of the Concentration of GHE Gases since the Beginning of the Industrial Era.

Gas	Concentration/ppm	
	1800	1987
Carbon dioxide	275	358
Methane	0.7	1.8
Dinitrogen oxide	0.29	0.32
Chlorofluorocarbons	0.0	0.0006

2. How can the concentration of GHE gases in the atmosphere be limited ?

Scientists are facing the problem of how to contain the increase in concentration of these gases in order to avoid serious negative consequences for life on our planet.[2] Several strategies for reaching this goal have been proposed and a severe cut down in the amount of each GHE gas emitted has been recently agreed upon at an international level.[3] Indeed, the goal of limiting the emission of certain gases (CFC) is quite a realistic objective, considering the use that is made of them (refrigerators and air conditioning, propellents, plastics) and the fact that they could be substituted by less dangerous species or else different technologies might be used. On the other hand, the problem of limiting carbon dioxide emission is much more serious as this species is generated in most metabolic and combustion processes, all essential to man's energy needs and strictly related to the living standards.

As reported in the introduction, the total amount of carbon dioxide emitted comes from several sources. The figures comment on the real dimension of the problem of controlling the carbon dioxide emission.

A cut down of 20 % in the emission of carbon dioxide by the year 2010,as agreed at the 1987 Montreal Conference, is equivalent to a diminution of 40 % on the 6.8 Gt (as carbon) emission forecast for that year. This poses the serious problem of finding valid strategies to be operative in the short- and medium-term.
An interesting comprehensive analysis of the technologies we have at hand has been made at the IEA-OECD Expert Seminar (Paris,12-14 April 1989).[4]

A few short-, and medium-term strategies for controlling carbon dioxide emission can be envisaged. They can be categorized as follows :

2.1. Increases in the efficiency of those energy producing proces ses that use carbon based fuels.

As a practical example we can consider : amelioration of the combustion process and recovery of heat from flue gases.
This will bring about a drop in the use of combustibles and, consequently, a drop in carbon dioxide emission.

2.2. The utilization of the energy so produced with greater efficiency, reduction of any excess and avoidance of any misuse of energy.

Useful suggestions for the diminution of carbon dioxide emission are :
to increase the end-use efficiency for oil,coal,gas;
to improve the efficiency of energy generating plants;
to improve some industrial processes and essentially steel production;
to substitute industrial- and space-heating fuel oil by gas;
to substitute coal and oil by hydrogen;
to improve the efficiency of energy transfer;
to ameliorate building insulation so that the amount of oil used for heating could be reduced;.
to use electricity where and when necessary and to avoid its use for heating;.
to limit the speed of cars as this would result in a noticeable reduction of gasoline consumption and,thus,of carbon dioxide emission.

2.3. The possible use of alternative energy sources to complement the burning of carbon based fuels.

The use of nuclear fuel is,in this respect,of ecological relevance,but it poses other serious problems.
The utilization of other energy sources (geothermal,wind- and water-power,solar) are of great importance in order to reduce the emission of carbon dioxide without environmental pollution.

2.4. The production and use of energy from biomass.

This strategy may turn out to be quite important and it can also use carbon dioxide.

2.5. Reforestation.

The role of plants in fixing carbon dioxide sould not be underestimated.
Moreover,the cost of carbon dioxide fixing through plants is one of the lowest considering the technologies currently available,and thus makes this strategy a priority choice.

2.6. Recovery and disposal or utilization of carbon dioxide.

This technology appears to be quite expensive today.
It has been estimated that the cost of carbon dioxide removal ranges around 13 USA $ per t of carbon dioxide (see below). This cost must be taken into account when we decide about which strategy for reducing the carbon dioxide concentration in the atmosphere should have priority today.
Other technologies,for example reforestation,may cost 20 to 50 times less.
It is worth noting that,to date,recovery has been aimed at the disposal of carbon dioxide in the oceans or in natural wells and this contributes to make the net cost negative If carbon dioxide is utilized as a raw material in the chemical industry,then the whole economical aspect of its recovery must be reconsidered.

3. Recovery of carbon dioxide : is this a priority choice today?
The recovery of carbon dioxide can be accomplished in two different ways :

- from concentrated sources such as flue gases from power plants (ca. 20 % of carbon dioxide) and from industries (up to 99 % pure carbon dioxide).
- from diluted sources such as the atmosphere where carbon dioxide is present at a low concentration (0.03% in volume).

If we consider the carbon dioxide sources listed in §1, it can be seen that 60% of the total amount of emitted carbon dioxide comes from power plants (36% of the total) together with industrial processes (fermentation and other industrial activities, 24%) alone. These sources of "concentrated carbon dioxide" should be preferred if the recovery of carbon dioxide is to be economically viable. The carbon dioxide generated by transportation and by heating systems in residences, however, will continue to be emitted into the atmosphere as recovery is either not feasible or its cost is really too high. In this case, plants and algae could be used for fixing carbon dioxide from the atmosphere. The cost of the recovery of carbon dioxide from the flue gases of a power station has been evaluated considering both the use of (a) air [5] and of (b) pure dioxygen for burning carbon fuels.

(a) If air is used for burning carbon fuels,the recovery of carbon dioxide will require a technology for the separation of emitted CO_2 from dinitrogen, nitrogen- and sulfur-oxides (NOx and SOx).
This can be achieved using either :
(i) condensed phases as a trap for carbon dioxide (Table 4)
or
(ii) selective membranes.

TABLE 4.
Processes for Carbon Dioxide Recovery.

Trapping by condensed phases such as
- Basic water solutions
- Amines
- Alcohols and glycols
- Other organic solvents (organic carbonates,ethers)
- Basic oxides
- Molecular sieves

Separation using selective membranes

The first technology is by far the most used nowadays,the second is very promising and may be the technology of the future, once the technological problems related to the dinitrogen separation have been solved.
Among the condensed phases used for carbon dioxide separation, monoethanolamine ($HOCH_2CH_2NH_2$, MEA) is the most common .In these conditions the actual cost of recovery ranges around 9.4 USA $ per ton of carbon dioxide The utilization of selective membranes can reduce

the cost of carbon dioxide recovery,because of the lower cost of both recovery plants and chemicals.

(b) When dioxygen is used for the combustion of carbon based fuels, we must consider the cost of air distillation and, in order to be able to use the same combustion plants we use today, dioxygen should be mixed with emitted carbon dioxide prior to being used in the combustion. The flue gases obtained in this way consist essentially of carbon dioxide accompanied by impurities of NOx and SOx derived from fossil fuels. In this case carbon dioxide can be compressed directly into pipelines, with or without SOx and NOx elimination, according to its end use. Despite the air distillation cost, the cost of carbon dioxide recovery in this case is estimated to be also lower than that of the process that uses air for fuel combustion.

If the end fate of the carbon dioxide thus recovered is its disposal in natural gas fields or in the oceans, to these costs we must add the cost of a pipeline, of carbon dioxide pressurizing and pumping.

A reasonable forecast of the cost is 3.8 USA $ per ton of carbon dioxide when a 600 MW sized power plant is considered (such a plant would dispose 450 t of carbon dioxide per hour).[6]

The total cost for recovery and disposal ranges around 13.2 USA $ per ton of carbon dioxide and is a net negative cost, as there is no way of recovering the total investment, maintenance, depreciation and interest.

Considering the cost of electricity produced, the recovery and disposal of carbon dioxide will cause an increase in the price of 0.01 USA $ per kWh.

The question arises : who will pay this cost? Should it be shared by the electricity producer and the consumer? Is a tax reduction policy advisable?

These are political and social problems that concern governments and local administrators and we will not discuss those here.

However,these considerations allow a rough evaluation of the economi- cal impact of the recovery and disposal of carbon dioxide to be made and may help us to understand why they have not been extensively practiced up to now.

The limitation of the use of carbon based fuels or else the better utilization of the energy produced by these fossil fuels are,thus,the first choice strategy for the short term,but they will not cover the total cut down of carbon dioxide emission agreed upon for 2010.

To conclude,if the recovery and disposal of carbon dioxide is to be practiced only if and when economically advantageous,we must expect

that they will be operative only when the negative effects of dischar-
ging carbon dioxide into the atmosphere have become more negative
than the cost of the recovery process. We are already at the very limit!
In any case,we have a means for making the cost of carbon dioxide
removal more profitable : this is to find ways of utilizing the recovered
carbon dioxide.

4. *The utilization of carbon dioxide.*

In these days only a few industrial processes based on carbon dioxide
are operated and these can be divided into two main categories :
i) Technological uses of carbon dioxide:
ii) The conversion of carbon dioxide (fixation of the entire molecule
    or reduction of carbon dioxide to other C1 molecules).

TABLE 5
Technological Applications of Carbon Dioxide (ca. 8 Mt per year) .

- Waste water treatment.
- Addition to beverages and drinks.
- Food packaging and freezing.
- Refrigerators.
- Extraction of active components from natural products.
- Extraction of oil.
- Welding,moulding.
- Fire-estinguishers.
- Sprays.

Tables 5 and 6 list the present uses of carbon dioxide.
The total amount used ranges around 100 Mt per year and the lifetime
of the end products (if any) of the processes, i.e. the length of time before they release the carbon dioxide once again into the ecosystem
varies considerably.

TABLE 6
Utilization of Carbon Dioxide in Synthetic Chemistry (ca. 90 Mt per year).

- Synthesis of urea (Fertilizers,animal feeding additive,resins).
- Organic carbonates (Polycarbonates,carbamates,other chemicals).
- Inorganic carbonates (First and second group elements,other salts
  used as pigments and in the glass industry etc,).
- Pharmaceuticals (Aspirin,others).
- Additive in the synthesis of methanol.

Table 7 indicates the sources of the carbon dioxide so used. Surprisingly, a considerable amount of carbon dioxide is extracted from natural wells,which can give quite pure (> 99%) carbon dioxide.

TABLE 7.
Industrial Sources of Carbon Dioxide

- Natural wells (high purity,>99%).
- Industrial processes (high purity,>99%).
- Fermentation processes (high purity,>99%).
- Cement manufacture.
- Fuels (coal,natural gas,oil,wood) burning.

The extraction of carbon dioxide can be in some cases the main source, depending on its availability in the country concerned. It is worth noting that the cost of extraction from natural wells ranges around 13.8 USA $ per ton. A considerable amount of carbon dioxide is also recovered from fermentation and other industrial processes that can give very pure CO_2. Its cost is roughly comparable with that of the extracted carbon dioxide. These figures need some comment as they appear to be comparable with the cost of the recovery of carbon dioxide from flue gases and the question might arise : why extract carbon dioxide from natural wells if we may recover it from flue gases at a cost which is very close to what we pay for the extraction?. The cost evaluation in the case of carbon dioxide recovery from flue gases has been made,as we have already seen,for a recycle of several hundreds ton of carbon dioxide per hour. This amount is something like 1.5 orders of magnitude higher than the usual productivity of a conventional extraction or treatment plant and this difference in size might push the two prices towards the same limit. But it must be pointed out that while extraction costs are for an operating technology, the recovery costs are evaluated for an untried technology and this means that they might be higher ort lower. Anyway, it is clear that the recovery of carbon dioxide from flue gases might enter into practice if we solve the problem of its final fate. At present production rate we cannot think about recovering it for use. But it seems obvious that if we find a way for using or converting large masses of carbon dioxide, then the cost of its recovery from flue gases (9.4 USA $ per ton, equivalent to 34.5 USA $ per ton of carbon) can be considered as the cost of a "bulk starting material". If the utilization of carbon dioxide is to be considered seriously, then we must analyze a few key points in order to evaluate the practicability of such hypothesis.

The reference points are:

1) The lifetime of the products obtained from carbon dioxide.

2) The amount of carbon dioxide we must use for the utilization to be an effective tool for carbon dioxide level control.

3) The source of the energy required in the conversion processes.

We shall analyze these points in order to have a clear picture of the potential application of the recovery and utilization of carbon dioxide in the short- and medium-term.

4.1 The lifetime of the products obtained from carbon dioxide.

The life of chemicals obtainable from carbon dioxide is one of the fundamental parameters for carbon dioxide utilization. As we have seen before, the final aim of the recovery of carbon dioxide can be either its disposal in oceans or natural wells or its recycling. Disposal in natural fields can be considered as a practice that eliminates carbon dioxide from the environment forever. In this respect, disposal in natural wells should be preferred to disposal in the oceans.

The fixation of carbon dioxide into chemicals poses the point that when we use that species we convert it into carbon dioxide. For this reason the "lifetime" of chemicals and goods obtainable from carbon dioxide is an essential parameter.

It is obvious that the fixation into chemicals having a lifetime of the order of decades or centuries can be considered as a solution to the problem of the carbon dioxide control.

Chemicals of the type of calcium carbonate are real "carbon dioxide fixing agents" as they can last for an unlimited time, at least if standard ambient conditions (298 K and 0.103 MPa) are considered. Table 8 gives the "lifetime" for some species obtained from carbon dioxide. "Lifetime" is here considered as the average time required for carbon dioxide release from a species when it comes into its typical utilization.

TABLE 8
Lifetime of Species Obtained from Carbon dDoxide

Classification	Lifetime	Example
Imperishable	>> Centuries	Calcium carbonate
Long life	Decades - centuries	Polymers
Short life	Months - year	Agrochemicals (urea,herbicides)
Very short life	< Hours	Pharmaceuticals, fuels, etc.

It is evident that to fix carbon dioxide into chemicals that have a life time of the order of "years" is not a definitive solution and cannot make a substantial contribution to the problem of the control of carbon dioxide concentration in the atmosphere,even if considerable amounts of carbon dioxide are used. This is the case in the synthesis of urea and of similar products (carbamates used as agrochemicals : pesticides, herbicides etc). These chemicals are back converted into carbon dioxide in a short time : apparently they are not useful tools for the control of carbon dioxide level. Nevertheless, it would be of great economic interest to develop new carbon dioxide based synthetic procedures for obtaining such chemicals especially if these procedures fulfill the following requirements:
- The use of carbon dioxide as the starting material in place of those toxic or more expensive species that are presently used;
- The use of more direct synthetic processes;
- The adoption of less drastic operative conditions.
In any case this would allow energy to be saved and would ensure safer working conditions in industry and laboratories.
The utilization of carbon dioxide in the synthesis of polymers such as polyuretanes and polycarbonates is of interest for the lifetime of these macromolecules. It is worth to note that some polyurethanes have good insu lating properties and they might contribute in two concurrent ways to the limitation of carbon dioxide emission in the atmosphere: in a direct way as carbon dioxide is fixed in the molecule;and in an indirect way as they might be used as insulators in buildings and contribute,thus, to reducing the amount of fossil fuels used for residence heating.
An alternative utilization of carbon dioxide is as a "hydrogen carrier" or as the starting material for the synthesis of fuels. In the first case carbon dioxide can be converted into formic acid,HCOOH, which is easily converted back to hydrogen and carbon dioxide in the presence of Pd . This peculiar utilization might be of interest also in the case of a "hydrogen based economy" as it would contribute to the solution of two of the key problems dealing with hydrogen : its storage and trans portation. To store and transport hydrogen as formic acid would reduce both the cost and risks.
The conversion of carbon dioxide into fuels poses the problems of the source of energy to be used for the conversion and of the amount of carbon dioxide to be used. In this case,the point of great interest is the rate of production of such fuels rather than the "lifetime of the product".

In fact,as far as fuels are considered they will not be evaluated for their "storage" ability, but we must think of the utilization of carbon dioxide in terms of a "man made cycle" (Fig. 1) that cooperates with the "natural cycle".

If the process becomes operative,then we should have the possibility of converting several Mt of carbon dioxide per year.

Fundamental parameters in the artificial cycle are,thus, : the rate of conversion of carbon dioxide and the source of the energy required in the reduction process.

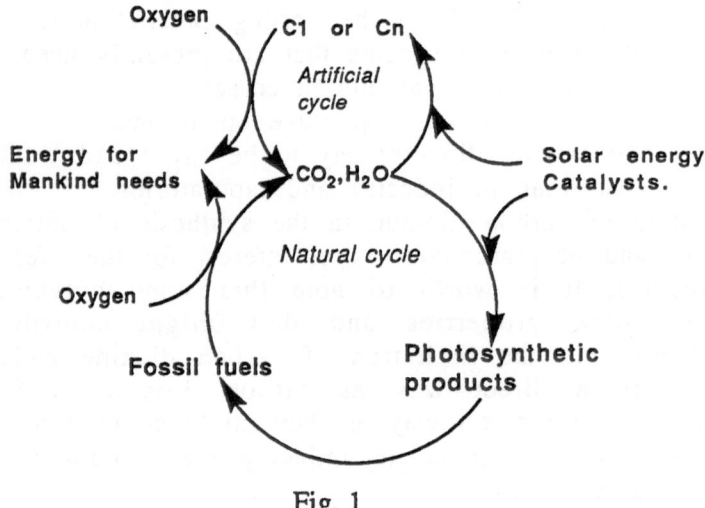

Fig. 1

4.2 How much carbon dioxide we should convert ?

To perform such an artificial photosynthetic cycle most probably would assure the conversion of huge amounts of carbon dioxide. The question of the amount of carbon dioxide we can fix or convert into other C1 or Cn molecules is the second key point to be considered if the utilization of carbon dioxide is to become a practicable technology for the control of the level in the atmosphere.

In fact,one of the proposals of the Montreal Conference was to dimi-nish by 40% the amount of carbon dioxide expected to be emitted in 2010, that means removing 2.7 Gt of carbon. If we pursue the hypothesis of the utilization, we should be able to fix an average amount of 270 Mt of carbon per year from now to 2010 this corresponds approximately to 1 Gt of carbon dioxide per year). At the moment we use $ca.$100 Mt of carbon dioxide per year, but the majority of it is used in technological applications and in the synthesis of chemicals with a short lifetime (see for example urea).

Table 9 gives an idea of the present trend and future expectation for the utilization of carbon dioxide together with an indication of the classes of chemicals obtainable.

TABLE 9
Estimated Amount of Carbon Dioxide used in Relation to the Classes of Species Obtained : Present Trend and Future Expectation.

Species	Present*	Future*
Carboxylated products	> kt	>> kt
Agrochemicals (ureas, carbamates etc.)§	> Mt	>> Mt
Polymers	> kt	>> Mt
New materials		>> Mt
Fuels °	> Mt	>> Mt

* Values are given per year. Each sign > indicates a factor of ten. § Carbon dioxide can substitute for phosgene in the synthesis of carbamates. ° At present carbon dioxide is used essentially as an additive for carbon monoxide in the synthesis of methanol and in Fischer Tropsch processes.

The main objective remains a shift to the synthesis of long lifetime chemicals and of fuels. The utilization of carbon dioxide in the synthesis of carboxylated products and of agrochemicals would mean releasing carbon dioxide into the atmosphere with a delay of only a few months. Nevertheless,this does not mean that we must not pursue the strategy of using carbon dioxide in the synthesis of fine chemicals. In fact, to succeed in this strategy would mean making operative more safe industrial processes by avoiding the use of toxic species, such as phosgene, that is used in the synthesis of carbamates and carbonates with a current amount of several Mt per year in the western world.[7] Still, even if fine chemicals cannot be synthesized in an unlimited amount using the process based on carbon dioxide, it is of interest from an economical point of view since such chemicals are of much greater value.[8]

4.3 Energetics of the processes based on carbon dioxide and the source of energy.

Carbon dioxide is, with water, the end product of metabolic pathways and of most combustion processes. Nevertheless, it cannot be considered an unreactive species and, in fact, it undergoes several reactions under very mild conditions provided that an electron rich species is available as the energy source. However, all the reactions in which carbon dioxide is involved can be considered as endoergic. The free energy content can be of the order of a few kJ or of hundreds of kJ, depending on the variation of the oxidation state that takes place in the process. In fact, all reactions must be considered as reduction reactions from the formal oxidation state of +4 to lower values. Table 10 gives values for the energy required in a few reduction processes in the presence or absence of hydrogen.

TABLE 10
Thermodynamic Data for Carbon Dioxide Conversion in other C1 Molecules., $\Delta G°/kJ$. (All species in the gas phase).

CO_2		=	CO	$+ 1/2\ O_2$		257.15
CO_2	$+\ H_2$	=	$HCOOH$			58.66
CO_2	$+\ 2H_2$	=	CH_2O	$+\ H_2O$		55.73
CO_2	$+\ H_2$	=	CO	$+\ H_2O$		28.49
CO_2	$+\ 3H_2$	=	CH_3OH	$+\ H_2O$		3.9
CO_2	$+\ 4H_2$	=	CH_4	$+\ 2H_2O$		-113.9

It is worth noting that there is a strong entropic effect in these reactions. In fact,for example,for the synthesis of methanol the $\Delta G°$ is -9.04 kJ when both methanol and water are considered in the liquid phase.Formic acid and methanol are, in my opinion, the most interesting spe cies as they can act as hydrogen carriers and can be considered, thus, as energy vectors. In this respect formaldehyde does not play a key role. Conversely, it is very important for the industrial applications it finds in the polymer industry that use considerable amounts of this species (several Mt per year in the 1980s). The synthesis of methane from carbon dioxide is an exergonic process but needs hydrogen. This process is accomplished in nature by anaerobic bacteria (Methanogenic bacteria) that can use hydrogen (or protons plus electrons) as the reducing agent.

This conversion is of interest for its biotechnological implications as it is used in the anaerobic digestion of various types of wastes to produce methane and is related to the biodegradation of wastes and to carbon recovery.

The carbon dioxide reduction to other C1 molecules requires, thus, energy and/or hydrogen. It seems obvious to think to the possibility of coupling this reaction with the water splitting reaction that generates hydrogen and oxygen from water. The overall reaction, after the utilization of dihydrogen in the synthesis of the hydrogenated C1 species, would be mimic of the photosynthetic reduction of carbon dioxide.

$$CO_2 \ + \ H_2O \ \xrightarrow{\text{Solar energy}} \ 1/n \ (CH_2O)_n \ + \ O_2$$

The utilization of water as the source of hydrogen and electrons is of fundamental importance for the carbon dioxide conversion could be considered as an operative strategy.

5. Can we use plants and algae for the carbon dioxide level control?

A fascinating approach to the control of carbon dioxide concentration in the atmosphere is the utilization of selected plants for carbon dioxide fixation. It is known that RuBisCO (Ribulose-Bisphosphate-Carboxy lase-Oxygenase), the most abundant enzyme in nature, present in green leaves of both C3 and C4 plants, is able to catalyze both the carboxy lation and oxidation of ribulose with a 50 % efficiency for each process. This enzyme is at present the target of intensive research by several research groups who are investigating the site-directed mutagenesis. The final purpose of this long-term research is to get a better RuBisCO that would fix carbon dioxide with a greater efficiency with respect to the natural one. The result would be highly efficient crop plants obtained through technological manipulation of the enzyme. This would allow plants to reduce the atmospheric carbon dioxide concentration more efficiently than observed at present.

This is a long-term strategy and it is difficult to make a forecast when or if it will be operating. At present, we have some algae which are able to fix carbon dioxide at a rate much higher (light efficiency utilization is *ca.* 5 times higher) than that commonly found in plants.

6 *The catalysts for carbon dioxide conversion.*

All reactions aimed at use of carbon dioxide need energy and a catalyst for the activation and fixation of the cumulene. Transition metal systems are good candidates for this role.

To date they have been used as chemical, electro-chemical, photo-chemical, photo-electro-chemical catalysts in a number of reactions[1a] that can be categorized as :

1) Carbon dioxide reduction (thermal, electrochemical and photochemical) to carbon monoxide.
2) Carbon dioxide reduction (chemical, electrochemical, photoelectroche mical) to other C1 molecules.
3) Carbon dioxide transfer to organic substrates (synthesis of carboxy lated products : lactones, carbonates, carbamates etc.).
4) Synthesis of polymers (polycarbonates, polyurethanes).
5) Synthesis of ureas and polyureas.
6) Other reactions.

The extension of the use of carbon dioxide from the actual level to that considered as appropriate for the utilization of carbon dioxide to become an operative strategy for the control of its level into the atmosphere requires a long term program and a great investment, both in terms of funds and of intellectual efforts,for developing new catalysts having a higher turnover number and higher selectivity,new ma terials and new technologies.

The utilization of solar energy seems to be a must in this case.

7. *Conclusions.*

Scheme 1 represents an integrated process concerning carbon dioxide : its recovery and possible modes of conversion.

It seems evident that an interdisciplinary approach is necessary : it does not seem likely that only one type of scientist will have the right solution at hand. The organization of a "task force" for solving the problem of carbon dioxide recovery and utilization seems to be likely as this is a strategy of great intellectual significance and can give a positive contribution to the solution of the problem of carbon dioxide level control. At present carbon dioxide is used in the synthesis of chemicals and in other industrial applications, but we are really far away from having explored all possibilities and exploited the potentiality.

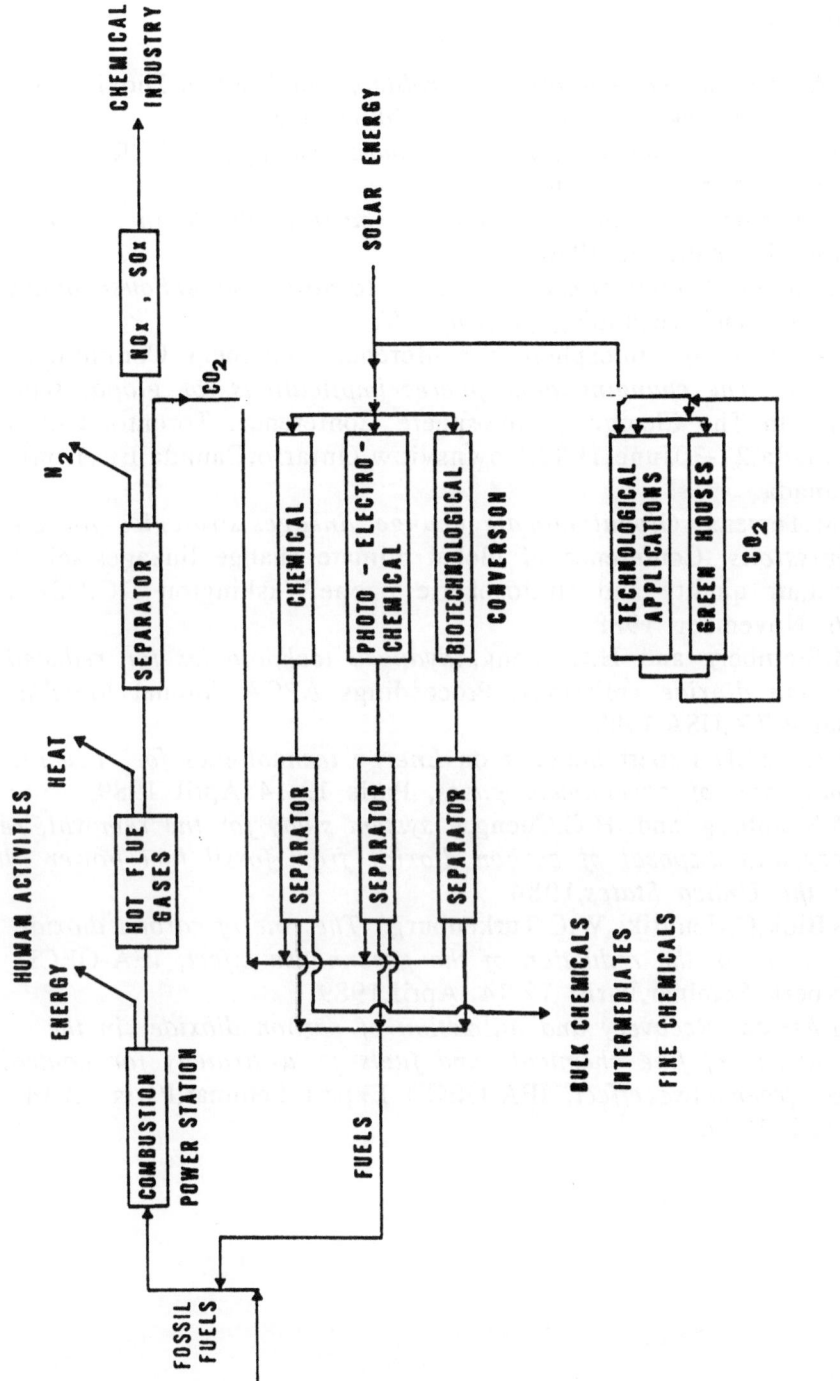

Scheme 1

18

References.

1 a) M.Aresta,*The carbon dioxide problem* , in "Carbon Dioxide as a
      Source of Carbon", Reidel Publ.,1986,p 1-22.
   b) I.M.Smith, *Carbon dioxide and climatic change*, IEACR/O7,London,
      1988,IEA coal research.
   c) M.Shepard,*The greenhouse effect : earth's climate in transition.*,
      EPRI Journal,USA 1986.
   d) J.Mathews,*Global climate change : toward a greenhouse policy.*,
      Science and Technology,3 ,1987, 57.
2)   The changing atmosphere : conference Statement Committee
      (1988). *The changing atmosphere: implications for global secu
      rity*. In The Changing Atmosphere Conference, Toronto, Ontario,
      Canada,27-30June,1988.Downsview,Ontario,Canada,Environment
      Canada.
3a)   I.M.Torrens, *Global climate change linkages;strategies for control
      emissions*. Conference of global climate change linkages;acid
      rain,air quality and stratospheric ozone,Washington DC,USA,15-
      16 November 1988.
   b) M.Steinberg and H.C.Cheng,*Advanced technologies for reduced
      carbon dioxide emissions*, Proceedings APCA Annual Meeting,
      Dallas,TX,USA,1988.
4)   IEA-OECD Expert Seminar on *Energy technologies for reducing
      emissions of greenhouse gases*, Paris 12-14 April 1989.
5)   M.Steinberg and H.C.Cheng,*A system study for the removal, reco
      very and disposal of carbon dioxide from fossil fuel power plants
      in the United States*,1984
6)   K.Blok,C.Hendriks,W.C.Turkenburg, *The role of carbon dioxide
      recovery in the reduction of the greenhouse effect*, IEA-OECD
      Expert Seminar,Paris 12-14 April,1989.
7)   M.Aresta, *Recovery and utilization of carbon dioxide :in the
      synthesis of fine chemicals and fuels : a strategy for controlling
      the greenhouse effect*, IEA-OECD Expert Seminar,Paris 12-14
      April 1989.

MARKET AND APPLICATION OF MERCHANT CO_2

PAOLO GRAZIANO
PERGINE S.p.A.
VIALE SPARTACO LAVAGNINI 42
50129 FIRENZE (ITALY)

INTRODUCTION

Over four hundred years have passed since Van Helmont discovered Carbon Dioxide, and just over a hundred years since it was first used in industry.

In fact, in 1888 Liquid Carbonic of Chicago, first used dry ice to cool horsedrawn vehicles delivering ice cream.

Since then world consumption of CO_2 has developed and has now reached an estimated value of 8 million tons; this consumption only refers to the so-called "merchant CO_2", i.e. CO_2 sold in thanks as a subcooled liquid, in cylinders or as a dry ice; it does not include the so-called "captive" CO_2, which is self-produced and reused as raw material in the production of chemical products, essentially urea.

Merchant CO_2 counts for approximately 20% of total consumption.

1) MARKET

The world market for merchant CO_2 is estimated at 8 million tons subdivided as follows:
- North America $4,5 \times 10^6$ tons
- Western Europe $1,5 \times 10^6$
- Eastern Europe 1×10^6
- Far East $0,5 \times 10^6$
- Developing Countries $0,5 \times 10^6$

The most precise data regard North America and Europe: it is very difficult to make an estimate regarding developing countries because in these countries the CO_2 is produced by individual bottling companies with their own oil burning CO_2 producer plants.

19

M. Aresta and J. V. Schloss (eds.),
Enzymatic and Model Carboxylation and Reduction Reactions for Carbon Dioxide Utilization, 19–22.
© 1990 *Kluwer Academic Publishers.*

In developing countries nearly all the CO_2 is used in the mineral water sector and/or for soft drinks.

In technically more advanced countries, CO_2 is used for widely different purposes. In North America the market is divided up as follows :
- Food freezing and chilling   40%
- Beverage carbonation   20%
- Chemical production   10%
- Metallurgy   10%
- Others   20%

In Western Europe, 45% óf CO_2 consumption is destined for mineral waters and soft drinks.

In Italy the market is at present divided up as follows :
- Beverage carbonation   26%
- Waste water treatment   23%
- Food freezing   13%
- Welding   10%
- Others   28%

Forecasts up to 1995 show a further reorganization of mineral waters of around 20%.

The present per-capita consumption is as follows :
- North America   18   kg per-capita
- Western Europe   3.6   "   "
- Italy   2.2   "   "

The market has a growth trend of between 5 and 10% per year and is being held back by a lack of raw material. Merchant CO_2 takes only a small fraction of the total carbon dioxide demand, = 20%, the other part is for urea and some inorganic carbonates.

2) SOURCES OF CO_2

Approximately 70% of merchant CO_2 comes from ammonia and/or hydrogen plants following the classic steam-reforming procedure, with subsequent gas-water shift conversion; the remaining 30% is almost exclusively obtained from mining.

In Italy the ratio between mined CO_2 and CO_2 of chemical origin is around 50/50.

The main Italian minerary fields are situated in Tuscany and North Latium; the CO_2 is very pure, the only impurity of any importance being

H_2S.

The typical features of dry unrefined CO_2 obtained from deposits in Tuscany are as follows :

- CO_2 97-98%
- Inerts   2-3%
- Hydrocarbons   200-500 ppm
- H_2S 20-200 ppm

The pressure varies between 0 and 8-20 bars, and the temperature between 15 and 30° C.

A typical CO_2 purification and liquefaction plant consists of a section for the elimination of sulphur compounds which can be carried out either through activated carbon or absorption in a solution (normally copper sulphate), drying on aluminia beds followed by compression at 20 bars and liquefaction.

There is another interesting production method that will soon be used in Italy, based on the recovery of the CO_2 obtained as a byproduct of the treatment of waste water from TiO_2 plants; these sulphate based acid muds are neutralized with finely ground $CaCo_3$, forming a gypsum and realising almost pure CO_2 .

3) CO_2 APPLICATIONS

Carbon dioxide is an extremely versatile fluid and can be used in many different sectors by exploiting its physical features as a gas which is much heavier than hair, and its chemical and thermodynamic features as a vehicle for refrigeration.

CO_2 has traditionally had its most interesting applications in the food sector, in the gassing of mineral waters and soft drinks, the freezing and chilling of foodstuffs and the creation of controlled atmospheres, which make it possible to preserve the quality of food over a length of time.

One particular sector being studied at present is of exceptional interest and regards the use of CO_2 in the disinfection of wheat and foodstuffs in general.

This technique dates back to Roman times: in fact the Romans used to store wheat in caves where, due to fermentation , the atmosphere was rich in CO_2, with the subsequent protection of the wheat from attack. by insects.

Recently, the increasing problems caused by the use of pesticides and the residues of chemical products used for disinfection, such as phos-

phine and methyl bromide, have induced researchers to experiment with
CO_2 ; the infected foodstuffs are kept for a period varying from 3 to
10 days in a CO_2 atmosphere of between 60 and 80%. This technique is already being widely used in Australia and the United States, whilst it is
only at the initial stages in Europe.

This application represents a link between traditional applications in
the foodstuffs sector and the ecological and environmental conservation
applications being prepared today :

- the replacement of chlorofluorocarbons both as aerosol propellants as
well as polymer expanders,
- the replacement of sulphuric acid in the treatment of alkaline waste
water,
- the nobilitation of waste products from manufacturing industries, by
converting them into fertilizers and construction materials.

Lastly, it is worthy remembering the use of CO_2 in supercritical conditions as a solvent: there are already many varied industrial applications, such as the decaffeinization of coffee and the extraction of hops
in beer-making.

THEORETICAL STUDIES OF CARBON DIOXIDE ORGANOMETALLIC
REACTIVITY

A. Dedieu[*], C. Bo and F. Ingold

Laboratoire de Chimie Quantique
UPR 139 du CNRS
Université Louis Pasteur, Strasbourg (France)

Despite its inherent stability and its relatively inertness carbon dioxide has now been found to be potentially active in a large variety of organometallic reactions [1] and many investigations, both experimental and theoretical, have been performed in order to assess the various mechanisms by which it can be activated and/or react. Theoretical studies of this type have first focused on the coordination of CO_2 in its complexes, leading to a better understanding of the factors governing carbon dioxide bonding. Credit should be given here to some pioneering studies [2 − 4] particularly those of Sakaki et al. [4] from which a general picture of the bonding of CO_2 to transition metal complexes, involving both orbital and electrostatic interactions has emerged [4c] : In ML_nCO_2 complexes the η^2 coordination mode is favored by a high lying doubly occupied d_π orbital (with the additional requirement that one has no doubly occupied orbital of d_σ type also lying high in energy). The $\eta^1 − C$ coordination mode on the other hand is favored by a high lying doubly occupied orbital of d_σ type but may suffer from a quite unfavorable electrostatic interaction between the positively charged carbon atom of CO_2 and the metal atom which is most often positively charged . For that reason a high oxidation state of the metal is not desirable if one wants to achieve a $\eta^1 − C$ coordination .

The most recent theoretical studies undertaken either in our group or in the group of Sakaki are now more directed towards the organometallic reactivity of carbon dioxide [5 − 9], trying to delineate reaction ‑pathways either for the carbon

M. Aresta and J. V. Schloss (eds.),
Enzymatic and Model Carboxylation and Reduction Reactions for Carbon Dioxide Utilization, 23–42.
© 1990 Kluwer Academic Publishers.

dioxide attack at coordinated ligands, or for processes involving attack of coordinated CO_2. They usually rely on ab initio calculations carried out first at the SCF level but also at the MC-SCF, CI or Moller-Plesset (MP) levels in order to include electron correlation effects. We will review here our own work, our goal being to show how such studies can improve our understanding of the mechanisms of the reactions involving CO_2 or can add new perspective to the CO_2 activation problem. For complete information about the computational details (basis sets, geometries) we refer the reader to our original publications [5 − 7] .

THE CO_2 INSERTION INTO THE Cr-H BOND OF HCr(CO)$_5^-$

The first study of this type undertaken in our group was devoted to the insertion of an incoming CO_2 into the metal hydride bond of HCr(CO)$_5^-$. The reaction was known to occur quite readily [10 − 13] but there were still some uncertainties regarding its intimate mechanism. Experimental studies of Darensbourg and coworkers had shown that the insertion of CO_2 into the Cr-H bond of HCr(CO)$_5^-$ took place through a dissociative type mechanism [10 − 13] , involving probably the dissociation of an equatorial CO ligand. One might wonder however about the driving force for this mechanism since the insertion into the Cr − CH$_3$ bond of the analogous CH$_3$Cr(CO)$_5^-$ system was found to occur through an associative or an associative interchange type mechanism [14,15] . There was some hypothesis [11,12] that the dissociation of the ancillary CO equatorial ligand might be triggered by the formation of a [Cr(CO)$_5$H...CO$_2$]$^-$ adduct between HCr(CO)$_5^-$ and CO_2 acting as a Lewis base and a Lewis acid respectively, but no definite proof had been given. In fact for this reaction as for the CO_2 insertion into the Cu(I)-H bond which has been analyzed by Sakaki and Ohkubo [8] , the reaction path (or at least structures relevant to the reaction path) had not been characterized either structurally or electronically.

SCF calculations were first carried out for the [Cr(CO)$_5$H...CO$_2$] adduct, leading after a geometry optimization of the [Cr...HCO$_2$] unit to the structure sketched in 1 , Cr-H = 1.88 Å , C-H = 1.15 Å , O-C-O = 134.5°, Cr-H-C = 170° [6] .

1

The adduct was computed to be more stable than the separated reactants by 8.4 kcal/mol, once the so-called basis set superposition error had been taken into account through the counterpoise method [16]. It is known that this corrective procedure leads generally to an underestimation of the interaction energies and the value of 8.4 kcal/mol is probably a lower bound of the actual SCF interaction energy. One may nevertheless worry about the effect of neglecting the electron correlation. Somewhat analogous calculations carried out for acid base adducts of main group elements (such as Al$_2$H$_7^-$ or B$_2$H$_6^-$) and which also involve a bridging hydrogen indicate that the stabilization energy is not changed by more than a few kcal/mol when electron correlation is taken into account (the stabilization is in fact slightly increased) [17,18] . For the HCr(CO)$_5^-$ system however CAS SCF calculations which include non dynamical correlation effects reduce the negative charge of the hydride, thus making it slightly less prone to interact with CO$_2$ [19]. Calculations of this type which were carried out for the acid- base adduct and for the two reactants separated by 50 Å indeed decrease the stabilization energy by 3.1. and 3.6 kcal/mol for the σ and π correlation effects respectively [6] . On the other hand MP2 calculations which also account for the dynamical correlation effects have been performed by Sakaki and Ohkubo [8] for the CO$_2$ insertion into the Cu(I)-H bond of HCu(PH$_3$)$_2$. They point to a slight change of the energetics of the reaction up to the transition state (in the product region they decrease the exothermicity). The changes are more marked for the

HCu(PH$_3$)$_3$ reaction [8] . One may nevertheless reasonably suspect that some stabilization energy should remain for the [Cr(CO)$_5$H...CO$_2$]$^-$ adduct.

This stabilization and the $\eta^1 - C$ coordination mode have been traced to two types of interaction [6]. There is first a donor-acceptor interaction (see **2**) between the s$_H$ + d$_{z^2}$ doubly occupied orbital and the empty π^* orbital of CO$_2$ which is more

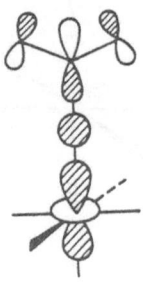

2

localized on the central carbon atom than on the terminal oxygen atoms. This is similar to what is observed in CO$_2$ transition metal complexes with a $\eta^1 - C$ coordination mode where the two electron interaction is between a doubly occupied d$_\sigma$ orbital and the π_{CO_2} orbital. The two electron stabilizing interaction is probably weaker in [Cr(CO)$_5$H...CO$_2$]$^-$ because of a less directional character of the s$_H$ + d$_{z^2}$ orbital. But in addition to the orbital type interaction, one finds in the [Cr(CO)$_5$H...CO$_2$]$^-$ adduct an <u>attractive</u> electrostatic interaction which has no counterpart in the $\eta^1 - C$ coordinated carbon dioxide complexes [4c,6] . Here the negatively charged hydride (-0.298 according to a Mulliken population analysis, see Table 1) interacts attractively with the positively charged (+ 0.728) carbon atom of CO$_2$.

It appeared to us that both the geometry and electronic distribution of the adduct **1** was very much like those of the formate product, whatever the isomer either **3a** or **3b** is : in **1** the optimized C-H bond length, 1.15 Å , is close to the value expected for the formate ligand [20] , whereas the Cr-H bond length is elongated by about 0.2 Å . The atomic charges of the HCO$_2$ unit in the adduct and in the formate isomers are also quite similar, see Table 1. All these features therefore suggest that the end of the

Table 1. Atomic charges for the reactants and product of the CO_2 insertion reaction into the Cr-H bond of $HCr(CO)_5^-$

	$HCr(CO)_5^-$ [a]	CO_2 [b]	$Cr(CO)_5(O_2CH)^-$ (3a)[c]	$Cr(CO)_5(O_2CH)^-$ (3b)[c]
Cr	- 0.184	-	+ 0.176 (+ 0.186)	+ 0.231
C_{eq}	+ 0.307	-	+ 0.322 (+ 0.316)	+ 0.297
O_{eq}	- 0.405	-	- 0.390 (- 0.384)	- 0.381
C_{ax}	+ 0.290	-	+ 0.298 (+ 0.298)	+ 0.297
O_{ax}	- 0.415	-	- 0.409 (- 0.402)	- 0.399
H	- 0.298	-	- 0.028 (- 0.036)	- 0.058
C	-	+ 0.728	+ 0.494 (+ 0.490)	+ 0.482
O_b	-	- 0.364	- 0.687 (- 0.683)	- 0.684
O_t	-	- 0.364	- 0.573 (- 0.579)	- 0.533

[a] Experimental geometry. [b] with the geometry of the adduct. [c] For a Cr-CO bond length of 2.0 Å (the values in parentheses refer to the optimized Cr-CO bond length of 2.18 Å).

3a   3b

insertion process should involve a gradual shift from **1** to **3a** or **3b**. There are of course various pathways for this rearrangement and the Scheme 1 sketches the two most plausible ones. The first pathway involves the dissociation of an equatorial CO

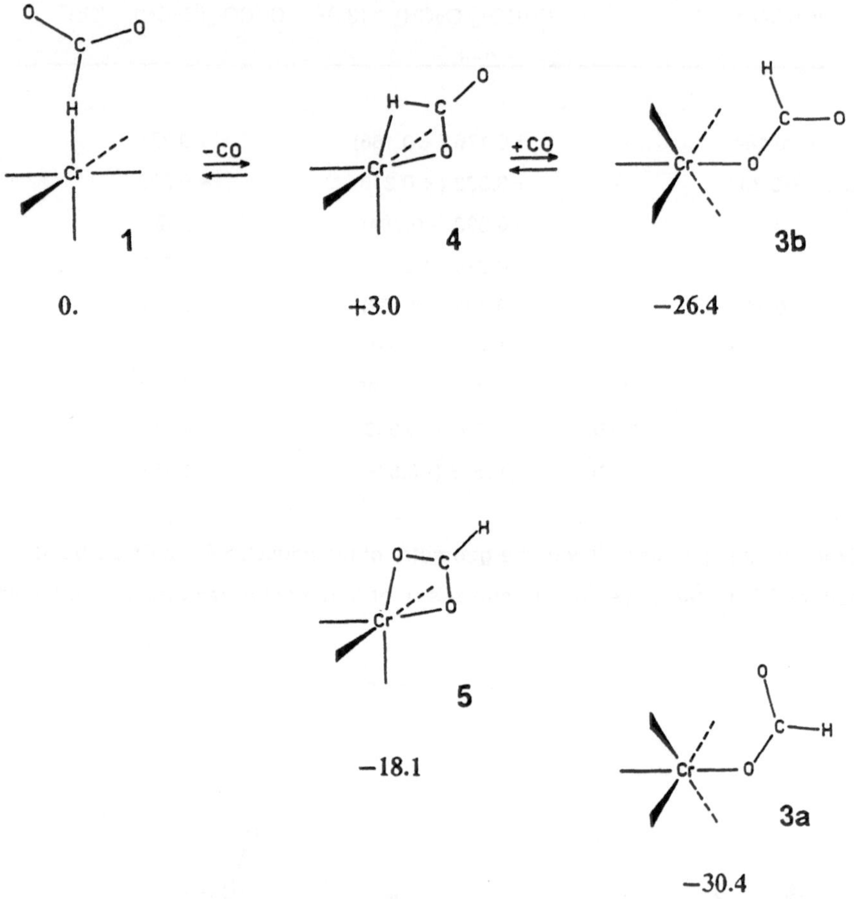

1

0.

−CO

4

+3.0

+CO

3b

−26.4

5

−18.1

3a

−30.4

Scheme 1 (SCF relative energy values are in kcal/mol)

ligand as experimentally proposed [10 – 13], the bonding of an oxygen atom to give the η^2 structure **4** in which both the hydrogen and oxygen atoms are bound to the chromium atom and finally the relaxation of **4** to **3b** followed by the rotation of CO_2 around the C-O bond to yield the thermodynamically most stable isomer **3a**. This rotation does not require very much energy, 2.3 kcal/mol according to our calculations. As Sakaki showed for the $Cu(PH_3)_n(OCHO)$ systems [8] the greater stability of **3a** over

3b is probably due to a greater electrostatic attraction of the $Cr^{\delta+}$ atom with the oxygen atom than with the hydrogen atom (the respective charges of O and H being -0.53 and -0.06, see Table 1).

The second pathway would involve a rotation of CO_2 while the hydrogen in **4** is pulled off, leading directly to the η^2 structure **5** in which the two oxygen atoms are linked to the chromium atom, and a final relaxation to **3a**.

That the CO_2 coordination to the hydride labilizes the cis carbonyl ligand is best seen from the decrease of the computed Cr-CO bond dissociation energy from 29.4 kcal/mol in $HCr(CO)_5^-$ to 23.6 kcal/mol in $[Cr(CO)_5H...CO_2]^-$ (the dissociation of the trans CO ligand is more endothermic, 28.1 kcal/mol instead of 23.6 kcal/mol). But a more salient feature is the rather small destabilization of **4** with respect to **1** , 3.0 kcal/mol. Although we cannot assess at this stage of our calculations whether or not **4** is a transition state or close to a transition state, we can reasonably conclude that this structure is most likely involved in the insertion process. We traced the relatively low destabilization of **4** to a quite strong two electron stabilizing interaction **6** between the empty d_π orbital which is of b_2 symmetry in the C_{2v} $Cr(CO)_4$ fragment left by

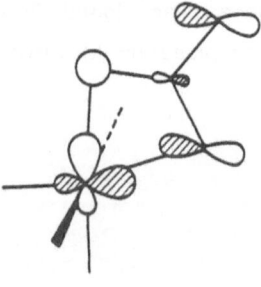

6

the dissociation of the cis carbonyl ligand and a doubly occupied valence orbital of the incipient HCO_2^- formate ligand. This latter orbital is characterized by an out-of-phase combination between the p components on the oxygen atoms and the s component on the hydrogen atom. We also noticed in the course of our study [6] that for the insertion of CO_2 into the metal-methyl bond of the $CH_3Cr(CO)_5^-$ analog, the orbital to consider in the incipient acetate ligand $CH_3CO_2^-$ would now have a p component on the methyl carbon atom interacting in an out-of-phase manner with the p components on the oxygen atoms. The interaction similar to **6** would then be much weaker, therefore providing no decisive stabilization of the $[\eta^2-(CH_3), O]$ structure and making the ancillary CO dissociation unnecessary. This may explain why an associative type mechanism prevails for the insertion into the metal-methyl bond.

We have already mentioned the possibility of another dihapto structure, **5**, for the incipient formate ligand. According to our calculations **5** is more stable than **4** by 21.1 kcal/mol (see the Scheme 1). **5** + CO is more stable than the adduct **1** by 18.1 kcal/mol, but still higher in energy than the two product isomers **3a** and **3b** (by 12.3 and 8.3 kcal/mol respectively). This makes **5** as a likely candidate for an intermediate in the reaction pathway, but we cannot definitively conclude on this issue without a full geometry optimization and a determination of the force constants matrix.

It is interesting to note at this point that new experimental investigations have now started in the Darensbourg group and that the corresponding results nicely complement the information obtained from our theoretical study: **4** is best seen as a transition state and **5** as an intermediate on the insertion reaction pathway [21]

How does this insertion of CO_2 into the Cr-H bond of $HCr(CO)_5^-$ compares to reactions of the same type for other transition metal hydrides? We rationalized in our original publication [6] why no ancillary ligand dissociation was involved in the decarboxylation reaction of square planar d^8 $ML_3(HCO_2)$ complexes **7** such as

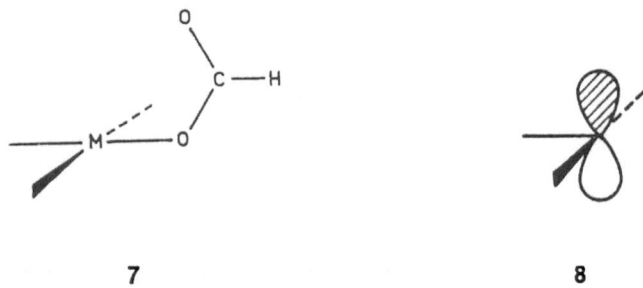

7   **8**

$(PPh_3)_3Rh(HCO_2)$ or various $HPtL_2(HCO_2)$ systems [20a,22,23]. The rationale lies in the existence in such systems of a low lying empty orbital of b_2 symmetry, see **8**, which can assist, as in **6**, the η^2 H,O structure (or eventually the η^2 O,O structure). The Cu(triphos)(HCO_2) system on the other hand does not decarboxylate at room temperature. This is again consistent with our explanation, since the C_{3v} Cu(triphos)$^+$ fragment lacks a low lying empty orbital of b_2 symmetry. Indeed

Sakaki and Ohkubo [8] found in the course of their study of CO_2 insertion into the Cu-H bond of $HCu(PH_3)_3$ that the η^2 structure is much less stable than the η^1 structure. Their study also allows a quite useful comparison with ours: As in the $HCr(CO)_5^-$ case, the reaction with $HCu(PH_3)_2$ and $HCu(PH_3)_3$ is driven by the charge transfer interaction from the metal complex to CO_2 and by the electrostatic attraction between the hydride and the positive carbon atom. The overall reaction is also exothermic, the energy difference between the two formate isomers being greater in the copper complexes than in the chromium complex (13 - 14 kcal/mol instead of 4 kcal/mol). This can be traced to the fact that the hydrogen atom of the formate ligand is positively charged in the copper complex, therefore inducing an electrostatic repulsion when H lies near the Cu atom (we have seen that in the Cr system, the hydrogen atom bears a slight negative charge giving rise to a weak electrostatic attraction). The same reason may be invoked to explain that the rotation leading to the most stable isomer occurs without any barrier in the copper complex.

At this point one may wonder about the appropriateness of the SCF method to treat CO_2 reactivity problems. We have seen that the inclusion of non dynamical effects seemed to decrease the stabilization energy of the $[Cr(CO)_5H...CO_2]^-$ adduct by about 6.7 kcal/mol. The study of Sakaki and Ohkubo [8] provides some information about the inclusion of dynamical correlation effects through MP2 type calculations: they decrease the overall exothermicity of the reaction with either $HCu(PH_3)_2$ or $HCu(PH_3)_3$. As far as the energy barrier is concerned the results seem to be less general since the barrier is barely affected in the case of $HCu(PH_3)_2$ but significantly raised in the case of $HCu(PH_3)_3$. A more comprehensive study of the correlation effects is apparently in progress [24] and it should shed some useful light on this subject. But one may already conclude that going beyond the Hartree-Fock approximation does not change the overall qualitative picture that one gets from a SCF study. One should not however extend this conclusion to any reactivity problem. The coupling reaction between CO_2 and ethylene mediated by Ni(0) centers illustrates, as we shall see now, the need to go in some instances beyond the SCF approximation.

THE COUPLING REACTION BETWEEN CO_2 and C_2H_4 AT Ni(0) CENTERS

Hohberg and coworkers recently found that CO_2 and a variety of alkenes could oxidatively couple at Ni(0) centers to yield Ni(II) metallalactones. [25]

$$C_2H_4 + CO_2 + Ni^0 + 2DBU \rightarrow (DBU)_2Ni(C_3H_4O_3)$$

$$Ni^0 = [Ni(cod)_2], \text{ cod} = 1,5\text{cyclooctadiene}$$

The reaction took place quite easily, but no reaction mechanism could be assessed. A X-ray crystal structure of the metallalactone product revealed a square planar co-ordination of the Ni atom with cis imino N atoms [25a] There are of course many reaction pathways that one could envision for this reaction. We restricted ourselves to three possible coupling channels which require a prior formation of a bis(ligand)Ni(0) complex [7]. The first channel involves a concerted approach of C_2H_4 and CO_2 to the Ni metal. In the second one C_2H_4 coordinates first to the metal,

i) $L_2Ni + C_2H_4 + CO_2 \longrightarrow$

ii) $L_2Ni(C_2H_4) + CO_2 \longrightarrow$

iii) $L_2Ni(CO_2) + C_2H_4 \longrightarrow$

giving rise to a three coordinate Ni(0) complex which then undergo an electrophilic attack of CO_2 on the coordinated C_2H_4. Alternatively in the third channel CO_2 could coordinate first and then be attacked by an incoming ethylene. The first pathway is in fact reminiscent of the coupling of two ethylene at NiL_2 systems which was shown by Hoffmann et al. [26] to be symmetry forbidden provided that C_{2v} symmetry is retained. Here the replacement of one ethylene by a carbon dioxide molecule relieves the strict forbidness but some avoided crossings remain and the reaction is still quite unfavorable *(vide infra)*. We therefore focused mainly on the two other pathways and considered in a first stage structures which are relevant to these pathways, namely $NiL_2(\eta^2 - C_2H_4)$, **9**, $NiL_2(\eta^1 - C_2H_4)$, **10**, $NiL_2(\eta^2 - CO_2)$, **11**, and $NiL_2(\eta^1 - CO_2)$, **12**.

Although the η^1 complexes are probably not true intermediates, they provide a measure of the deformation energy of the $NiL_2(C_2H_4)$ or $NiL_2(CO_2)$ skeletons on going from the η^2 complexes to the metallalactone. We then compared the approach of CO_2 and C_2H_4 to **10** and **12** respectively, optimizing for each Ni-C bond distance the most important geometrical parameters [9,27].

The Table 2 gives the relative SCF energies of the $NiL_2(C_2H_4) + CO_2$ and $NiL_2(CO_2) + (C_2H_4)$ systems in the conformations **9** - **12** and for three sets of ligands L, namely L = NH_3, $NH = CH_2$, and PH_3 [29]. Interestingly bis-phosphine complexes are found to stabilize the η^2 ethylene complex relative to the η^2 CO_2 complex whereas bis-amine or bis-imine complexes stabilize preferentially the η^2 CO_2

Table 2. SCF relative energies (in kcal/mol) of the $NiL_2(CO_2)(C_2H_4)$ isomers (with a L-Ni-L angle standardized to 90°)

	NH_3	$NH(=CH_2)$	PH_3
$L\diagdown_{Ni}\diagup^L$ + CO_2 + C_2H_4	0.	0.	0.
$L\diagdown_{Ni}\diagup^L$ H_2C-CH_2 + CO_2	-55.2	-61.6	-34.1
L—Ni—L $H=C-C$ + CO_2	-13.9	-10.7	+9.7
$L\diagdown_{Ni}\diagup^L$ $O-C_{O}$ + C_2H_4	-59.8	-73.5	-23.6
O—Ni—L $O-C$ + C_2H_4	-33.5	-46.1	+11.0

complex. But the most interesting feature of this Table lies in the fact that the deformation energy to overcome in order to go from the η^2 structure to the η^1 structure is much less for CO_2 complexes than for C_2H_4 complexes, thus already suggesting a better propensity of these systems to undergo a C_2H_4 attack on coordinated CO_2. One has to worry however about the near degeneracy correlation effects: Ni complexes, especially the low valent ones, are usually characterized by a manifold of low lying electronic configurations which have to be taken into account in the calculation [30] . This is best achieved with a MC-SCF type calculation. We therefore carried out CAS-SCF calculations for the bis amine complexes. As shown in the Table 3 the η^2 structures are of similar stability but the deformation energy to

Table 3. CAS- SCF relative energies (in kcal/mol) of the $Ni(NH_3)_2(CO_2)(C_2H_4)$ isomers

	NH_3
$L\diagdown_{NI}\diagup L$ + CO_2 + C_2H_4	0.
$L\diagdown_{NI}\diagup L$ H_2C-CH_2 + CO_2	-21.9
$\overset{L}{\underset{\underset{H}{H}-C}{\overset{\mid}{NI}-L}}$ + CO_2	+7.1
$L\diagdown_{NI}\diagup L$ $O-C_O$ + C_2H_4	-22.9
$O-NI-L$ $O-C$ + C_2H_4	-21.6

reach the η^1 structure **12** is now very small, the CAS-SCF energy difference between **12** and **11** being only 1.3 kcal/mol. An additional interesting result is provided by a natural orbital analysis of the corresponding CAS-SCF wave function which points to a strong diradical character for **12**. The orbital occupations of the the bonding orbital between $d_{x^2-y^2}$ and π_{CO_2} (the HOMO in the SCF ground state wavefuction) and its antibonding counterpart (which is the LUMO) are 1.288 and 0.714 respectively. Since these two orbitals are equally spread over the the metal and the CO_2 ligand, the $\eta^1 - O$ bonded structure **12** is probably best seen as 1,3 diradical centered on the nickel and carbon atoms, and most probably with a singlet ground state since the triplet state was computed to lie 4.1 kcal/mol above the singlet state.

The relatively low energy of $NiL_2(\eta^1 - CO_2)$ and its strong diradical character suggest that the coupling reaction might well involve the attack of C_2H_4 on coordinated CO_2 rather than the electrophilic attack of CO_2 on coordinated C_2H_4. (It may also explain the very easy coupling reaction observed with O_2 [31]). We checked this hypothesis by carrying out CAS SCF calculations either for the approach of C_2H_4 to coordinated $\eta^1 - CO_2$ or for the approach of CO_2 to coordinated $\eta^1 - C_2H_4$ [27,28]. The first pathway is found to be much more favorable, as sketched in the Scheme 2. Having C_2H_4 at 3 Å of the Ni atom (which corresponds roughly to a C-C bond length of 2.5 Å requires only 11.1 kcal/mol from the separated $Ni(NH_3)_2(\eta^2 - CO_2)$ and C_2H_4 molecules. A second point not shown on the Scheme and corresponding to Ni-C = 2.5 Å (and C-C = 2 Å i.e. typical of the C-C bond length in transition states of coupling reactions) is found to be destabilized by 15.3 kcal/mol only. On the other hand, when CO_2 approaches $Ni(NH_3)_2(C_2H_4)$, the destabilization for the point corresponding to Ni-C = 3 Å (and C-C = 2.5 Å) is much higher, 36.1 kcal/mol with respect to $Ni(NH_3)_2(\eta^2 - C_2H_4) + CO_2$. This is again a strong indication of the preference for the C_2H_4 attack on coordinated CO_2 over the CO_2 attack on coordinated C_2H_4.

We also carried out a single calculation for the concerted approach of C_2H_4 and CO_2 on NiL_2 [28]. With both C_2H_4 and CO_2 distant by 3 Å from the Ni atom, the destabilization with respect to $Ni(NH_3)_2 + CO_2 + C_2H_4$ is quite high, 26 kcal/mol. This

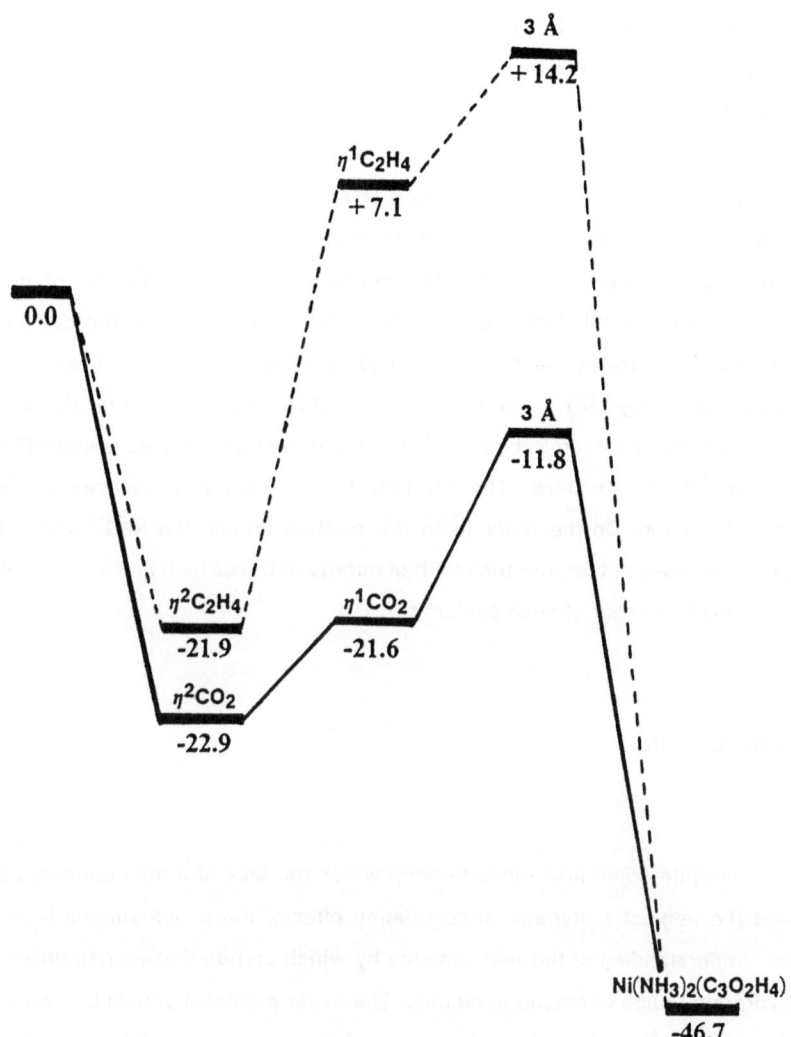

Scheme 2: Relative CAS-SCF energies of structures relevant to the CO_2 approach to $Ni(NH_3)_2(C_2H_4)$ (- - -) and to the C_2H_4 approach to $Ni(NH_3)_2(CO_2)$ (-----). The zero of energy refers to the separated $Ni(NH_3)_2 + CO_2 + C_2H_4$ systems and the values are in kcal/mol.

feature was expected from our previous considerations on the forbidness or quasi forbidness of the reaction and allows us to discard this reaction pathway as an operative one.

The η^2 and $\eta^1 - O$ structures are of course not the only ones that one can think of for $NiL_2(CO_2)$ complexes. We considered other geometries in connection with the fluxional behaviour claimed for the Ni bisphophine complexes [32 – 34] The single resonnance observed in ^{31}P NMR spectra either in solution [32] or in liquid CO_2 [33] has been attributed to a η^1 coordination mode of CO_2 at the carbon atom, to a fast rotation around the $Ni - (\eta^2 - CO_2)$ bond or to a process involving a partially dissociated CO_2 [34] . The CAS SCF calculations carried out for the $Ni(NH_3)_2(CO_2)$ isomers shown on the Scheme 3 [28] point to a strong destabilization (46.8 kcal/mol) of the $\eta^1 - C$ structure. The rotation of $\eta^2 - CO_2$ also induces an appreciable destabilization. On the other hand the rotation around the Ni-O bond in the $\eta^1 - O$ structure does not require too much of energy (9.0 kcal/mol) and might account for the fluxional behaviour of such systems.

CONCLUSION

Despite some limitations (among which the lack of a full geometry optimization and the neglect of dynamical correlation effects) the above studies have increased our understanding of the mechanisms by which carbon dioxide can insert into metal hydrogen bonds or couple to olefins. The experimentalist should be aware of the fact that the results which have been reported here are qualitative or at most semi-quantitative. Nevertheless they have led us to propose new intermediate structures relevant to the corresponding reaction paths. New experimental data and also more refined calculations (including dynamical correlation effects) should allow to test the validity of these proposals. In that respect the recent experimental studies of Darensbourg and Wiegreffe which have been designed towards the spectroscopic identification of such structures in the CO_2 insertion reaction [21] illustrate nicely the interplay that can exist between theory and experiment in the field of mechanistic investigations of carbon dioxide activation and reactivity.

Scheme 3 (L = NH₃).

REFERENCES AND NOTES

[1]   For reviews, see for instance: a) D. J. Darensbourg, R. A. Kudaroski, Adv. Organomet. Chem. **22**, 129 (1983);   b) D. A. Palmer, R. Van Eldik, Chem. Rev. **83**,51 (1983);   c) R. Ziessel, Nouv. J. Chim. **7**, 613 (1983);   d) D. J. Darensbourg, C. Ovalles, CHEMTECH **15**, 636 (1985);   e) D. Walther, Coord. Chem. Rev. **79**, 135 (1987);   f) A. Behr, Angew. Chem. **100**, 681 (1988); Angew. Chem. Int. Ed. Engl. **27**, 661 (1988);   g) P. Braunstein, D. Matt, D. Nobel, Chem. Rev. **88**, 747 (1988).

[2]   C. Mealli, R. Hoffmann, A. Stockis, Inorg. Chem. **23**, 56 (1984).

[3]   T. Ziegler, Inorg. Chem. **25**, 2721 (1986).

[4]   a) S. Sakaki, K. Kitaura, K. Morokuma, Inorg. Chem. **21**, 760 (1982);   b) S. Sakaki, A. Dedieu, J. Organomet. Chem. **314**, C63 (1986);   c) S. Sakaki, A. Dedieu, Inorg. Chem. **26**, 3278 (1987);

[5]   A. Dedieu, V. Branchadell, Inorg. Chem. **26**, 3966 (1987).

[6]   C. Bo, A. Dedieu, Inorg. Chem. **28**, 304 (1989).

[7]   A. Dedieu, F. Ingold, Angew. Chem. in press.

[8]   a) S. Sakaki, K. Ohkubo, Inorg. Chem. **27**, 2020 (1988);   b) S. Sakaki, K. Ohkubo, Inorg. Chem. **28**, 2583 (1989);

[9]   S. Sakaki, T. Aizawa, N. Koga, K. Morokuma, K. Ohkubo, Inorg. Chem. **28**, 103 (1989).

[10] D. J .Darensbourg, A. Rokicki, M. Y. Darensbourg, J. Am. Chem. Soc. **103** , 3223 (1981).

[11] D. J. Darensbourg, A. Rokicki, Organomet. **1**, 1685 (1982).

[12] M. Y. Darensbourg, R.Bau, M. W. Marks, R. R. Burch Jr., J.C. Deaton, S. Slater, J. Am. Chem. Soc. **104**, 6961 (1982).

[13] D. J. Darensbourg, M. J. Pala, J. Am. Chem. Soc. **107**, 5687 (1985).

[14] D. J. Darensbourg, R. Kudaroski Hanckel, C. G. Rauch, M. Pala, D. Simmons, J. Am. Chem. Soc. **107**, 7463 (1985).

[15] D. J. Darensbourg, G. J. Grotsch, J. Am. Chem. Soc. **107**, 7473 (1985).

[16] For a review see: W. Kolos, Theor. Chim. Acta, **51**, 219 (1979).

[17] C. Hoheisel, W. Kutzelnigg, J. Am. Chem. Soc., **97**, 6970 (1975).

[18] R. A. Chiles, C. E. Dykstra, Chem. Phys. Lett., **92**, 471 (1982).

[19] A. Dedieu, V. Branchadell, ACS Symp. Ser., **394**, 58 (1989).

[20] (a) A. Immirzi, A. Musco, Inorg. Chim. Acta **22**, L35 (1977); (b) D. J. Darensbourg, C. S. Day, M. B. Fischer, Inorg. Chem. **20**, 3577 (1981); (c) D. J. Darensbourg, M. B. Fischer, R. E. Schmidt Jr., B. J. Baldwin, J. Am. Chem. Soc. **103**, 1297 (1981); (d) D. M. Grove, G. Van Koten, H. J. C. Hubbels, R. Zoet, A. L. Spek, J. Organomet. Chem. **263**, C10 (1984); (e) C. Bianchini, C. A. Ghilardi, A. Meli, S. Midollini, A. Orlandinni, Inorg. Chem. **24**, 924 (1985).

[21] D. J. Darensbourg, H. Pickner Wiegreffe, Inorg. Chem., in press.

[22] S. H. Strauss, K. H. Whitmire, D. F. Shriver, J. Organomet. Chem. **174**, C59 (1979).

[23] R. S. Paonessa, W. C. Trogler, J. Am. Chem. Soc. **104**, 3520 (1982).

[24] see the reference 52 of reference 8b

[25] a) H. Hoberg, Y. Peres, C. Krüger, Y.- H. Tsay, Angew. Chem. **99**, 799 (1987); Angew. Chem. Int. Ed. Engl. **26**, 771 (1987); b) H. Hoberg, Y. Peres, A. Milchereit, S. Gross, J. Organomet. Chem. **345**, C17 (1988).

[26] R. J. McKinney, D. L. Thorn, R. Hoffmann, A. Stockis, J. Am. Chem. Soc. **103**, 2595 (1981).

[27] The geometry of the $Ni(CO_2)(C_2H_4)$ entity was optimized at the SCF level, except for the O-C-O angle in the $[Ni(NH_3)_2(\eta^1 - CO_2)----(C_2H_4)]$ system for which a CAS-SCF optimization of the O-C-O angle was found to be necessary [28]. Conversely for the $Ni(NH_3)_2(\eta^1 - CO_2)$ system alone the SCF and CAS-SCF optimizations of the O-C-O angle yielded quite similar values, 130.3° and 136.5° respectively, the CAS-SCF energy difference between these two points amounting to 1.2 kcal/mol only. The values quoted in the Table 2 [7] therefore refer to the SCF value of the O-C-O angle.

[28] F. Ingold, A. Dedieu, to be published.

[29] The $NH = CH_2$ ligand was chosen as a better model of the DBU ligand as compared to NH_3.

[30] See for instance: a) M. R. A. Blomberg, U. B. Brandemark, P. E. M. Siegbahn, K. B. Mathisen, G. Karlstrom, J. Phys. Chem. **89**, 2171 (1985); b) P. O. Widmark, B. O. Roos, P. E. M. Siegbahn, J. Phys. Chem. **89**, 2180 (1985).

[31] M. Aresta, C. F. Nobile, J. Chem. Soc. Dalton Transactions 708 (1977).

[32] M. Aresta, E. Quaranta, I. Tommasi, J. Chem. Soc., Chem. Comm. 450 (1988).

[33] M. G. Mason, J. A. Ibers, J. Am. Chem. Soc. **104**, 5153 (1982).

[34] M. Aresta, E. Quaranta, I. Tommasi, R. Gobetto, NATO-ASI Summer School on Enzymatic and Model Carboxylation and Reduction Reactions, Ginosa (Italy) 1989.

The Organometallic Chemistry of Carbon Dioxide Pertinent to Catalysis

Donald J. Darensbourg
Department of Chemistry
Texas A&M University
College Station, Texas 77843
United States of America

ABSTRACT. The mechanistic aspects of the C-X bond forming reactions resulting from CO_2 insertion into M-X bonds, where X = H, R, OR, and NR_2, must be elaborated if we are to be successful in developing catalysts for processes leading to useful chemicals derived from carbon dioxide. Of the three reactions involving the coupling of CO_2 with coordinated ligands (M-H, M-R, M-OR) which are of primary interest in our research, only the C-C bond formation process is generally irreversible.

Our present understanding of the mechanism of the carbon-carbon bond forming process is the most detailed. Since carbon dioxide insertion into anionic metal-hydrides and metal-alkoxides is often a rapid process, limited mechanistic information for these processes is available. In contrast, the reverse process, decarboxylation, generally occurs on a timescale which is amenable to kinetic analysis by isotopic labelling experiments such as depicted in the equation below.

$$M\text{-}OC(O)H \ + \ {}^*CO_2 \ \text{---}> \ M\text{-}O^*C(O)H \ + \ CO_2$$

The mechanistic view emerging from these studies is that the decarboxylation reaction is strongly dependent on the nature of the X group in the [MOC(O)X] moiety. That is, the X group must have an appropriate orbital for interaction with the metal center during the decarboxylation process. This in turn requires significant interaction between CO_2 and X group during the C-X bond forming reaction. A detailed analysis of the carboxylation/decarboxylation pathways will be presented. These insertion reactions can play pivotal roles in the mechanisms of the syntheses of organic products from carbon dioxide. Such processes include the production of alkyl formates, carboxylic acids, lactones, alkylpolycarbonates, etc.

Finally, the utilization of organometallic chemistry to develop better heterogeneous metal catalysts for the production of chemical feedstocks from CO_2, e.g. CH_4 and CH_3OH, will be discussed. One of the essential problems to be dealt with in heterogeneous catalysts is the preparation and stabilization of very small metal particles supported on oxide carriers. In this regard metal carbonyl clusters in and on solid metal supports such as silica, alumina, and magnesia, serve as good sources of highly dispersed, low-valent metals for catalysis. Our efforts in the area of CO_2 methanation employing ruthenium carbonyl clusters on alumina and in zeolite supercages will be summarized.

M. Aresta and J. V. Schloss (eds.),
Enzymatic and Model Carboxylation and Reduction Reactions for Carbon Dioxide Utilization, 43–64.
© 1990 *Kluwer Academic Publishers.*

44

1. Introduction

The activation of carbon dioxide by transition metal complexes has been extensively studied, both experimentally and theoretically.[1] Central reactions in this chemistry are the insertion of CO_2 into M-X bonds, where X = H, C, O, and N. (eq. 1-4). We are presently investigating the mechanistic aspects of these reaction processes and will herein describe our current level of understanding. Comparisons of the pathway of the carbon-carbon bond forming process in transition metal chemistry with the well known analogous chemistry involving organolithium reagents will be presented. Furthermore, the role of these reaction types in both homogeneous and heterogenous catalytic processes leading to useful chemicals will be elaborated.

$$[M]-H + CO_2 \quad ---> \quad [M] O_2CH \qquad (1)$$

$$[M]-R + CO_2 \quad ---> \quad [M]O_2CR \qquad (2)$$

$$[M]-OR + CO_2 \quad ---> \quad [M]O_2COR \qquad (3)$$

$$[M]-NR_2 + CO_2 \quad ---> \quad [M]O_2CNR_2 \qquad (4)$$

Insertion of CO_2 into the Metal-Hydride Bond.

The reaction of anionic group 6 (Cr, Mo, W) transition metal hydrides with carbon dioxide to afford metalloformates occurs readily at ambient temperature and reduced pressures of carbon dioxide.[2] This insertion process is referred to the normal pathway (**Scheme 1**). There are no documented cases of CO_2 insertion into the metal hydride bond to provide the alternative, metallocarboxylic acid, isomer (referred in **Scheme 1** as abnormal).[3] Recent theoretical studies ascribe this preference to an unfavorable electrostatic interaction and poorer orbital overlap in the latter process.[4] Nevertheless, these complexes can be prepared **via** direct nucleophilic attack

Scheme 1

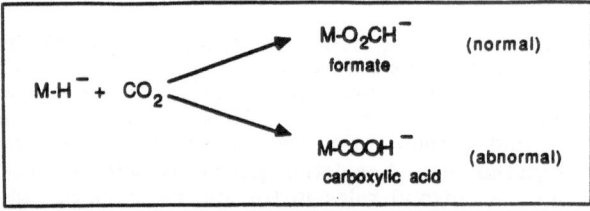

of hydroxide ion at a metal bound carbon monoxide ligand (eq. 5).[5] Reaction (5) is reversible as evidenced by oxygen atom exchange processes when $^{18}OH^-$ is employed

$$M-CO + OH^- \rightleftharpoons M-COOH^- \qquad (5)$$

where the metal derivative becomes enriched in oxygen-18 labelled CO.[6]

Presently in the literature there is only one kinetic study of CO_2 insertion into a metal-hydride bond. This investigation involves a hydride derivative much less hydridic than the anionic group 6 metal hydrides, i.e., fac-Re(bpy)(CO)$_3$H (bpy = 2,2'-bipyridine).[7] Reaction (6) was demonstrated to be first-order in both metal complex and CO_2 concentrations with activation parameters ($\Delta H^{\neq}$ = 12.8 kcal/mol and $\Delta S^{\neq}$ = -33.0 e.u.) indicative of an I_a or A mechanism. The transition state for the insertion reaction is proposed to have significant charge transfer character (1), as revealed by the second order rate constant varying from 1.97×10^{-4} M^{-1}-sec^{-1} in THF to 5.44×10^{-2} M^{-1}-sec^{-1} in CH$_3$CN at constant temperature. Reaction (6) displays an inverse isotope effect (k_H/k_D = 0.53) which is also consistent with notable C-H bond formation in the transition state.

$$\textbf{fac-Re(bpy)(CO)}_3\textbf{H} + CO_2 \longrightarrow \textbf{fac-Re(bpy)(CO)}_3\textbf{O}_2\textbf{CH} \qquad (6)$$

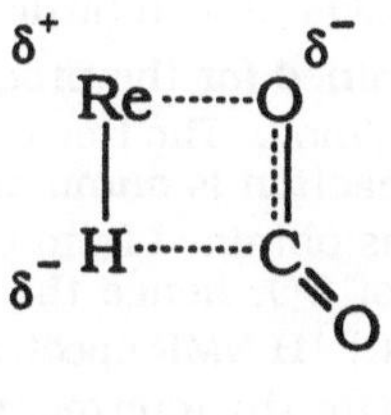

1

Rate data for the decarboxylation reaction involving a Re(I) derivative, eq 7, similarly reveal the importance of breaking the C-H bond in the transition state (2).[8] In this instance a kinetic isotope effect (k_H/k_D = 1.55) was determined. The activation parameters ($\Delta H^{\neq}$ = 26.8 kcal/mol and $\Delta S^{\neq}$ = -6.3 e.u) were determined for the decarboxylation process which was shown not to be retarded by added PPh$_3$ concentration and to occur with retention of configuration about the metal center.

$$CpRe(NO)(PPh_3)O_2CH \longrightarrow CpRe(NO)(PPh_3)H + CO_2 \qquad (7)$$

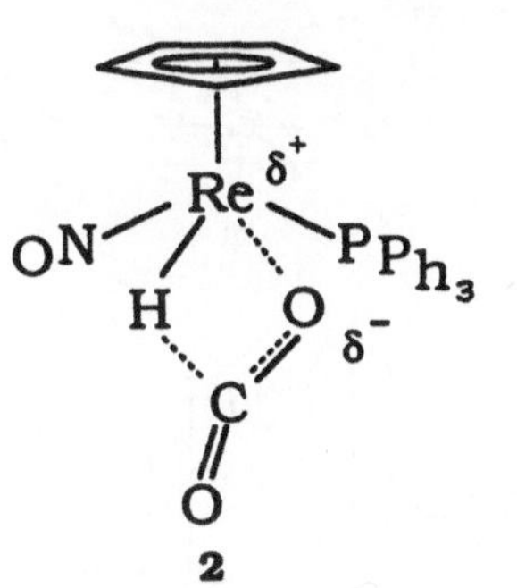

2

As previously mentioned the insertion of CO_2 in to the metal-hydride bond of $HM(CO)_5^-$ (M = Cr, Mo, W) derivatives occurs very rapidly at ambient temperature. Hence in an effort to gain knowledge about the transition state of this process we have investigated the less facile reverse reaction, decarboxylation. This has been accomplished **via** determining the rate parameters for the $^{13}CO_2$ exchange reaction defined in eq. (8).[9] This process is readily monitored by means of ^1H NMR, where the formate resonance at ∂ 8.28 ppm is split by ^{13}C (J_{C-H} = 186 Hz) upon $^{13}CO_2$ exchange. Concomitantly, in the ^{13}C NMR spectrum the formate signal at ∂167.3 ppm is enhanced.

$$[M]\text{-OC(O)-H} + {}^*CO_2 \rightleftharpoons [M]\text{-O}^*C(O)\text{-H} + CO_2 \qquad (8)$$

The decarboxylation reaction of $HCO_2Cr(CO)_5^-$ was observed to be first-order in metal complex, zero-order in CO_2, and retarded by addition of carbon monoxide. The activation parameters determined for the process were $\Delta H^{\neq}$ = 18.2 kcal/mol and $\Delta S^{\neq}$ = -14.3 e.u. ($\Delta G^{\neq}$ = 22.5 kcal/mol). The free energy profile for the decarboxylation reaction and the carboxylation reaction is summarized in **Scheme 2**, where the barrier for proceeding from **8** to **7** was obtained from CO substitution data.[10] Formation of species **6** and **7** require loss of CO, hence the origin of the inhibition by carbon monoxide. Low temperature (-80°C) ^1H NMR spectra of **3** in the presence of excess CO_2 has thusfar provided no evidence for the intermediacy of **5**. Intermediates of the type **7** have been spectroscopically characterized for the tungsten acetate analog.[10]

Scheme 2

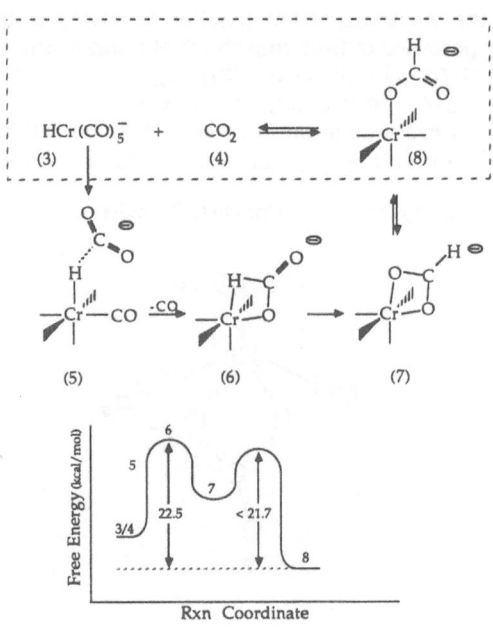

Although we have presented **5** as a transient along the reaction coordinate in **Scheme 2** and not as an intermediate, quantum chemistry computations indicate the bond enthaply for formation of **5** to be 8.4 kcal more favorable than $HCr(CO)_5^-$ and CO_2.[11] There is of course an unfavorable entropic term associated with formation of **5**. Nevertheless, the calculations of Bo and Dedieu[11] are consistent in all respects with the mechanistic details outlined in **Scheme 2.**

We have further investigated the mechanistic aspects of the decarboxylation reactions of square-planar $HM(O_2CH)(PCy_3)_2$ (M = Ni, Pd, Pt; Cy = cyclohexyl) derivatives (eq. 9).[9,12] In these processes the possibilities exist for the insertion of carbon dioxide into the H-M bond prior to or after CO_2 loss, i.e., the intermediacy of $(HCO_2)_2M(PCy_3)_2$ or $H_2M(PCy_3)_2$.

$$HM(O_2CH)(PCy_3)_2 + {}^*CO_2 \quad \text{--->} \quad HM(O_2{}^*CH)(PCy_3)_2 + CO_2 \qquad (9)$$

The rate of carbon dioxide exchange in $HNi(O_2CH)(PCy_3)_2$ was shown to be first-order in metal complex and zero-order in carbon dioxide. Unlike in the coordinatively saturated group 6 metal formates, where CO dissociation is a prerequisite to formation of the η^2-H, O formate species **6**, in the group 10 metal derivatives no ligand dissociation is necessary. This was evident in that the exchange reaction was not inhibited by the presence of excess tricyclohexylphosphine. In these d^8 metal derivatives there is a low-lying empty orbital for interaction in forming the $\eta2$–H, O structure (**9**).[11]

9

The activation parameters for reaction (9), where M = Ni, were determined to be $\Delta H^{\neq}$ = 22.1(9) kcal/mol. and $\Delta S^{\neq}$ = -5.(3) e.u. In the closely related process, reaction (10), similar activation parameters were found, i.e. $\Delta H^{\neq}$ = 23.0 (10) kcal/mol. and $\Delta S^{\neq}$ = -4(4) e.u. Hence we propose that reactions (9) and (10) proceed **via** a common dihydride intermediate.

Consistent with this proposal the palladium analog undergoes CO_2 exchange ca. 2200 times faster at 24°C than the nickel complex. This is anticipated since the suggested intermediate $H_2Pd(PCy_3)_2$ complex in the carbon dioxide exchange process is a known species,[13] whereas $H_2Ni(PCy_3)_2$ is unknown although there have been several serious attempts at its synthesis. In general there is an increase in the stability of the metallodihydride upon proceeding down the group 10 metal triad (**Figure 1**). Indeed Trogler and coworkers have reported $HPt(O_2CH)(PCy_3)_2$ to readily decarboxylate at ambient temperature to afford an equilibrium mixture with the $H_2Pt(PCy_3)_2$ complex, the latter species being predominant.[14]

Figure 1. Free-energy profile for CO_2 exchange process.

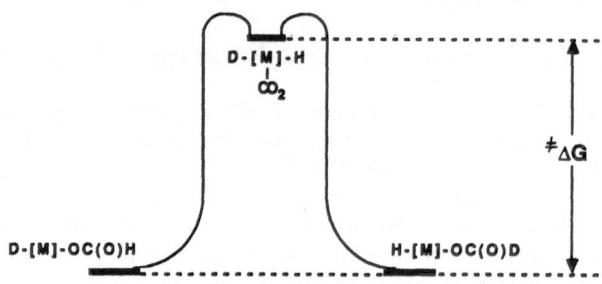

$^{\neq}\Delta G$: **nickel > palladium > platinum**

Insertion of CO_2 into Metal-Carbon Bonds.

Anionic alkyl and aryl complexes of group 6 metal carbonyls undergo irreversible CO_2 insertion forming metallocarboxylates as indicated in equation (11).[15] This process has been investigated in detail mechanistically for anionic tungsten-alkyl and -aryl complexes.[16]

$$RM(CO)_5^- + CO_2 \quad \text{--->} \quad RCO_2M(CO)_5^- \tag{11}$$

These reactions were shown to obey second-order kinetics, first-order in anionic metal substrate and first-order in carbon dioxide, with activation parameters indicative of an associative interchange (I_a) mechanism. That is, $\Delta H^{\neq}$ and $\Delta S^{\neq}$ were determined to be 10.2 kcal/mol and -43.3 e.u., respectively for the insertion of CO_2 into $CH_3W(CO)_5^-$ in THF to afford $CH_3CO_2W(CO)_5^-$. Hence, the rates of these reactions are considerably slower than the corresponding process involving the analogous metal-hydrides (**vide supra**). In one instance we have compared the rates of CO_2 insertion into M-H **vs.** M-R bonds directly, i.e., employing the complex $Ni(H)(Ph)(PCy_3)_2$. Carbon dioxide insertion occurred quantitatively and exclusively at the Ni-H bond to yield the $(HCO_2)Ni(Ph)(PCy_3)_2$ derivative.

The rate of reaction (11) was only slightly influenced by the nature of R, increasing along the series R = $C_6H_5 < C_2H_5 < CH_3$, with relative rates of 1:3.8:6.0.

Consistent with an I_a pathway the stereochemistry about the α–carbon atom is maintained during the carbon dioxide insertion reaction.[17] This was evident by the conversion of **threo**-$W(CO)_5CHDCHDPh^-$ with CO_2 to **threo**-$W(CO)_5O_2CCHDCHDPh^-$ (eq.12). A similar observation was noted for the phosphine substituted derivative **threo**-$W(CO)_4[PMe_3]CHDCHDPh^-$.

$$(12)$$

The importance of electron-donating ancillary ligands on the carboxylation process is amply demonstrated upon replacing CO ligands in $CH_3W(CO)_5^-$ by phosphorus-donor ligands. The rate of CO_2 insertion is greatly accelerated for **cis**-$CH_3W(CO)_4L^-$ complexes in the order $L = CO < P(OMe)_3 < PMe_3$. This rate enhancement was shown to be due to an increase in nucleophilic character at the metal center and not due to a weakening of the W-CH_3 bond on the basis of X-ray structural data of $CH_3W(CO)_5^-$ **vs.** **cis**-$CH_3W(CO)_4PMe_3^-$,[15b,16,18] where the W-CH_3 bond is shorter in the latter species. Furthermore, the neutral group 7 analogs, $RM(CO)_5$ ($M = Mn$, Re) do not react with carbon dioxide.

The proposed transition-state consistent with experimental findings for the carbon-carbon bond forming reaction involving carbon dioxide is illustrated in **10**, where the counterion (M^+) aids in stabilizing the negative charge buildup on the distal oxygen atom of the incipient carboxylate. The effect of the cation on the rate of CO_2 insertion was revealed in a kinetic study of the carboxylation reaction as a function of Na^+, free and encapsulated with kryptofix-2.2.1.[19] For example, the rate of CO_2 insertion was found to be an order of magnitude faster for uncomplexed sodium ions. The ferocity of the distal oxygen atom of the carboxylate ligand for ion-pairing with Na^+ is manifested in the solid-state structure of [Na-kryptofix-2.2.1][HCO_2W(CO)_5], where the shortest $Na \cdots O$ interaction involves the distal oxygen atom of the formate ligand. The low activation energy calculated for the highly exothermic carboxylation reaction of methyllithium similarly involves a transition state dominated by the lithium cation complexation to oxygen.[20]

10

Because these anionic metal-alkyl and -aryl carbonyl derivatives readily undergo CO insertion to provide acyl complexes (eq. 13), it is not a trivial process to establish that the metal center remains coordinatively saturated during the CO_2 insertion reaction. That is, reaction (13) complicates an investigation of the affect of added CO on the CO_2 insertion reaction. However, independent kinetic measurements of reactions (11) and (13) combined with composite rate studies revealed both processes to be occurring **via** separate pathways (**Scheme 3**),[16] i.e., two concurrent processes. **Table I** presents a comparative summary of mechanistic aspects of carbonylation **vs.** carboxylation reactions. The consequences of these mechanistic differences should be of major importance in catalytic processes designed to utilize carbon dioxide as a C_1 feedstock.

$$RM(CO)_5^- + CO \dashrightarrow RC(O)M(CO)_5^- \qquad (13)$$

Scheme 3

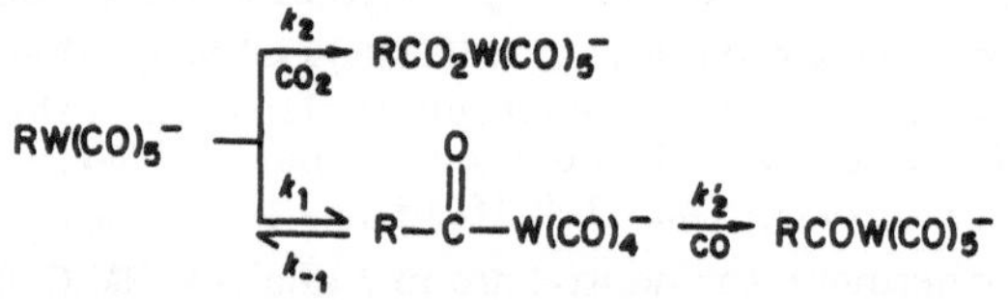

$$\frac{d[RW(CO)_5^-]}{dt} = \left\{ k_2[CO_2] + \frac{k_1 k_2'[CO]}{k_{-1} + k_2'[CO]} \right\} [RW(CO)_5^-]$$

$$= (k_2[CO_2] + k_1)[RW(CO)_5^-], \text{ for } k_2'[CO] \gg k_{-1}$$

Table I. Summary of Mechanistic Aspects of Carbonylation vs Carboxylation Reactions.

Reaction Variables	**Carboxylation**	**Carboxylation**[a]
Kinetic Order in CO_2 or CO	First-order in CO_2	Mixed-order in CO; independent of CO at high CO pressures
Nature of Metal	Third row more reactive than first row	First row more reactive than third row
R dependence	Small dependence on R group, alkyls faster than aryls	Reaction greatly retarded by electron-withdrawing R substituents

Table I. (continued)

Reaction Variables	Carboxylation	Carboxylation[a]
Ancillary ligands	Sterically nonencumbering phosphorus donor ligands greatly accelerate reaction	Little effect
Stereochemistry at α–carbon	Retention of configuration	Retention of configuration

[a]These observations have been extensively noted for $RMn(CO)_5$, and the more limited study on the group 6 anionic analogs reported herein is in complete agreement with these generalizations.

Insertion of CO_2 into Metal-Oxygen Bonds.

An investigation of carbon dioxide insertion reactions involving the analogous group 6 metal-alkoxide and -aryloxide derivatives (eq. 14) is dependent on the availability of these complexes.[21-23] Hence, much of our efforts in this area has entailed the development of synthetic methods for generating these species. There are two major problems which must be overcome in order to isolate monomeric $M(CO)_5OR^-$ derivatives. Firstly, reaction (15) is readily reversible, where the mononuclear metal alkoxides are unstable with respect to formation of the metal hy-

$$M(CO)_5OR^- + CO_2 \longrightarrow M(CO)_5O_2COR^- \qquad (14)$$

$$[M]H^- + R_2CO \longrightarrow [M]OCHR_2^- \qquad (15)$$

dride and corresponding aldehyde or ketone.[23] Secondly, these monomeric species have a great prosperity to aggregate to dimers or higher nuclearity clusters.[22,25] We have avoided the difficulty of the reversible nature of reaction (15) by employing aryloxides as ligands, and have overcome the second obstacle by performing the chemistry under an atmosphere of carbon monoxide.

The synthetic methodology for generating the mononuclear aryloxide complexes is summarized in **Scheme 4**. These complexes are unstable and decompose **via** loss of CO ligands to form binuclear $[M_2(CO)_6(OR)_3]^{-3}$ (M = W) or tetranuclear $[M_4(CO)_{12}(OR)_4]^{-4}$ (M = Cr, Mo, W) species. The stability of the $M(CO)_5OR^-$ complexes follows the trend W > Cr > Mo. Several of the derivatives have been characterized by X-ray crystallography, including $[Et_4N][W(CO)_5OPh \cdot 0.5H_2O]$ which has 1/2 a molecule of water in the crystal lattice (**Figure 2**).

Scheme 4

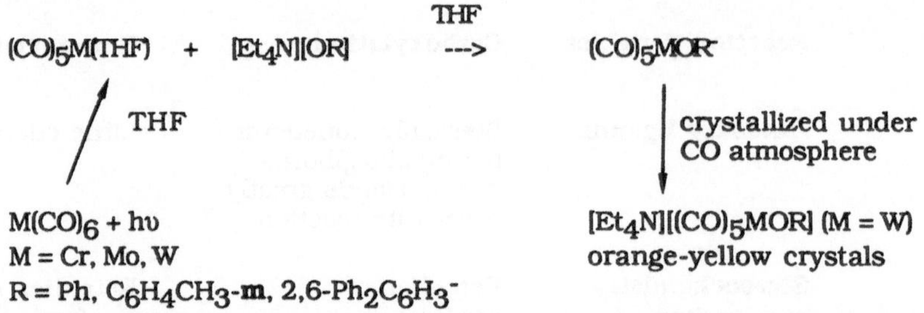

$$(CO)_5M(THF) + [Et_4N][OR] \xrightarrow{\quad THF \quad} (CO)_5MOR^-$$

(CO)$_5$M(THF) has an arrow from M(CO)$_6$ + hυ (THF)

(CO)$_5$MOR$^-$ crystallized under CO atmosphere

M(CO)$_6$ + hυ
M = Cr, Mo, W
R = Ph, C$_6$H$_4$CH$_3$-**m**, 2,6-Ph$_2$C$_6$H$_3^-$

[Et$_4$N][(CO)$_5$MOR] (M = W)
orange-yellow crystals

The focus of our efforts have been on the tungsten derivatives, since these are the most stable complexes. Consistent with the observation that W(CO)$_5$OPh$^-$ is stable with respect to cluster formation in a CO atmosphere, these complexes have been shown to undergo a facile CO ligand exchange reaction with ^{13}CO in solution (eq. 16). No further reactions with CO were noted, even at high pressures (> 500 psi) of carbon monoxide, either with respect to phenoxide displacement to afford W(CO)$_6$ or CO insertion to provide W(CO)$_5$C(O)OPh$^-$. This latter species has been independently synthesized by nucleophilic addition of OPh$^-$ to W(CO)$_6$ in tetrahydrofuran.

$$(CO)_5WOPh^- + {}^{13}CO \longrightarrow ({}^{13}CO)_5WOPh^- + CO \qquad (16)$$

As is evident in **Figure 2**, the metal bound phenoxide ligand can be involved in hydrogen-bonding with water in the solid state. An analogous interaction is seen in solution with alcohols or Na$^+$ ions (**11** and **12**) as is exhibited by the shift to higher frequencies of the υ(CO) vibrations in the infrared spectra. It would therefore be anticipated that other electrophiles should interact at the oxygen atom of the metal bound aryloxide ligand.

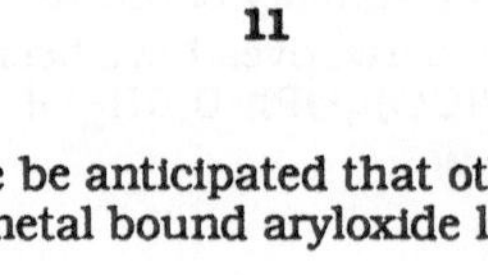

11 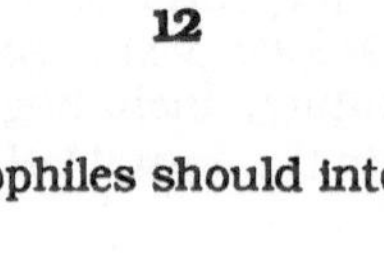   **12**

It would therefore be anticipated that other electrophiles should interact at the oxygen atom of the metal bound aryloxide ligand.

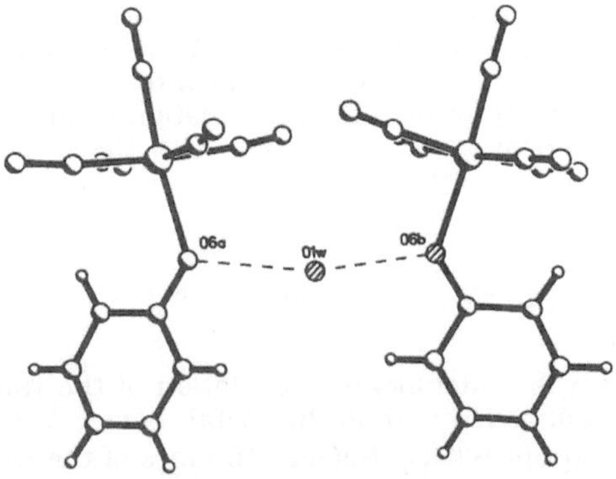

Figure 2. Hydrogen-bonding network in the $W(CO)_5OPh \cdot 0.5H_2O^-$ anions.

Relevant to the subject under discussion herein, in the absence of steric hindrance these metal aryloxides readily react with the electrophile carbon dioxide to afford metal arylcarbonates (eq. 15). Similar reactivity was noted for the analogous reactions with COS and CS_2, however, due to the increased M-S bond strengths in these derivatives the reactions were irreversible. Because of the reversible nature of these reactions, in general the carbonate derivatives were only stable in a CO_2 atmosphere. Although there are slight shifts to higher frequencies in the $\upsilon(CO)$ region of the infrared spectra upon CO_2 insertion into the M-O bond, these reactions are best monitored by ^{13}C NMR where the carbon resonance in free CO_2 at 125.9 ppm shifts to 159.8 ppm in the $-O_2\underline{C}OR$ ligand.

When the carbon dioxide insertion reactions of $W(CO)_5OPh^-$ was preformed in 2-10 atmospheres of carbon monoxide at ambient temperature, no $W(CO)_6$ was formed nor was the rate of carbon dioxide insertion altered significantly. Hence, the phenoxide moiety remains in the coordination sphere of the metal during the insertion process and CO dissociation from the metal center is not rate-determining. We suggest that CO_2 behaves as an electrophile, interacting initially at the lone pair of the bound phenoxide in much the same manner as Na^+ ions or alcohols, with subsequent formation of the ligated phenyl carbonate (**Scheme 5**).

Scheme 5

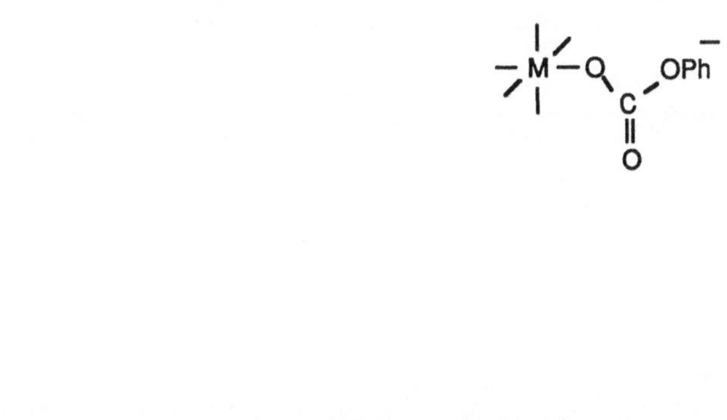

The reversibility of the carbon dioxide insertion reaction (**Scheme 5**) is consistent with our current view of the decarboxylation mechanism. That is, whether or not decarboxylation occurs appears to be strongly dependent on the nature of the X group in the [MOC(O)X] moiety. Decarboxylation is only observed when the X group has an appropriate orbital for interaction with the metal center during the decarboxylation process, e.g., when X = -H,[11] -CH$_2$CN,[26], -OR.

$$\begin{array}{c} M-O \\ \diagdown \\ \diagup \\ X \end{array} C=O \quad \longrightarrow \quad M-X \; + \; CO_2 \qquad (17)$$

Qualitatively, there is a significant retardation of the rate of carbon dioxide insertion into the metal-alkoxide bond as the metal center is made sterically more hindered. For example, in **cis**-W(CO)$_4$(L)OPh$^-$ the rate of the CO$_2$ or CS$_2$ insertion processes parallels an increase in the L ligands cone angle,[27] with CO >> P(OMe)$_3$ > PMe$_3$ > PPh$_3$. Similarly, when the oxygen atom is sterically blocked by bulky groups on the aryloxide, e.g., in the 2,6-diphenyl-C$_6$H$_3$O complex (**Figure 3**), no CO$_2$ insertion is observed even at pressures greater than 800 psi.[28]

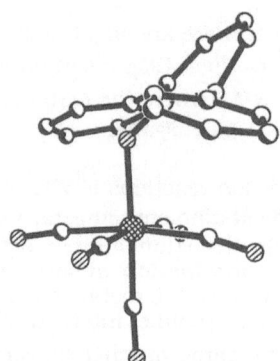

Figure 3. X-ray structure of the W(CO)$_5$OC$_6$H$_3$Ph$_2^-$ anion.

The addition of small quantities of water to THF solutions of the tungsten arylcarbonates leads to precipitation of bright-orange crystals of the carbonate species, W(CO)$_4$CO$_3^=$ (eq. 18). This complex has been characterized by infrared and ^{13}C NMR spectroscopies, and X-ray crystallography.

$$W(CO)_5O\overset{\overset{\displaystyle O}{\|}}{C}\text{-}OPh^- \; + \; H_2O \quad \dashrightarrow \quad (CO)_4W \overset{O}{\underset{O}{\diagdown\diagup}} C=O \quad ^{-2} \qquad (18)$$

Insertion into the Metal-Nitrogen Bond,

Most instances of CO_2 insertion into the $M-NR_2^-$ groupage (eq. 4) have been ascribed to proceed **via** an amine-catalyzed sequence of reactions (eq. 19-20).[29] This area will be discussed in detail in a chapter of these proceedings by Calderazzo and coworkers. However, most recently Cowan and Trogler[30] have reported experi-

$$Me_2NH + CO_2 \quad \rightleftharpoons \quad Me_2NCOOH \qquad (19)$$

$$M-NMe_2 + Me_2NCOOH \quad \longrightarrow \quad MO_2CNMe_2 + Me_2NH \qquad (20)$$

mental evidence based on isotopic studies that there is at least one case which proceeds by way of CO_2 insertion analogous to that previously discussed herein for the M-OR linkage.

Homogeneous Catalyzed Reactions of Carbon Dioxide

The production of methyl formate from the hydrocondensation of carbon dioxide in alcohols utilizing anionic group 6 metal hydrides as catalysts has been reported under rather mild reaction conditions (loading pressures of CO_2 and H_2, 250 psi each, and 125°C).[31] Reaction (1) represents a fundamental step in this process, with a subsequent reaction of the metalloformate complex with dihydrogen via a ligand-assisted heterolytic splitting mechanism leading to formic acid. HCOOH consequently rapidly reacts with methanol to provide methyl formate (see **Scheme 6**).

Scheme 6

Methyl Formate Synthesis from CO_2

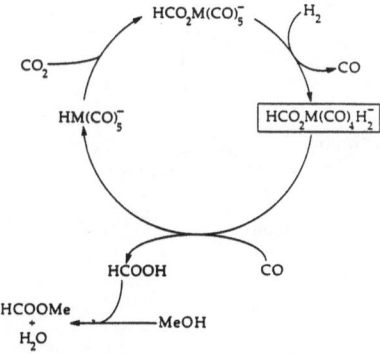

Consistent with **Scheme 6** the addition of CO retards alkyl formate production, strongly implying CO_2 as the primary source of the carboxylic carbon atom in HCOOR. This was verified by carrying out reactions in the presence of $HCO_2W(^{13}CO)_5^-$ which provided only $H^{12}COOR$ after short reaction times. However, in the absence of CO_2 and H_2 the anionic metal hydrides were observed to be effective catalyst precursors for converting CO and methanol into methyl formate. In this connection it is important to recall that a combination of reactions (21) and (22)

leads to reaction (23), i.e., the formation of methyl formate from the carbonylation of methanol.

$$H_2O(g) + CO(g) \rightleftharpoons H_2(g) + CO_2(g) \qquad (21)$$

$$CO_2 + H_2 + MeOH \longrightarrow HCOOMe + H_2O \qquad (22)$$

$$MeOH + CO \longrightarrow HCOOMe \qquad (23)$$

The carbonylation process (eq.23) involving group 6 metals as catalysts is more efficient at producing HCOOMe than the reaction proceeding through carbon dioxide (eq. 22).[32] For example, under comparable reaction conditions the turnover numbers per day are approximately 15 and 270, respectively for the carboxylation and carbonylation catalytic processes employing anionic tungsten carbonyl hydrides as catalysts. A catalytic cycle in accordance with experimental observations is represented in **Scheme 7.** Intermediates for both processes (**Schemes 6** and **7**) were in general identified by in situ FTIR studies using a cylindrical internal reflectance cell.[33] The methanol carbonylation process is greatly inhibited by carbon dioxide. This inhibition involves the reaction of the cocatalyst, OMe⁻, with CO_2 to yield methyl carbonate.

Scheme 7

Methyl Formate Synthesis from CO

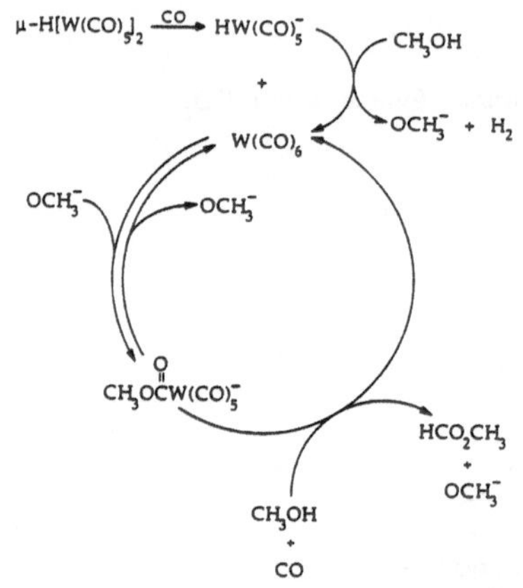

C-H bond activation in alkanes mediated by transition metal complexes is currently well-documented in the literature.[34] Very limited progress has been made in functionalizing the activated alkane. **Scheme 8** illustrates the catalytic cycle which encompasses both alkane and carbon dioxide activation, where the L_nM fragment is generated by thermal or photochemical labilization of a bound ligand.

Scheme 8

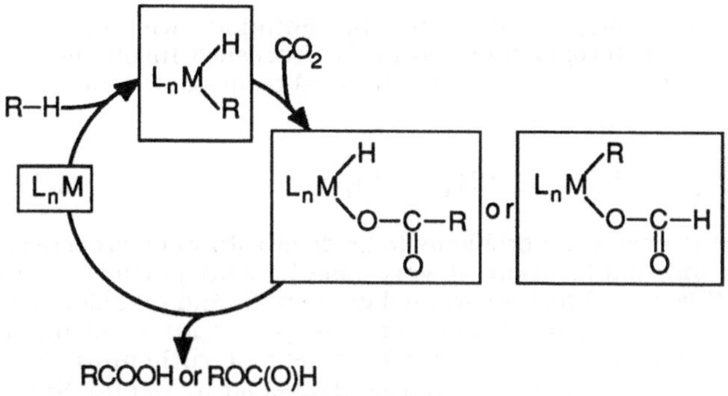

Based on our current knowledge of the relative barrier for carbon dioxide insertion into M-H **vs.** M-R bonds, production of the alkyl formate product would be anticipated. Alternatively, there is precedence for heterolytic activation of the alkane with concomitant addition of the nucleophile to carbon dioxide (eq. 24).[26,35]

$$HIr(PMe_3)_4 \;+\; CH_2(CN)_2 \;\dashrightarrow\; [H_2Ir(PMe_3)_4]^+[CH(CN)_2]^- \qquad (24)$$

$$\Big\downarrow CO_2$$

$$(CN)_2CHCO_2^-$$

Finally, the insertion of carbon dioxide into M-O bonds (reaction 3) is of paramount importance in the development of homogeneous catalysts for the copolymerization of carbon dioxide and epoxides (**Scheme 9**)[36]. Work in this area is currently being pursued in our laboratories.

Scheme 9

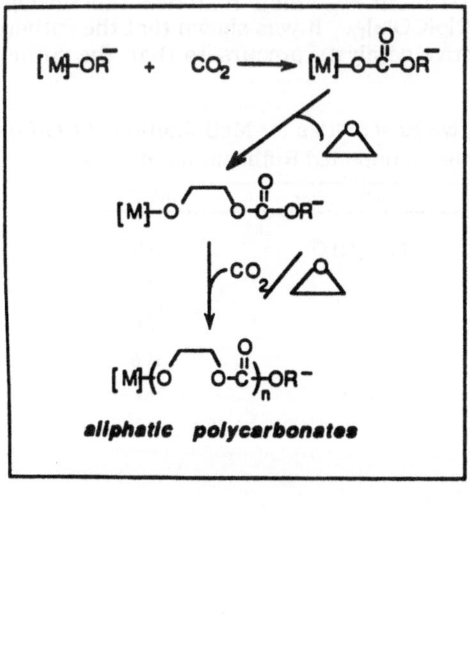

Heterogeneous Catalyzed Reactions of Carbon Dioxide.

In the remaining portion of this presentation I wish to change the focus from homogeneous to heterogeneous catalysis processes involving carbon dioxide. In particular an investigation of supported metal catalysts for the hydrogenation of CO_2 to methane, methanation (eq. 25). [37]

$$CO_2 + 4H_2 \longrightarrow CH_4 + 2H_2O \qquad (25)$$

One of the essential problems to be dealt with in heterogeneous catalysis is the preparation and stabilization of very small metal particles supported on oxide carriers. In this regard metal carbonyl clusters in and on solid metal supports such as silica, alumina, and magnesia, serve as good sources of highly dispersed, low valent metals for catalysis. A large variety of ruthenium clusters have been supported on alumina and activated as described in eq.(26).[38] The infrared $\upsilon(CO)$ spectra of all the supported clusters were found to be identical after activation, having a characteristic two band pattern with peaks at 2043 and 1963 cm^{-1}. The structure of these surface species has been proposed to be $[Ru(CO)_2X_2]_n$, where X is a surface oxygen atom of alumina and **n** is unknown.[38,39]

e.g. ,

$$H_4Ru_4(CO)_{12} \xrightarrow[\text{2. } H_2, 200°C]{\substack{\text{1. THF or hexane} \\ Al_2O_3}} (Ru(CO)_2X_2)_n \qquad (26)$$

X = Oxygen atoms on alumina
n = unknown

Catalysts prepared similarly from $RuCl_3$ produce a different surface species upon activation under hydrogen at 200°C followed by exposure to carbon monoxide. The infrared spectrum of this species exhibits a three band pattern with peaks at 2140, 2075, and 2013 cm^{-1}, which matches the $\upsilon(CO)$ infrared pattern of $[RuCl_2(CO)_3]_n$ or $[RuCl_2(CO)_4]_n$. It was shown that the ruthenium carbonyl clusters are generally more active catalysts precursors than the ruthenium chloride derived catalyst (**Table II**).

Table II. Relative Reactivities for Methanation of CO_2 using Alumina- Supported Ruthenium Catalysts

Catalyst Precursor	180 °C
$RuCl_3 \cdot 3H_2O$	1.0
$Ru(CO)_5$	4.7
$Ru_3(CO)_{12}$	12.6
$H_4Ru_4(CO)_{12}$	14.6
$CRu_6(CO)_{17}$	21.7

Although all the surface species produced by the various ruthenium carbonyl derivatives appear to be identical, it is obvious upon inspection of Table II that these vary significantly in catalytic activity toward methanation. Indeed a correlation is noted between metal cluster size and reactivity, with the higher nuclearity clusters exhibiting greater reactivity. This is best seen in a plot of metal dispersion as determined from O_2 chemisorption measurements and precursor cluster nuclearity (**Figure 4**). Since the metal particles are generally greater than 50 Å as determined by electron microscopy, it emerges that the degree of metal aggregation upon decarbonylation (or activation) is inversely proportioned to cluster nuclearity.

The surface reactions of the activated cluster catalysts were studied by diffuse reflectance FTIR spectroscopy in an effort to identify the surface species which leads to methane production.[40] **Figure 5** shows the diffuse reflectance spectra of the surface of an activated catalyst under a mixture of $CO_2:H_2$ of 1:2.5, as is used in catalysis. Under this atmosphere the original carbonyl bands maintain their band pattern but are shifted slightly to lower energies at 2036 and 1954 cm^{-1}. A formate species is also present on the surface under catalytic conditions as evidenced by the symmetric and antisymmetric $\upsilon(CO_2)$ stretches at 1591 cm^{-1} and 1375 cm^{-1} and also

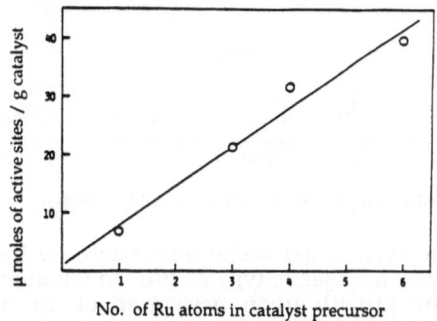

Figure 4. Plot of metal dispersion **vs.** cluster nuclearity.

the C-H stretch at 2904 cm^{-1}. At 150°C, no methane is observed; however, when the sample is heated to 200°C, methane production commences as is seen by the appearance of the band at 3016 cm^{-1}. This band can be assigned to the C-H stretching frequency of gaseous methane which is trapped in the DRIFTS cell, and so is observed in the IR band pattern. Methane production increases with increasing temperature, as can be seen by the amount of methane produced at 250°C in **Figure 5**.

Independent experiments indicate that when a catalyst pretreated with CO_2 and H_2 at 130°C to afford surface formates is evacuated and retreated with hydrogen, methane production occurred without an appreciable diminish of the infrared bands due to the formate species. Hence, hydrogenation of the surface formates does not appear to be the source of methane. This taken along with changes noted in the terminal CO region of the infrared spectra during catalysis leads us to propose a mechanism for methane formation which proceeds **via** metal carbonyl and RuO(ads) to RuC(surface species). Hydrogenation of RuO(ads) and RuC(surface species) ultimately result in methane and water production.

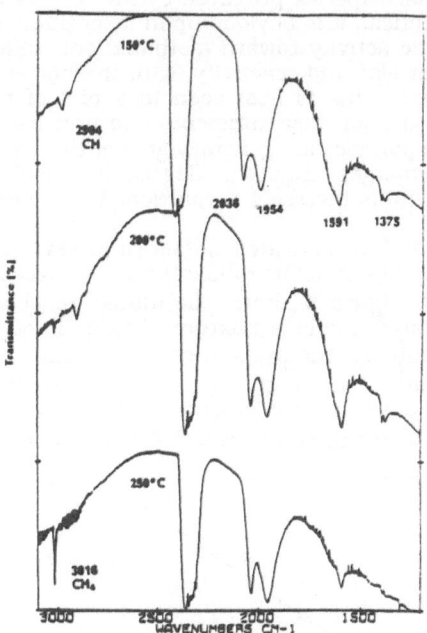

Figure 5. IR Spectra of Catalyst under CO_2 and H_2 Mixture.

In an effort to prepare even better metal dispersion, we have investigated a higher surface area support, i.e., a faujasite-type zeolite. In addition, the zeolite pores might limit metal crystallite growth upon activation of an intrazeolite metal carbonyl cluster. However, high nuclearity ruthenium carbonyl clusters are difficult to support on such zeolites from solution. Some success had been achieved by sublimation of $Ru_3(CO)_{12}$ onto Y-type zeolite.[41] In this instance it has been suggested that the ruthenium carbonyl molecules are located inside the supercages of the support. This does not seem possible since the $Ru_3(CO)_{12}$ cluster appears to be too large (9.2 Å) to fit through the zeolite pores (7.6 Å).

In our investigations we have employed two different procedures for supporting $Ru_3(CO)_{12}$ on zeolites. The first of these follows the previously reported vapor phase deposition onto the dry NaY support. The second preparation proceed through the $Ru(CO)_5$ monomer under high CO pressure. The size of the monomer (6.3 Å) should allow penetration of the zeolite pores. It has been shown that the 13 Å supercages of the NaY zeolite accommodates three molecules of the structurally similar $Fe(CO)_5$ molecule.[42] Therefore, clusterification to the stable $Ru_3(CO)_{12}$ upon release of CO pressure can occur inside the zeolite supercages. The location of the $Ru_3(CO)_{12}$ species in the two catalyst precursor preparations was determined to be outside the supercage for the vapor phase synthesis and inside the supercage for the high pressure synthesis (**Figure 6**) by their size selective reactivity with phosphorus donor ligands (PPh_3 **vs.** $P(OMe)_3$). That is, upon addition of solutions of PPh_3 (a ligand previously shown to be too large to penetrate the zeolite pores[43]) to the two differently prepared solid catalysts, only the vapor phase deposited catalyst reacted to provide $Ru_3(CO)_{12-n}(PPh_3)_n$ derivatives in solution with concomitantly little

$Ru_3(CO)_{12}$ remaining on the zeolite support. On the other hand the high pressure synthesized catalyst was found to only react with smaller phosphine ligands, such as $P(OMe)_3$, with no dissolution of the trinuclear cluster.[44]

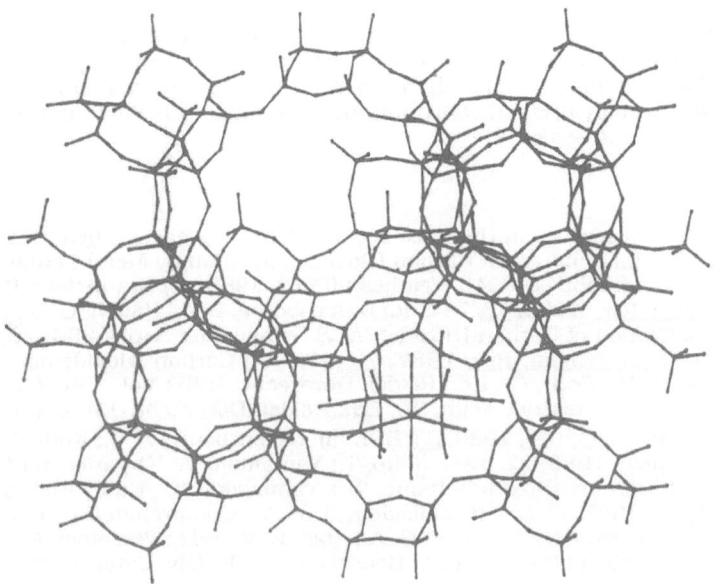

Figure 6. NaY with $Ru_3(CO)_{12}$ encapsulated in the bottom-right supercage.

The catalytic activities of the two catalyst precursors upon activation under hydrogen or helium at 200°C for 8-10 hrs were investigated. In both cases selectivity to methane was greater than 95%, with a trace of CO at temperatures at or above 300°C being the only other observable carbon-based product. In each case, the catalytic activity is compared to that of the previously discussed $Ru_3(CO)_{12}/Al_2O_3$ precursor. The best catalyst precursor for methanation of CO_2 is the one in which $Ru_3(CO)_{12}$ was deposited from the vapor phase onto the outer surface of the zeolite. For the zeolite supported catalysts the metal dispersion was greater for the helium activated (>60%) than for the hydrogen activated catalyst. Indeed in the case of the intrazeolite catalyst, helium activation appears to result in little metal aggregation, as is apparent in that the catalyst retains its yellow color upon activation, rather than decomposing to a grey solid (suggestive of the growth of metal particles) which is the case for the H_2-activated catalyst.

Contrary to the behavior of the vapor phase deposited $Ru_3(CO)_{12}$ catalyst, which suffered a slight loss in catalytic activity with time, presumably due to a buildup of inactive carbon on metal sites, the behavior of the intrazeolite ruthenium catalyst displayed an increase in activity with time under catalytic conditions. Nevertheless, during this time period the metal dispersion decreased significantly For example, an increase in catalytic activity of >300% was exhibited by a helium-activated system after 72 hrs of catalysis at 300°C. This is most likely due to the high volatility of the very stable ruthenium oxides which migrate and agglomerate outside the zeolite supercages,[45,46] where the ruthenium has been shown to be catalytically more active. The low activity initially manifested by the highly dispersed ruthenium metal in the intrazeolite catalysts is evidently a mass transport problem, even though the reactants and products are themselves all small entities.

Acknowledgements. The author is most grateful to the National Science Foundation, whose support has made possible his contributions to the research described herein. He is likewise extremely appreciative to all his colleagues mentioned in the references, whose many original contributions have made this such an exciting area of research in which to work.

References

1. For the pertinent reviews in this area see: (a) Behr, A. Angew. Chem. Int. Engl. Ed. **1989**, 27, 661. (b) Behr, A. in "Carbon Dioxide Activation by Metal Complexes"; VCH Verlagsgesellschaft mbH, D-6940 Weinheim (FRG), 1988. (c) Braunstein, P.; Matt, D.; Nobel, D. Chem. Rev. **1988**, 88, 747. (d) Darensbourg, D. J.; Bauch, C. G.; Ovalles, C., "Catalytic Activation of Carbon Dioxide", ACS Symposium Series **363**, 26, 1988. (e) Walther, D. Coord. Chem. Rev. **1987**, 79, 135. (f) Carbon Dioxide as a Source of Carbon; Aresta, M., Forti, G., Ed.; Reidel: Dordrecht, 1987; Vol. 206. (g) Behr, A. in "Catalysis in C₁ Chemistry"; Keim, W., Ed.; Reidel: Dordrecht, 1983. (h) Palmer, D. A.; van Eldik, R. Chem. Rev. **1983**, 83, 651. (i) Darensbourg, D. J.; Kudaroski, R. Adv. Organomet. Chem. **1983**, 22, 129. (j) Ito, T.; Yamamoto, A. "Organic and Bio-organic Chemistry of Carbon Dioxide"; Inoue, S.; Yamazaki, N.; Eds.; Kodansha, Ltd.; Toyoko, Japan, 1982, p. 79. (k) Sneeden, R. P. A. "Comprehensive Organometallic Chemistry"; Wilkinson, G.; Stone, F. G. A.; Abel, E.W.; Eds.; Pergamon Press: Oxford, 1982; Vol. 8, p. 225. (l) Eisenberg, R.; Hendriksen, D. E. Adv. Catal. **1979**, 28, 79. (m) Kolomnikov, I. S.; Grigoryan, M. Kh. Russ. Chem. Rev. **1978**, 47, 334.

2. (a) Darensbourg, D. J.; Rokicki, A.; Darensbourg, M. Y. J. Am. Chem. Soc. **1981**, 103, 3223. (b) Darensbourg, D. J.; Rokicki, A. Organometallics **1982**, 1, 1685. (c) Slater, S. G.; Lusk, R.; Schumann, B. F.; Darensbourg, M. Y. Organometallics **1982**, 1, 1662.

3. (a) Volpin, M. E.; Kolomnikov, I. S. Organomet. React. **1975**, 5, 313; Pure Appl. Chem. **1973**, 33, 567 (b) Kolomnikov, I. S.; Stepovska, G.; Tyrlik, A.; Volpin, M. E. Zh. Obshch. Khim. **1972**, 42, 1652; J. Gen. Chem. USSR (Engl. Transl.) **1972**, 42, 1645.

4. Sakaki, S.; Ohkubo, K. Inorg. Chem. **1989**, 28, 2583.

5. For a recent review of this area see: Ford, P. C.; Rokicki, A. Adv. Organomet. Chem. **1988**, 28, 139.

6. Darensbourg, D. J.; Baldwin, B. J.; Froelich, J. A. J. Am. Chem. Soc. **1980**, 102, 4688 and references therein.

7. Sullivan, B. P.; Bruce, M. R. M.; O'Toole, T. R.; Bolinger, C. M.; Megehee, E.; Thorp, H.; Meyer, T. J. ACS Symposium Series.**363**, 26, 1988.

8. Merrifield, J. H.; Gladysz, J. A. Organometallics **1983**, 2, 782.

9. Wiegreffe, P. W. Ph.D. Dissertation, 1988, Texas A&M University, College Station, TX.

10. Darensbourg, D. J.; Wiegreffe, H. P. Inorg. Chem. in press.

11. Bo. C.; Dedieu, A. Inorg. Chem. **1989**, 22, 304.

12. Darensbourg, D. J.; Darensbourg, M. Y.; Goh, L. Y.; Ludvig, M.; Wiegreffe, P. J. Am. Chem. Soc. **1987**, 109, 7539.

13. Kudo, K.; Hidai, M.; Uchida, Y. J. Organomet. Chem. **1973**, 56, 413.

14. Paonessa, R. S.; Trogler, W. C. J. Am. Chem. Soc. **1982**, 104, 3529.

15. (a) Darensbourg, D. J.; Rokicki, A. J. Am. Chem. Soc. **1982**, 104, 349. (b) Darensbourg, D. J.; Kudaroski, R. J. Am. Chem. Soc. **1984**, 106, 3672.

16. Darensbourg, D. J.; Kudaroski, R.; Bauch, C. G.; Pala, M.; Simmons, D.; White, J. N. J. Am. Chem. Soc. **1985**, 107, 7463.

17. Darensbourg, D. J.; Grotsch, G. J. Am. Chem. Soc. **1985**, 107, 7474.

18. Darensbourg, D. J.; Bauch, C. G.; Rheingold, A. L. Inorg. Chem. **1987**, 26, 977.

19. Darensbourg, D. J.; Pala, M. J. Am. Chem. Soc. **1985**, 107, 5687.

20. Kaufman, E.; Sieber, S.; Schleyer, P. v. R. J. Am. Chem. Soc. **1989**, 111, 4005.

21. Darensbourg, D. J.; Sanchez, K. M.; Rheingold, A. L. J. Am. Chem. Soc. **1987**, 109, 290.

22. Darensbourg, D. J.; Sanchez, K. M.; Reibenspies, J. H. Inorg. Chem. **1988**, 27, 821.

23. Darensbourg, D. J.; Sanchez, K. M.; Reibenspies, J. H.; Reibenspies, A. L. J. Am. Chem. Soc. **1989**, 111, 0000.

24. Tooley, P. A.; Ovalles, C.; Kao, S. C.; Darensbourg, D. J.; Darensbourg, M. Y. J. Am. Chem. Soc. **1986**, 108, 5465.

25. (a) McNeese, T. J.; Cohen, M. B.; Foxman, B. M. Organometallics **1984**, 3, 552. (b) McNeese, T. J.; Mueller, T. E.; Wierda, D. A. ; Darensbourg, D. J.; Delord, T. J. Inorg. Chem. **1985**, 24, 3465.

26. Behr, A.; Herdtweck, E.; Herrmann, W. A.; Keim, W.; Kipshagen, W. J. Chem. Soc., Chem. Commun. **1986**, 1262.

27. Tolman, C. A. Chem. Rev,. **1977**, 77, 313.

28. Darensbourg, D. J.; Mueller, B. manuscript to be submitted for publication.

29. Chetcuti, M. J.; Chisholm, M. H.; Folting, K.; Haitko, D. A.; Huffman, J. C. J. Am. Chem. Soc. **1982**, 104, 2138.

30. Cowan, R. L.; Trogler, W. C. J. Am.. Chem. Soc. **1989**, 111, 4750.

31. (a) Darensborg, D. J.; Ovalles, C. J. Am. Chem. Soc. **1984**, 106, 3750. (b) Darensbourg, D. J.; Ovalles, C. J. Am. Chem. Soc. **1987**, 109, 330.

32. (a) Darensbourg, D. J.; Gray, R. L.; Ovalles, C.; Pala, M. J. Mol. Catal. **1985**, **29**, 285. (b) Darensbourg, D. J.; Gray, R. L.; Ovalles, C. J. Mol. Catal. **1987**, **41**, 329.

33. Darensbourg, D. J.; Gibson, G., in: "Experimental Organometallic Chemistry", A. L. Wayda and M. Y. Darensbourg, Eds., American Chemical Society, Washington, D. C. (1987).

34. (a) Shilov, A. E. "Activation of Saturated Hydrocarbons by Transition Metal Complexes", D. Reidel, Dordrecht, 1984. (b) Crabtree, R. H. Chem. Rev., **1985**, **85**, 245. (c) Wenzel, T. T.; Bergman, R. G. J. Am. Chem. Soc., 1986, 108, 4856. (d) Janowicz, A. H.; Periana, R. A.; Buchanan, J. M.; Kovac, C. A.; Stryker, J. M.; Wax,

35. Behr, A.; Herdtweck, E.; Herrmann, W. A.; Keim, W.; Kipshagen, W. Organometallics **1987**, **6**, 2307.

36. Inoue, S. Chemtech 1976, **6**, 588.

37. Darensbourg, D. J.; Ovalles, C.; Bauch, C. G. Rev. Inorg. Chem. **1985**, **7**, 315.

38. Darensbourg, D. J.; Ovalles, C. Inorg. Chem. **1986**, **25**, 1603.

39. Kuznetsov, V. L.; Bell, A. T. J. Catal. **1980**, **68**, 374.

40. Darensbourg, D. J.; Mangold, D. J. unpublished observations.

41. Goodwin, J. G.; Naccache, C. J. Mol. Catal. **1982**, **14**, 259.

42. Bein, T.; Jacobs, P. A. J. Chem. Soc., Faraday Trans. **1983**, **79**, 1819.

43. Herron, N.; Stucky, G. D.; Tolman, C. A. Inorg. Chim. Acta **1985**, **100**, 135.

44. Darensbourg, D. J.; Gibson, G. unpublished results.

45. Verdonck, J. J.; Jacobs, P. A.; Genet, M.; Poncelet, G. J. Chem. Soc., Faraday I, **1980**, **76**, 403.

46. Pederson, L. A.; Lunsford, J. H. J. Catal. **1980**, **61**, 39.

CATALYTIC INCORPORATION OF CO_2 FOR THE SYNTHESIS OF ORGANIC COMPOUNDS

Pierre H. Dixneuf, Christian Bruneau and Jean Fournier.
Unité de recherche CNRS DO415, Campus de Beaulieu, Université de Rennes, 35042 Rennes (France).

ABSTRACT. The direct incorporation of CO_2 into alkynes in the presence of nucleophiles has been investigated. Vinylcarbamates and dienylcarbamates can be regioselectively produced in one-step from terminal alkynes or isopropenylacetylene, CO_2 and secondary amines with ruthenium catalysts. Reactions of propargyl alcohol derivatives and CO_2 can lead selectively to β-oxopropylcarbamates, with ruthenium catalysts, and to α-methylene cyclic carbonates or functional α-methylene oxazolidinones with phosphines catalysts.

Introduction

Carbon dioxide constitutes a natural source of carbon and an alternative C_1 synthetic unit, with a potential for the functionalization of substrates and new organic compounds syntheses. Due to its high stability, the non-natural incorporation of CO_2 into unsaturated hydrocarbons requires activation which can be provided by transition metal centers (1).

Since the first characterization of a $M(\eta^2\text{-}CO_2)$ complex $Ni(CO_2)(PCy_3)_2$ (2) a variety of metal-CO_2 complexes have been prepared (3-8). Examples of insertion of CO_2 into M-H, M-C, M-O or M-N bonds are now well-documented (9,10). However, the direct involvement of metal-CO_2 complexes, or CO_2 insertion products into metal-atom bonds, in catalytic syntheses remains unusual (11,12).

For the catalytic incorporation of CO_2 to an unsaturated substrate, recent results show that the activation of the unsaturated hydrocarbon itself at a coordinatively unsaturated metal center, and in the presence of CO_2, appears very successful. For example, the activation of dienes by palladium(0) complexes in the presence of CO_2 has afforded a catalytic route to unsaturated lactones (13-15). The catalytic ring-opening of methylenecyclopropanes with palladium-phosphine derivatives and CO_2 provides a five-membered ring lactone synthesis (16) which can reach 80% yield (17). Allene itself also reacts with CO_2 when activated by palladium

65

M. Aresta and J. V. Schloss (eds.),
Enzymatic and Model Carboxylation and Reduction Reactions for Carbon Dioxide Utilization, 65–77.
© 1990 *Kluwer Academic Publishers.*

(18) or rhodium (18,19) catalysts to produce lactone and unsaturated esters. Ethylene has been transformed using rhodium catalyst and high pressure of CO_2 under reducing conditions into propionic acid (20).

Catalytic co-oligomerization of alkynes and CO_2 leading to pyrones has been initially performed by Inoue using Nickel(0)-diphosphine complexes (21,22) and improved using electron-rich phosphines (23). It has been extended by the selection of hindered phosphines or bidentate ligands to the transformation of diynes into bicyclic pyrones (24,25).
    The systematic study of the interaction between electron-rich nickel(0) intermediate, alkynes and CO_2, or isocyanates, has led H. Hoberg to show that oxidative coupling at the nickel center of one alkyne and CO_2 initially occurs (26) (Eq. 1).

$$'L_nNi(0)' \xrightarrow[CO_2]{R-C \equiv C-R} \quad \longrightarrow \qquad \text{Eq.1}$$

Analogous key-steps have been shown to take place for stoechiometric coupling of CO_2 with allenes, alkenes, dienes or even imine C=N and ketone C=O bonds (26). Oxidative coupling of alkyne and CO_2 at a Ni(0) site is also expected to take place in the electrochemical carboxylation of alkynes to give unsaturated carboxylic acids (27,28).

From these results one has learnt that the incorporation of CO_2 into an unsaturated substrate may not necessarily require the activation of CO_2 itself, but may be achieved by the coordination of the hydrocarbon molecule to a coordinatively unsaturated metal center allowing the oxidative coupling with CO_2. This key-step is expected to be favoured by low valent metal complexes or very electron-donating ligands.

Another general process for the incorporation of CO_2 appears to consist in the activation towards a nucleophilic CO_2-adduct of an unsaturated molecule, A=B, such as alkyne or alkene at an electrophilic metal center. Carbonates, formates or even carboxylates can be considered as nucleophilic CO_2-adducts. (Eq. 2).

$$M(A=B) \quad + \quad {}^{-}O\underset{O}{\overset{Y}{\bigvee}} \quad \longrightarrow \quad M\text{-}A\text{-}B\text{-}O\underset{O}{\overset{Y}{\bigvee}}$$

$$\underset{L}{\uparrow} \qquad\qquad\qquad\qquad\qquad\qquad \underset{L}{\uparrow}$$

In such a process, one coordination site has to be vacant on the metal allowing the fixation of the unsaturated molecule (A=B). The coordination of the CO_2-adduct is likely not to take place. This key-step involving the oxygen-carbon bond formation (Eq.2) is expected to lead to the synthesis of useful chemicals for agriculture, such as carbonate, carbamate or urea derivatives. These compounds are usually made by multi-step syntheses starting with phosgene.

We have attempted to activate various alkynes by ruthenium(II) precursors in the presence of carbamates and carbonates in the search of processes using CO_2 as a substitute for phosgene. The main aspects of our results will be presented here.

1 - Catalytic synthesis of vinylcarbamates

Vinylcarbamates, or enol carbamates, are suitable precursors for agriculture chemicals or polymerization. General methods of formation of alkylcarbamates derive from isocyanates or chloroformates. Catalytic carbonylation of amines (29) or nitroarenes (30) and alkylation of ammonium carbamates (31) or phosphacarbamates (32) also provide routes to alkylcarbamates. However, various carbamate salts have been used for the preparation of interesting derivatives such as alkylcarbamates (33), 2-hydroxyethylcarbamates (34), carboxylato and carbonatoanhydrides (35), isocyanates (36) or ureas (37). None of these reactions is suitable for the access to vinylcarbamates. Enol carbamates were previously obtained by multi-step syntheses from phosgene derivatives: dehydrohalogenation of halogenoalkylcarbamates (38,39), addition of amines to vinylchloroformates (40-42) or coupling of isocyanates with alkylidene carbenes (43).

We have considered the possibility of the addition to activated alkynes of ammonium carbamates, the protected forms of carbamic acids, resulting from the addition of a secondary amine to CO_2 (44). Our first attempts were achieved with $Ru_3(CO)_{12}$ as a catalyst precursor (45). Phenylacetylene (10 mmol) was reacted with diethylamine (20 mmol) and CO_2 (5 MPa) in toluene, THF or acetonitrile in the presence of $Ru_3(CO)_{12}$ (0.2 mmol). At 125-140°C for 20 h, 15 to 36 % yields of carbamates were formed, although the conversion of the alkyne was good. A mixture of the three isomers **2Z, 2E** and **3** (minor) was obtained (45) (Eq.3).

Eq.3

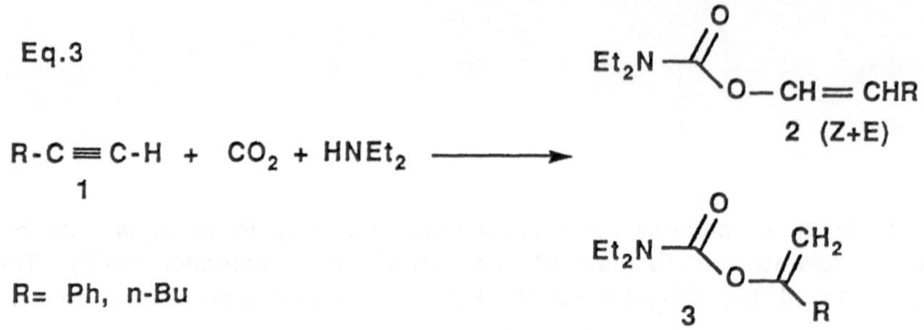

$$R\text{-}C \equiv C\text{-}H \ + \ CO_2 \ + \ HNEt_2 \ \longrightarrow$$

1

R= Ph, n-Bu

$$Et_2N \overset{O}{\underset{}{\parallel}} C - O - CH = CHR$$

2 (Z+E)

$$Et_2N \overset{O}{\underset{}{\parallel}} C - O - \overset{CH_2}{\underset{R}{\parallel}}$$

3

The efficiency of the reaction is largely increased by using mononuclear ruthenium precursors (46). With complexes such as $RuCl_2(PMe_3)(C_6Me_6)$, $RuCl_2(PMe_3)(p\text{-cymene})$, $RuCl_2(py)_2(norbornadiene)$, $RuCl(PMe_3)_2(p\text{-cymene})^+BF_4^-$ at 125°C the yield in carbamates increased to 50-67 %. With phenylacetylene or hex-1-yne only the isomers **2** were obtained, besides the formation as a by-product of the dimer of the alkyne. A stereoselectivity in the formation of the Z-isomer was observed. The reaction is <u>regiospecific</u>: the carbamate adds to the terminal carbon of the alkyne. Whereas carboxylic acids can add to diphenylacetylene (47,48) the addition of carbamates to diphenylacetylene was not observed in the presence of ruthenium and thus this reaction appears to be specific of terminal alkynes.

Acetylene itself leads to simple vinylcarbamates especially when cyclic secondary amines are used. $Ru_3(CO)_{12}$ presents a low activity (49) whereas $RuCl_3,3H_2O$ or $(RuCl_2(norbornadiene))_n$ in acetonitrile at 80-100°C appears to be the best catalyst precursors (50) (Eq. 4).

Eq.4

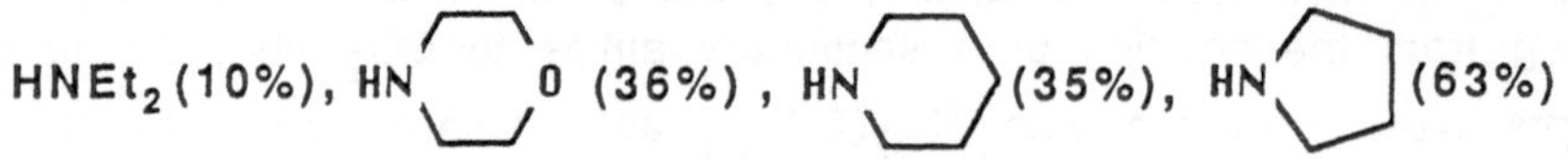

$$H\text{-}C \equiv C\text{-}H \ + \ CO_2 \ + \ R_2NH \ \longrightarrow \ H_2C = CH - O \overset{}{\underset{O}{\overset{\parallel}{C}}} - NR_2$$

$HNEt_2$ (10%), $HN\!\!<\!\!\rangle\!\!O$ (36%), $HN\!\!<\!\!\rangle$ (35%), $HN\!\!<\!\!\rangle$ (63%)

The reaction can be extended to vinylacetylene derivatives. Isopropenylacetylene in the presence of $RuCl_3,3H_2O$ in acetonitrile affords 2-methyl-3-butadienyl carbamates in 37% ($HNEt_2$) or 50% (BuNHMe) yield with a good stereoselectivity (Z/E isomers 80/20) (51) (Eq. 5).

Eq.5

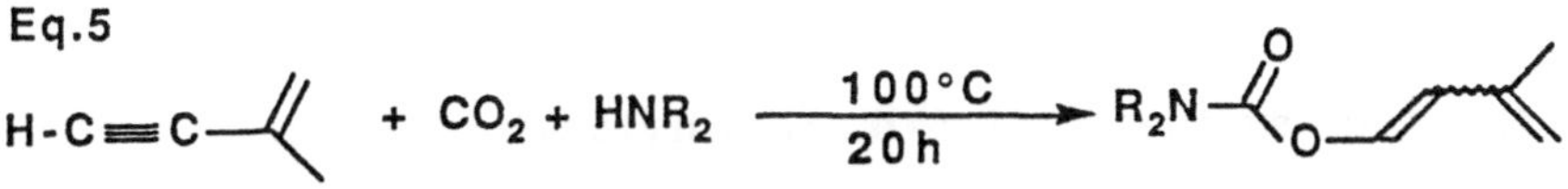

$$H-C\equiv C-\diagup\diagdown + CO_2 + HNR_2 \xrightarrow[20h]{100°C} R_2N-\overset{O}{\underset{\|}{C}}-O-\diagup\diagdown$$

These experiments show that (i) the formation of vinylcarbamates is restricted to <u>secondary amines</u>, (ii) mononuclear complexes catalyse the addition of ammonium carbamates to <u>terminal alkynes</u> only, and (iii) the addition is <u>regioselective</u> with (carbamate)oxygen-carbon(1) bond formation.

To explain this reaction, an active species, specific of terminal alkynes has been suggested. It is known that $M(\eta^2 HC\equiv CR)$ complexes readily rearrange into metal vinylidene intermediates $M=C=CHR$ presenting an electrophilic "carbene" carbon as any heteroallenes (52). Therefore a catalytic cycle has been suggested (Scheme 1) (44).

Scheme 1

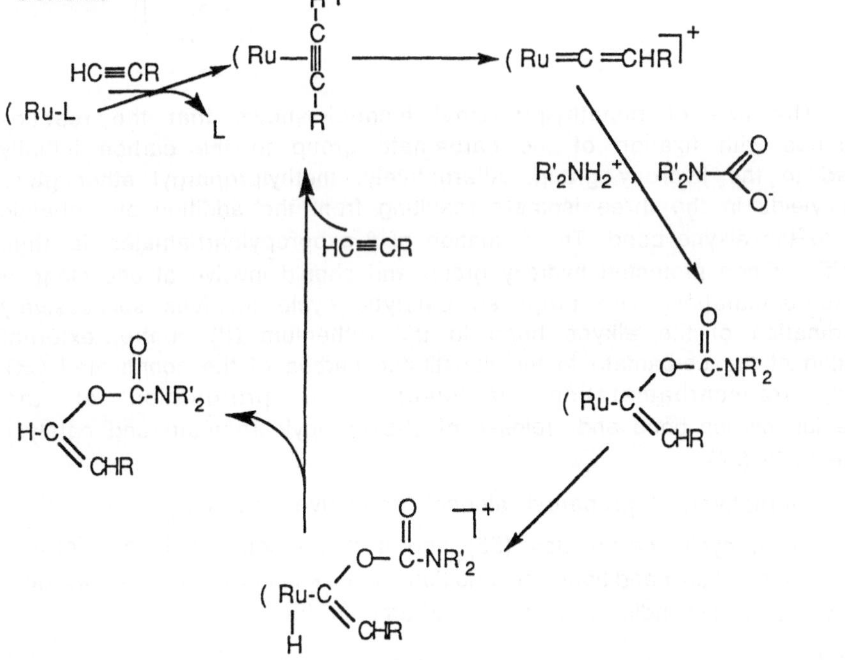

This one-step synthesis of vinylcarbamates, based on CO_2, is easy to perform and presents avantages over multi-step syntheses starting from phosgene.

2 - Catalytic synthesis of β-oxopropylcarbamates

Among cheap terminal alkynes, the activation by ruthenium catalysts of propargyl alcohol derivatives towards ammonium carbamates has been investigated. Propargyl alcohol and secondary amines in acetonitrile under 5 MPa of CO_2, do not give vinylcarbamates but lead to β-oxopropylcarbamates. The reaction is catalysed either by $Ru_3(CO)_{12}$ (53) or preferentially by $[RuCl_2(norbornadiene)]_n$ under milder conditions 50-70°C than with previous terminal alkynes yields when R=H : $HNEt_2$(40%), piperidine (38%) or pyrrolidine (24%) (54) (Eq. 6).

Eq.6

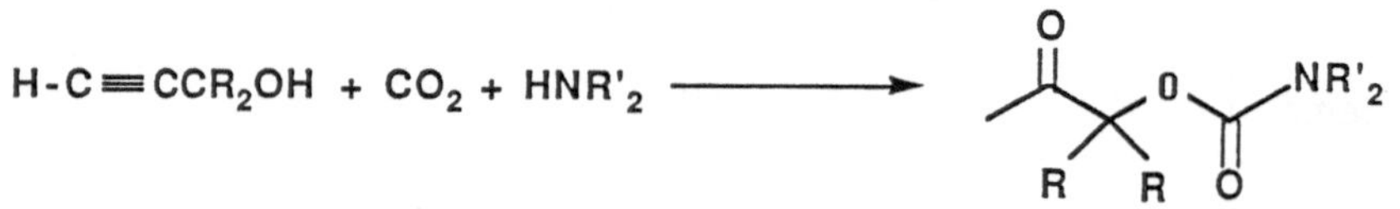

$$H-C\equiv CCR_2OH + CO_2 + HNR'_2 \longrightarrow$$

The use of dimethylpropargyl alcohol shows that the reaction proceeds with fixation of the carbamate group to the carbon initially linked to the hydroxy group. Alternatively, methylpropargyl ether gives poor yields in the three isomers resulting from the addition of carbamic acid to the alkyne bond. The formation of β-oxopropylcarbamates is thus specific of non protected hydroxy group and should involve at one stage a transcarbamatation. The proposed catalytic cycle involves successively coordination of the alkyne bond to the ruthenium (II) centre, external addition of the carbamate to the substituted carbon of the coordinated C≡C bond, transcarbamatation, tautomerisation, protonation of the ruthenium-carbon bond and release of β-oxopropylcarbamate and catalyst (Scheme 2) (54).

Alternatively, propargyl alcohol derivative and CO_2 may lead to α-methylene cyclic carbonates (55) and if these compounds are formed under the reaction conditions the addition of a secondary amine would give rise to the corresponding β-oxopropylcarbamates (Eq. 7).

Eq.7

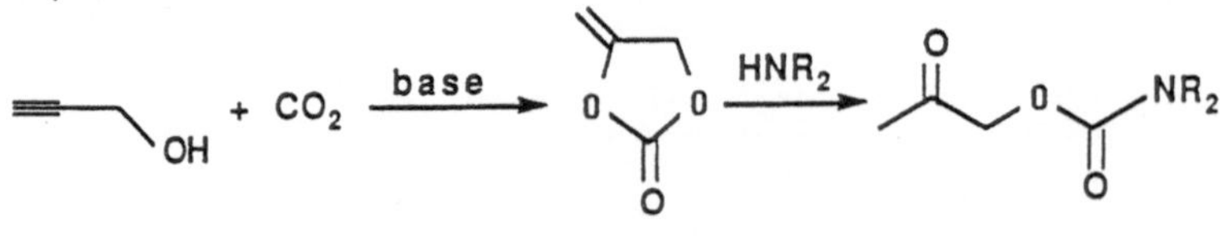

Scheme 2

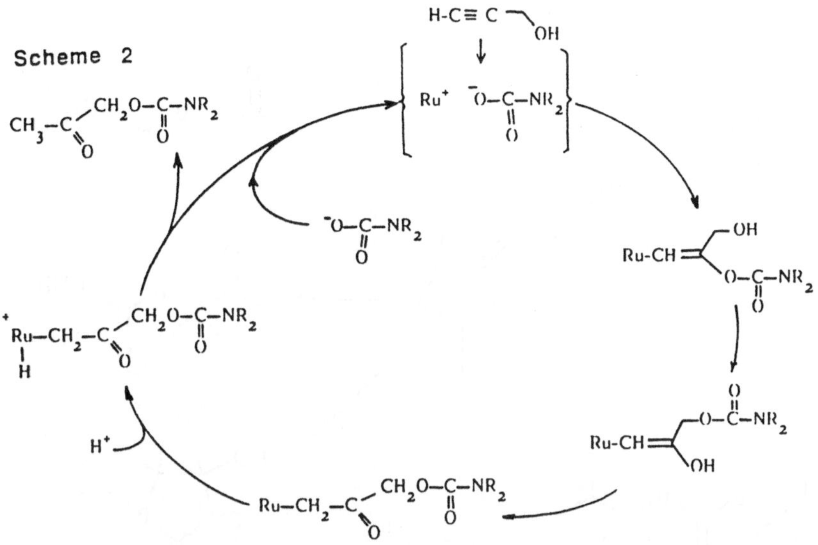

3 - Synthesis and use of α-alkylidene cyclic carbonates

Cyclic carbonates are potentially precursors for the synthesis of functional carbamates, esters or oxazolidinones (1). They are usually made by insertion of CO_2 into the C-O bond of oxiranes in the presence of Lewis acids or nucleophiles (56). α-methylene cyclic carbonates can be prepared from CO_2 and α-ethynyl alcohols by transition metal catalysts (57-60).

We have found a new and efficient way to obtain α-methylene cyclic carbonate **A** in one step from disubstituted propargyl alcohol and CO_2 in the presence of catalytic amount of phosphine (61) (Eq. 8).

Eq.8

$$\mathrm{H\text{-}C{\equiv}CCR_2OH} + \mathrm{CO_2} \xrightarrow[\text{5 MPa, 100°C, 8h}]{\mathrm{PR_3\ (cat)}}$$

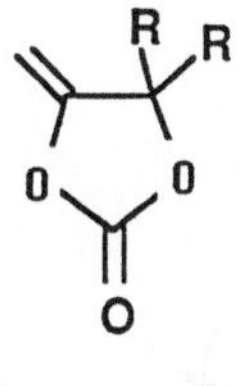

A

It is likely that the phosphine plays the role of deprotonating reagent to form a propargyl carbonate which can then cyclise into the α-methylene cyclic carbonate **A**.

The easy access to compound **A** (R=Me) (98%) has allowed its use for the access to a variety of functional carbamates and carbonates by simple reactions with amines and alcohols according to Scheme 3.

Scheme 3

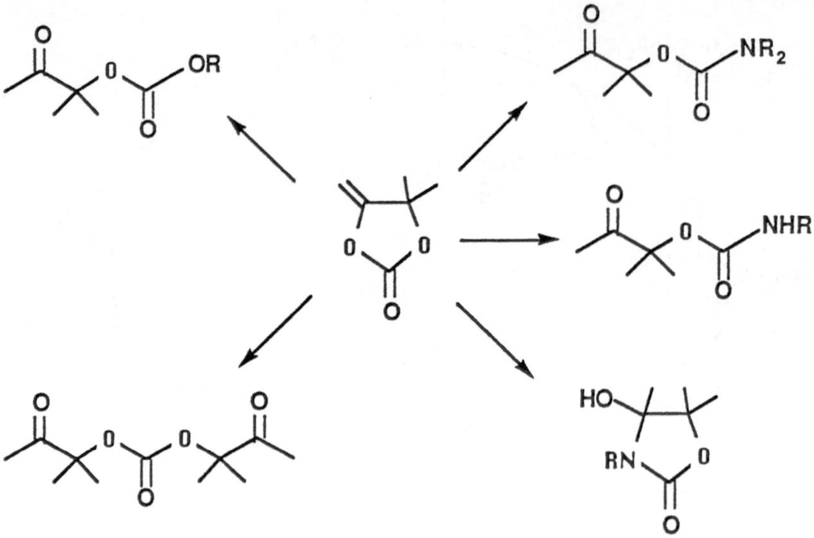

The phosphine catalysed synthesis of cyclic carbonates (61) has been used for the access to functional cyclic carbonates starting from unsaturated propargyl alcohols derivatives (62) (Eq. 9, 10).

Eq.9

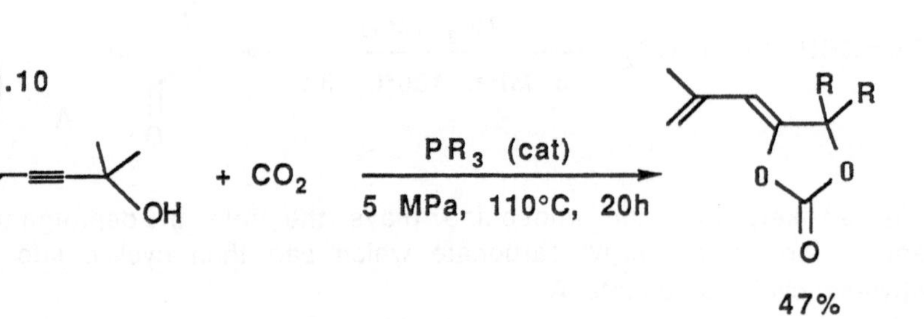

54%

Eq.10

47%

4 - Synthesis of oxazolidinones

Oxazolidinones constitute an important class of compounds as active biological agents or as polymer precursors (63). Methods of synthesis of oxazolidinones usually involve cyclisation of β-aminoalcohols or reaction of oxiranes with isocyanates (63). By contrast, α-methylene cyclic oxazolidinones may result directly from CO_2, amine and propargyl alcohol derivatives with copper catalysts (64) or from CO_2 and propargylamines (65). We have shown that the phosphine catalysed transformation of propargyl alcohols into cyclic carbonates (61) (Eq. 8) can be adapted for the synthesis of α-methylene oxazolidinones containing a functional group bounded to the nitrogen atom. Thus, 5,5-dimethyl-4-methylene-oxazolidin-2-ones can be obtained in one step directly from functional primary amines, dimethylpropargyl alcohol and CO_2 in the presence of catalytic amounts of tributylphosphine (Eq.11).

Eq.11

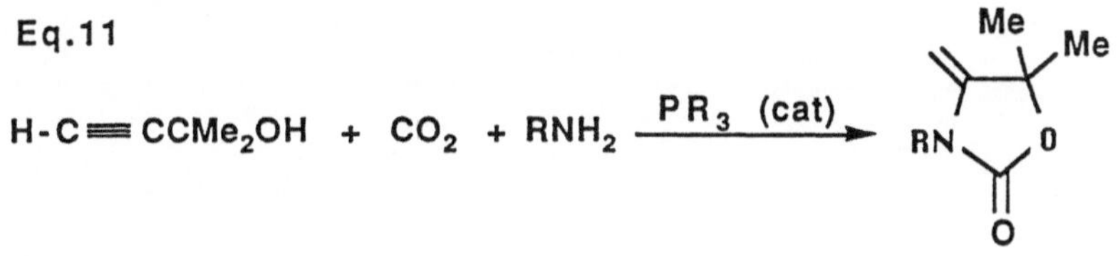

$$H-C\equiv CCMe_2OH \ + \ CO_2 \ + \ RNH_2 \ \xrightarrow{PR_3 \ (cat)}$$

R = C_6H_{11} (62%); $CH_2=CHCH_2$ (67%);   Ph (63%)

This synthesis can be extended to the preparation in good yields of bridged bis oxazolidinones such as **B** in the reaction of 1,3-diaminopropane with CO_2 and dimethylpropargyl alcohol . (Eq.12).

Eq.12

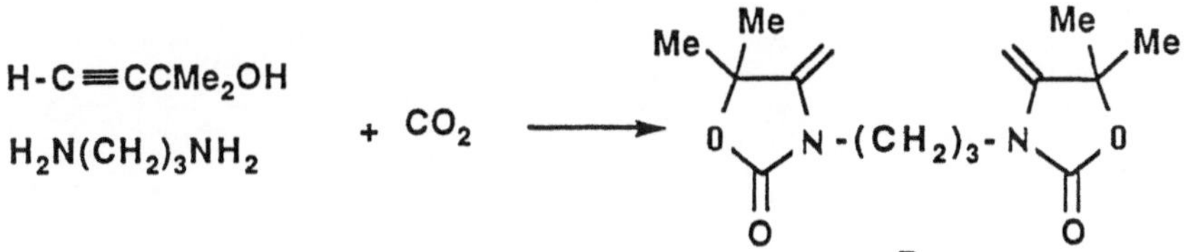

$$H-C\equiv CCMe_2OH$$
$$H_2N(CH_2)_3NH_2 \ + \ CO_2 \ \longrightarrow$$

B

Conclusion

Terminal alkynes and propargyl alcohol derivatives appear to be suitable substrates for the catalytic incorporation of CO_2 to produce unsaturated and functional carbamates, carbonates and oxazolidinones. These reactions involve either ruthenium catalysed activation of the terminal alkyne bond towards ammonium carbamates or the phosphine catalysed intramolecular cyclisation of alkynyl intermediates. These syntheses show that CO_2 can be used directly as phosgene substitute for the preparation in good yield of functional chemicals.

REFERENCES

1. M. Aresta and G. Forti, *Carbon Dioxide as a source of carbon*, NATO ASI Series C, vol. 206, **1987**.
2. M. Aresta, C.F. Nobile, V.G. Albano, E. Forni, and M. Manessero, *J. Chem. Soc., Chem. Commun.*, **1975**, 636; M. Aresta and C.F. Nobile, *J. Chem. Soc., Dalton trans.*, **1977**, 708.
3. G. Facchinetti, C. Floriani and P.F. Zanazzi, *J. Am. Chem. Soc.*, **1978**, *100*, 7405.
4. J.C. Calabrese, T. Herskovitz, and J.B. Kinney, *J. Am. Chem. Soc.*, **1983**, *105*, 5914.
5. R. Alvarez, E. Carmona, E. Gutierrez-Puebla, J.M. Marin, A. Monge, and M.L. Poveda, *J. Chem. Soc. Dalton trans.*, **1984**, 1326. R. Alvarez, E. Carmona, J.M. Marin, M.L. Poveda, E. Gutierrez-Puebla, and A. Monge, *J. Am. Chem. Soc.*, **1986**, *108*, 2286.
6. G.S. Bristow, P.B. Hitchcock, and M.F. Lappert, *J. Chem. Soc., Chem. Commun.*, **1981**, 1145.
7. S. Gambarotta, C. Floriani, A. Chiesi-villa, and C. Guastini, *J. Am. Chem. Soc.*, **1985**, *107*, 2985.
8. H.G. Alt, K.-H. Schwind and M.D. Rausch, *J. Organomet. Chem*, **1987**, *321*, C9.
9. D. Darensbourg, and R.A. Kudaroski, *Adv. Organomet. Chem.*, **1983**, *22*, 129.
10. A. Behr, *Angew. Chem. Int. Ed. Engl.*, **1988**, *27*, 661.
11. D. Walther, *Coord. Chem. Rev.*, **1987**, *79*, 135.
12. P. Braunstein, D. Matt, and D. Nobel, *Chem. Rev.*, **1988**, *88*, 747.
13. Y. Inoue, Y. Sasaki, and H. Hashimoto, *Bull. Chem. Soc. Jpn.*, **1978**, *51*, 2375.
14. A. Musco, C. Perego, and V. Tartari, *Inorg. Chim. Acta.*, **1978**, *28*, L147.
15. P. Braunstein, D. Matt, and D. Nobel, *J. Am. Chem. Soc.*, **1988**, *110*, 3207.
16. Y. Inoue, T. Hibi, M. Satake, and H. Hashimoto, *J. Chem. Soc., Chem. Commun.*, **1979**, 982.
17. P. Binger, and H-J. Weintz, *Chem. Ber.*, **1984**, *117*, 654; Ger. Offen. DE 3 403 793, **1985**.
18. A. Döhring, and P.W. Jolly, *Tetrahedron Lett.*, **1980**, *21*, 3021.
19. M. Aresta, E. Quaranta, and A. Ciccarese, C_1 *Mol. Chem.*, **1985**, *1*, 293.
20. A. Lapidus, S.D. Pirozhkov, and A.A. Koryakin, *Bull. Acad. Sci. USSR., Engl. Transl.*, **1978**, 2513.
21. Y. Inoue, Y. Itoh, and H. Hashimoto, *Chem. Lett.*, **1977**, 855; **1978**, 633.
22. Y. Inoue, Y. Itoh, H. Kazama, and H. Hashimoto, *Bull. Chem. Soc. Jpn*, **1980**, *53*, 3329.
23. D. Walther, H. Schönberg, E. Dinjus, and J. Sieler, *J. Organomet. Chem.*, **1987**, *334*, 377.
24. T. Tsuda, S. Morikawa, R. Sumiya, and T. Saegusa, *J. Org. Chem.*, **1988**, *53*, 3140.

25. T. Tsuda, S. Morikawa, and T. Saegusa, *J. Chem. Soc., Chem. Commun.*, **1989**, 9.
26. H. Hoberg in Ref. 1, p. 275.
27. E. Labbé, E. Dunach, and J. Périchon, *J. Organomet. Chem.*, **1988**, *353*, C51.
28. E. Dunach, and J. Périchon, *J. Organomet. Chem.*, **1988**, *352*, 239.
29. S.I. Murahashi, Y. Mitsue, and K. Ike, *J. Chem. Soc., Chem. Commun.*, **1987**, 125. H. Alper, G. Vaspollo, F.W. Hartstock, M. Mlekuz, *Organometallics*, **1987**, 6, 2391. S. Fukuoka, M. Chono, and M. Khono, *J. Chem. Soc., Chem. Commun.*, **1984**, 399.
30. S. Cenini, M. Pizzotti, C. Crotti, F. Porta, and G. La Monica, *J. Chem. Soc., Chem. Commun.*, **1984**, 1286.
31. Y. Yoshida, S. Ishii, and T. Yamashita, *Chem. Lett.*, **1984**, 1571.
32. M. Aresta, and E. Quaranta, *J. Org. Chem.*, **1988**, *53*, 4153.
33. T. Tsuda, K. Watanabe, K. Miyata, H. Yamamoto, and T. Saegusa, *Inorg. Chem.*, **1981**, *20*, 2728.
34. F. Kojima, T. Aida, and S. Inoue, *J. Am. Chem. Soc.*, **1986**, *108*, 391.
35. D. Belli Dell'Amico, F. Calderazzo, and U. Giurlani, *J. Chem. Soc., Chem. Commun.*, **1986**, 1000.
36. A. Belforte, D. Belli Dell'Amico, and F. Calderazzo, *Chem. Ber.*, **1988**, *121*, 1891.
37. Y. Morimoto, Y. Fujiwara, H. Taniguchi, Y. Hori, and Y. Nagano, *Tetrahedron Lett.*, **1986**, *27*, 1809.
38. R.A. Olofson, G.P. Wooden, and J.T. Marks, *Eur. Pat. 104984, Chem. Abstr.*, **1984**, *101*, 190657 u. B.R. Franco-Filipasic, and R. Patarcity, *Chem. Ind.*, **1969**, *8*, 166.
39. C.G. Overberger, H. Ringsdorf, and N. Weinshenker, *J. Org. Chem.*, **1962**, *27*, 4331. M. Shimizu, E. Tanaka, and H. Yoshioka, *J. Chem. Soc., Chem. Commun.*, **1987**, 136.
40. R.A. Olofson, B.A. Bauman, and D.J. Vancowicz, *J. Org. Chem.*, **1978**, *43*, 752.
41. R.A. Olofson, R.C. Schnur, L. Bunes, and J.P. Pepe, *Tetrahedron Lett.*, **1977**, 1567. R.A. Olofson, and R.C. Schnur, *Ibid.*, **1977**, 1571.
42. F.E. Kung, U.S. Pat. **1945**, 2377085. L.H. Lee, *J. Org. Chem.*, **1965**, *30*, 3943.
43. P.J. Stang, and G.H. Anderson, *J. Org. Chem.*, **1981**, *46*, 4585.
44. R. Mahé, Y. Sasaki, C. Bruneau, and P.H. Dixneuf, *J. Org. Chem.*, **1989**, *54*, 1518.
45. Y. Sasaki, and P.H. Dixneuf, *J. Chem. Soc., Chem. Commun.*, **1986**, 790.
46. R. Mahé, P.H. Dixneuf, and S. Lecolier, *Tetrahedron Lett.*, **1986**, 6333.
47. M. Rotem, and Y. Shvo, *Organometallics*, **1983**, *2*, 1689.
48. C. Ruppin, and P.H. Dixneuf, *Tetrahedron Lett.*, **1986**, *27*, 6323.
49. Y. Sasaki, and P.H. Dixneuf, *J. Org. Chem.*, **1987**, *52*, 314.
50. C. Bruneau, P.H. Dixneuf, and S. Lecolier, *J. Mol. Catal.*, **1988**, *44*, 175.
51. C. Bruneau, J.M. Joumier, and P.H. Dixneuf unpublished results.
52. M.I. Bruce, and A.G. Swincer, *Adv. Organomet. Chem.*, **1983**, *22*, 59.

53. Y. Sasaki, and P.H. Dixneuf, *J. Org. Chem.*, **1987**, *52*, 4389.
54. C. Bruneau, and P.H. Dixneuf, *Tetrahedron Lett.*, **1987**, *28*, 2005.
55. Y. Sasaki, *Tetrahedron Lett.*, **1986**, 1573.
56. U. Petersen in *Methoden der Organischen Chemie*, Houben-Weyl, E$_4$, **1983**, 95. A. Behr in *Catalysis in C$_1$ Chemistry*, W. Keim Editor, D. Reidel, **1983**, 169.
57. Y. Inoue, J. Ishikawa, U. Taniguchi, and H. Hashimoto, *Bull. Chem. Soc. Jpn.*, **1987**, *60*, 1204.
58. Y. Inoue, K. Ohuchi, and S. Imaizumi, *Tetrahedron Lett.*, **1988**, *29*, 5941.
59. P. Dimroth, E. Schefczik, and H. Pasedach, *Ger. Patent.*, **1963**, DBP 1145632.
60. K. Schneider, *Ger. Patent.*, **1986**, DE 3433 403 A1.
61. J. Fournier, C. Bruneau, and P.H. Dixneuf, *Tetrahedron Lett.*, **1989**,*30*, 3981.
62. J. Fournier, C. Bruneau, and P.H. Dixneuf unpublished results
63. M.E. Dyen and D. Swern, *Chem. Rev.*, **1967,** *67*, 197. V.A. Pankratov, Ts. M. Frenkel' and A.M. Fainleib, *Russian Chem. Rev.*, **1983**, *52*, 576.
64. P. Dimroth, H. Pasedach, and E Schefczik, *Ger. Pat.*, **1964**, DBP 1151507.
65. P. Dimroth and H. Pasedach, *Ger. Pat.*, **1964**, DBP 1164411.
T. Mitsudo, Y. Hori, Y. Yamakawa and Y. Watanabe, *Tetrahedron Lett.*, **1987**, *28*, 4417.

Photoreduction of carbon dioxide to formate and photo-oxidation of carbon monoxide to carbon dioxide by the use of transition metal complexes and visible light

Raymond Ziessel
Laboratoire de Chimie Organique Physique (URA 422 du CNRS)
Institut Le Bel, Université Louis Pasteur
4, rue Blaise Pascal - 67000 STRASBOURG,
France.

ABSTRACT. Visible light irradiation of solutions containing $[Ru(bpy)_3]^{2+}$ (bpy = 2,2' - bipyridine) as photosensitizer and cis-$[Ru(bpy)_2(CO)(X)]^{n+}$ (X = Cl, H, n = 1 or X = CO, n = 2) or cis-$[Ru(bpy)_2(S)_2]^{n+}$ (S = CH_3CN, DMF, n = 2 or S = Cl, n = 0) or cis-$Ru(bpy)(CO)_2(Cl)_2$ as homogeneous catalysts mediate CO_2 reduction to $HCOO^-$.

The photochemical process, which consumes triethanolamine (election donor), has a quantum yield of 15% and was studied by ^{13}C-nmr spectroscopy using labelled CO_2 in order to determine the origin of the $HCOO^-$. The process consists of two catalytic cycles : a photochemical one for the ruthenium-trischelate and a darkreaction pathway for the ruthenium-bisbpy or ruthenium-monobpy complex. Reductive quenching of $[Ru(bpy)_3]^{2+*}$ excited state by the tertiary amine gives $[Ru(bpy)_3]^+$ which reduces the CO_2 activation catalyst to Ru(I) and futher to Ru(O). Cyclic voltametry confirms the formation of a doubly reduced complex and a strong electrocatalytic current based on the second reduction wave.

$[(\eta^5\text{-}Me_5C_5)Ir(L)X]^+$, L = bpy or phen, (phen = 1,10 - phenanthroline), X = Cl, H complexes are shown to be active homogeneous catalysts for the light-driven oxidation of CO to CO_2 under extremely mild conditions (visible light, 1 atm. CO, pH = 7.0, room temperature). The photochemical process consumes CO, H_2O and light and produces catalytic amounts of both CO_2 and H_2 with turnover numbers up to 10 h^{-1}.

INTRODUCTION

There are many reasons why chemists are interested in CO_2 and CO activation processes. CO_2 is a natural and abundant source of raw material and is considered a major atmospheric pollutant, mostly involved in the greenhouse effect which may ultimatly affect the temperature of our planet [1]. CO is used in many important industrial processes like carbonylation, hydroformylation, Fisher-Tropsch reactions... There are also some fundamental reasons for studying CO_2 activation processes : CO_2 is an inert molecule with carbon in its highest oxidation state, its activation thus being difficult to achieve. Consequently very few efficient, catalytic systems involving CO_2 are known today. CO_2 could either be reduced to C_1 compounds or incorporated into organic molecules to create exciting new derivatives. Photochemical systems are of interest for conversion and storage of solar energy as well as in modelling natural

79

M. Aresta and J. V. Schloss (eds.),
Enzymatic and Model Carboxylation and Reduction Reactions for Carbon Dioxide Utilization, 79–100.
© 1990 *Kluwer Academic Publishers.*

photosynthesis. Electrochemical systems could be used to convert and to store electricity. A great deal of attention has been recently devoted to photochemical or electrochemical systems capable of activating CO_2 [2], involving a number of different strategies :

(i) **electrochemical systems** use an electrode to either directly reduce CO_2 or to reduce a soluble transition metal complex which in it's reduced state reacts futher with CO_2 ;
(ii) **photoelectrochemical systems** use an illuminated semi-conductor electrode in order to gain in applied potential, CO_2 being reduced in the same way as mention ed previously ;
(iii) **purely photochemical systems** may be considered as either **heterogeneous** involving the use of semi-conductor suspensions [3], or as **homogeneous** involving the use of aqueous solutions of metal ions [4], organic dyes [5] or of transition metal complexes. Examples of transition metal catalysts for the reduction of CO_2 can be grouped into five classes :

1.- macrocyclic cobalt and nickel complexes [6]
2.- soluble phthalocyanines [7] or porphyrin complexes [8]
3.- metal clusters such as iron-sulfur [9a] or ruthenium carbonyl [9b]
4.- rhodium [10] and palladium [11] phosphine complexes
5.- polypyridine complexes of cobalt [12], rhenium [13, 14], rhodium and iridium [15], ruthenium [15a, 16-18] and osmium [19].

The use of this latter group of complexes has recently attracted much attention. Most of these complexes efficiently reduce CO_2 to CO or $HCOO^-$. The majority of the mechanistic data on these reductions has come from studies on complexes containing polypyridyl ligands. These ligands, for example bpy or phen, have the unique properties of stabilizing metals in a wide range of oxidation states while at the same time they are "electron reservoirs" capable of storing electrons at potentials between -0.7 and -1.7 V by using vacant Π^* orbitals. The thermodynamics of CO_2 reduction, as well as the properties of the photosensitizers, used during our studies have been previously discussed in detail [20]. It is however of importance to stress that polyelectronic reactions involving 2 to 8 electrons, needed much less energy per electron transferred, than a monoelectronic pathway (see Fig. 1). Most of the homogeneous systems already known generate CO and $HCOO^-$ in reduction processes involving two electrons (Fig. 1), however one can envisaged the design and synthesis of complexes able to store and transfer more than two electrons. In most of the systems which produced traces of CH_2O, CH_3OH or CH_4, their origin has not been clearly determined. As will be mention ed in part B of this paper, complexes which are not active in CO_2 reduction reactions might be active in a reverse reaction, like the oxidation of CO (eq. 1).

$$CO + H_2O \qquad CO_2 + H_2 \quad (1)$$

Photoreduction of CO_2 to CO and photooxidation of CO to CO_2 are related by the so-called water-gas-shift-reaction (WGSR), which can be considered as a simple model of CO_2 activation. It represents therefore, a first approach to the reductive transformation of CO_2 into a fuel or other organic product, somewhat similar to processes involved in natural photosynthesis.

Our goals in the field of CO_2 activation are :
- the search for photo- or electrochemical systems which operate, selectively, at high turnover numbers near the thermodynamic potential ;
- the design and synthesis of new catalysts which can control the course of CO_2

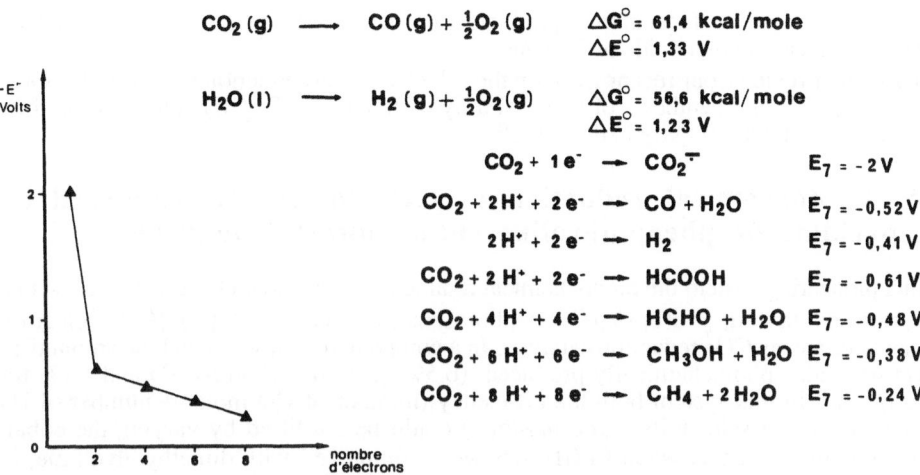

Figure 1. Redox potential of CO_2 reduction versus the number of electrons involved in the reaction. Potentials are given versus the normal-hydrogen-electrode, at pH = 7.

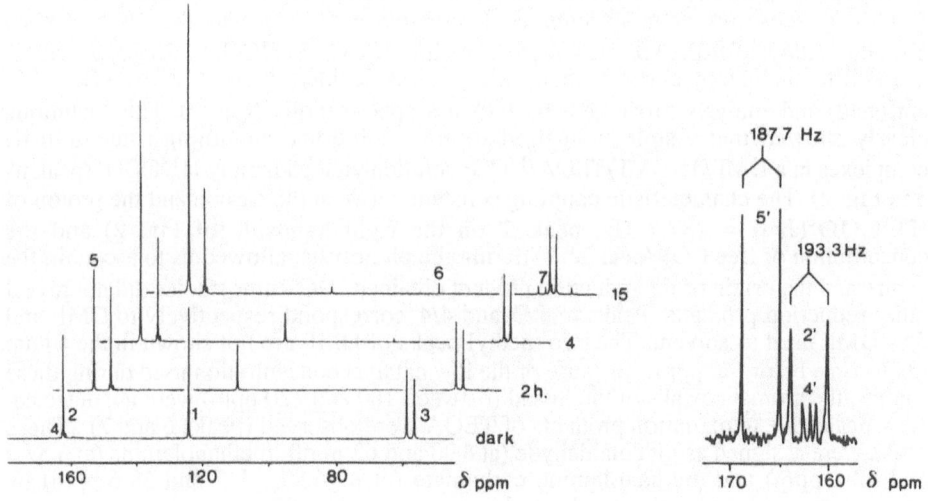

Figure 2. ^{13}C-nmr spectra of $H^{13}COO^-$ photogeneration by visible light irradiation of $[Ru(bpy)_3]^{2+}/TEOA/^{13}CO_2$ in DMF/d_7-DMF solutions : left hand-side proton decoupled spectrum (180 to 40 ppm) ; right hand-side without proton decoupling (170 to 160 ppm).

reduction toward the formation of products such as formaldehyde, methanol or oxalate, rather than carbon monoxide or formate.

This paper reviews our recent work on the selective photoreduction of CO_2 to $HCOO^-$, on the design and synthesis of a new family of Ir(III) and Rh(III) complexes and on their photocatalytic activity in the WGSR.

A. Photochemical reduction of CO_2 to $HCOO^-$ catalyzed by bipyridine or phenanthroline ruthenium(II) complexes.

In a pioneering system the photochemical reduction of CO_2 to CO was achieved with a system containing $[Ru(bpy)_3]^{2+}$ (as photosensitizer) and $[Co(bpy)_3]^{2+}$ (as electron mediator and as CO_2 reduction catalyst). In a competitive pathway catalytic amounts of H_2 were also photochemically produced (6.5% quantum yield for H_2 and 1.2% for CO)[12a]. In this system both the efficiency (number of CO moles + number of H_2 moles) and the selectivity (ratio n_{CO}/n_{H2}) could be modified by varying the cobalt ligands. For instance, when Co(II) ions were complexed with dimethylglyoxime, in place of bpy or phen derivatives, only H_2 production was observed with 16% conversion of light energy [6b]. However no ligand variation allowed us to reduce CO_2 to formate . Only non-catalytic traces of formate were detected.

In a second specific system based on the single fac-Re(bpy)(CO)$_3$Cl complex (acting both as photosensitizer and as CO_2 reduction catalyst in its reduced state), high yields of CO (14% quantum yield) were photochemically produced, in the presence of an excess of chloride anions [13]. In the absence of this excess, traces of fac-[Re(bpy)(CO)$_3$HCOO] were isolated and characterized by X-ray diffraction [21].

In light of these previous systems, an interesting extension was the search for an efficient catalyst for the specific photoreduction of CO_2 to $HCOO^-$. Since formate is difficult to analyze by classical techniques, it was decided to use labelled CO_2 (90% enriched) and analysis carried out by ^{13}C-nmr spectroscopy (Fig. 2). This technique clearly showed that visible light-irradiation of solutions containing ruthenium(II) complexes in a DMF/d7-DMF/TEOA/$^{13}CO_2$ solution yielded mainly H^{13}COO$^-$ (peak n° 5 in Fig. 2). The characteristic coupling constant between the carbon and the proton of H^{13}COO$^-$(J_{CH} = 187.7 Hz, peak 5' on the right handside of Fig. 2) and the consumption of free CO_2 (peak n° 1) during the photolysis, allowed us to ascertain the nature and the origin of the reduction product obtained. ^{13}C - nmr studies did not reveal other reduction products. Peaks n° 2/2' and 4/4' correspond respectively to DMF and d7 - DMF used as solvent. The two methyl peaks of DMF are not shown in the figure (they come below 40 ppm). Because of the low catalyst concentration used during these studies, the aromatic peaks of the ligand (between 160 and 120 ppm) were not detected. New peaks due to oxidation products of TEOA were observed (peaks 6 and 7) . These peaks were assigned as glycolaldehyde (at δ 90 and 62 ppm), diethanolamine (at δ 57.9 and 52.1 ppm) and diethanolamine carbamate (at δ 162.4, 59.7 and 50.6 ppm) by comparaison with authentic samples and literature data [13, 22].

Formate generation using the simple $[Ru(bpy)_3]^{2+}$ system, in the presence of water.

It was obvious from previous studies [23] that Ru-trisbpy itself (monoelectronic reducing agent) could not be the active CO_2 reduction catalyst. This led us to suppose that this complex plays the role of light absorber and as catalyst precursor. In order to prove this, the photocatalysis was followed by ^{1}H-nmr spectroscopy (Fig. 3). No dark reaction was observed, but an increase in H^{13}COO$^-$ (doublet at 8.52 ppm with J_{CH} =

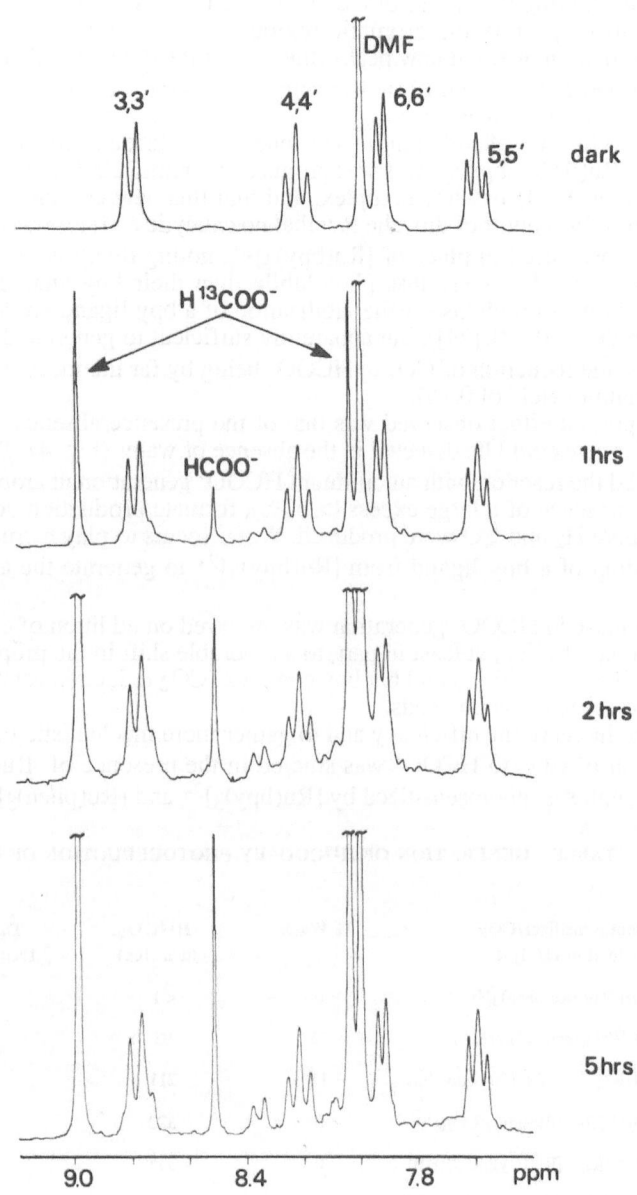

Figure 3. ^{1}H-nmr spectra (9.2 to 7.4 ppm) of a solution containing d$_7$-DMF/TEOA/H$_2$O (3:1:1), ^{13}CO$_2$ and [Ru(bpy)$_3$]$^{2+}$, in the dark and after 1, 2 and 5 hrs irradiation with visible light (250 W non focused halogen lamp).

irradiation time. During the course of the photolysis the colour of the solution changed from orange to red, while the aromatic region of the proton nmr spectra (Fig. 3) showed the formation of small new peaks (the area of the doublet at 8.36 ppm being ca 15% of the triplet at 8.22 ppm of the starting $[Ru(bpy)_3]^{2+}$ complex), indicating the formation of one or more new species. These observations, along with previous studies on the photolabilisation of ruthenium (II)-trichelate in various solvents [24] led us to suppose that a Ru(II) bis-bpy species was produced by photolabilisation of a bpy ligand from the starting Ru(II) tris-bpy complex, and that this new complex was the active catalytic species. In support of this, the fact that no catalytic activity was observed when $[Ru(phen)_3]^{2+}$ was used in place of $[Ru(bpy)_3]^{2+}$, noting that, due to ligand rigidity Ru(II)-trisphen complexes are less photolabile then their bpy analogues [25]. The photochemical quantum yields for the labilisation of a bpy ligand are low in DMF or H_2O (of the order of 0.001 [24]), but apparently sufficient to generate the catalytically active species, the reduction of CO_2 to $HCOO^-$ being by far the most efficient process (maximal quantum yield of 0.15).

The most important effect observed was that of the presence/absence of water in the system ; no formate could be detected in the absence of water (Fig. 4). The presence of water enhanced the reaction with an optimum $HCOO^-$ generation at around 15% added water. In the presence of a large excess (> 50%), formate production decreased while significantly more H_2 and CO were produced. Water seems to play a crucial role in the photolabilisation of a bpy ligand from $[Ru(bpy)_3]^{2+}$ to generate the active catalytic species.

A marked increase in $HCOO^-$ generation was observed on addition of excess bpy (see Table). This could be due, at least in part, to a favorable shift in the proportions of tris-bpy complex (photosensitizer) and bis-bpy complex (CO_2 reduction catalyst) present in the solution as the reaction proceeds.

In the order to increase the efficiency and to gather more mechanistic information, the photoreduction of CO_2 to $HCOO^-$ was studied in the presence of Ru(II) bis-bpy or mono-bpy complexes, photosensitized by $[Ru(bpy)_3]^{2+}$ and $[Ru(phen)_3]^{2+}$ complexes.

TABLE : GENERATION OF $H^{13}COO^-$ BY PHOTOREDUCTION OF $^{13}CO_2$

Photosensitizer/CO2 reduction catalyst	% Water	$H^{13}CO_2^-$ (μmoles)	Turnover frequency (h-1)
$[Ru(bpy)_3]^{2+}$ or $[Ru(phen)_3]^{2+}$	0	< 1	-
$[Ru(bpy)_3]^{2+}$ / generated in situ	15	93	10
$[Ru(bpy)_3]^{2+}$ / " / + 100 equiv. bpy	15	211	22
$[Ru(bpy)_3]^{2+}$/cis - $[Ru(bpy)_2(CO)(Cl)]^+$	0	322	200
$[Ru(bpy)_3]^{2+}$/cis - $[Ru(bpy)_2(CO)(H)]^+$	0	335	200
$[Ru(phen)_3]^{2+}$/cis - $[Ru(bpy)_2(CO)(Cl)]^+$	0	60	36
$[Ru(phen)_3]^{2+}$/cis - $[Ru(bpy)_2(CO)(H)]^+$	0	71	43
$[Ru(bpy)_3]^{2+}$/cis - $[Ru(bpy)_2(CO)_2]^{2+}$	0	142	85
$[Ru(bpy)_3]^{2+}$/cis - $[Ru(bpy)_2(DMF)_2]^{2+}$	0	110	66
$[Ru(bpy)_3]^{2+}$/cis - $Ru(bpy)_2(CO)_2(Cl)_2$	0	258	156

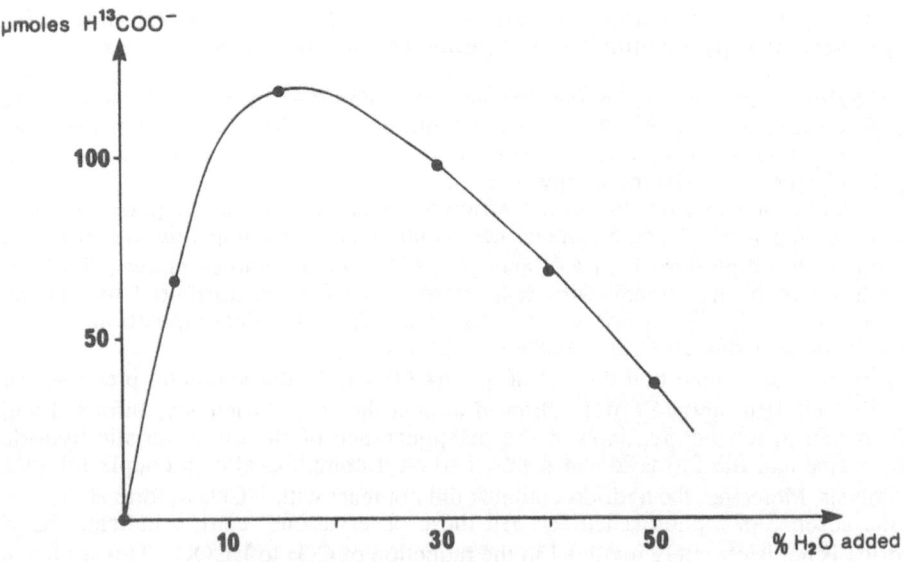

Figure 4. Amount of $H^{13}COO^-$ produced from $^{13}CO_2$ after 2 h photolysis, as a function of the amount of H_2O added ; the solution contains $[Ru(bpy)_3]^{2+}$, DMF/d_7-$DMF/TEOA/H_2O$ (3:1:1:1).

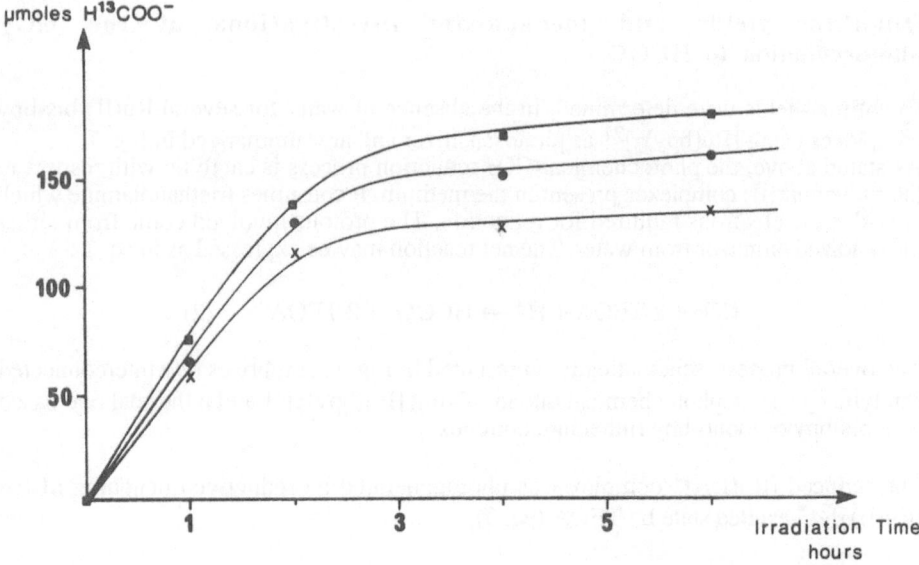

Figure 5. Photogeneration of $H^{13}COO^-$ from $^{13}CO_2$ as a function of time, using $[Ru(bpy)_3]^{2+}$ alone (X), $[Ru(bpy)_3]^{2+} + cis$-$[Ru(bpy)_2(CO)(H)]^+$ (■), $[Ru(bpy)_3]^{2+} + cis$-$[Ru(bpy)_2(CO)(Cl)]^+$ (●), in DMF/d_7-$DMF/TEOA/H_2O$ (3:1:1:1) solution.

Formate generation using the mixed system of ruthenium tris-bpy with ruthenium bis-bpy or mono-bpy complexes, in the absence of water.

As previously mentioned, in the absence of water the ruthenium(II)-trischelate complexes do not generate formate during photolysis (see Fig.4 and Table). The most efficient systems involved complexes where a CO group was coordinated to either a Ru(II) bis-bpy or a Ru(II) mono-bpy species.

The addition of water decreased the efficiency and stability of the photochemical systems (compare Fig. 5 and 6). Under these conditions water competitively reacts with the reduced complex to form CO and H_2 [26]. Kinetic studies showed $HCOO^-$ generation to be non-linear with respect to time. This is attributed to : i) the consumption of CO_2 (closed systems) and ii) some decomposition of the photosensitizer and/or of the active catalytic species.

It is of interest to note that cis - $[Ru(bpy)_2(CO)(Cl)]^+$, the synthetic precursor of complex cis-$[Ru(bpy)_2(CO)H]^+$ showed almost the same efficiency. Infra-red and proton nmr spectroscopy, showed the disappearance of the characteristic hydride resonnance and the formation of a new carbonyl complex, after a couple minutes photolysis. Moreover, the hydrido-complex did not react with $^{13}CO_2$ to form $H^{13}COO^-$ in the absence of a photosensitizer. All these observations seem to indicate that a hydride is not necessarily involved in the reduction of CO_2 to $HCOO^-$. This is also in good agreement with literature data in which hydroxycarbonyl-bpy complexes of ruthenium [17], iridium and rhodium [2b, 15] have been proposed as intermediates in this kind of reduction (see mechanistic section below). A high activity was also observed using cis-bisbpy-bissolvent complexes (see Table). This was an important observation because cis-$[Ru(bpy)_2(DMF)_2]^{2+}$ might be generated in situ by photolabilisation of bpy and followed by coordination of two DMF molecules.

Quantum yields and mechanistic investigations of the CO_2 photoreduction to $HCOO^-$.

Quantium yields were determined, in the absence of water for several Ru(II) bis-bpy complexes using $[Ru(bpy)_3]^{2+}$ as photosensitizer and are summarized in Fig. 7.

As stated above, the photochemical CO_2 reduction process is catalytic with respect to the ruthenium(II) complexes present in the medium. It consumes triethanolamine which provides the electrons required for reduction. The protons involved come from either the oxidized amine or from water. The net reaction may be expressed as in eq. 2 :

$$CO_2 + 2 \text{ TEOA} + H^+ \rightarrow HCOO^- + 2 \text{ TEOA}^+. \quad (2)$$

The overall process, schematically represented in Fig. 8, comprises two interconnected catalytic cycles : a photochemical one involving $[Ru(bpy)_3]^{2+}$ and a thermal one based on a bis-bpy or mono-bpy ruthenium complex.

The reduced $[Ru(L)_3]^+$ complex was photogenerated by reductive quenching of the $[Ru(L)_3]^{2+*}$ excited state by TEOA (eq. 3).

$$[Ru(bpy)_3]^{2+*} + \text{TEOA} \rightarrow [Ru(bpy)_3]^+ + \text{TEOA}^+. \quad (3)$$

From a Stern-Volmer plot of the luminescence intensity changes ($\lambda_{emission}$ = 607 nm) with addition of increasing amounts of quencher (Fig. 9), a rate constant of 1.7×10^5

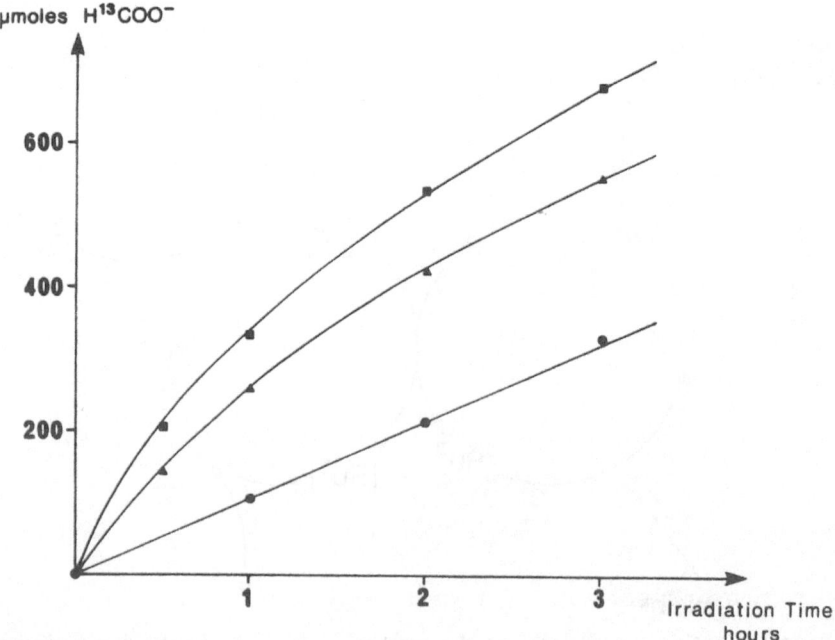

Figure 6. Photogeneration of $H^{13}COO^-$ from $^{13}CO_2$ as a function of time, using $[Ru(bpy)_3]^{2+}$ (as photosensitizer), and cis-$[Ru(bpy)_2(DMF)_2]^{2+}$ (●), cis-$Ru(bpy)(CO)_2(Cl)_2$(▲), or cis-$[Ru(bpy)_2(CO)(Cl)]^+$ (■) (as carbon dioxide reduction catalysts) in a DMF/d$_7$-DMF/TEOA (3:1:1) solution.

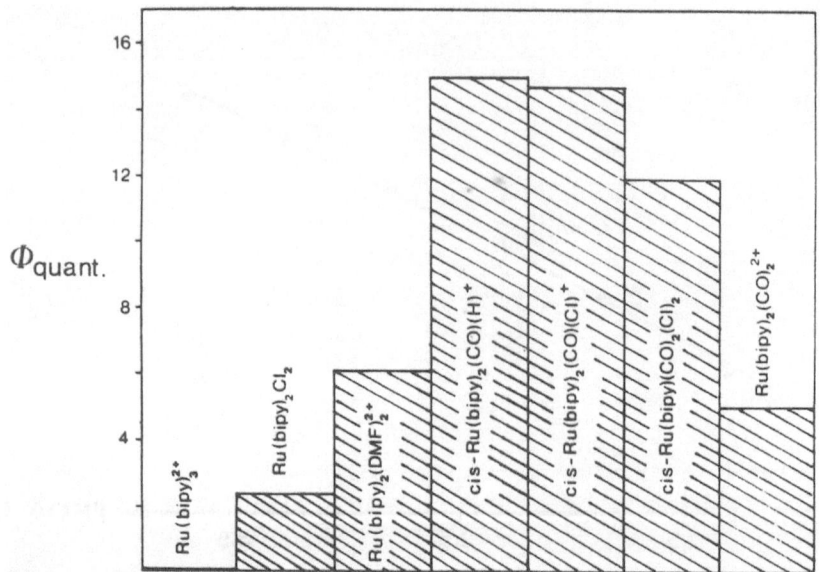

Figure 7. Quantum yields for the photoconversion of CO_2 to $HCOO^-$ using various ruthenium bis-bpy or mono-bpy complexes.

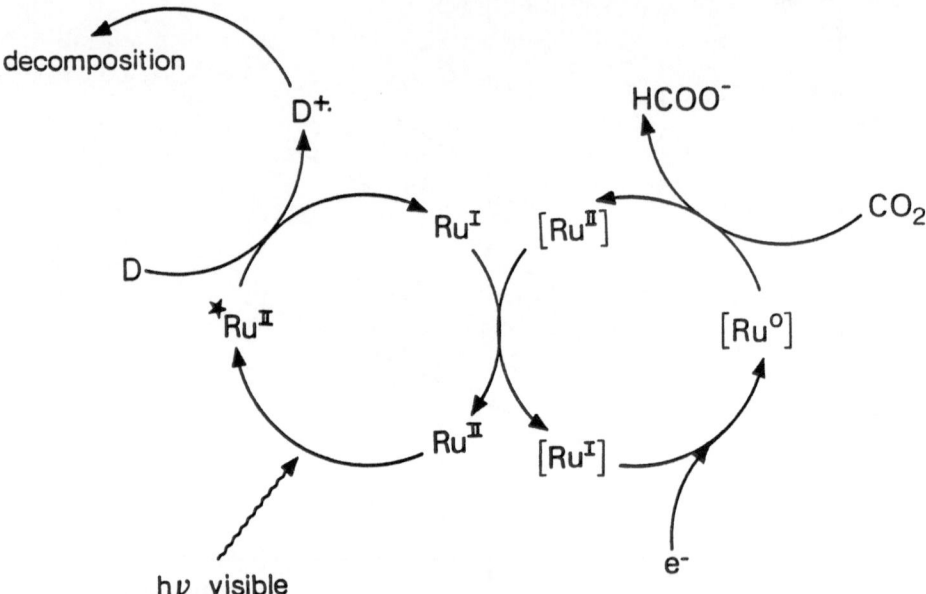

Figure 8. Schematic representation of the processes occuring in the generation of formate by photoinduced reduction of carbon dioxide, using the mixed catalytic system and following a pathway of reductive quenching of the excited state of the ruthenium photosensitizer ; it involves a photosensitizer cycle (ruthenium-trischelate) (left) and a CO_2 catalytic reduction cycle (ruthenium bis- or monochelate) (right) ; D represents TEOA (electron donor) ; the ligands on the metal ions are not included.

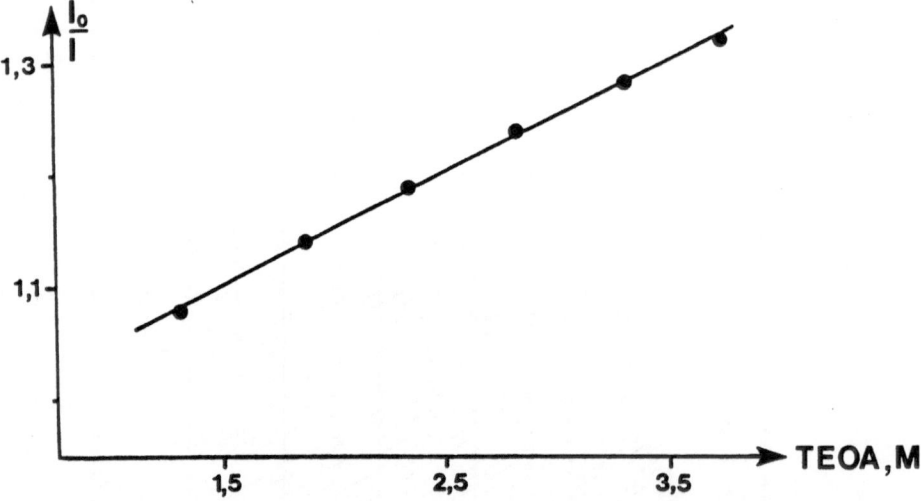

Figure 9. Stern-Volmer plot of the reductive quenching of $[Ru(bpy)_3]^{2+*}$ fluorescence by TEOA, at 30°C in DMF, under argon.

$M^{-1}s^{-1}$ for the quenching reaction (eq. 3) was determined. Flash photolysis experiments on $[Ru(bpy)_3]^{2+}$ in the presence of a tertiary amine have also indicated the formation of $[Ru(bpy)_3]^+$ [27].

The cyclic voltammogram (c.v.) of $[Ru(bpy)_2(CO)(Cl)]^+$ in an **argon satured DMF** solution (Fig. 10) showed two successive one-electron reductions. One reversible reduction at -1.39 V and the other irreversible at -1.64 V vs SCE (satured colomel electrode). The peak separations between the cathodic and anodic waves of the redox reactions at $E_{1/2}$ = -1.39 and -1.64V are 60 and 80 mV, respectively, at a sweep rate of 400 $mV.s^{-1}$, using a platinum electrode. This confirms that the two-electron reduction of $[Ru(bpy)_2(CO)(Cl)]^+$ is followed by a slow chemical reaction, induced by the high reactivity of the apparent 20-electron configuration of the ruthenium atom. Moreover, the c.v. of $[Ru(bpy)_2(CO)(Cl)]^+$ in **CO_2 satured DMF** (dashed line of Fig. 10) showed that little or no current enhancement occurs at the first reduction peak. A strong catalytic current begins to flow around -1.5 V and continues through the potential region characteristic of the second two electron reduction of $[Ru(bpy)_2(CO)(Cl)]^+$; this being formally a Ru(O) complex. Removal of CO_2 from the solution by bubbling argon through it resulted in complete disappearance of the strong cathodic current to give the c.v. of the complex in argon-satured DMF solutions. Constant potential electrolysis of solutions containing $[Ru(bpy)_2(CO)(X)]^+$, X = Cl, H, in freshly distilled, CO_2 satured DMF/H_2O solutions at -1.65 V resulted in sustained electrocatalytic reduction of CO_2 to HCOOH. These results are similar to previous studies[13] performed on fac-Re(bpy)(CO)$_3$Cl in which a two-electron reduction of CO_2 to CO, involves the anion $[Re(bpy)(CO)_3]^-$. This has also been clearly shown by cyclic voltametry (see Fig. 11). Studies on photo-and electrochemical cis-$[Ru(bpy)_2(CO)_2]^{2+}$ catalyzed CO_2 reduction, led to the proposition that this complex was reduced stepwise by two moles of $[Ru(bpy)_3]^+$ to give a coordinatively unsatured, formally Ru(0), complex after the release of one CO molecule [17]. Similarly in the present system the same active species $[Ru(bpy)_2(CO)]^0$ (18e$^-$ species) could be generated by double reduction of the starting cis - $[Ru(bpy)_2(CO)(Cl)]^+$ and release of a chloride anion (Fig. 12). The reduction of CO_2 to HCOO$^-$ might proceed through the formation of a metallocarboxylic species (eq. 4), which after chloride coordination would give formate and the starting ruthenium-bis-bpy complex (eq. 5). Such a complex $[Ru(bpy)_2(CO)(COO^-)]^+$ has been identified and isolated during the water-gas-shift-reaction catalyzed by $[Ru(bpy)_2(CO)(Cl)]^+$ [28]. It could be formed during catalysis, by nucleophilic attack of the Ru(0) complex on the CO_2 molecule (eq. 4).

$$[Ru(bpy)_2(CO)]^0 + CO_2 + H_2O \rightarrow [Ru (bpy)_2(CO)(COOH)]^+ + OH^- \quad (4)$$
$$[Ru(bpy)_2(CO)(COOH)]^+ + Cl^- \rightarrow [Ru(bpy)_2(CO)(Cl)]^+ + HCOO^- \quad (5)$$

Monomethoxycarbonyl ruthenium complexes have been obtained by reaction of ruthenium(0) clusters with methoxide anion in methanol [29]. Hydroxyl-carbonyl complexes of platinum were prepared by nucleophilic attack of OH$^-$ on a carbonyl ligand [30] or by insertion of CO into a hydroxy platinum complex [31]. Hydroxycarbonyl-bpy complexes of ruthenium [17], iridium and rhodium [15] have been proposed as intermediates in the reduction of CO_2 to HCOO$^-$.
A classical mechanism involving the formation of a hydride (eq. 6) followed by CO_2 insertion and release of the coordinated formate (eq. 7 and 8) has been ruled out because of the inertness of the cis-$[Ru(bpy)_2(CO)(H)]^+$ toward $^{13}CO_2$.

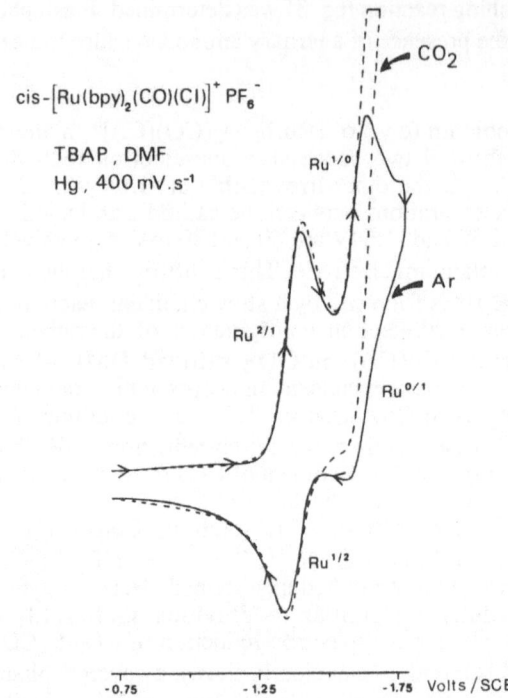

cis-$[Ru(bpy)_2(CO)(Cl)]^+$ PF_6^-

TBAP, DMF
Hg, 400 mV.s^{-1}

Figure 10. Cyclic voltamogram of cis-[Ru(bpy)$_2$(CO)(Cl)]$^+$ PF$_6^-$ in DMF with tetrabutylammonium perchlorate as supporting electrolyte using a hanging Hg electrode taken at a sweep rate of 400 mVs^{-1}. The solid line corresponds to c.v. performed under argon and dashed line under CO$_2$.

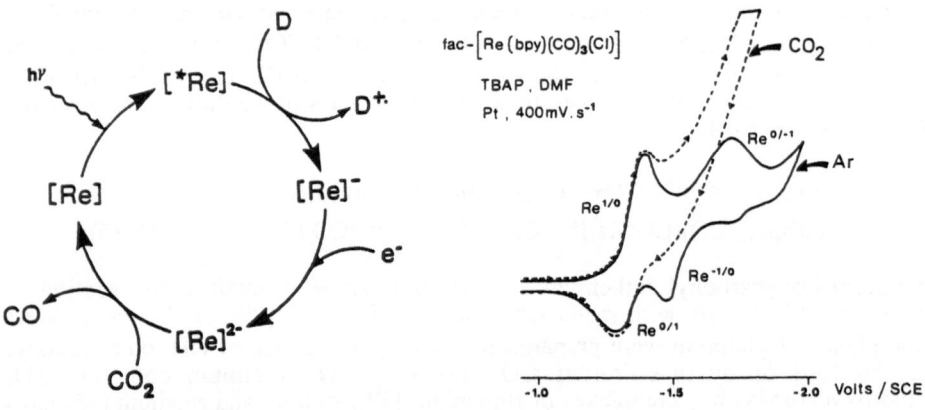

Figure 11. Left hand-side : schematic representation of the photochemical CO generation process using fac-Re(bpy)(CO)$_3$Cl as CO$_2$ activation catalyst. Right hand-side : c.v. of fac-Re(bpy)(CO)$_3$Cl in the same conditions as in Fig. 10, using a platinum electrode. The solid line corresponds to c.v. performed under argon and dashed line under CO$_2$.

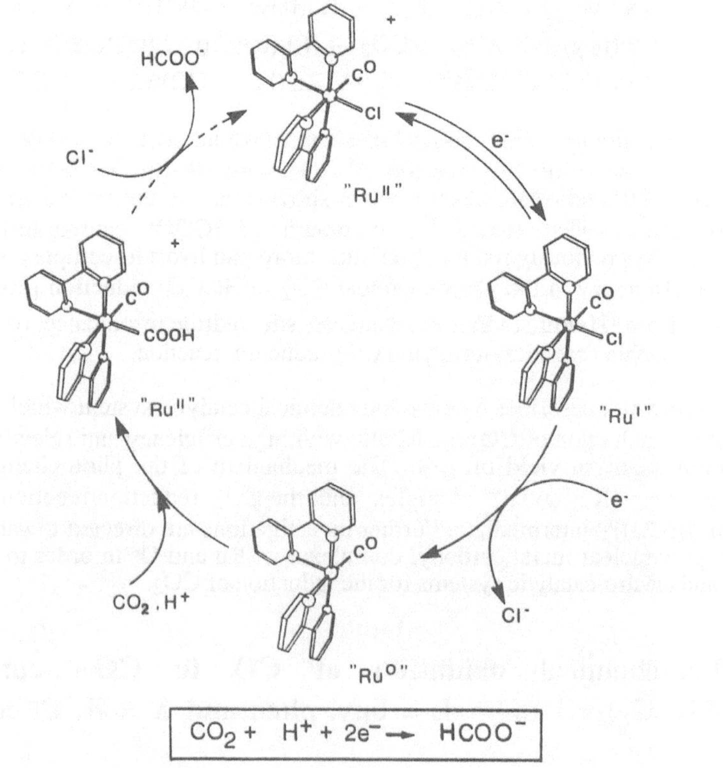

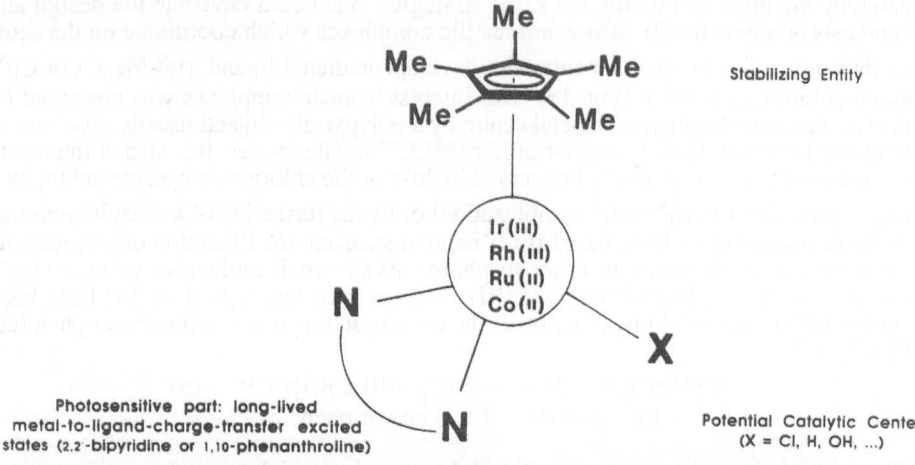

Figure 12. Schematic representation of a possible catalytic ruthenium cycle for the reduction of CO_2 to $HCOO^-$.

Figure 13. Schematic representation of the synthesis of new photocatalysts based on cyclopentamethylcyclopentadienyl, polypyridine and anion ligands. The global charges of the resulting complexes are not shown.

$$[Ru(bpy)_2(CO)]^0 + H_2O \rightarrow [Ru(bpy)_2(CO)(H)]^+ + OH^- \quad (6)$$
$$[Ru(bpy)_2(CO)(H)]^+ + CO_2 \rightarrow [Ru(bpy)_2(CO)(OOCH)]^+ \quad (7)$$
$$[Ru(bpy)_2(CO)(OOCH)]^+ + Cl^- \rightarrow [Ru(bpy)_2(CO)(Cl)]^+ + HCOO^- \quad (8)$$

Other reasons ruling out this classical mechanism would include the observation that no thermal or photochemical reaction of CO_2 with $[Ru(bpy)_2(CO)(H)]^+$ occured. However, FT-IR and NMR spectroscopies showed that decomposition of the complex, with formation of near stoechiometric amounts of $HCOO^-$ occured during long term photolysis. As previously mentionned, the chloro and hydrido-complexes have almost the same efficiency in the photochemical CO_2 to $HCOO^-$ reduction process (Fig. 7). Moreover from ^1H-nmr studies, the characteristic hydride resonnance ($\delta = -11.3$ ppm) disappears within secondes during the CO_2 reduction reaction.

The present work describes a new photochemical catalytic system which performs the dielectronic reduction of CO_2 to $HCOO^-$ with high efficiency and selectivity, giving a maximum quantum yield of 15%. The mechanism of the photochemical reaction involves the $[Ru(bpy)_3]^+$ complex and the CO_2 reduction reaction involves the $[Ru(bpy)_2(CO)]^O$ intermediate. Further investigations are directed towards the use of related polynuclear metal carbonyl complexes of Ru and Os in order to develop new photo-and electro-catalytic systems for the reduction of CO_2.

B. Photochemical oxidation of CO to CO_2 catalyzed by $[(\eta^5\text{-}Me_5C_5)Ir(L)X]^+$ L = bpy, phen and X = H, Cl complexes

In the previous section, we have seen that the coordination of different ligands (bpy, phen, carbonyl, chloride or hydride...) on a mononuclear metal centre provides a rich coordination chemistry with interesting photophysical properties, as well as unexpected catalytic qualities. Following the same strategies, one could envisage the design and synthesis of a new family of organometallic complexes which coordinate on the same metal centre a bpy or phen, a pentamethylcyclopentadienyl ligand ($\eta^5\text{-}Me_5C_5$ or Cp^*) and a chloride or hydride (Fig. 13). Our interest in such complexes was prompted by the fact that complexation of a metal centre by a polypyridine ligand usually gives rise to long lived metal-to-ligand-charge-transfer (MLCT) excited states. It is also of interest to point out that vacant sites may be created by loss of the chloride or hydride anion, by a ring-slip mechanism ($\eta^5 \rightarrow \eta^3$ complexation) or by the formation of a sesquibipyridine. Both the possibility of forming a MLCT excited state and the liberation of a vacant site on the metal are required in order to photoactivate small molecules using a single complex. Iridium (III) and rhodium (III) complexes of this type (Fig. 14) have been synthesized in high yield by reaction of the corresponding dimer with bpy or phen (eq. 9) [32, 33]

$$[Cp^*MCl_2]_2 + 2\,LL \rightarrow 2\,[Cp^*M(LL)Cl]^+ \, Cl^- \quad (9)$$
$$M = Ir \text{ or } Rh \quad LL = bpy \text{ or } phen$$

The iridium hydrido-complexes **2** and **4** were obtained by reaction with sodium cyanoborohydride (eq. 10).

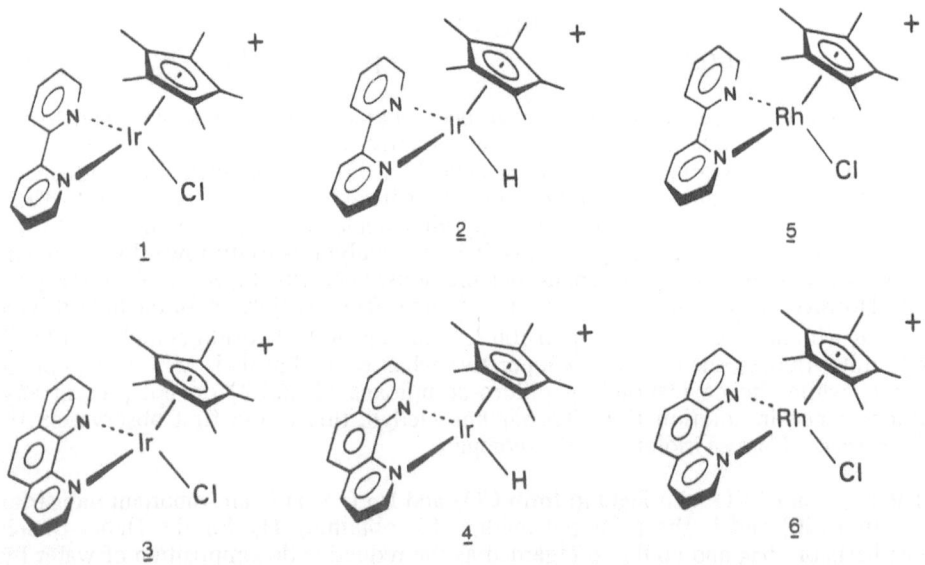

Figure 14. Complexes synthesized as described in Fig. 13. Complexes **1**, **3**, **5** and **6** were isolated as the chloride salts and complexes **2** and **4** as the tetraphenylborate salt.

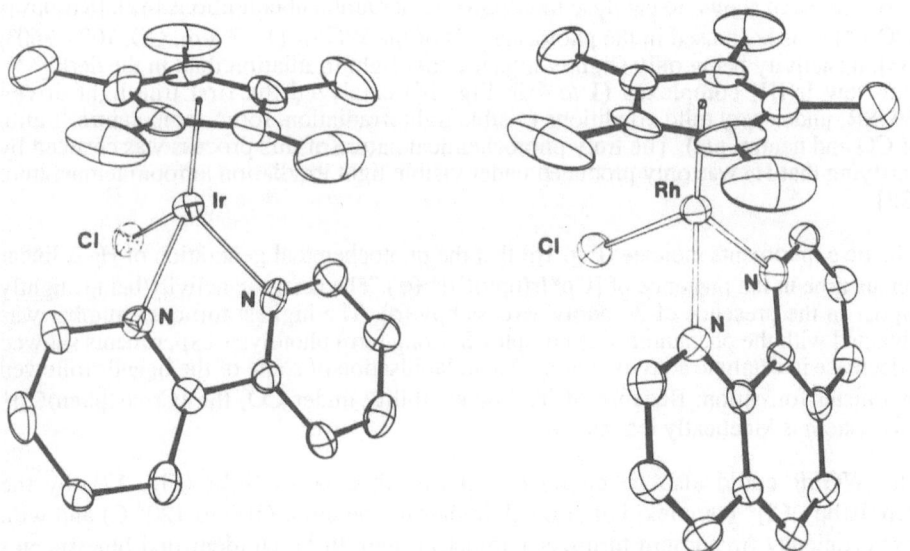

Figure 15. Molecular geometry for complexes **1** (left hand-side) and **6** (right hand-side). The ellipsoids are scaled to represent the 50% probability surface. Hydrogen atoms are omitted for clarity.

$$[Cp^*Ir(LL)Cl]^+ \quad \xrightarrow{NaBH_3CN} \quad [Cp^*Ir(LL)H]^+ \quad (10)$$

Two of these complexes (**1** and **6**) have been characterised by single crystal X-ray diffraction and exhibit (Fig. 15) a characteristic three-legged "piano-stool" arrangement [33]. During this work, related mononuclear [34, 35] and dinuclear [36] ruthenium complexes were synthesized, and complex **5** as its chloride salt was used in hydride transfer reactions involving protons [37], pyridine nucleotides, or cyclohexanone [38]. None of these complexes (Fig. 14) display any catalytic activity toward CO_2, when photolyzed or electrolyzed in conditions previously described (section A above, [12, 13]. However the iridium complexes (**1** to **4**) are effective catalysts for the light-driven WGSR with turnover numbers up to $10h^{-1}$. Solutions of the hydrido-complexes (**2** and **4**) exhibit fluorescence at room temperative when excited at their MLCT absorption bands while the corresponding chloro-complexes (**1** and **3**) do not present any fluorescence in solution [39]. To our knowledge, this is the first observation of fluorescence from an iridium-hydrido complex.

The reaction of CO with H_2O to form CO_2 and H_2 (eq. 1) is an important industrial reaction [40] and is the principal method for obtaining H_2 for the Haber-Bosch synthesis of NH_3 and could be regarded as the reductive decomposition of water by CO. A few homogeneous systems capable of thermally catalyzing the WGSR based on Ru, Rh or Ir complexes of polypyridine ligands have been reported. For example, iridium complexes of aromatic diimines have been shown to thermally catalyse the WGSR (1 atm. CO, 100-150°) with turnover numbers close to $100h^{-1}$ [41]. Recently ruthenium (II) complexes of the monoprotonated form of ethylenediaminetetraacetic acid have been shown to catalyse the WGSR under ambient conditions [42]. [Ru(bpy)$_2$ (CO)Cl]$^+$ has been used in the photocatalysis of the WGSR (1 - 3 atm. CO, 100 - 160°) [43], its activity being only slightly higher under light irradiation than in the dark [44]. Our new Ir(III) complexes (**1** to **4** in Fig. 14) catalyzed the first true light-driven WGSR, under very mild conditions (visible light irradiation, room-temperature, 1 atm. of CO and neutral pH). The truly photochemical nature of this process was checked by verifying that H_2 was only produced under visible light irradiation at room temperature [32].

Kinetic experiments indicate (Fig. 16) that the photochemical generation of H_2 is linear versus time in the precence of [Cp*Ir(bpy)Cl]$^+$ (●). The catalytic activity being slightly higher in the presence of 25 equiv. excess bpy (o). The highest turnover number was obtained with the phenanthroline complex **3**. Long term photolysis experiments showed a decrease in catalytic activity mainly due to labilisation of some of the ligands followed by cluster formation. Because of it's low solubility under CO, the [Cp*Ir(phen)Cl]$^+$ (❑) system is kinetically less stable.

The WGSR could also be catalysed, **in the absence of light** (Fig. 17), by the [Cp*Ir(bpy)Cl]$^+$ complex, but at much higher temperature (100 to 120° C) and with less efficiency (maximum turnover number around $3h^{-1}$). Unidentified blue species were formed under CO at these high temperatures and **an induction period** was observed (Fig. 16). This indicates that the active species, in the thermal system, was formed slowly from [Cp*Ir(bpy)Cl)]$^+$ but was obtained instantaneously in the photochemical system (no induction period was observed (Fig. 15).

A marked pH effect for both the photochemical(X) and thermal (●) systems was observed (Fig. 18). In the former an optimum efficiency, around pH 7, was obtained,

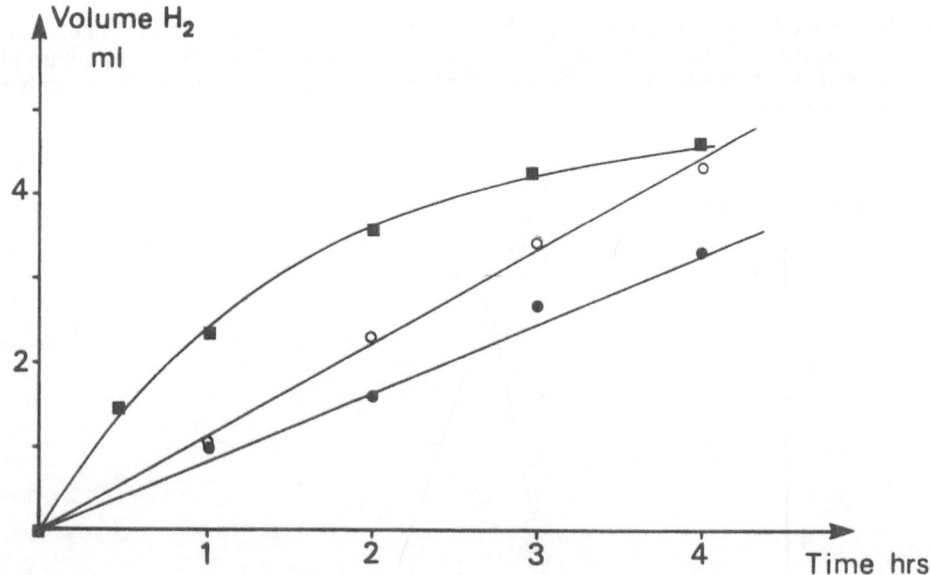

Figure 16. Photogeneration of H_2 from H_2O, as a function of time, using $[Cp^*Ir(L)Cl]^+$ as homogeneous catalysts ; L = phen and 25 equiv. excess phen (■) ; L = bpy (●) ; L = bpy and 25 equiv. excess bpy (O), at pH = 7.0 and 1 atm. CO.

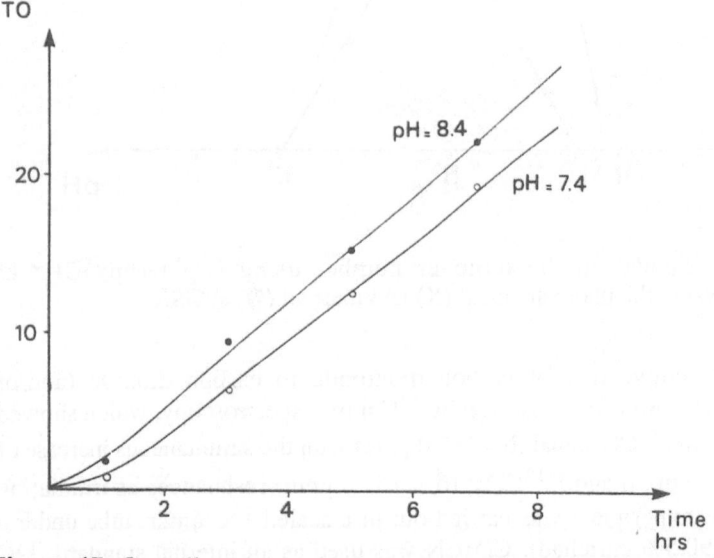

Figure 17. Overall turnover number measured as a function of time in the thermal generation (120°C, 1 atm. CO) of H_2 from H_2O, using $[Cp^*Ir(bpy)Cl]^+$ as homogeneous catalyst.

while in the latter system this value shifted to 8.4. Since CO_2 formation is likely to be accelerated in a basic media and H_2 formation favoured at low pH, the optimum pH observed near neutrality is an indication that both CO oxidation and H_2 release are both kinetically important.

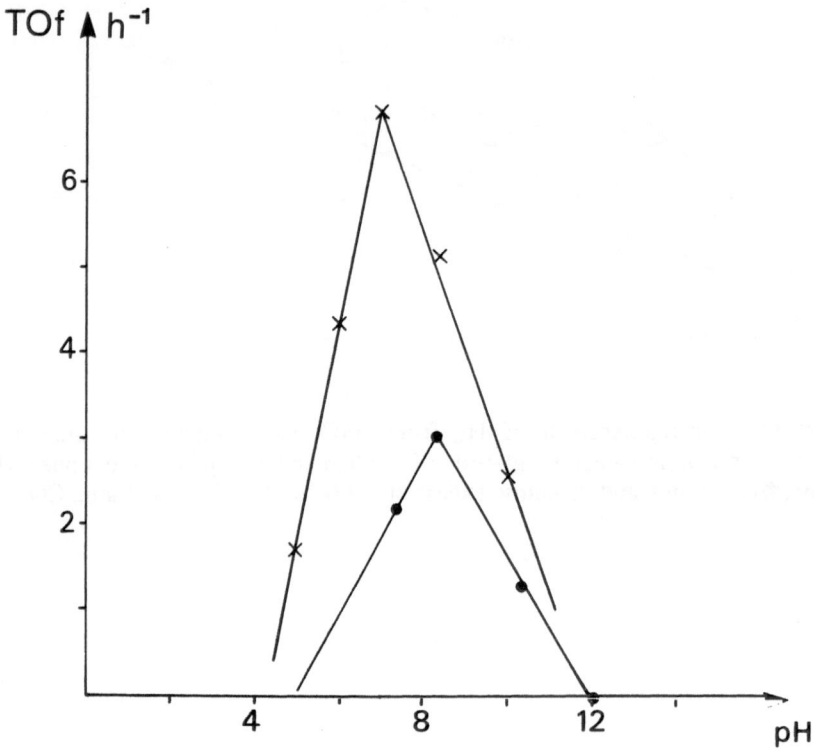

Figure 18. pH influence on the turnover number using $[Cp^*Ir(bpy)Cl]^+$ as homogeneous catalyst in the photochemical (X) and thermal (●) WGSR.

The photochemical conversion of carbon monoxide to carbon dioxide (and/or hydrogenocarbonate) was clearly observed by ^{13}C n.m.r. spectroscopy, which showed the disappearance of the ^{13}CO signal ($\delta = 182.0$ ppm) with the simultaneous increase of the $^{13}CO_2$ ($\delta = 123.0$ ppm) and $H^{13}CO_3^-$ ($\delta = 157.8$ ppm) resonances as irradiation was continued. This experiment was carried out in a sealed ^{13}C n.m.r. tube under a labelled CO atm. (90.5% enriched). CD_3CN was used as an internal standard. ^{13}C n.m.r. spectroscopy also showed the appearance of a new resonance at $\delta = 177.9$ ppm. We believe that this signal, as well as the deepening of the yellow colour of the solution, are due to the formation of an iridium (III)-carbonyl compound. No formate formation could be detected during the WGSR.

The presence of triphenylphosphine or pyridine totally inhibited the reaction, implying that the vacant site, propably created by chloride loss plays an important mechanistic role. A possible mechanism for this reaction could involve a loss of chloride and the formation of a photoactive Ir(III)-carbonyl species. The MLCT excited-state might favour the nucleophilic attack of water on the coordinated CO followed by the classical WGSR mechanism [43, 45] : (i) formation of a hydroxy-carbonyl species ; (ii) decarboxylation with hydride formation ; (iii) protonation of the hydride to give H_2 ; and finally (iv) coordination of the chloride anion. The $[Cp*Ir(bpy)H]^+$ complex **2**, a possible intermediate for the formation of H_2, is also an active catalyst for the photochemical WGSR (TO $\approx$ 3h^{-1}).

In conclusion, association of an aromatic diimine, a pentamethylcyclopentadienyl ligand, and a chloride or a hydride on an iridium (III) centre allowed the synthesis of new photosensitizers and new catalysts for the photoactivation of small molecules. Work is in progress to ascertain the nature of the catalytic species and to elucidate the photochemical and thermal steps of the process. The substitution effect (donor/acceptor) on the bpy ligand, the use of labile anions (triflate, hydroxide...) in place of chloride and the influence of various experimental parameters (catalyst concentration, CO pressure...) on the catalytic activity are currently being studied. Preliminary results on activation energy measurement seem to indicate that the reductive elimination of H_2 is the rate limiting step. Spectroscopic studies (UV-visible absorption, FT-IR, ^{1}H- and ^{13}C-nmr) will also help in the full understanding of this new catalytic process.

Acknowledgements

K. Watson is gratefully acknowledged for her kind participation in the preparation of the manuscript and for helpful discussions. We are also grateful to the Centre National de la Recherche Scientifique for supporting this work.

98

REFERENCES

[1] R. ZIESSEL, Paper presented at the OECD-IEA Expert Seminar on Energy Technologies for Reducing Emissions of Greenhouse Gases, Paris 12-14 April 1989.

[2] a) M. ARESTA and G. FORTI (Eds), in "Carbon dioxide as a source of carbon : Biochemical uses ; Nato ASI Series C, 206 (1987) b) S. INOUE and N. YAMAZAKI (Eds.), in "Organic and bioorganic chemistry of carbon dioxide", Halsted Press ; NY 1982 ; c) B.P. SULLIVAN, M.R.M. BRUCE, T.R. O'TOOLE, C.M. BOLLINGER, E. MEGEHEE, H. THORP and T.J. MEYER, in "Catalytic activation of carbon dioxide" ; M.W. AYERS (Ed.) ; ACS Symposium Series 363, Chapter 6, 52 ; American Chemical Society : Washington, D.C., 1988.

[3] T. INOUE, A. FUJISHIMA, S. KONISHI and K. HONDA, Nature, **277** (1979) 637 ; A. HENGLEIN and M. GUTIERREZ, Ber. Buns Phys. Chem., **87** (1983) 852.

[4] N. GETOFF, Z. Naturfosch, B, **17** (1962) 87 ; ibid **18** (1963) 169.

[5] S. TAZUKE and N. KITAMURA, Nature, **275** (1978) 301.

[6] a) B. FISCHER and R. EISENBERG, J. Am. Chem. Soc., **102** (1980) 7361 ; b) J. HAWECKER, J.M. LEHN and R. ZIESSEL, Nouv. J. Chim., **7** (1983) 271 ; c) A.H.A. TINNEMANS, T.P.M. KOSTER, D.H.M.W. THEWISSEN, A. MACKOR, Rec. Trav. Chim. Pays-Bas, **103** (1984) 288 ; d) M. BELEY, J.P. COLLIN, R. RUPPERT and J.P. SAUVAGE, J. Am. Chem. Soc., **108** (1986) 7461 ; e) J.L. GRANT, K. GOSWAMI, L.O. SPREER, J.W. OTVOS and M. CALVIN, J. Chem. Soc. Dalton Trans., (1987) 2105 ; J.P. COLLIN, A. JOUAITI and J.P. SAUVAGE, Inorg. Chem., **27** (1988) 1986.

[7] K. HIRATSUKA, K. TAKAHASHI, H. SASAKI and S. TOSHIMA, Chem. Lett., (1977), 1137.

[8] S. MESHITSUKA, M. ICHIKAWA and K. TAMARU, J. Chem. Soc., Chem. Comm., (1974) 158 ; K. TAKAHASHI, K. HIRATSUKA, H. SASAKI and S. TOSHIMA, Chem. Lett., (1979) 305 ; J.Y. BECKER, B. VAINAS, R. EGER and L. KAUFMAN, J. Chem. Soc., Chem. Comm., (1985) 1471 ; C.M. LIEBER and N.S. LEWIS, J. Am. Chem. Soc., **106** (1984) 5033.

[9] a) **Iron-sulfur clusters** : M. TEZUKA, T. YAJIMA, A. TSUCHIYA, Y. MATSUMOTO, Y. UCHIDA and M. HIDAI, J. Am. Chem. Soc., **104** (1982) 6835 ; M. NAKAZAWA, Y. MIZOBE, Y. MATSUMOTO, Y. UCHIDA, M. TEZUKA and M. HIDIA, Bull. Chem. Soc. Jpn, **59** (1986) 809 ; b) **Ruthenium-carbonyl clusters** : B.H. Chang, J. Organomet. Chem., **291** (1987) C31.

[10] S. SLATER and J.H. WAGENKNECHT, J. Am. Chem. Soc., **106** (1984) 5367.

[11] D.L. DUBOIS and A. MIEDANER, J. Am. Chem. Soc., **109** (1987) 113.

[12] a) J.M. LEHN and R. ZIESSEL, Proc. Natl. Acad. Sci. USA, **79** (1982) 701 ; R. ZIESSEL, J. HAWECKER and J.M. LEHN, Helv. Chim. Acta, **69** (1986) 1065 ;

b) F.R. KEENE, C. CREUTZ and N. SUTIN, Coord. Chem. Rev., **64** (1985) 247.

[13] J. HAWECKER, J.M. LEHN and R. ZIESSEL, J. Chem. Soc., Chem. Comm., (1983) 536 ; J. HAWECKER, J.M. LEHN and R. ZIESSEL, J. Chem. Soc., Chem. Comm., (1984) 328 ; J. HAWECKER, J.M. LEHN and R. ZIESSEL, Helv. Chim. Acta, **69** (1986) 1990.

[14] B.P. SULLIVAN and T.J. MEYER, J. Chem. Soc.Chem. Comm., (1984) 1244 ; C. KUTAL, M.A. WEBER, G. FERRAUDI and D. GEIGER, Organometallics, **4** (1958) 2161 ; B.P. SULLIVAN and T.J. MEYER, Organometallics, **5** (1986) 1500 ; H. HUKKANEN and T.T. Pakkanen, Inorg. Chim. Acta, **114** (1986) L43; C. KUTAL, A.J. CORBIN and G. FERRAUDI, Organometallics, **6** (1987) 553.

[15] a) C.M. BOLLINGER, B.P. SULLIVAN, D. CONRAD, J.A. GILBERT, N. STORY and T.J. MEYER, J. Chem. Soc., Chem. Comm., (1985) 796 ; b) C.M. BOLLINGER, N. STORY, B.P. SULLIVAN and T.J. MEYER, Inorg. Chem., **27** (1988) 4582.

[16] J. HAWECKER, J.M. LEHN and R. ZIESSEL, J. Chem. Soc., Chem. Comm. (1985) 56.

[17] a) H. ISHIDA, K. TANAKA and T. TANAKA, Chem. Lett., (1985) 405 ; (1987) 1035 ; (1988) 339 ; b) H. ISHIDA, K. TANAKA and T. TANAKA, Organometallics, **6** (1987) 181 ; H. ISHIDA, H. TANAKA and T. TANAKA, J. Chem. Soc., Chem. Comm., (1987) 131.

[18] R. MAIDAN and I. WILLNER, J. Am. Chem. Soc., **108** (1986) 8100 ; D. MANDLER and I. WILLNER, J. Am. Chem. Soc., **109** (1987) 7884 ; I. WILLNER, R. MAIDAN, D. MANDLER, H. DURR, G. DÖRR and K. ZENGERLE, J. Am. Chem. Soc., **109** (1987) 6080.

[19] M.R.M. BRUCE, E. MEGEHEE, B.P. SULLIVAN, H. THORP, T.R. O'TOOLE, A. DOWNARD and T.J. MEYER, Organometallics, **7** (1988) 238.

[20] R. ZIESSEL, in ref. 2a, p. 113-138

[21] J. GUILHEM, C. PASCARD, J.M. LEHN and R. ZIESSEL, J. Chem. Soc. Dalton, (1989) in press.

[22] D. BARTH, P. RUBINI, J.J. DEQUECH, Nouv. J. Chim., **7** (1983) 563.

[23] K. KALYANASUNDARAM, Coord. Chem. Rev., **46** (1982) 998.

[24] J. VAN HOUTEN and R.J. WATTS, J. Am. Chem. Soc., **98** (1976) 4853 ; Inorg. Chem., **17** (1978) 3381. P.E. HOGGARD and G.B. PORTER, J. Am. Chem. Soc., **100** (1978) 1457 ; W.M. WALLACE, P.E. HOGGARD, Inorg. Chem., **18** (1979) 2934. M. GLERIA, F. MINTO, G. BEGGIATO and P. BORTOLUS, J. Chem. Soc. Chem. Comm. (1978) 285 ; B. DURHAM, J.M. WALSH, C.L. CARTER and T.J. MEYER, Inorg. Chem., **19** (1980) 860 ; R.F. JONES and D.J. COLE-HAMILTON, Inorg. Chim. Acta, **53** (1981) L3.

[25] B. DURHAN, J.V. CASPAR, J.K. NAGLE and T.J. MEYER, J. Am. Chem.

Soc., **104** (1982) 4803 ; S. TACHIYASHIKI, N. NAGAO and K. MIZUMACHI, Chem. Letters, (1988) 1001.

[26] J.M. LEHN and R. ZIESSEL, J. Organometal. Chem., (1989) in press.

[27] C.P. ANDERSON, D.J. SALMON, T.J. MEYER and R.G. YOUNG, J. Am. Chem. Soc., **99** (1977) 1980.

[28] H. ISHIDA, K. TANAKA, M. MORIMOTO and T. TANAKA, Organometallics, **5** (1986) 724.

[29] D.C. GROSS and P.C. FORD, Inorg. Chem., **21** (1982) 1702.

[30] T. YOSHIDA, Y. UEDA, S. OTSUKO, J. Am. Chem. Soc., **100** (1979) 3941.

[31] T.G. APPLETON and M.A. BENNETT, J. Organomet. Chem., **55** (1973) C88.

[32] R. ZIESSEL, J. Chem. Soc., Chem. Commun., (1988) 16.

[33] M.T. YOUINOU and R. ZIESSEL, J. of Organometal. Chem., **363** (1989) 197.

[34] M.O ALBERS, D.J. ROBINSON and E. SINGLETON, J. Organomet. Chem., **311** (1986) 207.

[35] U. KOELLE and J. KOSSAKOWSKI, J. Chem. Soc., Chem. Commun., (1988) 549.

[36] M.O. ALBERS, D.C. LILES, D.J. ROBINSON and E. SINGLETON, J. Organomet. Chem., **323** (1987) C39.

[37] U. KOLLE and M. GRATZEL, Angew. Chem. Int. Ed. Engl., **26** (1987) 567.

[38] R. RUPPERT, S. HERRMANN and E. STECKHAN, Tetrahedron Lett., **28** (1987) 6583.

[39] D. SANDRINI, M. MAESTRI and R. ZIESSEL, Inorg. Chim. Acta, in press.

[40] R.M. LAINE and E.J. CRAWFORD, J. Mol. Catal. **44** (1988) 357.

[41] J.P COLLIN, R. RUPPERT and J.P. SAUVAGE, Nouv. J. Chim., **9** (1985) 395.

[42] M.M. TAQUI KHAN, S.B. HALLIGUDI and S. SHUKLA, Angew. Chem. Int. Ed. Engl., **27** (1988) 1735.

[43] D. CHOUDHURY and D.J. COLE-HAMILTON, J. Chem. Soc., Dalton Trans., (1982), 1885.

[44] K. TANAKA, M. MORIMOTO and T. TANAKA, Chem. Lett., (1983), 901.

[45] P.C. FORD, Acc. Chem. Res., **14** (1981) 31 and references therein ; T. YOSHIDA, T. OKNOI, Y. LEDA and S. OTSUKA, J. Am. Chem. Soc., **103** (1981) 3411.

ELECTROCHEMICAL SYNTHESES INVOLVING CARBON DIOXIDE

Giuseppe Silvestri, Salvatore Gambino, Giuseppe Filardo
Dipartimento di Ingegneria Chimica dei Processi e dei
Materiali,
Università di Palermo,
Viale delle Scienze,
90128 Palermo,
Italy.

ABSTRACT. The recent developments of electrosynthetic processes
involving carbon dioxide are reviewed. Both processes involving carbon
dioxide alone and electrocarboxylations are taken under consideration.
The first of these two fields, mainly projected towards the production
of basic chemicals such as methanol or carbon monoxide, is at present at
a stage of fundamental research. Some interesting developments have
recently appeared in the literature, concerning the use of electrode
materials, or of transition metal complexes with particular catalytic
activity. Electrocarboxylations are at a stage of more applicative
development: some processes have reached the stage of pilot plant, and
other have interesting perspectives of application. A considerable
effort has been devoted to the development of processes in undivided
cells, with anodic reactions involving the oxidation of oxalates or
formates, or the dissolution of sacrificial anodes. Specially designed
electrochemical reactors have also been developed. The proper
application field for electrocarboxylation processes appears to be the
production of fine chemicals .

1. INTRODUCTION

The concern of the general opinion on the problems raised by the
increase of concentration of carbon dioxide in the atmosphere, and by
the future needs of organic compounds in a perspective of crude
shortage, has promoted a wide range of chemical investigations on this
compound. In the electrochemical field these researches are at present
in a rapid and enth usiastic expansion, encouraged by a series of
promising results, in view of the reduction of carbon dioxide to
compounds finding industrial use on a large scale. Beside these
ambitious and long term goals there are other more specific targets
placed in the fine chemistry field, where electro-organic chemistry is
particularly adapted for giving important contributions. In this second
case the electrochemical insertion of carbon dioxide into organic

101

M. Aresta and J. V. Schloss (eds.),
Enzymatic and Model Carboxylation and Reduction Reactions for Carbon Dioxide Utilization, 101–127.
© 1990 *Kluwer Academic Publishers.*

compounds belonging to quite a large number of classes is involved. The recent developments of these two fields will be reviewed here, taking them separately under consideration, as they are directed towards different scientific and applicative areas, and are at significatively different development levels.

2. ELECTROCHEMICAL ACTIVATION OF CARBON DIOXIDE ALONE.

For more than a century, after the first communication in 1870 (1), only formic acid (or formate anion), of the products listed in Scheme 1, was described as obtainable by electroreduction of carbon dioxide in

---

Scheme 1. ELECTROCHEMICAL REDUCTION OF CARBON DIOXIDE.

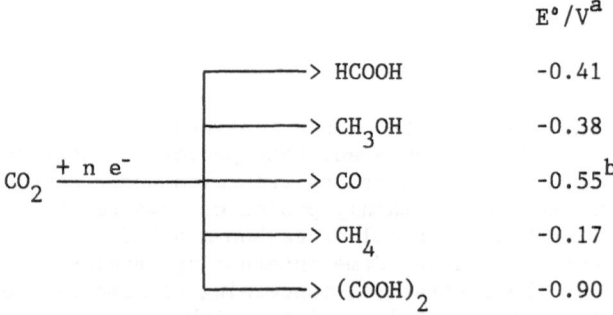

$E°/V^a$

> HCOOH	-0.41
> CH_3OH	-0.38
> CO	-0.55^b
> CH_4	-0.17
> $(COOH)_2$	-0.90

a) vs. NHE, pH = 7.
b) for the reduction to CO + CO_3^{2-}.

---

aqueous media. The formation, on gold electrodes, of substantial amounts of carbon monoxide in aprotic media was reported in 1967 (2). Five years later the synthesis of oxalate in aprotic media was reported by Baizer (3).

Since then, with the contribution of an increasing number of research groups, the list of the products obtained by electroreduction of carbon dioxide enlarged to methanol, to hydrocarbons (mainly methane, but even ethylene can be obtained with appreciable current yields), to fatty acids such as propanoic and butanoic, to hydroxyacids and higher dicarboxylic acids, such as glycolic, or tartaric acid. Examples of these "one pot" syntheses are reported in the literature, in a wide variety of experimental conditions (aqueous or aprotic media, different metals acting as cathodes, use of catalytic systems based on organometallic or coordination compounds of transition metals, organic or inorganic charge transfer mediators, different supporting electrolytes and so on). In some cases the reaction pathway, going through some key intermediates, is rather easily understandable. In other cases, complicated and still under investigation phenomena of adsorption of carbon dioxide or of its reduction derivatives, and of solvent molecules, or of hydride species, are involved in the process.

The field is at the stage of fundamental research, and is slowly moving towards the development of some particularly promising lines.

The final goal of this research is the setting up of large scale processes leading to some of the products listed in Scheme 1. Carbon monoxide and methanol represent the more appealing derivatives, as are involved in many processes of great industrial importance (Scheme 2). Nevertheless a series of realistic considerations has to be taken into account in the evaluation of the distance between the present level of development of these researches and the far perspective of industrial applications. Some preliminary considerations can be made on the more promising results obtained so far and on possible directions in which

Scheme 2 - SOME DERIVATIVES OF METHANOL AND CARBON MONOXIDE

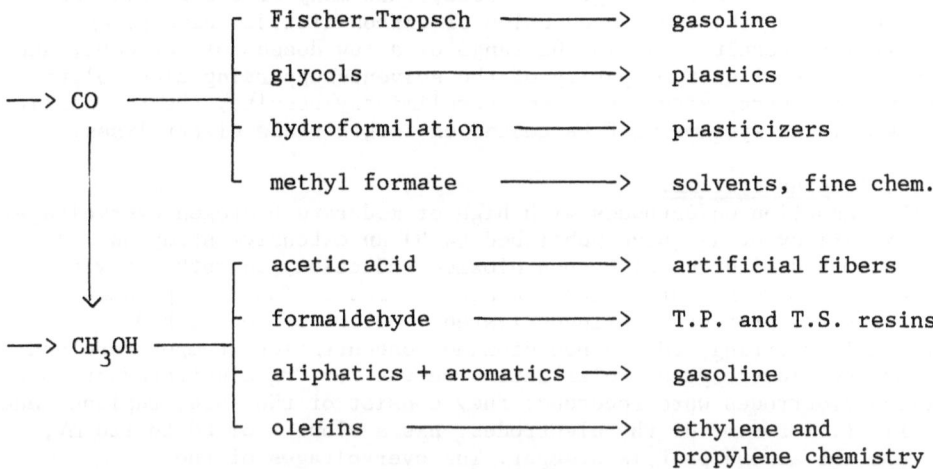

PRESENT SOURCES : NATURAL GAS - CRUDE OIL - COAL - OIL SHALE - TAR

SANDS - BIOMASS

this research could develop in order to remain more focused on applicative perspectives. High stability of solvent and supporting electrolyte, cathodic potentials nearest as possible to the thermodinamic value, catalytic systems not being deactivated by the products of the reduction (the formation of carbon monoxide leads often to the formation of stable metalcarbonyls), appear to be preliminary requisites for thinking of future developments. Water, of course, is the most suitable solvent, therefore in this part of the review, mainly focused on the synthetic aspects of carbon dioxide electrochemistry, the main attention will be devoted to electrochemical systems active in this medium.

The electrochemical reduction of carbon dioxide alone has been faced following two main routes : a) the use of cathodic materials having some particular catalytic activity, and b) the use of catalytic

systems based on transition metal coordination compounds or on organic
agents of homogeneous charge transfer.

These two routes will be taken under consideration in the following
sections.

2.1   The Reduction of Carbon Dioxide on Metals.

The catalytic activity of a cathode surface may be related to the
pretreatment of the electrode or to the modifications induced by the
electrolysis. Consequently the formation or the disappearance of
hydrido species, or of the bare metal, or the deposition of other metals
present as traces in the electrolytic medium, may deeply change the
behaviour of the electrode, depressing the selectivity towards a
particular derivative after the circulation of a limited amount of
charge. It is not surprising, therefore, that many of the metals tested
as cathodes for the electroreduction of carbon dioxide, have given
satisfactory results only in the range of a few dozens of coulombs, and
in conditions of extreme purity of the solvent-supporting electrolyte
system. Of course, after these very preliminary results, the reliability
of the related systems must be tested on larger scale electrolyses.

2.1.1. Aqueous systems.
i. The reduction on cathodes with high or moderate hydrogen overvoltage.
Vassiliev et al. have published (4,5) an extensive study on
mechanism and kinetics of carbon dioxide reduction, on cathodes with
different hydrogen overvoltage, in aqueous media. The main parameters
affecting the process were investigated : nature of the cathode,
potential, stirring, pH, carbon dioxide concentration, temperature and
supporting electrolytes. Polarization curves for CO_2 electroreduction on
various electrodes were recorded: they consist of two Tafel regions, and
the first, for some of the electrodes, has a slope from 80 to 120 mV,
whereas the second part is steeper. The overvoltages of the
electroreduction of carbon dioxide in the two regions depend on the
metal, but in different ways. In the first region the overvoltage
increases in the order :
$$Cu < Sn < Ga < Sb < Bi < In < Zn < Cd < Pb < Hg.$$
Variations are moderate from Cu to Bi, and larger for the rest of the
sequence. In the second region of the Tafel slope overvoltage increases
in the order:
$$Bi < Sb < Ga < Cu < Zn < In < Sn < Pb < Cd < Hg,$$
with marked differences between the terms of the sequence. The
dependence on the cathodic material of the CO_2 reduction indicates that
in both regions adsorbed particles take part in the rate determining
step. It can be deduced from this fundamental report that metals with
moderate overvoltage are indicated for the optimization of the reduction
of carbon dioxide.

From the preparative point of view, the best conditions up to now
reported for the reduction to formate (6) are those in which
amalgamated copper is used, in diaphragm cells to hinder the anodic
oxidation of formate itself.

The behavior of synthetic systems based on high hydrogen
overvoltage cathodes was investigated also by Ito (7,8) and by Hori (9).

Some technological developments of this process appeared recently in the literature, according to which carbon dioxide reduction is performed in cells equipped with gas diffusion electrodes. The electroactive materials dispersed into the hydrophobic PTFE matrix of the electrode were lead, indium and tin in a report from the BP laboratories (10). Excellent performances were obtained by the use of lead electrodes, with cathodic current densities up to 115 mA cm^{-2}, and current efficiencies of nearly 100% vs formate, at the quite negative cathode potential of -1.8 V (vs. SCE). The electroactive metals were obtained by reduction with hydrogen of the corresponding salts in the already prepared PTFE dispersion. Conflicting results were reported by Furuya (11): with apparently the same electrode material high yields of carbon monoxide were obtained, whereas electrodes obtained by incorporation of lead powder gave only formic acid (40%) and hydrogen (60%) as the products. So different results, which can probably be explained with different states of aggregation of the electroactive metal inside the matrix, require further investigations.

The use of tetraalkylammonium supporting electrolytes on mercury or on carbon electrodes in aqueous media leads to different results, due to the deep modifications induced in the structure of the electrified interface by the bulky organic cations. Eggins et al. (12) reported that in those conditions, at pH 9, oxalate is obtained at a cathodic potential of 0.9 V, whereas at more negative potentials glyoxylate is obtained, although with the coproduction of relevant amounts of formate.

ii. Molibdenum and Rutenium. A strong input to the use of new cathodic materials for the electroreduction of carbon dioxide came from heterogeneous catalysis: both Mo and Ru were known as active towards carbon dioxide respectively in the dissociative adsorption and in the conversion of carbon dioxide to reduced products. Once tested by Frese, who pioneered this new approach, these metals proved to be active also in the electrochemical reduction in aqueous media. Molibdenum gave, by electroreduction in aqueous sodium sulphate at pH 4.2, up to 84% faradic yield of methanol, at a cathodic current density of 0.12 mA cm^{-2}, and at a cathodic potential around -0.8 V (vs. SCE) (13,14). Electroplated rutenium cathodes gave, according to the known gas phase catalytic activity of this metal, methane (30% faradic yield), together with 25% methanol and 45% carbon monoxide (15,16) at a cathodic current density in the range 0.13 - 0.39 mA cm^{-2}, and at a cathodic potential of -0.55 V (vs SCE). The preliminarity of these results appears evident from the limited amount of charge passed in both experimental approaches (< 30 C), and from the extremely low current densities used.

iii. Copper. This metal can be at present considered one of the most promising cathodic materials for the reduction of carbon dioxide to hydrocarbons. Current yields for the reduction to methane and ethylene strongly depend on the temperature and on the purity of the metal surface (17 - 19). The synthetic runs reported in the first papers related to the use of this metal, gave faradic efficiencies of 50% for methane, 12% for ethylene, 3% for carbon monoxide, 10% for formate, and 24% for hydrogen, on 99.999% copper in 0.5M KHCO$_3$, at constant cathodic current density of 5 mA cm^{-2} at 16°C, but the ethylene yield dropped to one third of the preceeding value as the purity of copper changed from

99.999 to 99.99.

A faradic efficiency of 33% for the reduction of CO_2 to methane has been reported quite recently on commercial copper (20). The interesting approach of the "in situ" electrodeposition of copper, in order to keep under control the modifications of the electrode surface in the course of the electrolysis has been proposed by Cook et al. (21,22) in the frame of a study on the reaction mechanism and on the influence of the main operative variables on the catalytic activity of the metal. A further contribution to the elucidation of the reaction mechanism, related to the information from heterogeneous catalysis, is given by Frese et al. (23), who observed a noteworthy influence of the chemical preparation of the copper surface on the faradic yields to hydrocarbons.

iv. Silver and gold. Synthetic investigations on these metals in aqueous media showed that carbon monoxide is the derivative obtained with the highest current yield (up to 93%) (17,19,24). A study on the electrochemical behaviour of carbonaceous residues formed on gold in the presence of carbon dioxide has been recently published (25). The influence of the pretreatment of the surface on the interaction between CO_2 and H atoms in the potential range of hydrogen evolution was investigated.

v. Platinum. Only chemisorbed particles, at a reduction state lower than formic acid, are produced by electrochemical reduction of carbon dioxide in aqueous media on platinum (5). The desorption of these carbonaceous species is very slow and makes difficult any practical utilization of the reaction. Vassiliev reported a few years ago (26) that the reduction of carbon dioxide in aqueous solution saturated with sodium carbonate in diaphragmless cells, leads to formate and formaldehyde on platinum cathodes, possibly through the intermediation of percarbonate anions.

2.1.2.   <u>Non-aqueous and partially protic systems.</u> The first step of the reduction of carbon dioxide, whatever is the electrolytic medium, is its reduction to the corresponding anion radical $CO_2^-\cdot$ (4). In anhydrous DMF, with Et_4NClO_4 0.1M, the thermodynamic potential of the couple

$$CO_2 + e^- \underset{\longleftarrow}{\overset{\longrightarrow}{}} CO_2^-\cdot$$

was determined by fast cyclovoltammetry (27) as -2.40 V (vs. SCE). A slightly more negative value (-2.49 V vs. SCE), but with Bu_4NClO_4, was very recently obtained approaching the equilibrium conditions with homogeneous solution electron transfer agents (28).

The reductive coupling to oxalate and the contemporary formation of carbon monoxide and carbonate anion was rationalized in 1978 by Saveant (29), according to Scheme <u>3</u>.

$$CO_2^-\cdot \quad \overline{\underset{}{}} \begin{cases} +CO_2^-\cdot \longrightarrow \quad {}^-OOC\text{-}COO^- \\ +CO_2 \longrightarrow \quad O{=}C\text{-}O\text{-}\underset{\underset{O^-}{|}}{C}{=}O \quad -{+}e^- \longrightarrow \quad CO + CO_3^{2-} \end{cases}$$

Scheme <u>3</u>

Experimental results are in agreement with this substantially homogeneous reaction mechanism: the dimerization of the anion radical is
favoured by high current densities, whereas higher concentrations of
carbon dioxide address the synthesis towards carbon monoxide and
carbonate. Further investigations on the mechanism of this reduction on
different cathodic materials led Vassiliev (30) to the formulation of a
different reaction pathway, depending on the interactions between
the various cathodes and carbon dioxide, and related to the cathodic
potential. According to this second approach, the formation of oxalate
comes from a further reduction of an intermediate dimeric anion radical
$(CO_2)_2^{-\cdot}$ whereas carbon monoxide and carbonate arise from the disproportionation of two $CO_2^{-\cdot}$ species adsorbed on the electrode.

Vianello has very recently communicated the preliminary results of
research still in progress (28), in which the reduction is performed
either on the electrode or in the bulk of the electrolytic medium by
means of organic homogeneous charge transfer agents. The synthetic
results **show** that when the cathode is directly involved in the
reduction, it has a determining influence on the selectivity of the
reaction. As an example, in the same experimental conditions, a platinum
cathode gave 3.5 % oxalate and 90.8 % carbon monoxide faradic yield
whereas a mercury cathode gave 76% oxalate and 9.7 carbon monoxide
yield. Nevertheless on mercury current density and carbon dioxide
concentration have a detectable influence on the selectivity.

Further investigations are necessary for a better understanding of
this intriguing matter.

The reduction of carbon dioxide in dimethylsulfoxide with
tetraalkylammonium salts, under superatmospheric pressure, was
investigated by Ito (31) using several cathodic materials . Also in
these conditions, the nature of the electrode, and the composition of
the electrolytic media, are determinant on the selectivity of the
reaction: with Pb cathodes the main product is oxalic acid, but in the
presence of tetraethylammonium salts also higher carboxylic acids are
obtained in detectable amounts. With tetrabutylammonium salts only
traces of higher acids are obtained. Indium, tin and zinc electrodes
gave as the main derivative carbon monoxide.

Some years ago the electrochemical synthesis of oxalic acid was
brought to a pilot scale in a process with sacrificial zinc anodes and
zinc cathodes by the Dechema Institut (32).

In media of low proton availability products of further reduction
of oxalic acid, such as glycolic (33a) and glyoxylic acid , and of
further dimerization of these intermediates **to** tartaric acid, were
obtained with not particularly encouraging selectivities (33b).

2.2. Carbon Dioxide Reduction Catalyzed by Transition Metal Complexes.

Numerous complexes of transition metals have been proposed in the
literature as precursors of catalysts for the reduction of carbon
dioxide: metal cations arising from the first, second, seventh and
eighth group of the transition metals, and ligands belonging to several
classes (tetraazamacrocycles both [16] such as porphyrins and
phtalocyanins, and [14], with different substituents and at various
degrees of unsaturation, phosphines, bi- or ter-pyridines, carbon

monoxide). Square planar, often distorted, and octahedral complexes result from the association of the above mentioned cations and ligands. As a general consideration, it can be said that there are several examples of square planar complexes active in aqueous media, whereas the larger part of octahedral complexes is stable, and active, only in aprotic media. The general considerations already made at the beginning of this chapter apply also to these systems, so it will be given here major emphasis to systems active in aqueous media, and particularly to those already experienced on a synthetic scale.

The parts of this field that have been reviewed very recently, will not be considered again here in detail, addressing the reader to the related references.

In the following, the catalytic systems will be grouped on the basis of the ligands. For each class of compounds, the behaviour of aqueous and non-aqueous systems will be considered separately.

2.2.1. <u>Square planar complexes.</u>
i. Metal-phtalocyanines (M-Pc). The first note on the use in aqueous media of the divalent cations of Co, Ni, Mn, Pd, Cu, Fe phtalocyanines in the electrochemical reduction of carbon dioxide appeared in 1974 (34). The higher reactivity of Co(II)-Pc and Ni(II)-Pc towards the reduction of CO_2 was deduced from the electroanalytical behaviour of the various complexes, acting heterogeneously deposited on graphite rods. Further reports (35 - 37) confirmed the higher reactivity of Co(II)-Pc in various conditions, although the identification and the dosage of the derivatives appeared quite controversial. The reaction mechanism has been object of remarkable attention (38 - 40) as these complexes are highly stable and insoluble in water in a broad range of conditions, and appear highly promising for applicative purposes. The best large scale results on the use of metal-phtalocyanine complexes were obtained by incorporation of the complex itself in the PTFE hydrophobic matrix of gas diffusion electrodes (41). Co(II)-Pc complexes gave high yield of carbon monoxide, although at cathodic current density much lower than that attained in the already quoted similar experiments in which lead was incorporated in the same matrix. A strong influence of the metal cation was observed : Mn(II)-Pc, Cu(II)-Pc and Zn(II)-Pc gave predominant yields of hydrogen, whereas carbon dioxide was reduced to formic acid.

In this last paper (41) the main lines of the sequence of steps leading to the reduction of CO_2 are traced. The formation of a strong nucleophile Co(I)-Pc, which reacts with carbon dioxide leading to Co(III)Pc(CO_2^-), should be followed by protonation, then by a further monoelectronic reduction and proton transfer, leading, by dehydration, to CO, water and the Co(II)-Pc complex. The intermediaty of hydrido H-Co(I)-Pc⁻ species is discussed by Christensen et al. (40).
ii. Metal-porphirins. The carbon dioxide reduction mediated by metal--porphirin complexes refers for the larger part of examples to homogeneous systems, active both in aqueous and in aprotic media. The main derivative of the reduction with various Co-porphirin complexes (42 - 46) in aqueous media is carbon monoxide. Faradic yields are in general satisfactory.

In aprotic media the catalytic activity is dependent on both the metal involved and on the nature of the porphyrin ligand. Saveant (47) investigated the behaviour of Fe-porphyrins both planar and basket-shaped, bearing various substituents on the ring. The intermediation of an iron complex with the metal in the zero formal oxidation state is postulated on the basis of electroanalytical data. Octaethylporphirino complexes of Ag and Pd gave oxalic acid without carbon monoxide nor carbonate (48).

iii. Metal-[14]tetraazamacrocycle complexes. An intense research effort on metal complexes of various [14] tetraazamacrocyclic (teta) ligands was performed in the recent past. The field was recently reviewed by Eisenberg (49). Among the teta complexes investigated so far, those showing more enhanced activity are Ni(II) and Co(II) teta complexes, leading, in mixed acetonitrile water solvent, at cathodic potentials in the range from -1.5 to -1.6 V (vs. SCE), to a mixture CO/H_2 in ratios from 1 : 1 to 2 : 1, depending on the complex used. High current efficiencies, although with low turnovers, were observed in preliminary experiments (50,51). A careful investigation on this reaction and its mechanism was made by Pearce and Pletcher (52). Acceptable turnover numbers, but disappointing maximum cathodic current densities were observed with all these complexes. The reaction mechanism postulated on the basis of the cyclovoltammetric responses is reported in Scheme 4. The Authors formulate the hypothesis that the reaction does not produce formate anion, because the protonation is constrained to occur at an oxygen rather than at a carbon site, as the consequence of the formation of an intermediate ($[M(III)LCO_2^-Na^+]^{2+}$) in which the carbon dioxide molecule is linked to the metal through a Me-C bond.

An interesting approach to the use of these complexes has been proposed by Bailey et al. (53): it consists of the heterogeneization of a Ni-teta complex, $\{Ni[Me_4Bzo_2[14]tetraeneN_4]\}_n$, which was electropoly-merized on a Pt electrode and used for the reduction of carbon dioxide in acetonitrile/methanol mixture. The product was formate anion, obtained at a cathodic potential of -1.85 V (vs. SSCE).

$$[M(II)L]^{2+} + e^- \rightleftarrows [M(I)L]^+$$

$$[M(I)L]^+ + CO_2 + Na^+ \rightleftarrows [M(III)LCO_2^-Na^+]^{2+}$$

$$[M(III)LCO_2^-Na^+]^{2+} + H_2O \xrightleftharpoons{fast} [M(III)LCOOH]^{2+} + NaOH$$

$$[M(III)LCOOH]^{2+} + e^- \longrightarrow [M(II)LCOOH]^+$$

$$[M(II)LCOOH]^+ \xrightarrow{slow} [M(II)L]^{2+} + CO + OH^-$$

$$2\ NaOH + 2\ CO_2 \longrightarrow 2\ NaHCO_3 \qquad \text{Scheme 4 (Ref. 52)}$$

By far the most promising results, among those reported for this kind of complexes, in our opinion are those related to the Ni-cyclam complexes investigated by Sauvage et al. (54 - 56). These complexes are active in aqueous media, addressing the reaction towards the formation of carbon monoxide, and show high stability and relevant turnover

numbers (ca 10^3). A reaction mechanism going through the formation of
Ni(I) and Ni(III) species was postulated (Scheme 5). The catalytic
activity of the complex seems to be related to its adsorption on the
mercury electrode. The rather unexpected formation of formate anion in
low water content N,N-dimethylformamide was also reported (56). This
behaviour is possibly due to the different structure of the solvent cage

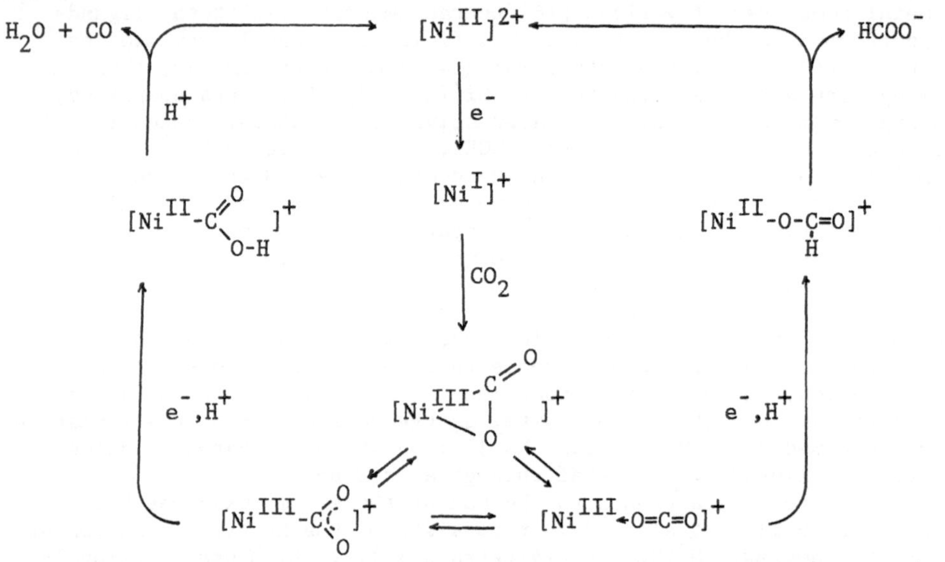

Scheme 5 (Ref. 56)

surrounding the initially formed Ni(III)CO$_2$ complex adsorbed on the
electrode, which induces the formation of Ni-formate, instead of Ni-car-
boxylate (see Scheme 5). No C-C bond formation nor enhancements in the
catalytic activity were observed when bis-cyclam complexes were used.

A recent paper (57) presents an electrochemical study on Ni(II) and
Co(II) complexes of tetraazamacrocycles bearing a pyridine functional
group. The electroanalytical data in organic solvents suggest a possible
catalytic activity of those complexes, but no synthetic data are until
now available.

iv. Phosphino complexes. To our knowledge no one of the several examples
of metal-phosphino complexes active in the electroreduction of carbon
dioxide reported in the literature is active in aqueous media.
Rh(diphos)$_2$Cl causes a shift towards more positive values of the
cathodic potential of ca. 700 mV for the electroreduction of carbon
dioxide, with respect to the uncatalyzed reduction (58). The process is
catalytic, and formate is obtained in acetonitrile, at the expenses of
the solvent deprotonation. Small yields of cyanoacetate, confirming the
involvement of the solvent, were observed.

The synthesis and caracterization of a new series of isoelectronic
complexes of the type [M(triphos)L](BF$_4$)$_2$, where M = Ni, Pd and Pt, and
L are various monodentate ligands, such as P(OMe)$_3$, CH$_3$CN, PEt$_3$,

$P(CH_2OH)_3$, PPh_3, has been object of a detailed study of research group from SERI (59,60). On the basis of cyclic voltammetry and bulk electrolyses, the Pd complexes are shown to catalyze the electrochemical reduction of carbon dioxide to CO in acidic acetonitrile solutions. A mechanism in which L reversibly dissociates from the complex, and an hydrido intermediate is formed before the insertion of carbon dioxide, is proposed. The rate limiting step of the catalytic cycle is the reaction of the coordinatively unsaturated palladium hydride intermediate with carbon dioxide. As seen in other catalytic cycles, also in this case the formation of a Pd-C bond preludes to the formation of carbon monoxide.

v. Metal-Shiff base complexes. A catalytic cycle substantially analogous to that reported in Scheme 4 was proposed for the electrochemical reduction of carbon dioxide catalyzed by Co(II)-salen complex (salen = 2,2'-[1,2-ethylene- -bis(nitrilomethylidyne)]bis phenol) (52). The catalytic activity of this square planar complex is comparable to that of the Ni(II)-teta complexes discussed in the same paper.

The catalytic electrochemical reduction of carbon dioxide using the Co(II)-salophen complex (salophen = 2,2'-[1,2-phenylenbis(nitrilo- -methylidyne)]bis phenol) in acetonitrile was recently reported (61). Products of the catalytic reduction are CO and carbonate anion. A positive influence of Li^+ cations on the efficiency of the catalysts was observed. Ethyl formate and traces of propionate anion are formed by addition of ethyl chloride to the reaction medium.

2.2.2. <u>Octahedral complexes.</u> Two very recent reviews (62,63) have already given a detailed account of this field, and the reader is addressed to them for a detailed description of the behaviour of each catalytic system. From the large amount of experimental data reported in the literature on polypyridyl complexes of the metals studied so far (Co, Ru, Rh, Re, Os), the following general features can be outlined:
a) carbon monoxide or formate are obtained, depending on the metal cation, and on the reaction medium;
b) several catalytic systems have been heterogeneized by insertion of a vinyl, or a pyrrole, group onto a bipy ligand, and electropolymerizing the resulting complexes on the electrode. Re (64 - 67) Rh and Ru (68), Co (69) complexes have been tested for this purpose;
c) good selectivities and good current efficiencies, with appreciable turnover numbers, are observed for many of the catalytic systems used;
d) several contemporary reaction pathways appear possible on the basis of the electrochemical behaviour of the systems. In some cases the intermediacy of species bearing a sesqui-bipyridine ligand has been hypothesized;
e) unfortunately, aprotic solvents do not seem easily replaceable with water, as the majority of the complexes tested so far are unstable in the presence of this solvent.

An interesting approach to the reduction of carbon dioxide consists in the use of a double catalytic cycle, based on the redox couple: Everitt's salt, $K_2Fe^{II}[Fe^{II}(CN)_6]$/Prussian blue $KFe^{III}[Fe^{II}(CN)_6]$. In this system the Everitt's salt, deposited onto an inert cathode, is oxidized to the Prussian blue, which is active in the reduction of a second redox catalyst, homogeneous, which in turn reduces carbon dioxide to methanol. Several metal complexes have been used as homogeneous catalysts, and for some of them good turnover number were observed. These systems are active in the presence of primary alcohols, therefore methanol could be produced in methanol itself (70). This catalytic cycle has been tested up to now only for limited amounts of charge passed (ca. 50 C): once demonstrated that it can be active for larger scale electrolyses, it could be considered as one of the possible directions of development of applicative interest.

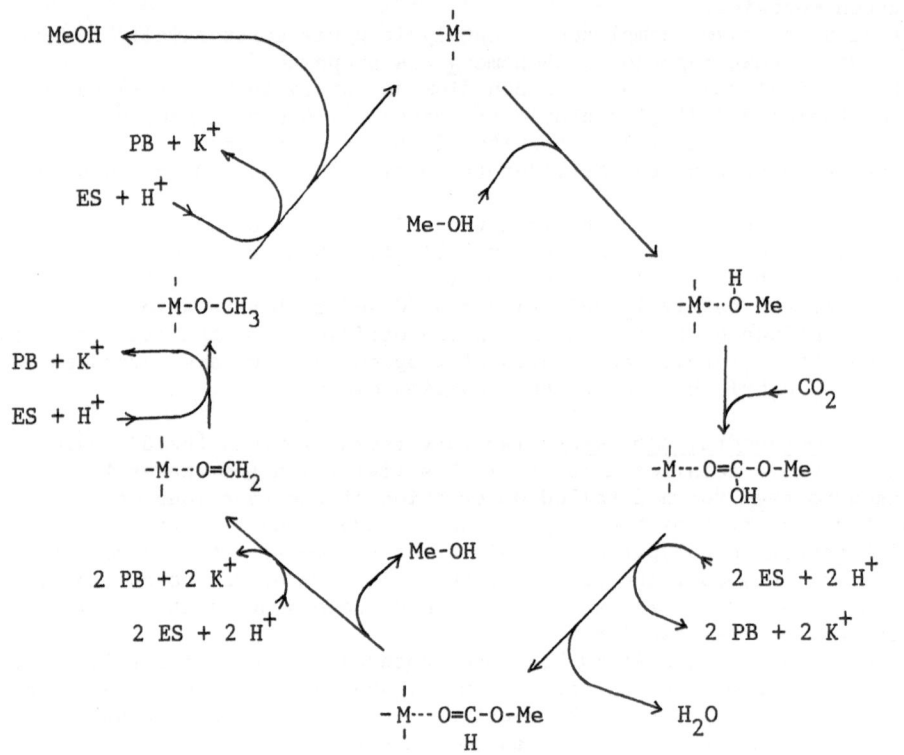

ES = Everitt's salt; PB = Prussian Blue; -M- = homogeneous catalytic system.

Scheme 6 (Ref. 70b)

The latest reports on the use of catalytic systems based on some metal-(bipy) complexes are of significant interest from the synthetic point of view and will be discussed both here and in the electrocarboxylation chapter.

The interest connected to the use of $[Ru(bipy)_2(CO)_2]^{2+}$ as catalyst for the electroreduction of carbon dioxide has been growing in the recent years. In the first papers (71,72) the production of both carbon dioxide and formate was reported: the ratio between the two products was dependent on the pH and on the composition of the medium, which was a saturated H_2O + DMF solution. The complex is active also in water, at pH 6, and at -1.5 V (vs. SCE) cathodic potential, 19% faradic yield of CO was observed. The yield of carbon monoxide increased up to 48 % in H_2O/DMF 1:1 at pH 6. No formic acid was detected in the electrolytic medium in both cases. At pH 9.5, and H_2O/DMF 9:1, 28% CO and 39 % formate faradic yield was reported.

The use of proton sources different from water, such as dimethyl-ammonium chloride or phenol, in acetonitrile, greatly increased the selectivity towards formate (>80% faradic yield)(73). Substantially the same system (i.e. Me_2NH + $Me_2NH_2^+$ Cl^- in MeCN) was shown that is also active for the production of N,N-dimethylformamide with a faradic yield of 21% (74). The reaction path leading to DMF is not clearly understandable: it could involve both the reaction of formate with dimethylammonium ions, or the reaction of carbon monoxide and the amine, promoted by bases present in the reaction medium.

The strongly basic intermediates involved in the catalytic cycles leading to CO or to formate can be used as deprotonating agents for organic molecules bearing activated hydrogen atoms, such as those in the alpha position of a carbonyl group (75). Thus acetophenone and cyclo-hexanone are carboxylated in the course of the reduction of carbon dioxide to both CO or HCO_2^-, leading to the corresponding carbanions in the alpha position, which react with CO_2 giving the anions respectively of benzoylacetic acid and cyclohexanone-2-carboxylic acid. The presence of molecular sieves seems to be necessary in order to sequestrate the water molecules produced in the course of the reaction.

Other catalytic systems, based on $Ni(bipy)_3$, catalyze the electrocarboxylation of alkynes, showing an unusual regioselectivity in the monoaddition of a carboxylate group. This and other catalyzed electrocarboxylations will be considered in the course of the following chapter.

3. ELECTROCARBOXYLATIONS

A considerable number of synthetic processes, by which one or more carboxylic groups are introduced into suitable organic starting materials, belongs to this important class of electroorganic reactions. The carboxylic group comes from a reductive reaction involving carbon dioxide and/or the organic co-starting material.

Several electrocarboxylations are at an advanced stage of development, both from the synthetic and the technological point of view. Some laboratories interested in applied electrochemistry are performing pilot scale tests for the synthesis of carboxylation derivatives used as pharmaceuticals.

The synthetic aspects of the electrocarboxylation processes were recently reviewed (76), and only the most recent results are shortly summarized here. The electrocarboxylations by reductive insertion of carbon dioxide will be grouped as follows:
a) e.c. by addition, taking place on starting materials bearing differ-ent kinds of unsaturations, such as alkenes, alkynes, polyenes, ketones, aldehydes, aromatics, imines, unsaturated heterocycles and so forth;

b) e.c. by substitution, taking place with starting materials bearing good leaving groups, such as halides, thioethers, thioacetals, acetates which are substituted by the carboxylate anion.

3.1. ELECTROCARBOXYLATION BY ADDITION

3.1.1.   Electrocarboxylation of unsaturated hydrocarbons.

The electrocarboxylation of styrene, taken as an example of an olefin having reduction potential slightly more positive than that of carbon dioxide, has been studied in detail. Experimental data suggest that the initial steps of the reaction mechanism should involve the reduction of the olefin, followed by a homogeneous electron transfer to carbon dioxide, forming the corresponding $CO_2^{-\cdot}$ anion radical. In anhydrous DMF phenylsuccinic acid with yields up to 85% is obtained (77). Different amounts of the monocarboxylic derivative are obtained, depending upon the proticity of the medium. The reaction can be performed at more positive cathodic potentials in the presence of organic mediators such as benzonitrile.

Some attention has been devoted to the electrocarboxylation of 1,3-butadiene, giving rise to a series of C_6 and C_{10} dicarboxylic acids, at present of industrial interest. The uncatalyzed reaction, described several years ago (see ref. 76), suffers of poor selectivity. Good yields of the 1,4-dicarboxylation derivative were obtained making use of $Fe_2Cp_2(CO)_4$ as catalyst in anhydrous THF (78). Two recent papers deal with the optimization of the uncatalyzed process (79). This last study was performed in diaphragmless cells, using tetraalkylammonium formates and oxalates as supporting electrolytes, as a continuation of a research started in the Shell Laboratories several years ago (80). According to that procedure, the electrolysis in the presence of oxalates or formates can be performed in diaphragmless cells with inert anodes, on which the anodic reaction is the oxidation of both anions to carbon dioxide, which of course is one of the reactants of the cathodic process. The cathodic products of carboxylation are therefore prevented from the oxidation, as it takes place at an anodic potential more positive than that of the oxidation of both oxalate and formate. By oxidation of oxalate $CO_2^{-\cdot}$ radical anions are formed, which could lead to a supplementary carboxylation of butadiene. The contradictory experimental results reported in the literature are discussed in ref. 79; the matter appears still controversial, and should be further investigated. The recent detailed study on the influence of electrolysis parameters on the selectivity of the electrocarboxylation of butadiene in the above system (79) has shown that the synthesis is strongly influenced by the cathodic material, with the best yield of C_6 derivatives obtained with Pb cathode, and of C_{10} derivatives with reticulated carbon. Current yields hardly exceeding 60% were observed, possibly due to a dissipative oxalate formation at the cathode and oxidation at the anode. No influence of the electrolyte flow rate and moderate influence of the concentration of the reactants, together with the influence of the cathodic material, are consistent with a chemistry taking place on the electrode surface, between adsorbed intermediates. The reaction has been investigated, in the second part of this study, in a bipolar fluidised bed cell constructed from vitreous carbon microspheres between two feeder electrodes (80). Current yields and selectivities for C_{10} are good in the early stages of the electrolysis, but deteriorate quite rapidly with charge passed, possibly because of unavoidable changes of the electrode surface.

The electrocarboxylation of alkynes was described for the first time by Wawzonek (81): from diphenylacetylene a mixture of dicarboxylated derivatives, including maleic, fumaric and succinic acids, was obtained with an overall yield around 40%. Recently Dunach and Perichon (82 - 84) have reported that a great improvement to the selectivity of this reaction comes from the use of nickel based catalytic systems. In fact, an unexpected regioselectivity in the electrocarboxylation of terminal alkynes was observed using some octahedral Ni(II) complexes as catalyst precursors in diaphragmless systems with sacrificial magnesium anodes (82,83). As an example, the electrocarboxylation of a series of terminal alkynes, in the presence of a Ni(0)bipy complex obtained in situ by reduction of $Ni(bipy)_3(BF_4)_2$, in acetonitrile or dimethylformamide, leads to a mixture of acrilic acids, and a fairly good selectivity towards the 2-addition, leading to alpha-substituted acrylic acids, is observed, with an overall yield of carboxylated derivatives ranging from 50 to 85 % (82). The same catalytic system is active towards the electrocarboxylation of disubstituted alkynes (84). In this case, symmetrically disubstituted alkynes yield with good selectivity the monocarboxylation derivative, a disubstituted acrylic acid. The Mg(II) species arising from the sacrificial anodes are necessary for the activation of the catalytic system. Interesting to note, the same $Ni(bipy)_3^{2+}$ complex is also active for the reduction of carbon dioxide alone in aprotic media (85).

3.1.2.   Electrocarboxylation of organic carbonyl compounds. By electrocarboxylation ketones and aldehydes lead to the corresponding alpha-hydroxycarboxylic acids (76). This reaction was described for the first time several years ago but only recently has been taken under consideration as an intermediate step for the production of alpha-aryl-propionic acids used as non steroidal anti-inflammatory agents (NSAI). In several of the above mentioned synthetic processes sacrificial anodes in undivided cells are used. This methodology constitutes, beside the already quoted use of easily oxidizable salts, a fruitful approach to the problem of the reaction at counter electrode in diaphragmless cells. This technology, for which the disposal of equivalent amounts of metal salts after the isolation of products has to be planned, presents several interesting aspects: a) the metal cations introduced stoichio-metrically into the reaction medium often have a catalytic effect on the carboxylation reaction (see the electrocarboxylation of alkynes), and b) always have a stabilizing effect on the carboxylate anions, hindering wasteful follow-up reactions; c) in many cases the solubility of the metal-carboxylate complex salts is different from that of the other side products (dimers, hydrogenation derivatives) and the separation procedures are quite easy.

To the production of NSAI two synthetic routes, respectively starting from the corresponding benzyl halide (discussed in the following) or from the ketone, can therefore be evaluated in the planning of an electrochemical synthesis. Depending upon the availability and the stability of starting materials and intermediates, a proper choice can me made on the more convenient process. If ketone is the starting material for the production of the halide, the route through its carboxylation appears to be more convenient, being the same yields and selectivities, as only two synthetic steps instead of three are involved, as it can be seen from Scheme 7.

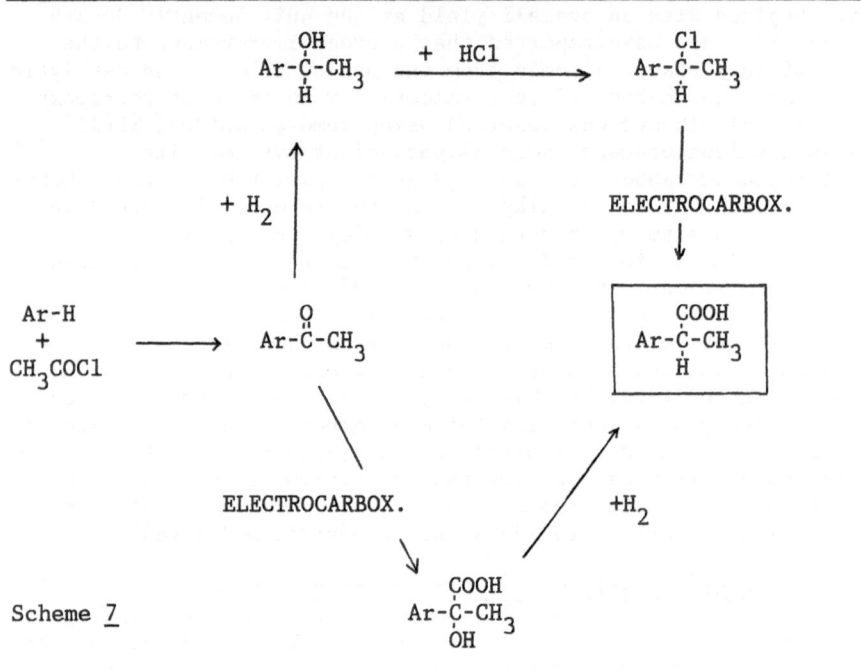

Scheme 7

As an example, the electrocarboxylation of aldehydes became possible with the use of the sacrificial anodes methodology, as the formation of complex salts with the hydroxocarboxylate anions hinders wasteful side reactions (86).

The sacrificial anodes methodology has been applied to the synthesis of intermediates to the production of 2-(6-methoxynaphtyl)-propionic acid (Naproxen) (87) and of 2-(4-i-butyl-phenyl)propionic acid (Ibuprofen)(88). The overall synthetic route to Naproxen appears to be competitive with the existing chemical methodologies. The starting material, 6-methoxyacetonaphtone, is cathodically reduced and carboxylated, in the presence of Al(III) species arising from the anodic reaction (scheme 8).

Main derivative of the electrosynthesis is the Al salt of the hydroxyacid, containing DMF as ligand (ca. 15% wt) and carbonate ions (ca. 10% wt). Side products, apart from traces of oxalate and of products of carboxylation of the aromatic ring, are pinacols deriving from the reductive dimerization of the ketone (reaction 2 of Scheme 7). The block diagram of the operations connected to the electrochemical stage of the synthesis of Naproxen is shown in Scheme 9.

cathode

$$Ar-\overset{\overset{O}{\|}}{C}-Me \quad + \quad 2\ e^- \quad + \quad CO_2 \quad \longrightarrow \quad \underset{Me}{\overset{Ar}{}}C\underset{CO_2^-}{\overset{O^-}{}} \qquad 1$$

$$2\ Ar-\overset{\overset{O}{\|}}{C}-Me \quad + \quad 2\ e^- \quad \longrightarrow \quad Ar-\underset{\underset{O^-}{|}}{\overset{\overset{Me}{|}}{C}}-\underset{\underset{O^-}{|}}{\overset{\overset{Me}{|}}{C}}-Ar \qquad 2$$

anode

$$Al \quad \longrightarrow \quad Al^{3+} \quad + \quad 3\ e^- \qquad 3$$

bulk

$$n\ Al^{3+} \quad + \quad m \ \underset{Me}{\overset{Ar}{}}C\underset{CO_2^-}{\overset{O^-}{}} \quad \longrightarrow \quad [Al^{3+}_n\ (\underset{Me}{\overset{Ar}{}}C\underset{CO_2^-}{\overset{O^-}{}}-)_m]^{(3n-2m)+} \qquad 4$$

Scheme 8   Ar = -2-(6-methoxynaphtyl)

The Al salt of the hydroxyacid is soluble in the reaction medium, and is precipitated from the effluent of the cell by addition of methyl-t-butyl--ether (MTBE).

Syntheses performed in a small scale batch pilot reactor gave 90% ketone conversion, and 90% product selectivity (as isolated acid). 2.7 KWh Kg^{-1} is the moderate energy consumption of the electrochemical stage, deduced from the operative parameters.

The influence of the nature of the cathode, and of the aluminium cations, on the electrocarboxylation of benzaldehyde, has been approached by cyclovoltammetry (89). The preliminary results are consistent with the hypothesis that the coordination of the carboxylated species on Al(III) could take place after the uptake of the first electron, and the addition of a carbon dioxide molecule on the resulting anion radical.

An optimization study is in progress, performed in a batch laboratory pilot reactor, on the carboxylation of acetonaphtone to the corresponding alpha-hydroxyacid (90). The system operates with a sacrificial aluminium anode in a batch reactor with cylindrical coaxial electrochemical cell. The molar ratio (MR) between the carboxylation derivative and the dimeric pinacol is influenced by the cathodic potential and ketone concentration. In particular, lower ketone concentrations and more negative potentials favour higher yields of the desired product. The negative influence the ketone concentration has on MR could be tentatively explained on the basis of a reaction pattern in which one of the cathodic intermediates (a radical anion or a dianion) could competitively react with the ketone, leading to the dimer, or with carbon dioxide, leading to the acid. Therefore at lower ketone concentrations the formation of the acid should result favoured. In both cases the reaction products are sequestered by the aluminium cations.

118

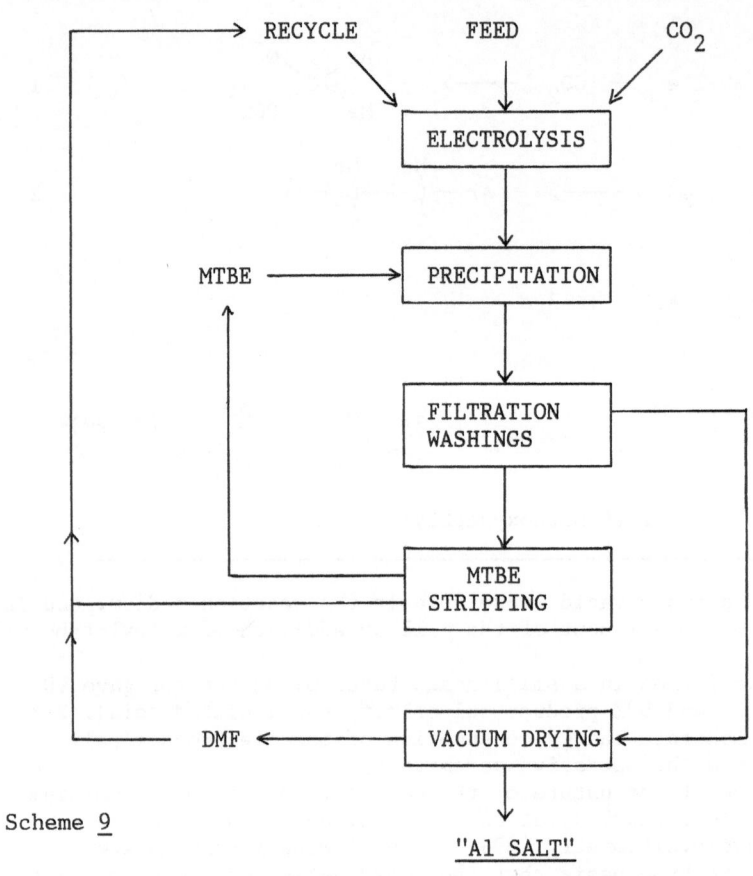

Scheme 9

"A1 SALT"

The results of a study on the mechanism of the reduction of 1,4-benzoquinone in the presence of carbon dioxide were published quite recently (91). A DISP1 mechanism (Scheme 10), according to the reaction scheme proposed by Saveant (92), is deduced from the results of double step chronopotentiometry : the monocarboxylated anion radical arising

$$A + e^- \xrightarrow{\longrightarrow}_{\longleftarrow} A^{-\cdot} \qquad 1$$

$$A^{-\cdot} + CO_2 \xrightarrow{k_1} [ACO_2]^{-\cdot} \qquad 2$$

$$[ACO_2^{-\cdot}] + A^{-\cdot} \longrightarrow [ACO_2]^{2-} + A \qquad 3$$

$$[ACO_2]^{2-} + CO_2 \longrightarrow A(COO)^{2-} \qquad 4$$

Scheme 10 (Ref. 91).   A = 1,4-benzoquinone

from reactions 1,2 , is furtherly reduced by the benzoquinone anion radical, and finally stabilized by a second carboxylation.

Synthetic small scale electrolyses in propylene carbonate gave good yield of 2,5-dihydroxybenzoic acid. Presumably the first carbon dioxide molecule is fixed onto the oxygen atoms of the reduced carbonyl group, and the second reacts with an anionic site on the ring originated by step 3.

Other examples of electrocarboxylations of carbonyl compounds which could gain some interest from the synthetic point of view were published by Simonet et al. (93). A quite surprising difference in reactivity towards the electrocarboxylation is shown by two classes of cyclic ketones having quite similar structure : indanones and tetralones. The electrochemical reduction of 2,2-disubstituted indanones in the presence of carbon dioxide gives only traces of carboxylation derivatives, and the presence of carbon dioxide seems to play only the role of enhancing the acidity of adventitious water; on the contrary, 2,2-disubstituted tetralones undergo the electrocarboxylation in the expected way, leading with moderate yields to the corresponding hydroxycarboxylic acid.

3.1.3. <u>Electrocarboxylation of the =C=N- unsaturation.</u> The synthetic potentialities of this reaction, leading the corresponding alpha-amino-carboxylic acids are at present under investigation both from the theoretical and applicative point of view. Quite a large number of Shiff bases has been tested for these reactions, which can give rise to several derivatives, as depicted in Scheme <u>11</u> (94).

scheme <u>11</u>

The first step of the reaction is, in the majority of cases, the reduction of the imine to the corresponding radical anion, whose structure, and spin and charge distribution are determinant for the product distribution. Quantum chemical HMO-, EHT-, and CNDO/2-calculations of the LUMO and free unpaired electron densities were correlated to the electrochemical data of 35 substituted imines by Komenda and Hess (95). A very good correlation between thoretical predictions and experimental results was found, suggesting that this interesting approach should be pursued also for other carboxylation processes.

The electrocarboxylation of the valuable pharmaceuticals 1,4-benzodiazepines, performed by Hess et al. with yields from 49 to 59% is a further example of the versatility of these processes (96).

The electrocarboxylation of imines with aluminium sacrificial anodes in diaphragmless cells has been also described (97). The presence of metal cations arising from the anodic reaction allows to work with higher concentrations of the starting material in comparison with the conventional electrochemical processes. This system appears to increase the selectivity towards the insertion of the carboxylic group on the carbon atom of the double bond of the imine.

3.2. ELECTROCARBOXYLATION BY SUBSTITUTION

3.2.1.   Electrocarboxylation of organic halides. Quite a large number of examples of electrocarboxylations of halides has been reported in the literature, due to the easy availabilty of halides as intermediates of organic synthesis, and the reliability of these electrochemical processes (76).

As it was shown in the case of ketones, the electrocarboxylation of halides has received large attention in the recent past, in view of the synthesis of several of the alpha-aryl-propionic acids to be used as anti-inflammatory agents. The electrocarboxylation of benzyl halides was described by Baizer (98), who found that it is followed by a nucleophilic substitution of the carboxylate anion on the unreacted halide, with formation of the corresponding benzyl-phenylacetate. This reaction, confirmed for the electrocarboxylation of benzyl iodide (99), does not seem to be of relevant importance for halides bearing bulky substituents, as recently shown by Fauvarque et al. (100). These Authors have recently published the results of a study on nickel catalyzed electrosyntheses of aryl-2 propionic acids. The results of this research were that when $NiCl_2dppp$ (dppp = 1,3-diphenylphosphino propane) associated to one equivalent of a labile co-ligand such as COD (COD = 1,5-cyclo-octadiene), was used as catalyst for the electrocarboxylation of some substituted benzyl halides, higher current densities and starting material concentrations can be used, if compared to uncatalyzed systems. Electrolyses were perfomed in undivided cells, with oxalates as supporting electrolytes. An interesting methodology for the monitoring of the progress of the electrosynthesis, based on HPLC, was also developed in the course of this research (101). It was commonly accepted that the mechanism of insertion of carbon dioxide onto a halogenated organic, catalyzed by Ni complexes, involved a Ni(0) - Ni(II) catalytic cycle, analogous to that accepted for the Ni catalyzed dimerization of halides (102). This last matter has been recently elucidated by Amatore and Jutand (103), who demonstrated that a more complex catalytic cycle, involving also Ni(I) and Ni(III) species, is active in the Ni-catalyzed dimerization of aryl halides. An analogous mechanism has been also demonstrated for the electrocarboxylation of halides (104).

A recent paper describes the electrocarboxylation of aryl halides in undivided cells with sacrificial magnesium anodes (105). The catalytic activity of Mg cations arising from the sacrificial anode is evident from the higher yields and selectivities towards the monocarboxylation of polyhalobenzenes, if compared to the results of conventional electrochemical systems.

The electrocarboxylation of 2-bromo-isobutyramides represents a new and useful synthetic way to ester-amides of 2,2-dimethylmalonic acid (106). When the electroreduction of 2-bromo-isobutyramides is performed in the presence of carbon dioxide the carboxylation is almost quantitative (Scheme 12). In order to obtain an isolable product, the unstable carboxylate anion was transformed into the corresponding ester by addition to the reaction medium of an alkylating agent.

A further example of interesting synthetic routes to carboxylic acids, deriving from mechanistic studies in which carbon dioxide is used

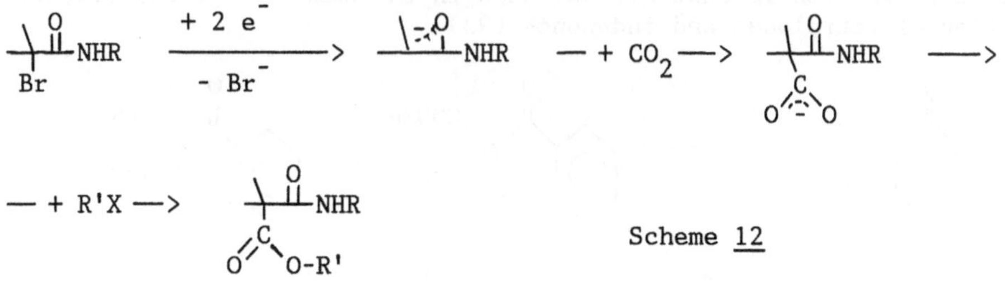

Scheme 12

as anion trap is found in the reduction of a series of substituted 3-halogeno-beta-lactams (Scheme 13).
The reductive cleavage of the C-halogen bond leads to an anionic intermediate, which reacts with carbon dioxide giving high yields of the

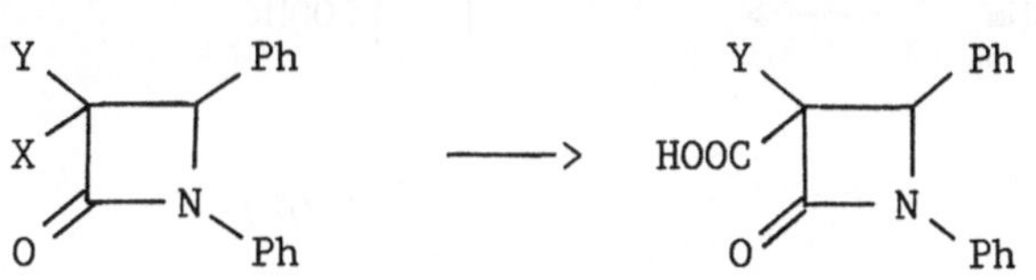

where: X = Br, Cl
Y = H or X

Scheme 13

corresponding 3-carboxy-beta-lactams, compounds which are not easily obtained by conventional chemical procedures (107).

3.2.2.   Carboxylation by substitution of the -SR group. The reductive substitution of -SR with a carboxylate anion was proposed several years ago by Matschiner et al. (108). The same research group has in the recent past enlarged the knowledge of those reactions of significant synthetic value (109), adding new examples of reactivity of organo-sulfur compounds bearing various substituents and different degrees of unsaturation (110,111). In the case of vinyl-ketones, the carboxylation for a structure like 1 involves the substitution of the halogen, whereas in the case of 2 it is the -SR group which is involved, and the corresponding diketo-carboxylic acid is obtained with acceptable yields.

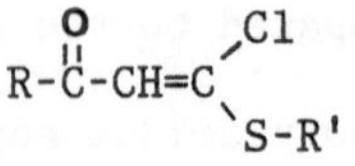

$$R-\overset{O}{\overset{||}{C}}-CH=C\overset{Cl}{\underset{S-R'}{\diagup}}$$

1

$$R-\overset{O}{\overset{||}{C}}-CH=C\overset{N<}{\underset{S-R'}{\diagup}}$$

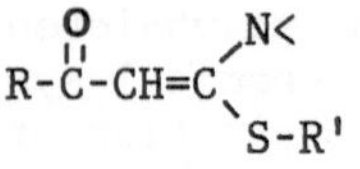

2

An interesting further example of carboxylative substitution of a thio-ether group involves 2-bis(methylthio)methylene-1-tetralone 3, which by electroreduction in the presence of carbon dioxide, leads to the espected carboxylate ion, which was trapped with methyl iodide as methylester 4. On the contrary, from 2-bis(methylthio)methylenecyclo-hexanone 5, in similar conditions, the addition product 6 is obtained (112). This difference in reactivity induced by side-substituents for some aspects recalls that already seen in the case of the electrocarboxylation of tetralones and indanones (93).

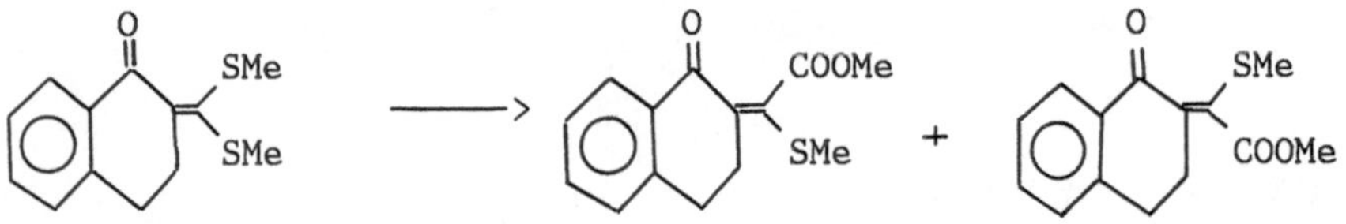

78%   13%

3   4

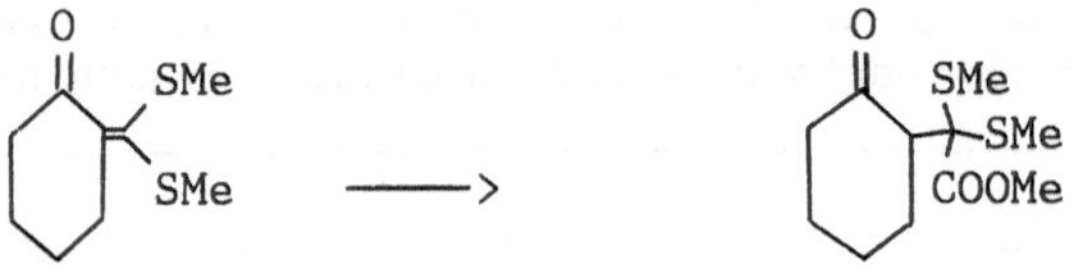

5   6 (40%)

4. TECHNOLOGICAL DEVELOPMENTS OF ELECTROCHEMICAL PROCESSES INVOLVING CARBON DIOXIDE.

In the course of this review some technological solutions to the problems raised by the performance of electrochemical processes involving a gaseous starting material have been presented.

Gas diffusion electrodes are of great importance in the fuel cells technology, and the expertise gained in this large field of applied research is transferred to all processes involving gaseous starting materials. As far as the electroreduction of carbon dioxide alone is considered, there are several examples of the use of this kind of electrodes, with incorporated various electroactive materials (10,11, 113 - 116).

Concerted cathodic and anodic processes are described in suitable electrochemical cells, for the transformation of carbon dioxide into different products, both in aqueous and organic media (32,117 - 120). Electrocarboxylations in fluidized bed electrolytic reactors have been tested on a pilot scale (52). The sacrificial anodes technology has been also developed, several models of cells have been patented (121 - 123), and some of them are under development.

REFERENCES.

1. Royer, M. E.; C. R. Hebd. Seances Acad. Sci. Fr.; 1870, 731, 70.
2. Haynes, L. V.; Sawyer, D. T.; Anal. Chem.; 1967, 332, 39.
3. Tyssee, D. A.; Wagenknecht, J. H.; Baizer, M. M.; Chruma, J. L.; Tetrahedron Lett.; 1972, 4809.
4. Vassiliev, Yu. B.; Bagotskii, V. S.; Osetrova, N. V.; Khazova, O. A.; Mayorova, N. A.; J. Electroanal. Chem. Interfacial Electrochem.; 1985, 271, 189
5. Vassiliev, Yu. B.; Bagotskii, V. S.; Osetrova, N. V.; Mikhailova, A. A.; J. Electroanal. Chem. Interfacial Electrochem.; 1985, 311, 189.
6. Udupa, K. S.; Subramanian, G. S.; Udupa, H. V. K.; Electrochimica Acta; 1971, 1593, 16.
7. Ito, K.; Ikeda, S.; Okabe, M.; Denki Kagaku; 1980, 247, 48.
8. Ikeda, S.; Takagi, T.; Ito, K.; Bull. Chem. Soc. Jpn.; 1987, 2517, 60.
9. Hori, Y.; Suzuki, S.; Bull. Chem. Soc. Jpn.; 1982, 660, 55.
10. Mahmood, M. N.; Masheder, D.; Harty, C. J.; J. Appl. Electrochem.; 1987, 1159-70, 17.
11. Furuya, N.; Matsui, K.; Motoo, S.; Denki Kagaku oyobi Kogyo Butsuri Kagaku; 1987, 787-8, 55.
12. Eggins, B. R.; Brown, E. M.; McNeill, E. A.; Grimshaw, J.; Tetrahedron Lett.; 1988, 945, 29.
13. Frese, K. W., Jr.; Leach, S. C.; Summers, D. P.; U.S. Pat. No.: 4609440 (1986). C.A. 105:199208
14. Summers, D. P.; Leach, S.; Frese, K. W., Jr.; J. Electroanal. Chem. Interfacial Electrochem.; 1986, 219-32, 205.
15. Frese, K. W., Jr.; Leach, S.; J. Electrochem. Soc.; 1985, 259-60, 132.
16. Frese, K. W., Jr.; Leach, S. C.; Summers, D. P.; U.S. Pat. No.: 4609441 (1986). C.A. 105:199209.
17. Hori, Y.; Kikuchi, K.; Suzuki, S.; Chem. Lett.; 1985, 1695.
18. Hori, Y.; Kikuchi, K.; Murata, A.; Suzuki, S.; Chem. Lett.; 1986, 897.
19. Hori, Y.; Suzuki, S.; Kenkyu Hokoku - Asahi Garasu Kogyo Gijutsu Shoreikai; 1986, 213, 48.
20. Cook, R. L.; MacDuff, R. C.; Sammells, A. F.; J. Electrochem. Soc.; 1987, 1873, 134.

21. Cook, R. L.; MacDuff, R. C.; Sammells, A. F.; J. Electrochem. Soc.; 1987, 2375, 134.
22. Cook, R. L.; MacDuff, R. C.; Sammells, A. F.; J. Electrochem. Soc.; 1988, 1320-6, 135.
23. Kim, J. J.; Summers, D. P.; Frese, K. W., Jr.; J. Electroanal. Chem. Interfacial Electrochem.; 1988, 223, 245.
24. Hori, Y.; Murata, A.; Kikuchi, K.; Suzuki, S.; J. Chem. Soc., Chem. Commun.; 1987, 728.
25. Alonso, C.; Gonzalez Velasco, J.; Arvia, A. J.; J. Electroanal. Chem. Interfacial Electrochem.; 1988, 183, 250.
26. Osetrova, N. V.; Vasiliev, Yu. B.; Bagotskii, V. S.; Sadkova, R.G.; Cherashev, A. F.; Khrushch, A. P.; Elektrokhimiya; 1984, 286, 20.
27. Lamy, E.; Nadjo, L., Saveant, J. M.; J. Electroanal. Chem. Interfacial Electrochem.; 1977, 403-07, 78.
28. Vianello, E. et al, Communication at the VIIth Heyrowsky Discussion, Liblice (Prague), 1989.
29. Amatore, C.; Saveant, J. M.; J. Am. Chem. Soc.; 1981, 5021-23, 103.
30. Vassiliev, Yu. B.; Bagotskii, V. S.; Khazova, O. A.; Mayorova, N. A.; J. Electroanal. Chem. Interfacial Electrochem.; 1985, 295, 189
31. Ito, K.; Ikeda, S.; Iida, T.; Nomura, A.; Denki Kagaku; 1982, 463, 50.
32. Fisher, J.; Lehman, Th.; Heitz, E.; J. Appl. Electrochem.; 1981, 743-50, 11.
33. a) Lamy, E.; Nadjo, L.; Saveant, J. M.; Nouv. J. Chim.; 1979, 21, 3.
    b) Kaiser, U; Heitz, E.;Ber. Bunsenges. Phys. Chem.; 1973, 818-23, 77.
34. Meshitsuka, S.; Ichikawa, M.; Tamaru, K.; J. Chem. Soc. Chem. Comm.; 1974, 158.
35. Hiratsuka, S.; Takahashi, H.; Sasaki, H.; Toshima, S; Chem. Lett.; 1977, 1137.
36. Kapusta, S.; Hackerman, N.; J. Electrochem. Soc.; 1984, 1511, 131.
37. Lieber, C. M.; Lewis, N. S.; J. Am. Chem. Soc.; 1984, 5033, 106
38. Tanabe, H.; Ohno, K.; Electrochim. Acta; 1987, 1121, 32.
39. Masheder, D.; Williams, K. P. J.; J. Raman Spectrosc.; 1987, 391, 18
40. Christensen, P. A.; Hamnett, A.; Muir, A. V. G.; J. Electroanal. Chem. Interfacial Electrochem.; 1988, 361, 241.
41. Mahmood, M. N.; Masheder, D.; Harty, C. J.; J. Appl. Electro chem.; 1987, 1223-7, 17.
42. Cao, X.; Huang, C.; Wang, M.; Gaodeng Xuexiao Huaxue Xuebao; 1983, 549, 4. C.A. 100:058659.
43. Cao, X.; Mu, Y.; Wang, M.; Luan, L.; Huaxue Xuebao; 1986, 220, 44. C.A. 104:195348.
44. Cao, X.; Mu, Y.; Wang, M.; Huang, C.; Gaodeng Xuexiao Huaxue Xuebao; 1986, 302, 7. C.A. 106:074777 .
45. Cao, X.; Zheng, G.; Gaodeng Xuexiao Huaxue Xuebao; 1987, 686, 8. C.A. 108:194575
46. Cao, X.; Zheng, G.; Teng, Y.; Gaodeng Xuexiao Huaxue Xuebao; 1988, 861-3, 9. C.A. 110:30379

47. Hammouche, M.; Lexa, D.; Saveant, J. M.; Momenteau, M.; J. Electroanal. Chem. Interfacial Electrochem.; 1988, 347, 2494.
48. Becker, J. Y.; Vainas, B.; Eger, R.; Kaufman, L.; J. Chem. Soc., Chem. Commun.; 1985, 1471.
49. O'Connell, C.; Hommeltoft, S. I.; Eisenberg, R.; NATO ASI Ser., Ser. C; "Carbon Dioxide as a Source of Carbon: Biochem. Chem. Uses", M. Aresta and G. Forti Eds., 1987, 33, 206.
50. Fisher, B.; Eisenberg, R.; J. Am. Chem. Soc.; 1980, 7361, 102.
51. Tinnemans, A. H. A.; Koster, T. P. M.; Thewissen, D. H. M. W.; Mackor, A.; Recl.: J. R. Neth. Chem. Soc.; 1984, 288, 103.
52. Pearce, D. J.; Pletcher, D.; J. Electroanal. Chem. Interfacial Electrochem.; 1986, 317, 197.
53. Bailey, C. L.; Bereman, R. D.; Rillema, D. P.; Nowak, R.; Inorg. Chim. Acta.; 1986, L45, 116.
54. Beley, M.; Collin, J. P.; Ruppert, R.; Sauvage, J. P.; J. Chem. Soc., Chem. Commun.; 1984, 1315.
55. Beley, M.; Collin, J.-P.; Ruppert, R.; Sauvage, J. P.; J. Am. Chem. Soc.; 1986, 7461, 108.
56. Collin, J.P.; Jouaiti, A.; Sauvage, J.P.; Inorg. Chem.; 1988, 1986-90, 27.
57. Che, C. M.; Mak, S. T.; Lee, W. O.; Fung, K. W.; Mak, T. C. W.; J. Chem. Soc. Dalton Trans.; 1988, 2153.
58. Slater, S.; Wagenknecht, J. H.; J. Am. Chem. Soc.; 1984, 5367, 106.
59. DuBois, D. L.; Miedaner, A.; J. Am. Chem. Soc.; 1987, 113-17, 109.
60. DuBois, D. L.; Miedaner, A.; ACS Symp. Ser.; 1988, 42, 363.
61. Gennaro, A.; Isse, A. A.; Vianello, E.; in "Recent Advances in Electroorganic Synthesis", S. Torii Ed., Stud. Org. Chem. (Amsterdam); 1987, 321, 30.
62. Mitchell R. M.; O'Toole, T. R.; Bolinger, C. M.; Megehee, E.; Thorp, H.; Meyer, T. J.; ACS Symp. Ser.; 1988, 52, 363
63. Ziessel, R.; NATO ASI Ser., Ser. C; "Carbon Dioxide as a Source of Carbon: Biochem. Chem. Uses", M. Aresta and G. Forti Eds., 1987, 113, 206.
64. Cabrera, C. R.; Abruna, H. D.; J. Electroanal. Chem. Interfacial Electrochem.; 1986, 101, 209
65. Meyer, T. J.; O'Toole, T. R.; Margerum, L. D.; Westmoreland, T. D.; Vining, W. J.; Murray, R. W.; Sullivan, B. P.; U.S. Pat. No.: 4711708; C.A. 109:013644.
66. Cosnier, S.; Deronzier, A.; Moutet, J. C.; J. Mol. Catal.; 1988, 381-91, 45
67. Cosnier, S.; Deronzier, A.; Moutet, J. C.; J. Electroanal. Chem. Interfacial Electrochem.; 1986, 315, 207.
68. Bolinger, C. M.; Sullivan, B. P.; Conrad, D.; Gilbert, J. A.; Story, N.; Meyer, T. J.; J. Chem. Soc., Chem. Commun.; 1985, 796.
69. Guadalupe, A. R.; Usifer, D. A.; Potts, K. T.; Hurrell, H. C.; Mogstad, A. E.; Abruna, H. D.; J. Am. Chem. Soc.; 1988, 3462, 110.
70.(a) Ogura, K.; Yoshida, I.; Electrochim. Acta; 1987, 1191-5, 32.
   (b) Ogura, K.; J. Electrochem. Soc.; 1987, 2749, 134.
71. Ishida, H.; Tanaka, K.; Tanaka, T.; Chem. Lett.; 1985, 405.

126

72   Ishida, H.; Tanaka, K.; Tanaka, T.; Organometallics; 1987, 181,6.
73. Ishida, H.; Tanaka, H.; Tanaka, K.; Tanaka, T.; J. Chem. Soc.,
     Chem. Commun.; 1987, 131.
74. Ishida, H.; Tanaka, H.; Tanaka, K.; Tanaka, T.; Chem. Lett.;
     1987, 597.
75. Tanaka, K.; Miyamoto, H.; Tanaka, T.; Chem. Lett.; 1988, 2033.
76. Silvestri, G.; NATO ASI Ser., Ser. C;"Carbon Dioxide as a Source of
     Carbon: Biochem. Chem. Uses", M. Aresta and G. Forti Eds.,; 1987,
     339, 206.
77. Gambino, S.; Gennaro, A.; Filardo, G.; Silvestri, G.; Vianello,
     E.; J. Electrochem. Soc.; 1987, 2172, 134.
78. Tkatchenko, I. B. M.; Ballivet-Tkatchenko, D. A.; El Murr, N.;
     Tanji, J.; Payne, J. D.; Fr. Pat. No.: 2542764; (1984). C.A.
     102:069341
79. Pletcher, D.; Girault, J. Tietje; J. Appl. Electrochem.; 1986,
     791-802, 16
80. Pletcher, D.; Girault, J. Tietje; Inst. Chem. Eng. Symp. Ser.
     Electrochem. Engineering; 1986, 13-21, 321-2, 98.
81. Wawzonek, S.; Wearring, D.; J. Am. Chem. Soc.; 1959, 2067 81.
82. Dunach, E.; Perichon, J.; J. Organomet. Chem.; 1988, 239-46, 352.
83. Labbe, E.; Dunach, E.; Perichon, J.; J. Organomet. Chem.; 1988,
     C51-C56, 353.
84. Dunach, E.; Derien, S.; Perichon, J.; J. Organomet. Chem.; 1989,
     C33-C36, 364.
85. Daniele, S.; Ugo, P.; Bontempelli, G.; Fiorani, M.; J.
     Electroanal. Chem. Interfacial Elecrochem.; 1987, 259, 219.
86. Silvestri, G.; Gambino, S.; Filardo, G.; Tetrahedron Lett.; 1986,
     3429-30, 27.
87.a) Filardo, G.; Silvestri, G.; Gambino, S.; Eur. Pat. No.: 189120
     (1986). C.A. 105:180522
   b) Maspero, F.; Piccolo, O.; Romano, U.; Gambino, S.; Eur. Pat. Appl.
     No.: 286944 (1988).
88.a) Wagenknecht, J. H.; U.S. Pat. No.: 4582577 (1986); C.A.
     105:050861.
   b) Wagenknecht, J. H.; U. S. Pat. No.: 4601797 (1986); C.A.
     106:024982.
89. Silvestri, G.; Gambino, S.; Filardo, G.; in" Recent Advances in
     Electroorganic Synthesis", S. Torii Ed. (Stud. Org. Chem.),
     Amsterdam; 1987, 287, 30.
90. Di Lorenzo, S.; Silvestri, G.; Filardo, G.; Gambino, S.;Chem. Eng.
     J.; 1989, 187, 40..
91. Bulhoes, L. O. de Sousa; Zara, A. J.; J. Electroanal. Chem.
     Interfacial Electrochem.; 1988, 159-65, 248.
92. Amatore, C.; Gareil, M.; Saveant, J. M.; J. Am. Chem. Soc.; 1988,
     4120, 102.
93. Orliac Le Moing, A.; Delaunay, J.; Lebouc, A.; Simonet, J.;
     Tetrahedron; 1985, 4483, 41.
94. Hess, U.; Bluemcke, C. O.; Z. Chem.; 1988, 144, 28
95.a) Komenda, J.; Hess, U.; Z. Phys. Chem. (Leipzig); 1984, 17, 265.
   b) Komenda, J.; Fiala, R.; Hess, U.; Z. Phys. Chem. (Leipzig); 1987,
     48, 268.
96. Hess, U.; Granitza, D.; Thiele, R.; Z. Chem.; 1988, 188-9, 28.

97. Silvestri, G.; Gambino, S.; Filardo, G; Gazz. Chim. It.; 1988, 643, 118.
98. Baizer, M. M.; Chruma, J. L; J. Org. Chem.; 1972, 1951, 37.
99. Koch, D. A.; Henne, B. J.; Bartak, D. E.; J. Electrochem. Soc.; 1987, 3062, 134.
100. Fauvarque, J. F.; Jutand, A.; Francois, M.; J. Appl. Electrochem.; 1988, 109-15, 18.
101. Fauvarque, J. F.; Jutand, A.; Francois, M.; Petit, M. A.; J. Appl. Electrochem.; 1988, 116, 18
102.a) Troupel,M.; Rollin, Y.; Sibille, S.; Fauvarque, J. F.; Perichon, J.; J. Chem. Res. Synop.; 1980, 26.
     b) Schiavon, G.; Bontempelli, G.; Corain, B.; J. Chem. Soc. Dalton Trans.; 1981, 1074
103. Amatore, C.; Jutand, A.; Organometallics; 1988, 2203, 7.
104. Amatore, C.; Jutand, A.; personal communication.
105. Heintz, M.; Sock, O.; Saboureau, C.; Perichon, J.; Troupel, M.; Tetrahedron; 1988, 1631, 44.
106. Maran, F.; Fabrizio, M.; D'Angeli, F.; Vianello, E.; Tetrahedron; 1988, 2351-8, 44.
107. Casadei, M. A.; Moracci, F. M.; Inesi, A.; J. Chem. Soc., Perkin Trans. 2; 1986, 419.
108. Ruettinger, H.H.; Rudorf, W. D.; Matschiner, H.; Electrochim. Acta; 1985,155, 30.
109. Janietz, S.; Ruettinger, H. H.; Matschiner, H.; J. Prakt. Chem.; 1988, 147-53, 330.
110. Ruettinger, H. H.; Matschiner, H.; Gollnow, W. D.; J. Prakt. Chem.; 1986, 539, 328.
111. Janietz, S.; Ruettinger, H. H.; Matschiner, H.; Z. Chem. 1987, 442, 27.
112. Schultz-von Itter, N.; Steckhan, E.; Tetrahedron; 1987, 2475, 43.
113. Cook, R. L.; Mc Duff, R. C.; Sammels, A. F.; J. Electrochem. Soc.; 1988, 1470, 135.
114. Brown, D. E.; Hall, S. M.; Mahmood, M. N.; Eur. Pat. No.: 81982; C.A. 99:060912.
115. Furuya, C.; Motoo, S.; Jpn.Pat. No.: 62280385; C.A. 108:228508.
116. Ang, P. G. P.; Sammells, A. F.; U.S. Pat. No.: 4673473; C.A. 107:105037 .
117. Ang, P. G. P.; U.S. Pat. No.: 4620906; C.A. 106:035102.
118. Morduchowitz, A.; Ang, P. G. P.; U.S. Pat. No.: 4608133; C.A. 105:215761.
119. Sammells, A. F.; U.S. Pat. No.: 4608132; C.A. 105:160917.
120. Ang, P. G. P.; Sammells, A. F.; Morduchowitz, A.; U.S. Pat. No.: 4595465; C.A. 105:087537.
121. Chaussard, J.; Eur. Pat. No.: 219367.
122. Cohen, S.; Eur. Pat. No.: 13215.
123. Silvestri, G.; Filardo, G.; Gambino, S.; Eur. Pat. No.: 283796; (1988). C.A 110:065730.

MAMMALIAN AND PLANT CARBONIC ANHYDRASES:

STRUCTURE, FUNCTION, MECHANISM AND CONTROL

Y. Pocker

Department of Chemistry, University of Washington
Seattle, Washington 98195, U.S.A.

Mammalian Carbonic Anhydrase

Carbonic anhydrase (carbonate hydrolyase, EC 4.2.1.1) is the first zinc metalloenzyme to have been discovered. Its detection by Meldrum and Roughton[1] confirmed an earlier observation by a group of physiologists and biochemists that the release of carbon dioxide from hemolyzed blood occurs more rapidly than that expected from the buffer catalyzed rate of conversion of bicarbonate to CO_2.

The physiological role of carbonic anhydrase (CA) is attributed to its catalysis of the interconversion of CO_2 and bicarbonate,[1,2] eq. 1:

$$CO_2 + H_2O \rightleftarrows HCO_3^- + H^+ \tag{1}$$

In addition to its intrinsic physiological importance, carbonic anhydrase is an extremely convenient enzyme for study.[3] It can be isolated by standard techniques of precipitation and column chromatography[4] or by affinity chromatography.[5] Isoenzymes I and II found in mammalian erythrocytes are stable in cold storage for many months; they are also stable in solution and can tolerate pH values in the range 5.5 to 12 for relatively extended periods. Because of its great catalytic power, robust constitution, and physiological importance, many new experimental techniques have first been applied to the high activity isoenzyme, CA-II.[6-10] Carbonic anhydrase proved often to be a test case in which the application of the principles of solution chemistry and enzyme catalysis were subjected to rigorous tests.[11-23]

Although carbonic anhydrase catalyzes, a reaction that is readily reversible under physiological conditions, most studies of the kinetic properties of the enzyme have focused on its CO_2 hydrase activity.[11-18] As a result of recent work on HCO_3^- dehydration,[11-14] carbonic anhydrase can now be seen to possess a striking degree of symmetry in kinetic properties of the forward and reverse reactions. Both substrates exhibit pH-independent Michaelis constants, K_m, coupled to complimentary sigmoidal variations in the turnover number k_{cat}. Carbonic anhydrase is most active towards CO_2 hydration at high pH and most active towards HCO_3^- dehydration at low pH. A single ionization governs the change in

129

M. Aresta and J. V. Schloss (eds.),
Enzymatic and Model Carboxylation and Reduction Reactions for Carbon Dioxide Utilization, 129–143.
© 1990 *Kluwer Academic Publishers.*

activity with a pK_a near 7 for the high activity isozyme, CA-II. Originally thought to be absolutely specific for CO_2 - HCO_3^- interconversion, the high specific-activity isozyme CA II from mammalian erythrocytes has been shown to possess a considerable degree of catalytic versatility, including hydration of numerous aldehydes,[24-26] pyruvic acid[27] and alkyl pyruvate esters,[28] as well as the hydrolysis of carboxylate esters,[29-40] carbonate esters,[30,41,42] alkyl pyruvate esters,[43] and phosphate esters.[44-47] In contrast to the interaction of diethyl p-nitrophenyl phosphate ("paraoxon") with the serine esterases, in which one molecule of p-nitrophenol is liberated during the inhibition of each enzyme molecule by the phosphate ester, carbonic anhydrase experiences dialkyl monoaryl phosphate triesters only as substrates. This is indeed not unexpected in view of the fact that diisopropyl phosphorofluoridate has been found not to inactivate the enzyme.[30,48]

The pH-rate profiles for the carbonic anhydrase catalyzed hydration of aliphatic aldehydes and the hydrolysis of esters are both sigmoidal below pH 9 and exhibit a close correspondence with the pH dependence of the CO_2 hydration reaction suggesting that these various catalytic properties of the enzyme are shared by the same active site.[6]

Isoenzymes of mammalian carbonic anhydrase exist as a single polypeptide chain of 260 or 261 amino acid residues containing a single zinc ion and a blocked N terminal group. Amino acid sequence variation among carbonic anhydrase isoenzymes accounts for their varying activity, response to chemical modification, and inhibition by monovalent anions and heterocyclic or aromatic sulfonamides.[5-10,49-57] The three isoenzymes of carbonic anhydrase found in human erythrocytes, designated HCA-I, HCA-II and HCA-III, are products of distinct gene loci.[58] However, the carbonic anhydrase multigene family almost certainly includes more than three gene loci based on well documented reports of additional distinct varieties.[58-60] These additional forms include: (i) a membrane-bound species in lung, CA-IV;[61] (ii) a variety associated with kidney tubules;[62,63] (iii) an enzyme found in saliva and secreted by parotid acinar cells;[64,65] and (iv) a mitochondrial form thought to participate in the urea cycle, gluconeogenesis and fatty acid synthesis, CA-V.[66,67] Tissue distribution studies of the well-characterized isoenzymes in mammals have revealed a wide occurrence of CA-II and a more restricted distribution of CA-I. CA-III is mainly found in red skeletal muscle and rodent liver.[56,68-71]

The active site of carbonic anhydrase is a conical cavity some 15Å deep formed by four strands of pleated β-sheets, designated γ, δ, ε and ξ, and a single zinc ion firmly bound to the protein by three histidyl residues, which arise from the central pleated sheet. N^δ_3 of His 94 and 96 are from the δ strand with the nitrogen in position 3, and N^ϵ_1 of His 119 is from the ε strand with the nitrogen in position 1 fixed to the zinc ion. This results in a somewhat distorted tetrahedral coordination, lying some 30° out of the imidazole ring plane.[72-77]

In two regions of the HCA molecule on either side of the large twisted β-structure, aromatic residues are packed together in an irregular way. About 37% of the residues are in β-structure. The ten central pleated β-sheet strands are antiparallel to each other and are twisted about a common axis by about 220° from the left edge to the right edge. Along most chains in the β-structure there is a right-handed twist of about 20° between the peptide planes of successive residues.[74-77] The active site on the antiparallel δ and ε strands is nearly buried at the bottom of the conical cavity, with the stability assisted by the ξ and κ strands, as shown in the x-ray diffraction studies of Nostrand et al.[74] Interactions between this κ-pleated β-sheet and the loops of the δ and γ strands at the bottom of the cavity stabilize the entire molecule provide proper alignment and orientation in the active site cavity and facilitate induced fit during enzymatic activity.[78a,b,c]

Numerous studies have explored the possibility that carbonic anhydrases participate in processes other than CO_2-bicarbonate interconversion, promotion of CO_2 diffusion, and control of pH. Though any additional physiologic roles of carbonic anhydrase are as yet uncertain, some features of enzymatic catalysis are experienced by all substrates undergoing hydration-dehydration and hydrolysis. For example, each of the carbonyl-containing substrates is subject to a common catalytic fate: nucleophilic attack by a zinc-aquo or zinc-hydroxo complex. In terms of this mechanism, zinc enzymes such as CAII appear to use the zinc ion to lower the pK_a of a coordinated water molecule and generate a coordinated, but still highly reactive, hydroxide at neutral pH. The hydrophobic environment within the active central cavity of carbonic anhydrase favors such a low pK_a. Specific evidence suggests that a mixed zinc complex involving protein ligands and water in the unique environment of the active site cavity lowers the pK_a of the water molecule while preserving a high degree of nucleophilicity.[6-14, 79,80]

Despite a detailed knowledge of the active site of carbonic anhydrase, rather little is known about the topology of the CO_2 and HCO_3^- binding site, the pathway of the proton transfer, or the organization of water molecules within the active site.[81] In addition, differences between carbonic anhydrase isoenzymes are also related, in part, to the acidity of active site histidine residues.[82-83] The transfer of a proton from zinc-bound H_2O to His-64 occurs via an extended water bridge involving three protons "in flight" during the protonation-deprotonation process.[84]

Energy minimization calculations based on x-ray crystallographic data from the National Protein Data Bank, and examination of the roles of water and zinc coordinated to histidyl groups during the reversible hydration of CO_2, are being performed. Such data are being used to generate real-time molecular graphics of the active site during catalysis that may shed light on CO_2 and HCO_3^- binding water organization and proton relay from and to the active site.

Plant Carbonic Anhydrase

Current interest in carbonic anhydrase in plants centers on its elusive role. Many models are based on the demonstration that carbonic anhydrase in or near artificial membranes can increase the flux of carbon dioxide through them;[85,86] others relate to the difficulty of explaining the measured rates of ribulose bisphosphate carboxylation in the alkaline stroma.[87] Periodically, carbonic anhydrase is invoked in discussions of regulatory mechanisms and fine tuning in photosynthesis. In addition, since in some systems carbonic anhydrase can be induced by environmental carbon dioxide,[88] there is another control mechanism yet to be elucidated. Finally, as indicated earlier, the isoenzymes of a mammalian carbonic anhydrase have provided a useful phylogenetic tree.[58] Similarly, plants also appear to have a variety of isoenzymes.[90,91]

The enzyme appears to be ubiquitous. It has been found in bacteria, cyanobacteria (blue-green algae), green, red and brown algae, and ferns and higher plants, including both monocotyledons and dicotyledons, and plants with C_3- and C_4-type photosynthetic pathways. The wide variation in level of enzymatic activity in a given plant is more likely to reflect differences in extractability and stability than inherent dissimilarities.[92] In *Chlorella* and cyanobacteria, the amount of enzyme varies depending on the CO_2 concentration in the growth medium. High CO_2 (>1%) appears to repress formation of the enzyme. The enzyme has been isolated and characterized from three microalgae, a fern and several higher plants, including two monocotyledons and five dicotyledons. The enzyme has been extracted from photosynthetic tissue in all cases, except for the root nodule tissue of the French bean. The dicotyledon type has been characterized from leaves of four species (spinach, parsley, pea and French bean). There is general agreement that the native enzyme from dicotyledon species is composed of a hexameric assembly of identical subunits. Each subunit has a MW of 26,500 to 30,000[93-97] and contains one zinc (II) ion.

The plant enzymes are characterized by a high content of cysteine residues, which may account for the observation by most workers that under aerobic conditions a sulfhydryl agent is required during extraction and handling of plant-enzyme preparations to maintain catalytic activity. The dicotyledon enzymes may be formally analyzed in terms of the Michaelis-Menten rate equation. The apparent pK_a of the activity linked group in the plant enzyme is around 7.7.

We deem it desirable to compare the kinetic properties of the dicotyledon enzyme with those of the erythrocyte enzyme with regard to both physiological substrates, CO_2 and HCO_3^-. As shown in the present paper, the simultaneous measurement of the kinetic parameters for these two substrates allowed us to assay the validity of the Haldane relationship, Eq. 2 (here K_{eq} represents the equilibrium constant between the two substrates):

$$k_{cat}^{CO_2} K_m^{HCO_3^-} / k_{cat}^{HCO_3^-} K_m^{CO_2} = K_{eq}/[H^+] \tag{2}$$

The enzyme-catalyzed hydration of carbon dioxide and dehydration of HCO_3^- were studied as a function of pH over a wide region of pH using several contiguous buffer sytems. The kinetic data for the forward and reverse processes, determined under identical conditions, were fitted to the Haldane relationship. The original expression given by Eq. 2 may be reexpressed after taking the log of both sides in the form given in Eq. 3 (here pK_{eq} represents minus the log of the equilibrium constant between the two substrates):

$$\log\left(k_{cat}^{CO_2} K_m^{HCO_3^-} / k_{cat}^{HCO_3^-} K_m^{CO_2} \right) = pH - pK_{eq} \tag{3}$$

The plotted points were found to define a line with a slope very near unity. The apparent pK_{eq} of 6.39 is in good agreement with a value of 6.35 determined at zero ionic strength.[99]

The inhibition of the enzyme-catalyzed hydration of CO_2 by both sodium azide and ethoxzolamide was studied at pH 7.0.[100] The corresponding plots of V_{enz} *vs.* $V_{enz}/[CO_2]$ at varying concentrations of inhibitor show a series of parallel lines that characterize the mode of inhibition as noncompetitive in both cases. Their respective inhibition constants, K_i, as determined from plots of $[I]/(1 - V_i/V_0)$ vs. V_0/V_i, are 3×10^{-5} M and 1×10^{-5} M, respectively. The inhibition of the enzyme-catalyzed dehydration of HCO_3^- by these two inhibitors was also studied at pH 7.0. The series of parallel lines produced by varying quantities of ethox-zolamide again characterize the mode of inhibition as noncompetitive. The inhibition constant, K_i, of $1 \times 10_{-5}$ M determined for ethoxzolamide is identical to that determined for the hydrase reaction. The pattern of lines produced by varying the concentration of azi de characterizes the mode of inhibition as competitive to a first approximation. The apparent inhibition constant K_i for azide is of the same order of magnitude as that determined for the hydration of CO_2, but is dependent upon substrate concentration. Such patterns have been observed with CA-II as well.[3a,b,100,101] Most of the plant carbonic anhydrases so far studied differ quite markedly in their response to the sulfonamide family of inhibitors compared to the isoenzymes from red blood cells, CA-I and II. Since the mode of action of these inhibitors is generally considered to be through association with the zinc ion in the enzyme molecule, it is possible that the structure of the active site is more constrained in the plant enzyme in comparison to that in isoenzymes I and II. The enzyme from some algae, however, is as sensitive to acetazolamide as the isoenzymes I and II from many animal species.[102a,b] A more recently discovered isoenzyme, CA-III, occurs in mammalian skeletal muscles; it exhibits 10^3-fold lower affinity for acetazolamide, but is nearly as anion-sensitive as the erythrocyte isoenzyme II. In contrast to the animal isoenzymes I and II, very little is known about the structure of the plant enzyme. The structure of the active site of plant carbonic anhydrase is unknown, though it is generally assumed to be similar to that of the animal enzyme. A zinc (II) ion is tightly bound to each subunit of the plant enzyme.[93-98] The lower sensitivity of most of the plant carbonic anhydrases to inhibition by sulfonamides suggests that the approach to the active site is narrower and the hydrophobic region smaller than in the animal isoenzymes I and II. However, since the algal enzymes are just as sensitive to acetazolamide as most animal carbonic anhydrases, the active site structure and

environment of the algal and high plant enzymes must differ. The Michaelis constant $K_m^{CO_2}$ for the spinach enzyme measured in dimethylimidazole buffer indicates a greater affinity of the plant enzyme for

CO_2 than is found with the mammalian isozymes I and II. Like the mammalian enzyme, $K_m^{CO_2}$ values for the plant enzyme are unaffected by pH in the range 7.5-9.0. However, in the presence of inorganic

phosphate, the plant enzyme shows a considerable increase in $K_m^{CO_2}$, especially at low pH[95,96,98] where the K_m is pH-dependent. It appears that the affinity of the spinach enzyme for CO_2 is dramatically

affected by phosphate-buffers in the pH range 6.0 to 7.4.[95,96] The monovalent buffer ion, $H_2PO_4^{2-}$ reduces the affinity of the enzyme for CO_2, a regulatory effect that is most pronounced when the activity-linked group in the enzyme is in its acidic form. The pH rate profiles of the turnover number k_{cat} are also sensitive to buffer components. Here, however, the k_{cat} for CO_2 hydration increases with increasing

phosphate concentration in a manner suggesting that the diamion $H_2PO_4^{2-}$ participates in proton transfer at high pH. The dehydration of HCO_3^- has characteristics opposite to those of CO_2 hydration,[98] in terms

of pH-rate profile and phosphate effects on the kinetic parameters $K_m^{HCO_3^-}$ and $k_{cat}^{HCO_3^-}$.

These different effects of phosphate and pH on K_m and k_{cat}, respectively, have the potential for regulation in plant biology[36] and are likely to be especially important in the chloroplast where phosphate plays an important role in the regulation of photosynthesis.

Recent Progress

a) Much of our current knowledge about carbonic anhydrase is based upon earlier studies of the two isozymes isolable from mammalian erythrocytes. The essential zinc (II) ion is tightly bound to the enzyme; yet, it is possible to remove it through the use of appropriate chelating agents.[103-107] The resulting apoenzyme is almost completely inactive; however, full activity can be restored by the addition of molar equivalents of either zinc (II) or cobalt (II) ions.[103-107] When the chromophoric paramagnetic cobalt (II) ion is substituted for the colorless, diamagnetic zinc of the spinach enzyme, it yields a catalytically active, spectrally distinct enzyme whose unique UV-VIS absorption, CD, MCD and EPR spectra reflect the unusual asymmetric environment that this enzyme creates for Co (II). These spectra are responsive to pH, buffer complexation, and the binding of regulators, inhibitors, substrates, pseudosubstrates and products. By combining a rapid scanning spectrometer and a low-temperature, stopped-flow instrument, we are able to execute simultaneous kinetic and spectral measurements at subzero temperatures and provide information that is otherwise inaccessible about mechanism and control.

b) Since carbonic anhydrase catalyzes the reversible hydration of CO_2, it is important to compare the activation energies, E_a, for the hydration of carbon dioxide by H_2O (14.3 kcal mol^{-1}), OH$^-$ (7.2 kcal mol^{-1}) and CA-II (6.0 kcal mol^{-1}), eq. 4,

$$EZnOH + CO_2 + H_2O \rightleftarrows EZn\overset{+}{O}H_2 + HCO_3^- \tag{4}$$

The attack by the zinc-hydroxo complex is a recurring mechanistic motif in the hydrase and esterase activities of carbonic anhydrase.[6-16] The sign and magnitude of the measured activation parameters of k_{cat} and k_{cat}/K_m is consistent with the idea of a rate-determining intramolecular proton transfer between a zinc-aquo complex and the proton shuttle group, histidine 64.[108-110] A two-water proton-relay mechanism appears to be a prominent feature of the deprotonation process that converts the zinc-aquo to a zinc-hydroxo complex.[108-112] There appear to be three protons in flight during the rate-determining intramolecular proton transfer step.[108] These protons are postulated to be contributed by two water molecules which form a bridge between the zinc-aquo complex and the postulated acceptor, His-64.[111,112] This hypothesis is reasonable considering that proton donor and acceptor are separated by 6 Å;[111,112] the zinc ion in carbonic anhydrase II lies at the bottom of a conical cavity nearly 15Å deep, while His-64 lies in the same cavity, but nearer the surface.[111]

It has recently been shown that when His-64, identified here as the internal base B_i and its conjugate acid B_iH^+, is replaced with alanine, which cannot transfer protons, enzyme activity is reduced, but only by ca. two-fold.[113] Consequently, this residue is not crucial for maintaining a high rate of proton transfer between active site and the reaction medium. It is likely, however, that due to the size difference between histidine and alanine side chains, the entry of a small external buffer molecule, B_e, into the active site is more facile in the "imperfect" mutant enzyme in comparison to the wild-type form. Indeed, the mutant Ala-64 enzyme is ideally suited to recognize and bind external buffers, especially imidazole, which are small enough to penetrate into the site originally occupied by a histidine residue[112] (compare eqs. 5 and 6).

$$B_e\!:B_i\!-\!EZn\overset{+}{O}H_2 \rightleftarrows B_e\!:{}^+HB_i\!-\!EZnOH \rightleftarrows {}^+HB_e\!:B_i\!-\!EZnOH \tag{5}$$

$$B_e\!:EZn\overset{+}{O}H_2 \rightleftarrows {}^+HB_e\!:EZnOH \tag{6}$$

Clearly, the alternative pathway involving direct proton transfer between the metal site and buffer can also contribute significantly to the overall catalytic rate, but the efficiency of this proton-shuttle depends on productive protein-buffer interactions and hence on the type of external buffer used and its mode of binding in the active site. It is relevant to point out that a detailed examination of buffer efficiency in CO_2 hydration and HCO_3^- dehydration reactions catalyzed by wild-type carbonic anhydrase II revealed the participation of two distinct families of buffers.[110] The mutant enzymes would be expected to be even more sensitive to the specific properties of buffer species with regard to their ability to fit into the active site in an orientation suitable for accepting (or donating) a proton from (or to) the zinc-coordinated water (or hydroxide ion).[114] Recently, the role of His-64 in the intramolecular proton-transfer has been studied with the aid of active site modifiers having different steric and hydrophobic requirements. These studies lead to the conclusion that while this side chain residue is not absolutely crucial for efficient catalysis it significantly enhances the catalytic efficiency of carbonic anhydrase[110,115] and renders its catalysis reasonably independent of the steric requirements of surrounding buffer molecules.

c) Liang and Lipscomb have recently carried out some very exciting and important theoretical calculations on the hydration of CO_2 by H_2O,[116] OH^-,[116] and carbonic anhydrase.[117] The most difficult aspect of modelling the enzymatic hydration—dehydration process is duplicating the binding of CO_2 and HCO_3^- to the enzyme.[118,119] This stems from the fact that synthetic models are far smaller than carbonic anhydrase and do not command the supramolecular modes of recognition that the enzymatic cavity possesses.[118] Furthermore, no zinc-hydroxo or cobalt-hydroxo model identified to date exhibits a nucleophilicity towards CO_2 which exceeds that of free OH^-, although in the enzyme, the active site zinc-hydroxo complex appears to be approximately 104-fold more reactive than hydroxide ion.[11,12,81,118] It has been suggested that during an enzymatic transformation, any water molecules that are not important to the catalytic behavior of an enzyme are removed from the active site[118,119]. This would imply that an important step associated with the broad catalytic versatility of carbonic anhydrase, i.e., the nucleophilic attack on C=O substrates, may occur in a hydrophobic millieu,[120] whereas in the rate-controlling proton-transfer step(s) associated with CO_2 - HCO_3^- turnover, an ordered array of water molecules is of functional importance.[108-111] In this regard, it should be noted that the side chain residues on one side of the active site cavity are predominantly hydrophilic, and on the other side hydrophobic. Furthermore, an additional enhancement in rate may be achieved if at the time that the zinc-hydroxo complex attacks CO_2, the zinc ion also plays the role, of a Lewis acid, by stabilizing the incipient bicárbonato complex.[116,119]

REFERENCES

1. Meldrum, N.U.; Roughton, E.J.W. *J. Physiol.* (London) **1933**, *80*, 1113-1119.

2. Edsall, J.T.; Wyman J. In *Biophysical Chemistry;* Academic Press, New York **1958**; Vol. 1, Ch. 10.

3. Edsall, J.T. In *Harvey Lectures*; Academic Press, New York, **1968**; Series *62*, pp 191-230.

4. a. Nyman, P.-O. *Biochim. Biophys. Acta* **1961**, *52*, 1-12.

   b. Rikli, E.E.; Ghazanfar, S.A.S.; Gibbons, B.H.; Edsall, J.T. *J. Biol. Chem.* **1964**, *239*, 1065-1078.

5. Khalifah, R.G.; Strader, D.J.; Bryant, S.H.; Gibson, S.M. *Biochemistry* **1977**, *16*, 2241-2247.

6. Pocker, Y.; Sarkanen, S. *Adv. Enzymol., Relat. Areas Mol. Biol.* **1978**, *47*, 149-274.

7. Lindskog, S. *Adv. Inorg. Biochem.* **1982**, *4*, 115-170.

8. Silverman, D.; Vincent, S.H. *CRC Crit. Rev. Biochem.* **1983**, *16*, 207-255.

9. Bertinti, I.; Luchinat, C. *Acc. Chem. Res.* **1983**, *16*, 272-279.

10. Coleman, J.E. *Ann. New York Acad. Sci.* **1984**, *429*, 26-48.

11. Pocker, Y.; Bjorkquist, D.W. *Biochemistry* **1977**, *16*, 5698-5708.

12. Pocker, Y.; Deits, T.L.; Tanaka, N. In *Advances in Solution Chemistry*; Bertini, I.; Lunazzi, L.; Dei, A. Eds.; Plenum Press: New York, **1981**, *103*, 3949-3951.

13. a. Pocker, Y.; Deits, T.L. *J. Am. Chem. Soc.* **1981**, *103*, 3949-3951.

    b. Pocker, Y.; Deits, T.L. *J. Am. Chem. Soc.* **1982**, *104*, 2424-2434.

14. Pocker, Y.; Deits, T.L. *J. Am. Chem. Soc.* **1983**, *105*, 980-986.

15. Lindskog, S.; Engberg, P.; Forsman, C.; Ibrahim, S.A.; Jonsson, B-H; Simonson, I.; Tibell, L. *Ann. New YorkAcad. Sci.* **1984**, *429*, 61-75.

16. Pocker, Y.; Deits, T.L. *Ann. New York Acad. Sci.* **1984**, *429*, 75-83.

17. Cook, C.M.; Allen, L.C. *Ann. New York Acad. Sci.* **1984**, *429*, 84-88.

18. Bertini, I.; Luchinat, C. *Ann. New York Acad. Sci.* **1984**, *429*, 89-98.

19. Koenig, S.H.; Brown, III R.D. *Ann. New York Acad. Sci.* **1984**, *429*, 99-108.

20. Navon, G.; Kushnir, T. *Ann. New York Acad. Sci.* **1984**, *429*, 112-113.

21. Khalifah, R.G.; Rogers, J.I.; Mukherjee, J.; Morley, P.J. *Ann. New York Acad. Sci.* **1984**, *429*, 114-128.

22. Henkens, R.W.; Merrill, S.P.; Williams, T.J. *Ann. New York Acad. Sci.* **1984**, *249*, 143-145.

23. Graf, G.; Ewert, D.H.; Mayo, B. *Ann. New York Acad. Sci.* **1984**, *429*, 149-151.

24. Pocker, Y.; Meany, J.E. *Biochemistry* **1965**, *4*, 2535-2541.

25. Pocker, Y.; Meany, J.E. *Biochemistry* **1967**, *6*, 239-246.

26. Pocker, Y.; Dickerson, D.G. *Biochemistry* **1968**, *7*, 1995-2004.

27. Pocker, Y.; Meany, J.E. *J. Phys. Chem.* **1970**, *74*, 1486-1492.

28. Pocker, Y.; Meany, J.E.; Davis, B.C. *Biochemistry* **1974**, *13*, 1411-1416.

29. Tashian, R.E.; Douglas, D.P.; Yu, Y.L. *Biochem. Biophys. Res. Commun.*, *14*, 256.

30. Malmstrom, B.G.; Nyman, P.O.; Starndberg, B.; Tilander, B. In *FEBS Symp. 1, Structure and Activity of Enzymes*; Goodwin, T.W.; Harris, T.J., Hartley, B.S. Eds.; Academic Press, New York, **1964**, p 121.

31. Pocker, Y.; Stone, J.T. *J. Am. Chem. Soc.* **1965**, *87*, 5497-5498.

32. Pocker, Y.; Stone, J.T. *Biochemistry* **1967**, *6*, 668-678.

33. Verpoorte, J.A.; Mehta, S.; Edsall, J.T. *J. Biol. Chem.* **1967**, *242*, 4221.

34. Thorslund, A.; Lindskog, S. *Eur. J. Biochem.* **1967**, *3*, 117.

35. Pocker, Y.; Storm, D.R. *Biochemistry* **1968**, *7*, 1202-1214.

36. Pocker, Y.; Watamori, N. *Biochemistry* **1971**, *10*, 4843-4852.

37. Pocker, Y.; Watamori, N. *Biochemistry* **1973**, *12*, 2475-2482.

38. Pocker, Y.; Beug, M.W. *Biochemistry* **1972**, *11*, 698-706.

39. Pocker, Y.; Bjorkquist, L.C.; Bjorkquist, D.W. *Biochemistry* **1977**, *16*, 3967-3973.

40. Pocker, Y.; Janjic, N. *J. Biol. Chem.* **1988**, *263*, 6169-6176.

41. Pocker, Y.; Guilbert, L.J. *Biochemistry* **1972**, *11*, 180-190.

42. Pocker, Y.; Guilbert, L.J. *Biochemistry* **1974**, *13*, 70-78.

43. Pocker, Y.; Meany, J.E.; Davis, B.C.; Arrigoni, J.; Stein, J.E. *J. Am. Chem. Soc.* **1978**, *100*, 2883-2885.

44. Pocker, Y.; Meany, J.E.; Jones, R.C. *J. Am. Chem. Soc.* **1982**, *104*, 4885-4889.

45. Pocker, Y.; Sarkanen, S. *Northwest Regional Meeting, Am. Chem. Soc.* **1973**, *28*, Abstr B 27.

46. Pocker, Y.; Sarkanen, S. *FEBS Meeting* **1975**, *10*, Abstr 782.

47. Pocker, Y.; Sarkanen, S. *Biochemistry* **1978**, *17*, 1110-1118.

48. Lieflander, M.; Zech, R.Z. *Physiol. Chem.* **1968**, *349*, 1466.

49. Coleman, J.E. *J. Biol. Chem.* **1967**, *242*, 5212.

50. a. Pocker, Y.; Stone, J.T. *Biochemistry* **1968**, *7*, 2936-2945.

    b. Pocker, Y.; Stone, J.T. *Biochemistry* **1968**, *7*, 3021-3031.

51. Whitney, P.L.; Folsch, G.; Nyman, P.-O.; Malmström, B.G. *J. Biol. Chem.* **1967**, *242*, 4206.

52. Whitney,. P.L.; Nyman, P.-O.; Malmström, B.G. *J. Biol. Chem.* **1967**, *242*, 4212.

53. Ko, B.P.N.; Yazgan, A.; Yeagle, P.L.; Lottich, S.C.; Henkens, R.W. *Biochemistry* **1977**, *16*, 1720-1725.

54. Kannan, K.K.; Petef, M.; Fridborg, K.; Cid. Dresner, H.; Lovgren, S. *FEBS Lett.* **1977**, *73*, 115-119.

55. Kannan, K.K.; Vaara, I.; Notstrand, B.; Lovgren, S.; Borell, A.; Fridborg, K.; Petef, M. In *Proceedings on the Symposium on Drug Action at the Molecular Level;* Roberts, G.K.C., Ed., Macmillan, London **1977**, pp 73-93.

56. Bertini, I.; Luchinat, C.; Scozzafava, A. *Structure and Bonding* **1982**, *48*, 46-92.

57. Tashian, R.E.; Hewett-Emmett, D.; Goodman, M. In *Isozymes: Current Topics in Biological and Medicinal Research*; Ratazzi, M.C.; Scandalios, J.G.; Whitt, G.S. Eds.; A.R. Liss, New York; **1984**, *Vol. 7*, pp. 79-100.

58. Hewett-Emmett, D.; Hopkins, P.J.; Tashian, R.E.; Czelusniak, J. *Ann. New York Acad. Sci.* **1984**, *429*, 338-358.

59. Venta, P.J.; Montgomery, J.C.; Wiebauer, K.; Hewett-Emmett, D.; Tashian, R.E. *Ann. New York Acad. Sci.* **1984**, *429*, 309-323.

60. Venta, P.J.; Montgomery, J.C.; Hewett-Emmett, D.; Wiebauer, K.; Tashian, R.E. *J. Biol. Chem.* **1985**, *260*, 12130-12135.

61. Whitney, P.L.; Briggle, T.V. *J. Biol. Chem.* **1982**, *257*, 12056-12059.

62. Wistrand, P.J. *Ann. New York Acad. Sci.* **1984**, *429*, 195-206.

63. Shephered, J.N.; Spencer, N. *Ann. New York Acad. Sci.* **1984**, *429*, 280-281.

64. Feldstein, J.B.; Silverman, D.N. *J. Biol. Chem.* **1984**, *259*, 5447-5453.

65. Murakami, H.; Sly, W.S. *J. Biol. Chem.* **1987**, *262*, 1382-1388.

66. Dodgson, S.J.; Forster, R.E.; Storey, B.T.; Mela L. *Proc. Nat'l. Acad. Sci. U.S.A.* **1980**, *77*, 5562-5566.

67. Dodgsom. S.J.; Forster, R.E.; Storey, B.T. *Ann. New York Acad. Sci.* **1984**, *429*, 516-524.

68. Holmes, R.S. *Eur. J. Biochem.* **1977**, *78*, 511-520.

69. Register, A.M.; Koester, M.K.; Noltmann, E.A. *J. Biol. Chem.* *1978*, 253, 4143-4152.

70. Carter, N.D.; Jeffrey, S.; Shields, A.; Edwards, Y.; Tiples, T.; Hopkinson, D.A. *Biochem. Genet.* **1979**, *17*, 837-854.

71. Carter, N.D.; Hewett-Emmett, D.; Jeffrey, S.; Tashian, R.E. *FEBS Lett.* **1981**, *128*, 114-118.

72. Liljas, A.; Kannan, K.K.; Bergsten, P.-C.; Waara, I.; Fridborg, K.; Strandberg, B.; Carlsson, U.; Jarup, L.; Lovgren, S.; Petef, N. *Nature, New Biol.* **1972**, *235*, 131-137.

73. Vaara, I.; Lovegren, S.; Liljas, A.; Kannan, K.K.; Bergsten, P.-C. *Adv. Exp. Med. Biol.* **1972**, *28*, 169.

74. Notstrand, B.; Vaara, IU.; Kannan, K.K. In *Isozymes: Molecular Structure, Markert, C.L., Ed.; Academic Press*, New York, **1975**, *Vol. 1*, pp 575-599.

75. Kannan, K.K.; Notstrand, B.; Fridborg, K.; Lovgren, S.; Ohlsson, A.; Petef, M. *Proc. Nat'l. Acad. Sci. U.S.A.* **1975**, *72*, 51-55.

76. Kannan, K.K. In *Biophysics and Physiology of Carbon Dioxide*; Bauer, C.; Gross, G.; Bartels, H., Eds.; Springer-Veriag, New York; **1980**, pp 216-225.

77. Kannan, K.K.; Ramanadham, M. *Ann. New York Acad. Sci.* **1984**, *429*, 49-60.

78. a. Koshland, D.E., Jr., *J. Cellular Comp. Physiol.* **1959**, *54*, Suppl. 1, 245-258.

   b. Koshland, D.E., Jr., *Cold Spring Harbor Symp. Quant. Biol.* **1963**, *28*, 473-482.

   c. Koshland, D.E., Jr., *The Enzymes* **1970**, *1*, 341-396.

79. Lindskog, S.; Coleman, J.E. *Proc. Nat'l. Acad. Sci. U.S.A.* **1973**, *103*, 1544-1550.

80. Sheridan, R.D.; Allen, L.C. *J. Am. Chem. Soc.* **1981**, *103*, 1544-1550.

81. a. Pocker, Y.; Bjorkquist, D.W. *J. Am. Chem. Soc.* **1977**, *99*, 6537-6543.

   b. Wooley, P. *J. Chem. Soc. Perkin II* **1977**, 318.

   c. Silverman, D.N.; Lindskog, S. *Acc. Chem. Res.* **1988**, *21*, 30-36.

82. Campbell, I.D.; Lindskog, S.; White, A.I. *J. Mol. Biol.* **1975**, *998*, 597-614.

83. Forsman, C.; Jonsson, B.H.; Lindskog, S. *Biochim. Biophys. Acta* **1983**, *748*, 300-307.

84. Pocker, Y.; Janjic, N.; Miao, C.H. *Fast React. Solution, Discuss R. Soc. Chem.* **1985**, *9*, 7-8.

85. Enns, T. *Science* **1967**, *155*, 54-56.

86. Broun, G.; Selegny, E.; Tran-Minh, C.; Thomas, D. *FEBS Lett.* **1970**, *7*, 223-226.

87. Werdan, K.; Heldt, H.W. *Biochim. Biophys. Acta* **1972**, *283*, 430.

88. Graham, D.; Atkins, C.A.; Reed, M.L.; Patterson, D.B.; Smillie, R.M. In *Photosynthesis and Photorespiration;* Hatch, M.D.; Osmond, C.B.; Slayter, R.O., Eds.; Wiley-Interscience, New York; **1971**, pp 267-274.

89. Atkins, C.A.; Patterson, B.D.; Graham, D. *Plant Physiol.* **1972**, *50*, 214-218.

90. Atkins, C.A.; Patterson, B.D.; Graham, D. *Plant Physiol.* **1972**, *50*, 218-223.

91. Graham, D.; Reed, M.L.; Patterson, B.D.; Hockley, D.G.; Dwyer, M.R. *Annual New York Acad. Sci.* **1984**, *429*, 222-237.

92. Everson, R.G. In *Photosynthesis and Photorespiration*; Hatch, M.D.; Osmond, C.B.; Slayter, R.O., Eds.; Wiley-Interscience, New York **1971**, pp 275-281.

93. Tobin, A.L. *J. Biol. Chem.* **1970**, *245*, 2656-2666.

94. Kisiel, W.; Graf. G. *Phytochemistry* **1972**, *11*, 113-117.

95. Pocker, Y.; Ng, J.S.Y. *Biochemistry* **1973**, *12*, 5127-5134.

96. Pocker, Y.; Ng, J.S.Y. *Biochemistry* **1974**, *13*, 5116-5120.

97. Kandel, M.; Gornall, A.G.; Cybulsky, D.L.; Kandel, S.I. *J. Biol. Chem.* **1978**, *253*, 679-685.

98. Pocker, Y.; Miksch, R.R. *Biochemistry* **1978**, *65*, 2030-2037.

99. Harned, H.S.; Davis, R., Jr. *J. Am. Chem. Soc.* **1943**, *65*, 2030-2037.

100. Pocker, Y.; Miksch, R.R. *Int. Congr. Biochem.* **1979**, *11*, 04-7-S60.

101. Pocker, Y.; Tanaka, N. *Science* **1978**, *199*, 907-909.

102. a. Okazaki, M. *Bull. Tokyo Gakugei Univ. Ser. IV* **1974**, *26*, 209-218.

  b. Bundy, H.F.; Cote, S. *Photochemistry* **1980**, *19*, 2531-2534.

103. Hunt, J.B.; Rhee, M.J.; Storm, C.B. *Anal. Chem.* **1977**, *79*, 614.

104. McCoy, Jr., L.F.; Wong, K.P. *Biopolymers* **1979**, *18*, 2893-2904.

105. Pocker, Y.; Fong, C.T.O *Biochemistry* **1980**, *19*, 2045-2050.

106. Pocker, Y.; Fong, C.T.O. *Biochemistry* **1983**, *22*, 813-818.

107. Lindskog, S. In *Zinc Enzymes*; Bertini, I.; Luchinat, C.; Maret, W.; Zeppezauer, M., Eds.; Birkhauser, Boston; **1986**, pp 307-316.

108. Venkatasubban, K.S.; Silverman, D.N. *Biochemistry* **1980**, *19*, 4984-4989.

109. Sen, A.C.; Tu, C.K.; Thomas, H.; Wynns, G.C.; Silverman, D.N. In *Zinc Enzymes*; Bertini, I.; Luchinat, C.; Maret, W.; Zeppezauer, M., Eds.; Birhauser, Boston; **1986**, pp 329-339.

110. Pocker, Y.; Janjic, N.; Miao, C.H. In *Zinc Enzymes*; Bertini, I.; Luchinat, C.; Maret, W.; Zeppezauser, M., Eds.; Birkhauser, Boston, **1986**; pp 341-356.

111. a. Eriksson, E.A.; Jones, T.A.; Liljas, A. In *Zinc Enzymes*; Bertini, I.; Luchinat, C.; Maret, W.; Zeppezauer, M., Eds.; Birkhauser, Boston; **1986**, pp 317-328.

    b. Eriksson, E.A.; Jones, T.A.; Liljas, A. *Protein: Struct., Fun ct., Genet.* **1988**, *4*, 274-282.

112. Pocker, Y.; Janjic, N.J. *J. Am. Chem. Soc.* **1989**, *111*, 731-733.

113. Forsman, C.; Behravan, G.; Jonsson, B.-H.; Liang, Z.-W.; Lindskog, S.; Ren, X.; Sandstrom, J.; Wallgren, K. *FEBS Lett.* **1988**, *229*, 360.

114. Pocker, Y.; Janjic, N. *Biochemistry* **1988**, *27*, 4114-4120.

115. a. Pocker, Y.; Miao, C.H. *Biochemistry* **1987**, *26*, 8481-8486.

    b. Tu, C.; Silverman, D.N.; Forsman, C.; Jonsson, B.-H.; Lindskog, S. *Biochemistry* **1989**, *28*, 7913-7918.

    c. Pocker, Y.; Miao, C.H., in preparation.

116. Liang, J.-Y.; Lipscomb, W.N. *J. Am. Chem,. Soc.* **1986**, *108*, 5051-5058.

117. Liang, J.-Y.; Lipscomb, W.N. *Biochemistry* **1987**, *26*, 5293-5301.

118. Williams, T.J.; Henkens, R.W. *Biochemistry* **1985**, *24*, 2459-2462.

119. Brown, R.S. In *Carbon Dioxide as a Source of Carbon*, NATO ASI Series; *Vol. 206*; Aresta, M.; Forti, G., Eds.; Reidel Publ.; Dordrecht, **1987**, pp 169-197.

120. a. Eklund, H.; Branden, C.-I. In *Zinc Enzymes*; Spiro, T.G., Ed.; Wiley; New York, **1983**; pp 123.

    b. Eklund, H.; Jones, A.; Schneider, G. In *Zinc Enzymes*; Bertini, I.; Luchinat, C.; Maret, W.; Seppezauer, M., Eds.; Birkhauser, Boston, **1986**, pp 377-392.

    c. Dewar, M.J.S.; Storch, D.M. *Proc. Nat'l. Acad. Sci. U.S.A.* **1985**, *82*, 2225.

CARBONIC ANHYDRASES: THE MECHANISTIC, SPECTROSCOPIC, AND MODEL STUDIES

R.S. BROWN
Department of Chemistry, University of Alberta
EDMONTON, Alberta, Canada T6G 2G2

ABSTRACT. The Carbonic Anhydrases (CAs) are Zn^{2+}-containing enzymes that occur ubiquitously in Nature. They have been extensively studied with respect to their ability to hydrate CO_2 and dehydrate HCO_3^-. The first part of the following discussion will focus on the recent studies with the enzyme that have in total, led to the formulation of a detailed mechanism of action, namely, the "zinc hydroxide" mechanism. The second part of the discussion deals with the simpler chemical models that mimic the spectroscopic properties of cobalt CA, and others which mimic the catalytic properties.

1. Introduction

The carbonic anhydrases are widely-occurring Zn^{2+}-metalloenzymes found in plants and animals [1]. The complete structures of the human CAI and CAII (formerly HCAB and HCAC respectively) have been elucidated by X-ray crystallographic studies [2]. The latter isozyme has been refined at 2.2 Å resolution [3], while CAI has been refined to 2Å [3b]. The active site of the latter [2c], shown in Fig. 1 is comprised of an essential Zn^{2+}-coordinated to three histidine imidazoles in a close to tetrahedral geometry; the fourth ligand is an H_2O or OH^- unit depending on the pH. Also in the active site and connected to the $Zn-OH_2$ by a water bridge, is His64, (His 200 for CAI). This His unit is conserved in isozymes I and II but is not present in a more recently studied and less active form, isozyme CAIII [4].

The only known function of the CA isozymes is to facilitate the interconversion of CO_2 and HCO_3^- as in eq. 1 so that they play key roles in diverse processes such as photosynthesis, physiological pH control, gas balance and calcification [1]. They will, however, catalyze other

$$H_2O + CO_2 \xrightleftharpoons{CA} H^+ + HCO_3^- \qquad (1)$$

processes involving nucleophilic attack of OH^- on electrophilic centers including hydration of aldehydes [5a], pyruvate [5b] and alkyl pyruvates [5c] as well as the hydrolysis of carboxylic, sulfonic, carbonic, and phosphoric esters [6]. The basic form of the enzyme (generated by ionization of an active site residue of apparent $pK_a \sim 7$) is active in CO_2 hydration, while the acidic form catalyzes the dehydration of HCO_3^-.

M. Aresta and J. V. Schloss (eds.),
Enzymatic and Model Carboxylation and Reduction Reactions for Carbon Dioxide Utilization, 145–180.
© 1990 *Kluwer Academic Publishers.*

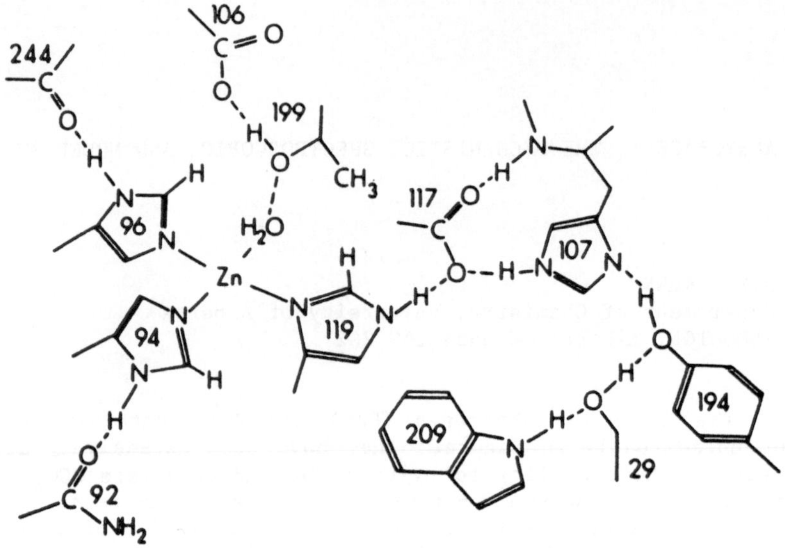

FIGURE 1. A stylized portion of the active site of human CAI showing the Zn^{2+}-environment. (Redrawn from [2c])

As given in Table 1, the three isozymes have different kinetic properties for CO_2 hydration; the high activity isozyme, CAII, is extraordinarily efficient and seems to fulfill the criteria for a perfectly evolved enzyme [7].

TABLE 1. Maximal Steady State Constants for the Hydration of CO_2 Catalyzed by the Basic Forms of CA Isozymes (from ref. [1g])

Isozyme	k_{cat} (s^{-1})	k_{cat}/K_M (M^{-1}s^{-1})
CAI (human)	2×10^5	5×10^7
CAII (human)	1.4×10^6	1.5×10^8
CAIII (feline, bovine)	2.5×10^3	3×10^5

The Zn^{2+}-ion of native CA can be removed by dialysis of the enzyme against 1,10-phenanthroline in acetate buffer (pH 5) [8a] or more rapidly by EDTA [8b]. Other divalent metals can be inserted into the apoenzyme [9] but of these, only Zn^{2+} and Co^{2+} restore significant catalytic activity (CoCA ~ 50% of native enzyme [8a]). CoCA has a pH-dependent visible absorption spectrum characteristic of a four and/or five coordinate environment about the metal [9,10]. Analysis of the pH dependence of the Co^{2+} visible absorption spectrum indicates the alkaline form of bovine CoCA (which is also catalytically active in CO_2 hydration) is predominantly distorted tetrahedral, while the acid form apparently consists of an equilibrium distribution of 4 and 5 coordinate forms, eq. 2 [10].

The pK_a responsible for the spectroscopic changes in CoCA was originally reported to be that of a single dissociating group (pK_a ~ 7) [9a,b]. More recent work in the presence of non-interacting buffers or ion free solutions indicates that the absorption vs pH profile is not as simple as that expected for a single dissociating group [10b,11].

$$
\begin{array}{ccc}
\overset{\displaystyle Im}{\underset{\displaystyle Im}{Im---Co}}\!\!\overset{\displaystyle OH_2}{\underset{\displaystyle OH_2}{\big\langle}}\!\!\!\!2+
& \underset{\displaystyle \rightleftharpoons}{\overset{\displaystyle -H_2O}{}}
& Im---\overset{\displaystyle Im}{\underset{\displaystyle Im}{Co}}\!\!2+\!\!-OH_2
\end{array}
\quad \underset{\displaystyle H_3O^+}{\overset{\displaystyle OH^-}{\rightleftharpoons}} \quad
Im---\overset{\displaystyle Im}{\underset{\displaystyle Im}{Co}}\!\!2+\!\!-OH^-
\qquad (2)
$$

acid forms   basic form

Two ionizing active site residues (Co^{2+}-OH_2 and His64-H^+) have been reported to be responsible for these observations [10b,11]. Finally, the Co absorption spectrum is markedly affected by the presence of added anions (which also inhibit the CO_2 hydration ability), a fact that indicates the inhibitory anions are coordinated to the metal [9,10]. Presented in Table 2 are the spectral properties of some anionic derivatives of CoCAII.

In the following, we will consider the spectroscopic properties of the Co^{2+}-substituted isozymes and the active site ionizations that control the absorption vs. pH profile. We will then turn to the native Zn^{2+}-enzymes with particular emphasis on the catalytic pathway for CO_2 hydration and HCO_3^- dehydration. Finally, we will briefly consider simpler tricoordinate M^{2+}-complexes that purport to mimic the spectroscopic and catalytic properties of the enzymes.

TABLE 2. Characteristic Coordination Environments for Various Complexes of CoCA[a]

4 Coordinate ($\varepsilon_{max} > 300$ M^{-1}cm^{-1})	4-5 Coordinate (200 M^{-1}cm^{-1} < ε_{max} < 300 M^{-1}cm^{-1})	5-Coordinate ($\varepsilon_{max} < 200$ M^{-1}cm^{-1})
bovine CoCAII (high pH)	bovine CoCAII (low pH)	human CoCAII (low pH)
human CoCAII (high pH)	HCO$_3^-$	SCN$^-$
CN$^-$	F$^-$	HSO$_3^-$
NCO$^-$	Cl$^-$	I$^-$
SH$^-$	Br$^-$	acetate
aniline	N$_3^-$	formate
imidazole (high pH)	imidazole (low pH)	oxalate
sulfonamides	benzoate	succinate

a. Ref. 10d.

2. Co^{2+}-Carbonic Anhydrases

a. ACID-BASE PROPERTIES

As can be judged from Fig. 2 [10b], the spectrum of CoCA from bovine and

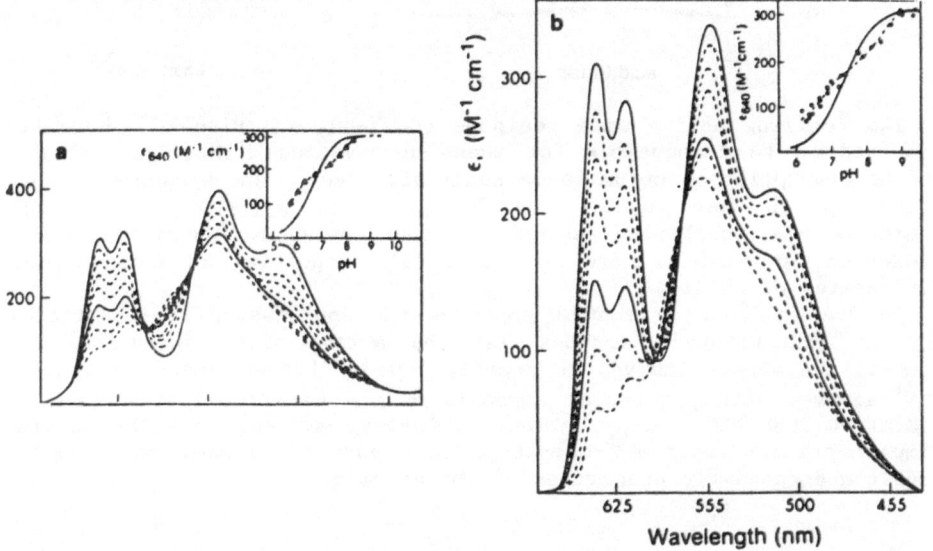

FIGURE 2. The absorption spectrum of Bovine CoCAI (left) and Human CoCAI (right) in unbuffered medium as a function of pH. (Redrawn from [10b])

human sources changes as a function of pH. At low pH ($<$ 6) the spectrum is that of an equilibrium of a 4 and 5 coordinate environment, which changes to a 4-coordinate environment with rising pH. The insets show that the ε_{640} vs pH profile is influenced by two ionizing groups associated with the active site. These have been identified as $Co^{2+}-OH_2$ and $His-H^+$ [10b,e,11]. Analysis of the profile can be made considering the pH dependent species given in Scheme 1 for the bovine CoCAII

Scheme 1 (His = His64; $OH_2 = Co^{2+}-OH_2$):

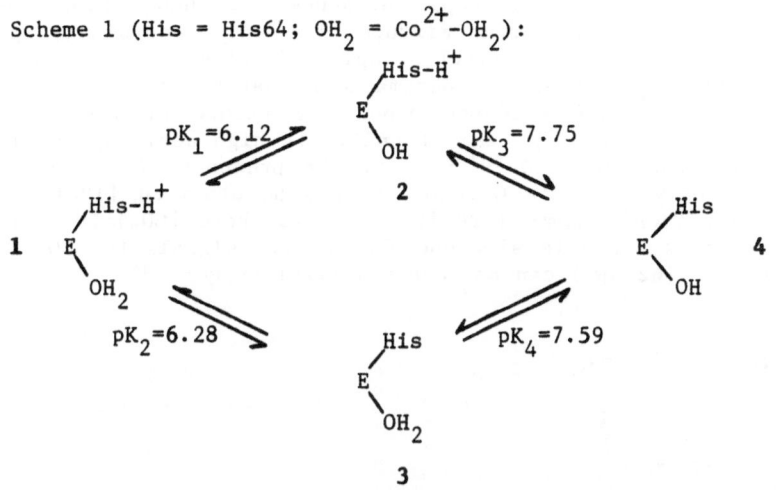

enzyme. In the absence of simplifying assumptions, the ε_{640} vs pH data can yield only two macroscopic ionization constants pK_a^1 (5.89) and pK_a^2 (7.97) [11]. However, if one assumes that molar absorbancies of species 1,3 and 2,4 are the same (probably justified since 2 and 4 are generated by ionization of the metal coordinated water, while 1 and 3 involve a non-coordinating $His-H^+$), then the four microconstants in Scheme 1 can be evaluated [11]. Simonsson and Lindskog [10e], also using bovine CoCAII in buffered media (Hepes, Mes) containing no $SO_4^=$, observed that the esterase activity (in hydrolysing p-nitrophenylacetate) and visible absorption spectrum vs pH plots had essentially identical inflection points corresponding to macroscopic pK_a values of 5.5 and 7.3 under their conditions. It is of note that the esterase activity of the native Zn^{2+}-enzyme also follows a pH/rate profile indicative of two active site ionizations, having macroscopic pK_a's of 4.8 and 7.1 [10e]. From the above, it seems relatively well established that the spectral and activity controlling ionizations in the cobalt enzyme are the same, and furthermore that the cobalt substitution produces a good model for the native enzyme.

The visible absorption spectrum of CoCAIII has been found by Engberg and Lindskog to be invariant with pH from 6-9 [10g]. In that range, the spectrum is virtually identical to that of the high pH form CoCAI and

II. This indicates that the pK_a of the Co-OH$_2$ in isozyme III must be considerably lower than 6, which is probably attributable to the fact that the active site region contains several basic amino acid residues (Lys-64, Arg-67, Arg-91) [4c] which, when protonated, would tend to electrostatically reduce the Co-OH$_2$ pK_a.

b. BINDING OF HCO$_3^-$

There is U.V.-visible spectroscopic evidence that HCO$_3^-$ binds to the metal of CoCA in an equilibrium mixture of 4 and 5-coordinate species (Table 2) [10a]. Another spectroscopic advantage of certain M^{2+}-substituted enzymes is that the paramagnetic nature of the ion can be used to provide ^{13}C-NMR evidence about the nature of the CO$_2$/HCO$_3^-$ binding sites [12]. In the case of CoCAII (a high activity form), the interconversion rate of ^{13}CO$_2 \rightleftharpoons H^{13}CO_3^-$ in the presence of the enzyme is so rapid that only a single C resonance can be observed [12a]. Using the less active CoCAI enzyme [12c,d] at pH 6.3 where [CO$_2$]≈[HCO$_3^-$], the enzyme promoted exchange is slow enough that two signals for CO$_2$ (δ 125 ppm) and HCO$_3^-$ (δ 162 ppm) can be observed [12c] (Figure 3). At pH 6.3

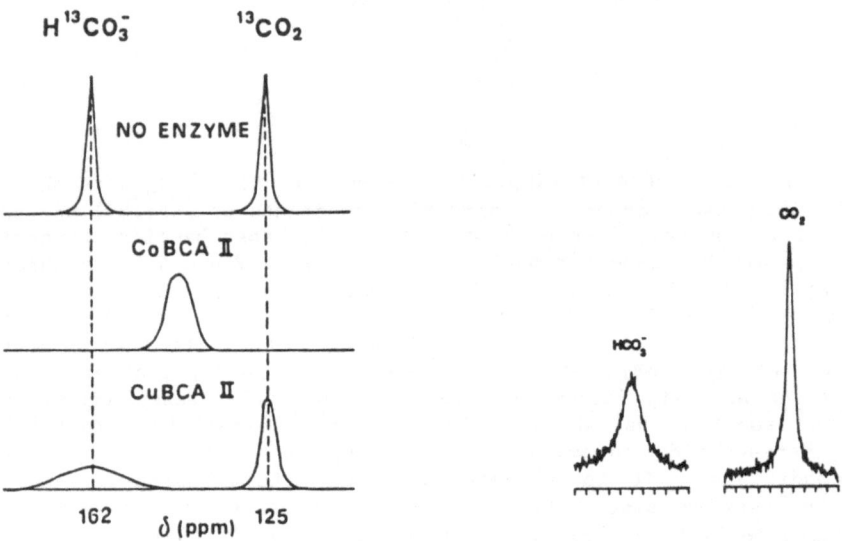

Figure 3. (Left) ^{13}C-NMR spectrum of CO$_2$ and HCO$_3^-$ in the presence of CoCAII and CuCAII (Redrawn from [12a])
(Right) ^{13}C-NMR spectrum of CO$_2$ and HCO$_3^-$ at 50.3 MHz in the presence of CoCA(I) at 14°C, pH 6.3; tick marks are 5 Hz apart. (Redrawn from [12c])

and T=14°C, the line width of the HCO_3^- resonance can be broadened to ~ 3 times that of CO_2. Line broadening can result from two effects; the chemical exchange between $CO_2 \rightleftharpoons HCO_3^-$, and paramagnetic effects due to Co(II). Broadening of the CO_2 line width is influenced by exchange, but the HCO_3^- line width is influenced by both [12c]. This gives evidence that HCO_3^- is bound within the inner coordination sphere of Co(II), in a way that is productive for catalyzed turnover the average distance of the ^{13}C being 3.22 Å from the metal. In a later study [13], Led and Neesgaard determined that in the Mn(II) substituted isozyme I (which retains ~ 7% of the $CO_2 \rightleftharpoons HCO_3^-$ activity of ZnCAI), the ^{13}C of the enzyme bound HCO_3^- is only 2.71 Å from the paramagnetic metal. This distance is compatible only with a structure in which the bicarbonate is bound bidentate as in eq. 3 (location of H is uncertain).

$$(3)$$

This type of complex is particularly thought provoking since it corresponds closely to one of the proposed pathways for $CO_2 \rightleftharpoons HCO_3^-$ interconversion promoted by the native enzyme.

3. Catalyzed Interconversion of $CO_2 \rightleftharpoons HCO_3^-$

a. RATE LIMITING PROTON TRANSFERS FROM THE ENZYME → MEDIUM

We turn here to a discussion of the mechanism of interconversion of CO_2 and HCO_3^- promoted by the native and M^{2+}-substituted isozymes. Considerable strides toward understanding of the details have been made by the groups of Silverman and Lindskog (summarized in ref. 1g). The bulk of the current information is consistent with the so-called $Zn-OH^-$ mechanism, a simplified model of which is given in eqs. 4,5. In effect,

$$EZnOH^- + CO_2 \rightleftharpoons EZn(OH^-)CO_2 \rightleftharpoons EZnHCO_3^- \rightleftharpoons EZnOH_2 + HCO_3^- \quad (4)$$

$$EZnOH_2 \rightleftharpoons H^+-EZnOH^- \rightleftharpoons EZnOH^- + H^+ \quad (5)$$

the evolved mechanism consists of two temporally separated events; the interconversion of $EZnOH^- + CO_2 \rightleftharpoons EZnHCO_3^-$ and a subsequent turnover step in which the protolysis of $EZnOH_2$ to regenerate $EZnOH^-$ occurs by release of a proton to the medium. In eq. 5, there is evidence that the proton transfer from the enzyme to the medium consists of two discreet steps (in the case of CAI and II); an intramolecular proton transfer and a subsequent intermolecular proton transfer to the medium. Under appropriate conditions, each of these can be rate-limiting. This leads to the conclusion that the enzyme promoted chemical interconversion ($CO_2 \rightleftharpoons HCO_3^-$) has a very low activation energy so that the most difficult step in the entire operation is in fact the dissociation of $H_2O \rightleftharpoons H^+ + OH^-$ in the enzyme active site.

We consider firstly the older evidence concerning the second step in eq. 5 involving intermolecular proton transfer from the enzyme to the medium. For the catalysed hydration of CO_2, both k_{cat} and k_{cat}/K_m have pH profiles approximated by a simple titration curve with $pK_a \sim 7$ and a maximal activity at high pH. There is a problem evident from the latter stages in eq. 5 since the maximum rate constant for the transfer of a proton from an active site residue having a $pK_a \sim 7$ to bulk H_2O should be $\sim 10^3 s^{-1}$. This is based on the assumption that the back reaction ($H_3O^+ + E\text{-}ZnOH^-$) is diffusion controlled ($\sim 10^{10} M^{-1} s^{-1}$) so that the k_{cat} value for isozyme II ($\sim 10^6 s^{-1}$) exceeds by 1000-fold the anticipated turnover rate constant. The problem is resolved considering that the medium's acceptor is not H_2O, but rather the buffer used for pH control [14]. Subsequent experiments indicated that below [buffer] ~ 10 mM, the enzyme catalyzed hydration rate of CO_2 decreased [15a] as did the catalyzed rate of exchange of ^{18}O between CO_2 and H_2O [15b]. Studies of CO_2 hydration and HCO_3^- dehydration in the presence of a series of buffers of different pK_a's indicated that the anticipated Eigen curves [16] for diffusion limited proton transfer from EEOH^- to B: (for CO_2 hydration) [17a] or BH^+ to EOH^- (for HCO_3^- dehydration) [17b] were indeed observed (Figure 4).

At sufficiently high [buffer], proton transfer between the enzyme and medium is no longer rate limiting. However, a solvent isotope effect (SIE) of $k_{cat}^{H_2O}/k_{cat}^{D_2O} \simeq 3.8$ is observed for CO_2 hydration catalyzed by CA(II) [18], while the SIE on k_{cat}/K_M is unity (Figure 5). The k_{cat} term includes all the steps in eqs. 4 and 5 including those where $EZnOH$ reacts with CO_2 to produce $EZnHCO_3^-$ and subsequently $EZnOH^-$. Since the latter step is not rate limiting at high [buffer], the large SIE of 3.8 can only be reconciled by some other process in which the degree of bonding to H changes in a rate contributing step. On the other hand, the k_{cat}/K_M ratio contains only the steps involving CO_2 binding to the enzyme and its interconversion to HCO_3^- up to the first irreversible step which is debinding of HCO_3^-. That the kinetic SIE on k_{cat}/K_M is unity can be rationalized by considering that none of the steps in eq. 4 involve processes where the degree of bonding to H changes markedly. Thus, some other step which follows debinding of HCO_3^- from the enzyme, and subsequently converts the enzyme into its basic form in order to turn over, is responsible for the large SIE on k_{cat}. This step is currently believed to be an intramolecular proton transfer (eq. 6)

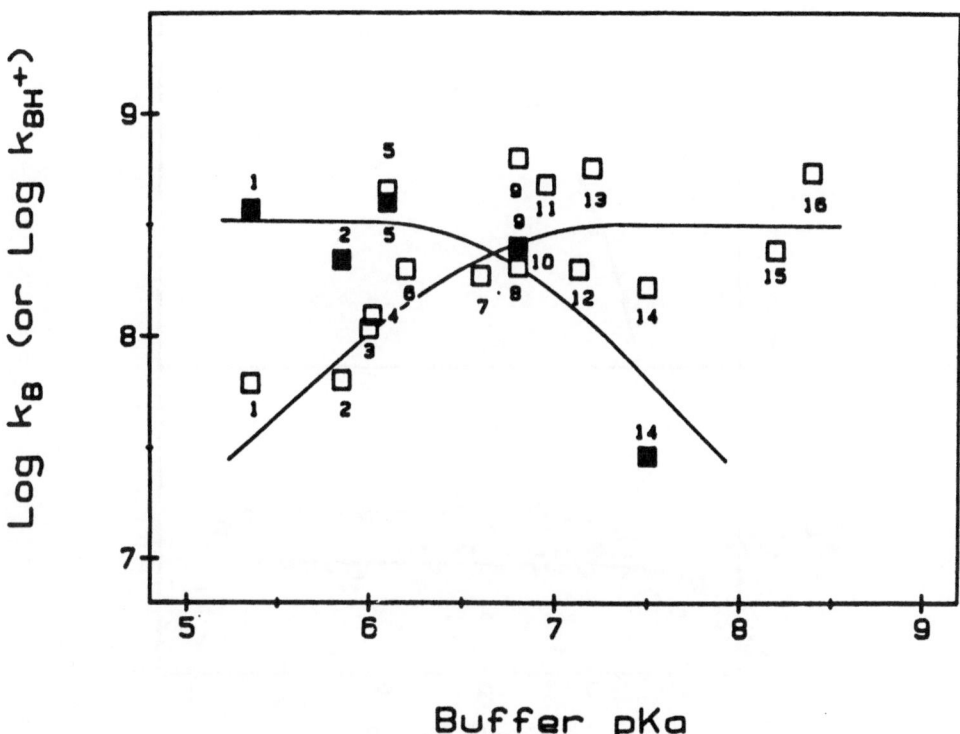

FIGURE 4. Brønsted plots for rate limiting proton transfer in CO_2 hydration and HCO_3^- dehydration between enzyme and external buffer: (1) malonate, (2) 3-picoline, (3) 2-picoline, (4) 4-picoline, (5) MES, (6) 3,5-lutidine, (7) 3,4-lutidine, (8) 2,4-lutidine, (9) ACES, (10) phosphate, (11) imidazole, (12) diethylmalonate, (13) N-methylimidazole, (14) HEPES, (15) 1,2-dimethylimidazole, (16) TAPS (Redrawn from [17b])

154

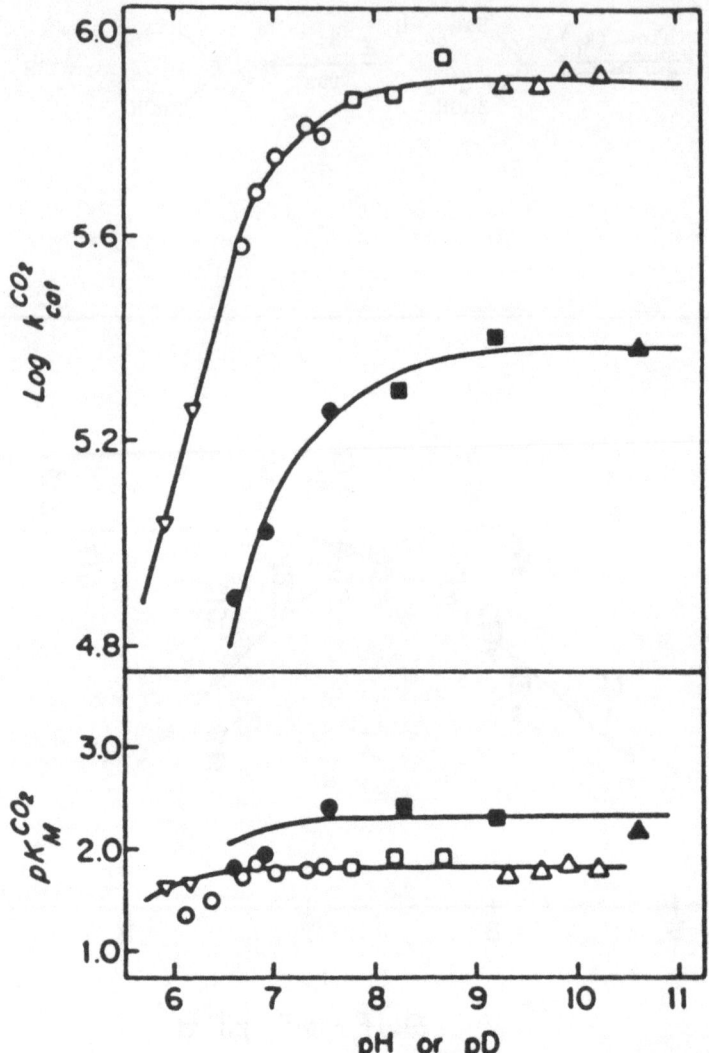

FIGURE 5. The pH dependence of k_{cat} and K_M for the hydration of CO_2 catalyzed by bovine CAII in H_2O (open symbols) and D_2O (filled symbols). T=25°C, μ = 0.1 (Na_2SO_4), buffers (0.02-0.005 M; (∇) 3-picoline, (O) phosphate, ($\square$) 1,2-dimethylimidazole, ($\triangle$) N,N-dimethyl-glycine. (Redrawn from [18b])

involved with the protolysis of Zn-OH$_2$. The rate constant (k_6) for this intramolecular proton transfer, is $\sim 10^6 s^{-1}$, a value that is consistent with the k_{cat}^{max} for enzyme promoted CO_2 hydration. In subsequent proton inventory studies Venkatasubban and Silverman [19] showed that the value of k_{cat}, (under [buffer] high enough that <u>intermolecular</u> proton transfer was not rate limiting), varied with the <u>mole</u> fraction of D_2O in a way that was consistent with two or more protons in flight. Since x-ray refinement [3] shows the intramolecular proton acceptor (His64) is $\sim$ 6Å away from the Zn^{2+}, it seems likely that the conversion of $EZnOH_2 \rightleftarrows H^+E$-$ZnOH^-$ is mediated by a water bridge as in eq. 7.

(7)

b. NON-ENZYMIC $CO_2 + H_2O \rightleftarrows H^+ + HCO_3^-$ PROCESS

Up to now we have only concentrated on the rate-limiting proton transfer steps required to regenerate the basic, $EZnOH^-$, form of the enyzme. The implication from the above is that the enzyme has evolved to the point where the steps required to interconvert $CO_2 \rightleftarrows HCO_3^-$ in the enzyme active site proceed with lower activation energies than the proton transfers. Prior to discussing the enyzmatic process, it is well at this point to consider the non-enzymic processes for hydration of CO_2 and the microscopic reverse reaction of dehydration of bicarbonate. Ostensibly, this is a simple process, but it is not understood in detail. At pH values above the pK_a of carbonic acid (pK_a = 3.76 [20]) eqn. 1 can be reformulated as eq. 8 where the values of k_{hyd} and k_{dehyd} at 25°C are 0.037 s^{-1} [20,21] and 4.1 x $10^4 M^{-1}s^{-1}$ [21] respectively. In D_2O, the corresponding values are 0.02 s^{-1} and 7.3 x 10^4 M$^{-1}s^{-1}$ [21]. The kinetic SIE's on the hydration ($k(H_2O)/k(D_2O) \simeq$ 2) and dehydration ($k(H_3O^+)/k(D_3O^+)$ = 0.56) are consistent with a reversible process involving a kinetically competent zwitterionic form of carbonic acid. The data available prior to 1985 were insufficient to distinguish between general acid (ga) or specific acid (sa) catalyzed pathways for dehydration. The conceptual distinction between the two is that in the ga-catalyzed pathway, the proton transfer from H_3O^+ to HCO_3^- is con-current with dehydration, while in the sa process, a preequilibrium production of $H_2^+OCO_2^-$ from H_3O^+ to $HOCO_2^-$ occurs prior to rate limiting dehydration. In the forward direction ($2H_2O + CO_2 \rightarrow H_3O^+ + HCO_3^-$), the solvent isotope effect of $\sim$ 2 and observed general base catalysis by dibasic $HPO_3^=$, is consistent with a concerted process [21]. General

$$\underset{\text{O}}{\overset{\text{O}}{\|}}\text{C} + 2H_2O \underset{k_{dehyd}}{\overset{k_{hyd}}{\rightleftharpoons}} \underset{\text{-O}}{\overset{\text{O}}{\diagdown}}C-OH + H_3O^+ \tag{8}$$

acid catalysis of the microscopic reverse reaction by $H_2PO_3^-$ is also observed, again consistent with a concerted process [21]. On the other hand, the $k(H_3O^+)/k(D_3O^+)$ of 0.56 for the dehydration supports a pre-equilibrium protonation of HCO_3^- followed by a rate controlling departure of H_2O. Nevertheless, an inverse isotope effect of this magnitude does not exclude a general acid pathway since examples are known where hydronium ion catalyzed hydration [22a] or hydrolysis reactions (which are also general acid catalyzed) exhibit $k(H_3O^+)/k(D_3O^+)$ values of 0.6-0.7 [23].

O'Leary and coworkers [25] reported carbon kinetic isotope effects for the dehydration of $H^{12}CO_3^-$ and $H^{13}CO_3^-$ in H_2O and D_2O media. The

value in D_2O (24°C) was found to be $k^{12}/k^{13} = 1.0178\pm0.0005$ while in H_2O (24°C) it was 1.0151 ± 0.0008 [25a]. Normal k^{12}/k^{13} kinetic isotope effects for decarboxylation reactions $R-CO_2^- \rightarrow R^- + CO_2$, (which can be assumed as approximate models for dehydration of HCO_3^-), are in the range of 1.03-1.06 [26]. Two points are of note. First, the k^{12}/k^{13} isotope effect on dehydration is > 1 which qualitatively indicates that breaking of a C-O bond occurs in a rate determining step. Second, the low value of the k^{12}/k^{13} ratio indicates the transition state is early. Taken together, the solvent isotope effect on k^{12}/k^{13} and the low value of the latter support the process in eq. 9 where the proton transfer from H_3O^+ to HCO_3^- and breakdown of $H_2\overset{+}{O}CO_2^-$ to products occur at similar rates [25].

$$H_3O^+ + HOCO_2^- \rightleftharpoons H_2O + \underset{\overset{|}{H}}{HO}-CO_2^- \longrightarrow 2H_2O + CO_2 \tag{9}$$

Above pH 8 (in the absence of enzyme), the dominant process for $HOCO_2^-$ production involves a direct nucleophilic attack of OH^- on CO_2. This is supported by $k_{OH^-}/k_{OD^-} = 6000 \ M^{-1}s^{-1}/9,500 \ M^{-1}s^{-1} = 0.63$ (25°C, $\mu = 0.1$) [21] with the more nucleophilic OD^- reacting fastest.

c. ENZYMATIC CO_2 HYDRATION AND HCO_3^- DEHYDRATION

As we have earlier stated, at high [buffer], the rate limiting step in eqs. 4 and 5 still involves an intramolecular proton transfer. By measuring the ^{13}C-line widths of a $CO_2 \rightleftharpoons HCO_3^-$ equilibrium mixture in the presence of human CAII, Simonsson et al [27] have determined that the intramolecular proton transfer is not directly involved in the catalytic interconversion of $CO_2 \rightleftharpoons HCO_3^-$. Silverman et al [28] have addressed this question by determining the enzyme promoted exchange of ^{18}O between CO_2 and HCO_3^- at chemical equilibrium [28,29]. The earlier studies [28] showed that catalyzed dehydration of $HCOO^{18}O^-$ labelled the active site with ^{18}O and that the dissociation of ^{18}O from the active site depends upon the presence of buffers. The subsequent work [29] investigated the rate of exchange between CO_2 and water and the rate of exchange between ^{12}C and ^{13}C-containing species of CO_2 in the presence of CAII. We deal with that study here, and following Silverman et al [29] define θ_{cat} as the rate constant for catalyzed exchange of ^{18}O between CO_2 and H_2O. ϕ_{cat} is defined as the rate constant for ^{18}O exchange between ^{12}C and ^{13}C-containing species of CO_2. The isotopic composition of CO_2 (both ^{12}C and ^{13}C forms containing 0, 1 or 2 ^{18}O) was determined by mass spectrometric analysis of the various molecular ions of CO_2 which was removed at various times from a reaction vessel containing CO_2 and HCO_3^- at chemical equilibrium and $\sim 2 \times 10^{-9}$ M enzyme. Given in Fig. 6 are

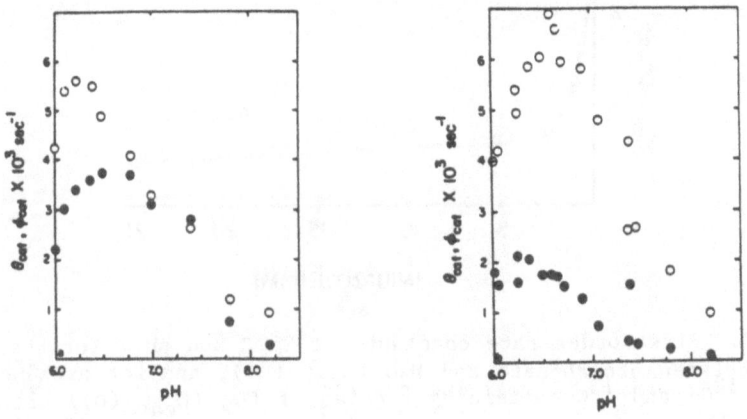

FIGURE 6. (Left) The rate constant θ_{cat} (O) for ^{18}O-exchange between CO_2 and H_2O, and ϕ_{cat} (O) for ^{18}O-exchange between ^{12}C- and ^{13}C-containing species of CO_2 at 25°C catalyzed by HCAII in the absence of buffers.

(Right) The same as the top figure but in the presence of 50 mM imidazole. (Redrawn from [29])

the observed values of Θ_{cat} and ϕ_{cat} as a function of pH in the absence and presence of 50 mM imidazole. Comparing Figs. 6a and 6b demonstrates that above pH 6.25, Θ_{cat} increases, and ϕ_{cat} decreases in the presence of imidazole buffer. Shown in Fig. 7 are the Θ_{cat} and ϕ_{cat} data at pH 7 as a function of [imidazole] [28b]. They can each be seen to reach plateau values at [imidazole] > 5 mM. The data indicate that increasing [buffer], (to provide concentrations of proton donor and acceptor that exceed $[H_3O^+]$ and $[OH^-]$), not only increases the value of the CO_2 hydration rate constant k_{cat} [14], but increases the rate of ^{18}O-exchange between H_2O and CO_2 (Θ_{cat}). ϕ_{cat} is seen to decrease (Fig. 7) with increasing imidazole. ϕ_{cat} arises because ^{18}O labels the active site of CAII during the dehydration of $HCOO^{18}O^-$, and resides there long

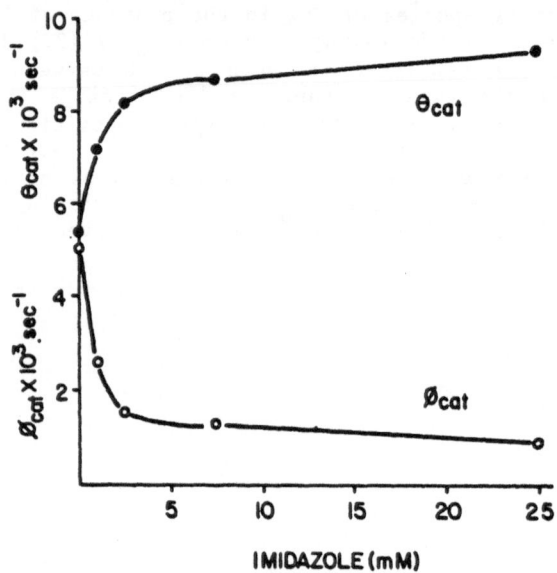

FIGURE 7. First order rate constants at 25°C and pH 7 for the exchange of ^{18}O between bicarbonate and H_2O (Θ_{cat} (0)), and for exchange of ^{18}O between ^{12}C- and ^{13}C-containing species of CO_2 (ϕ_{cat} (0)) catalyzed by bovine CAII as a function of [imidazole]. (Redrawn from [28b])

enough to regenerate HCO_3^- by catalytic rehydration of $^{13}CO_2$. A model fitting the data is given in Scheme 2. According to this model the increasing [buffer] influences the rate of protonation of the enzyme and thus increases the rate of loss of ^{18}O, (that labelled the active site), to the medium (Θ_{cat}). Correspondingly, since the buffer facilitates removal of ^{18}O from the enzyme and competes with step-1 of Scheme 2, it

Scheme 2

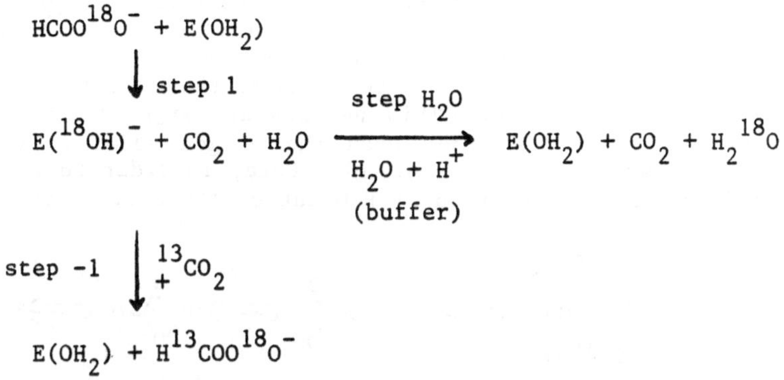

$$HCOO^{18}O^- + E(OH_2)$$

$$\downarrow \text{step 1}$$

$$E(^{18}OH)^- + CO_2 + H_2O \xrightarrow[\substack{H_2O + H^+ \\ (\text{buffer})}]{\text{step } H_2O} E(OH_2) + CO_2 + H_2^{18}O$$

$$\text{step -1} \downarrow \substack{^{13}CO_2 \\ +}$$

$$E(OH_2) + H^{13}COO^{18}O^-$$

diminishes the reformation of $H^{13}COO^{18}O^-$ from the enzyme catalyzed rehydration of $^{13}CO_2$ ($\emptyset_{cat}$). It is of note that both θ_{cat} and $\emptyset_{cat}$ have appreciable values even in the absence of external buffer. This implies that the enzyme has a rapid path for exchange from the active site that does not require protonation from the medium. One of the acceptable pathways for this exchange is the intramolecular proton transfer from His64-H$^+$ to E-Zn-OH$^-$ discussed in section 3a.

The catalytic properties of CAIII have been investigated. As indicated in Table 1, it is less active in the hydration of CO_2 (k_{cat} 2-3 x $10^{-3}s^{-1}$, feline, 25°C, 0.033 M NaSO$_4$) [4b] and does not contain an active site histidine. Nor does it exhibit a pH dependence for k_{cat} from pH 6-8. It also exhibits a $k_{cat}(H_2O)/k_{cat}(D_2O)$ solvent kinetic isotope effect of ~ 2.5 [4b] but k_{cat}/K_M exhibits a smaller, SIE of 1.1-1.5. These data suggest that the pK$_a$ of the ZnOH$_2$ in isozyme III is < 6 but that it follows a mechanistic scheme for CO_2 hydration similar to the more active isozymes I and II. Kararli and Silverman [30], using the ^{18}O-exchange method, have determined that the maximal rate of interconversion of CO_2 and HCO_3^- at chemical equilibrium has a value of > 10^4 s^{-1} between pH 6.0 and 7.1. However, since the steady state turnover for CO_2 hydration is 2 x 10^3 s^{-1} [4b], apparently proton transfer from the enzyme to the medium to regenerate EZnOH$^-$ limits the rate; this is consistent with the kinetic SIE with isozyme III of 2.5.

More detailed information about the CO_2 hydration with bovine CAIII has been provided using the ^{18}O-methodology of Silverman [31]. Silverman and Tu [31] reported two basic observations for ^{18}O-exchange catalyzed by CAIII at chemical equilibrium but before isotopic equilibrium. First, the ^{18}O loss from CO_2 was biphasic with the ^{18}O-content of CO_2 less than HCO_3^-. Second, the ^{18}O-distribution in CO_2 cannot be described by a simple binomial distribution as would be required if the isotopic distribution was random. These observations can be reconciled only by considering that HCO_3^- is more tightly bound to the enzyme than CO_2 or H_2O. Thus, the former dissociates more slowly than CO_2 or the processes involved in release of ^{18}O to the solvent. Also, the data require that more than one ^{18}O can be lost from a bound bicarbonate so that in its bound form there must be a scrambling of oxygens on the enzyme. This permits an oxygen to be removed from the intermediate complex other than the oxygen added to CO_2. A possible mechanism would be as in eq. 10. Note, in order to be consistent with the data, the dissociation of H_2O out of the active site must be

$$(10)$$

comparable to, or faster than, dissociation of either CO_2 or HCO_3^-. This aspect of the $CO_2 \rightleftharpoons HCO_3^-$ appears to be unique to CAIII since neither isozyme I or II exhibit these effects [31].

Information bearing on the relative rates of CO_2/HCO_3^- debinding and their chemical interconversion is provided by $^{12}C/^{13}C$ kinetic isotope effects (KIE) on HCO_3^- dehydration catalyzed by bovine CAII [26]. The carbon KIE is 1.0101 ± 0.0004 in H_2O and is 1.0107 ± 0.0007 in D_2O. This value is unaffected by substituting Co^{2+} for Zn^{2+} in the enzyme, nor is it affected markedly by increases in viscosity which would be anticipated to reduce diffusion of reactants to or from the active site [26a]. The lack of a D_2O effect on the carbon KIE in enzymatic dehydration contrasts the situation in the H_3O^+ promoted dehydration of HCO_3^- [25a] and requires for the former that the proton transfer to or from the enzyme occurs in a step temporally separated from the actual chemical interconversion of $HCO_3^- \rightarrow CO_2$. The lack of an effect of Co^{2+} or of added sucrose viscogens [32] on the isotope effect suggests that the diffusional processes are many fold faster than the rate limiting chemical interconversion. The low observed C KIE (1.01) suggests the transition state is very late with a large amount of $M(OH) \cdots CO_2$ bond-breaking. It is interesting to note that above pH 9, a formally similar high pH dehydration of $HOCO_2^- \rightarrow HO^- + CO_2$ in the absence of enyzme also exhibits a C KIE of 1.01 [26a].

Recent work of Rowlett et al has focussed attention on the activation parameters for CO_2 hydration catalyzed by both Zn^{2+} and Co^{2+} CAII in the absence of inhibitory anions at non-rate limiting [buffer] (10–50 mM) [33]. Given in Table 3 are the k_{cat}^{max} and k_{cat}^{max}/K_M values for both enzymes

TABLE 3. Rate and equilibrium constants derived from pH/rate profiles of CAII catalyzed CO_2 hydration, T=25°C, μ = 0.2 (Na_2SO_4), [Buffer] = 50 mM [33b].

| | Zn^{2+}-enzyme | |
	k_{cat}	k_{cat}/K_M
k_{max}	$(1.06\pm0.05)\times10^6 s^{-1}$	$(8.7\pm0.5)\times10^7 M^{-1}s^{-1}$
pK_a	6.87 ± 0.05	7.03 ± 0.05
	Co^{2+}-enzyme	
	k_{cat}	k_{cat}/K_M
k_{max}	$(3.05\pm0.12)\times10^5 s^{-1}$	$(8.8\pm0.4)\times10^7 M^{-1}s^{-1}$
pK_a	6.58 ± 0.05	7.23 ± 0.04

as well as the apparent pK_a's for the active site groups responsible for each term [33b]. For the Zn^{2+}-enzyme, the pK_a values obtained from the k_{cat} and k_{cat}/K_M profiles are similar and close to 7. The apparent K_M value is pH independent [18,33] which implies that the active site group having pK_a of 7 influences both k_{cat} and k_{cat}/K_M. By contrast, the Co^{2+}-enzyme exhibits an apparent K_M which is pH dependent, decreasing by ~ 3-fold in passing from pH 6-9. The pH dependence of K_M in the Co^{2+}-enzyme is manifested in a downward shift of ~ 0.6 units in the k_{cat} profile relative to k_{cat}/K_M. Given in Table 4 are the activation energies for Zn^{2+}- and Co^{2+}-CAII catalyzed CO_2 hydration [33a,b].

TABLE 4. Activation Energies for Zn^{2+}- and Co^{2+}-CAII catalyzed CO_2 hydration; 10 mM CHES buffer, μ = 0.2 (Na_2SO_4) [33b].

| | Zn^{2+}-Enzyme | | Co^{2+}-Enzyme | |
	k_{cat}	k_{cat}/K_M	k_{cat}	k_{cat}/K_M
E_a(k_{cal} Mol^{-1})	8.6 ± 0.3	1.2 ± 0.7	5.0 ± 0.5	-1.0 ± 0.3
ΔS^*(cal mol^{-1}K^{-1})	-3.5 ± 1.5	-20 ± 2	-18 ± 2	-27 ± 2
ΔH^*(k_{cal} mol^{-1})	8.1 ± 0.4	0.5 ± 0.7	4.4 ± 0.6	-1.7 ± 0.3

The data are interpreted within the M-hydroxide mechanism outlined in eqs. 4 and 5. In the native enzyme, both $Zn^{2+}-OH_2$ and $His64-H^+$ have similar pK_a's of ~ 7 [33,18]. Computer simulations indicate that as the pK_a values for $M^{2+}-OH_2$ and $His64-H^+$ are changed from ~ 7, the pK_a of $His64-H^+$ controls the pH dependence of k_{cat}, while the pK_a of the $M^{2+}-OH_2$ controls the pH dependence of k_{cat}/K_M [34]. Indeed, if the pK_a values of $M^{2+}-OH_2$ and $His64-H^+$ are similar, then the observed K_M for CO_2 will be pH independent [34]. However, lower value for the k_{cat} term is analyzed in terms of a lower pK_a for His64 and slightly higher pK_a for $Co^{2+}-OH_2$ than in the native enzyme.

The temperature dependence on k_{cat} is essentially localized in the intramolecular proton transfer outlined in eq. 5 and section 3a. The temperature effect on k_{cat}/K_M is predominantly associated with the first step in the catalysis or $EOH^- + CO_2 \rightarrow E-OH^- \cdot CO_2 \rightarrow E-\bar{O}CO_2H$ [33b]. By contrast, with CAI, no single step in eqs. 4,5 is uniquely controlling k_{cat} or k_{cat}/K_M [34a]. This may be anticipated to be the situation in highly evolved enzymes that have multi-step reactions (e.g., Figure 8) where the energy barriers for all steps are similar.

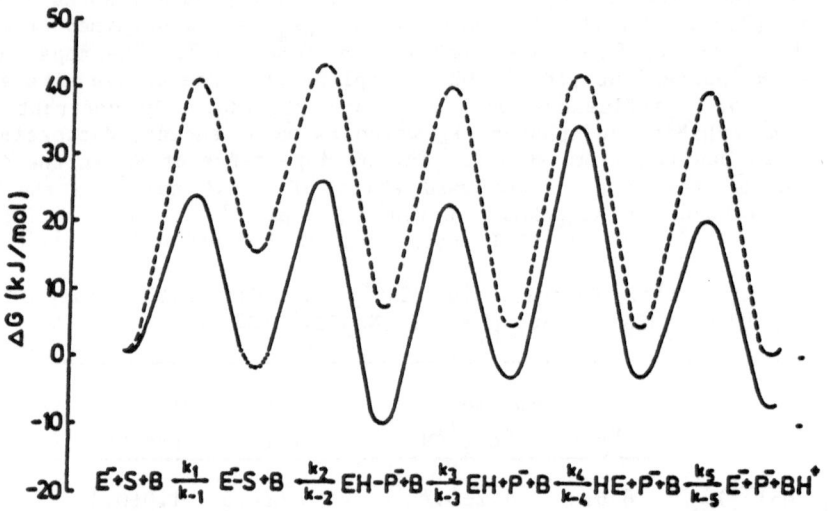

FIGURE 8. A hypothetical free energy profile for the pathway of eq. 4 and 5 catalyzed by human CAII. Fully drawn curve, standard states of 1 M; broken curve, 1 mM CO_2, 20 mM HCO_3^-, 10 mM each of BH^+ and B. (Redrawn from [1h])

4. Models for CA

The earlier work on simplified systems that mimic the spectroscopic and catalytic properties of CA have been extensively reviewed [35] and only those most relevant to the present discussion will be reiterated here. These fall into three main categories: spectroscopic, catalytic and theoretical, and we will deal with them in that order.

a. SPECTROSCOPIC MODELS FOR Co-CA

Octahedrally coordinated Co^{2+} gives rise to pink solutions having molar absorbancies (ε_{max}) of ~ 10 $M^{-1}cm^{-1}$ around 490-500 nm [36]. Solutions of tetracoordinate Co^{2+} are bright blue and have $\varepsilon_{max} > 300$ $M^{-1}cm^{-1}$ at 550-650 nm while five-coordinate Co^{2+} is typified by lighter blue solutions ($\varepsilon_{max} < 200$ $M^{-1}cm^{-1}$ at 530-630 nm) as well as a weak diagnostic transition ($\varepsilon_{max} < 50$ $M^{-1}cm^{-1}$) between 700 and 800 nm [10]. As shown earlier in Fig. 2, the spectrum of bovine and human CoCAI in unbuffered media changes as a function of pH. At low pH (5.8), the spectrum is characteristic of a 4-5 coordinate environment that changes to a 4 coordinate environment as the pH is raised. The absorbance vs pH profile for these gives evidence for two macroscopic pK_a's [10b,11] of 6.87 and 8.71 for CoCAI and 5.89 and 7.97 for CoCAII which are associated with the active site $Co-OH_2$ and His64 (for CAII) or His200 (CAI) residues.

Two independent studies with tridentate ligands mimic this behavior to some extent. Given in Fig. 9 is the spectrum of aquo [tris[(3,5-dimethyl-1-pyrazolyl)methyl]amine]:Co^{2+} perchlorate (1) as a function of pH which shows evidence for two spectral controlling ionizations (eq. 11) [37a].

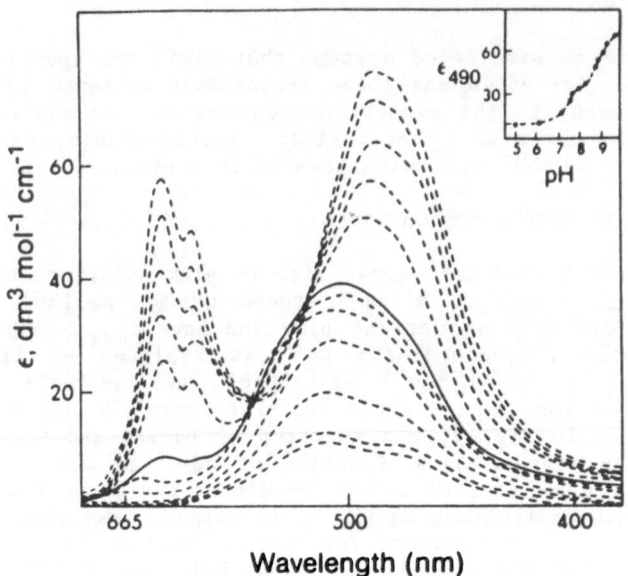

FIGURE 9. The absorption spectrum of aquo[tris(3,5-dimethyl-1-pyrazolyl) methyl]Co^{2+} as a function of pH. (Redrawn from [37a])

Overall, the high pH spectrum is reminiscent of the basic form of CoCA but the extinction coefficient is ~ $1/4$ that of the enzyme (~ 70 $M^{-1}cm^{-1}$ vs > 300 $M^{-1}cm^{-1}$) [37a]. The x-ray crystal structure of this perchlorate salt confirms the structure **2** and indicates apically bound N and OH$_2$ [37b]. However, the complex shows no activity in CO$_2$ hydration [37a].

Tris(2-imidazolyl)phosphines (**3**) have been synthesized and shown to bind divalent metals in a tridentate fashion [38]. One of these (**3c**) as its Co^{2+} chelate exhibits anion dependent absorption spectrum indicative of a 4-coordinate binding wherein the fourth ligand is the anion [38b] (Fig. 10). The absorption spectrum of **3c**:Co^{2+}Cl$^-$ is also pH dependent (Fig. 11) and is controlled by a ligand associated group having an

3a R=H

b R=CH$_3$

c R=CH(CH$_3$)$_2$

4

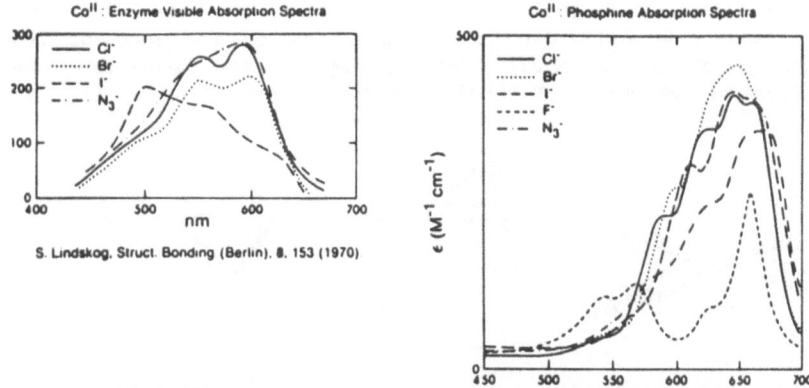

FIGURE 10. Anion dependent visible absorption spectra of 1 x 10^{-3} M **3c**:Co^{II} determined in 80% EtOH/H$_2$O solution saturated with NaX (X = I, Br, Cl, F) compared with those of CoCA. (Redrawn from [38b])

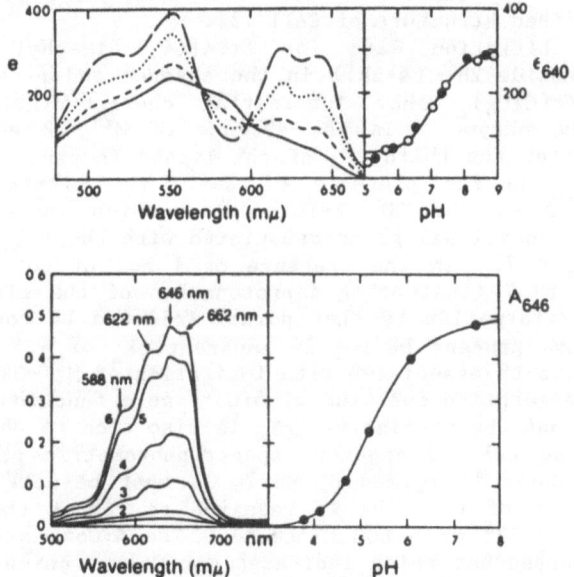

FIGURE 11. pH dependent absorption spectrum of CoCA (top). (Redrawn from [9a]) compared with that of **3c**:Co^{II} (1 x 10^{-3} M) in the presence of 10^{-1} M Cl$^-$ (80% EtOH/H$_2$O) (bottom), Redrawn from [38b])

apparent pK_a of ~ 5.6 (80% $EtOH/H_2O$). However, the tetracoordinate spectrum cannot be obtained in the presence of the non-coordinating anion ClO_4^-, so the ionization responsible for the spectrum cannot be attributable to $Co^{2+}-OH_2$. A likely process is given in eq. 12. X-ray

$$\text{"}pK_a\text{"} \approx 5.6$$

(12)

low pH form, octahedral
3c: $Co^{2+}Cl^-(OH_2)_n$

high pH form,
tetracoordinate

crystallographic analysis of 3c: $Zn^{2+}Cl_2^-$ indicates a tetra coordinate structure with one Cl^- bound to the metal and highly strained N–Zn–N bonds of ~ 95° [39]. These are ~ 15° less than the N–Zn–N bond angles found in the refined structure of CAII [3].

Quantitative titration data for tris[(4,5-dimethyl-2-imidazolyl)-methyl]phosphineoxide·$2H^+$ (4·$2H^+$) in the absence and presence of 1 eq. $Zn(ClO_4)_2$ or $Co(ClO_4)_2$ show interesting characteristics of direct relevance to the enzyme. In the absence of M^{2+}, 2 eq. of OH^- are consumed only after the third pK_a of the ligand is passed (pK_a^3 = 7.18, 80% $EtOH/H_2O$). In the presence of Zn^{2+}, the titration shows the consumption of 3 eq. of OH^- before pH 7 which indicates a formal titration of one additional group associated with the 1:1 complex having an apparent pK_a < 7. In the presence of 1 eq. of Co^{2+}, two OH^- are consumed before pH 7 (indicating deprotonation of the ligand and hence complete complex formation by that point) followed by consumption of a third OH^- in some process having an apparent pK_a of ~ 7.8. The third OH^- is believed to be associated with ionization of $M^{2+}-OH_2$.

The visible absorption spectrum of 4:Co^{2+} as a function of pH (Figure 12) also shows that the titrimetric pK_a is also tied to the formation of an intensely blue colored species, spectrophotometric pK_a ≃ 7.8 [40]. Thus while a complex is formed by pH 7, it must be octahedral and the further ionization of $Co^{2+}-OH_2$ is required to reduce the coordination number to 4 (ε_{max} 598 nm = 600 $M^{-1}cm^{-1}$). The 4:Co^{2+} spectrum is also markedly anion dependent which indicates monovalent anions such as Cl^-, I^-, Br^-, AcO^- associate directly with the metal. However, spectrophotometric determinations of 4:Co^{2+} as a function of pH in the presence of 0.2 M Cl^- clearly shows the formation of a unique 4(5) coordinate species which subsequently gives way on the addition of further OH^- to the same spectrum as was observed with only ClO_4^- as the added counterion. The appearance of the absorbance vs pH profile of 4:Co^{2+} in

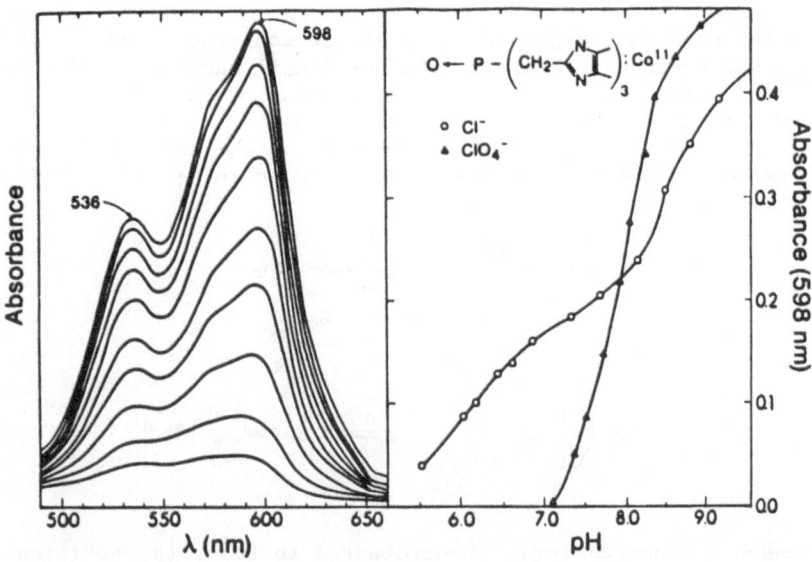

FIGURE 12. Visible absorption spectrum of **4**:Co^{2+} as a function of pH, 80% EtOH/H$_2$O. Redrawn from [40].

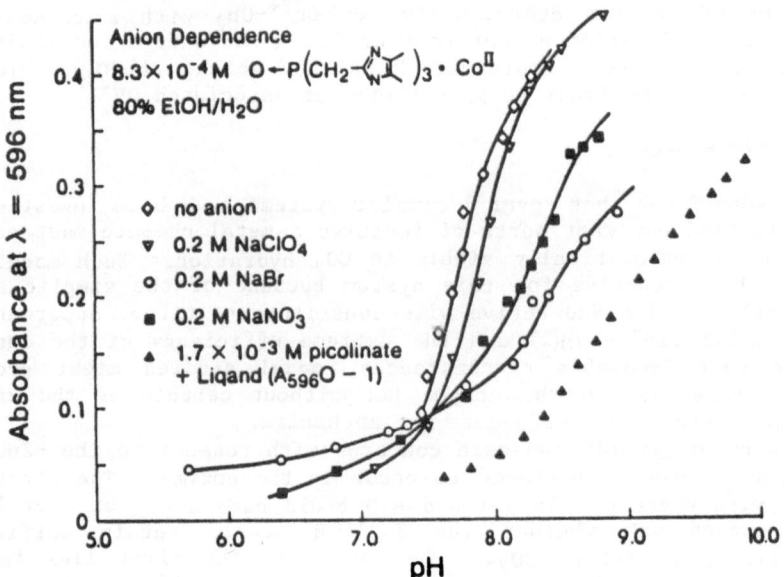

FIGURE 13. Spectrophotometric pH dependent titration curves of **6**:Co^{11} in the presence of various anions, 80% EtOH/H$_2$O. (Redrawn from [40])

the presence of 0.2 M Cl⁻ is also shown in Fig. 12 and is more complicated than that expected for a single ionizing group. As shown in Fig. 13, the spectrophotometric "pK_a" of $4:Co^{2+}$ has a general increasing trend in the presence of associating anions. A similar trend has been observed with CoCA and is reconciled in terms of a competition between OH⁻ and inhibitory anion for a common site on the metal [41]. Overall, the situation with $4:Co^{2+}$ is best visualized as in eq. 13. At low pH, **4**

$$4 : 2H^+ + Co^{2+} (H_2O)_6 \xrightleftharpoons{2OH^-} O=P\langle Im, Im, Im \rangle Co^{2+} (OH_2)_3$$

(13)

exists as a protonated form, uncoordinated to Co^{2+}, but addition of OH⁻ deprotonates $4:2H^+$ and allows full coordination to Co^{2+} by pH 7. However, this complex must be octahedral (OH_2 units occupying the residual metal sites) since no blue colors are evident at pH 7. Further addition of OH⁻ causes deprotonation of $Co^{2+}-OH_2$ with a concomitant reduction in coordination number to 4 or 5. Added anions coordinate to $4:Co^{2+}$ to form new 4(5) coordinate complexes which, at higher [OH⁻], give rise to the same basic complex having an associated OH⁻.

b. CATALYTIC MODELS

It is not surprising that several simpler systems have been investigated in order to find out what sorts of features a metal chelate must employ in order to be catalytically viable in CO_2 hydration. Such modelling appears to be attractive for this system because of the simplicity of the chemical reaction and active site constitutents (i.e., apparently a tris imidazolyl Zn^{2+} + OH_2) and the extreme efficiency of the enzyme. Under the most favorable circumstances, model studies might provide answers of relevance to the enzyme but without certain of the enzyme based complications that obfuscate the mechanism.

There were originally two main concerns with respect to the $-ZnOH^-$ + CO_2 → $-ZnHCO_3^-$ process believed to occur in the enzyme. The first was concerned with whether a Zn^{2+} bound H_2O could have a pK_a of 7 or less, and the second was whether the $Zn^{2+}-OH^-$ could retain sufficient nucleophilicity to attack CO_2. The answer to the first lies in two features embodied in the enzyme; a low coordinate $Zn^{2+}-$ and a somewhat hydrophobic active site. Both these features work to lower the pK_a of

$Zn^{2+}(OH_2)$ from 9 [42a] or 10 [42b,43] (for $Zn(OH_2)_x$ in H_2O), to values approaching 7. Five-coordinate **5a,b** are reported [43] to have pK_a values of 8.7 and 8.1 respectively while (presumably) 4-coordinate **5** has a pK_a of ~ 7 [44]. Also, as stated before, the pK_a of **4**:$Zn^{2+}-OH_2$ must

5a R=CH₃, R₁=H   **6**   **7a** R=H (tren)

  b R=R₁=H   **b** R=CH₃(Me₆ tren)

be lower than 7 (in 80% EtOH/H_2O) in order to account for the quantitative titration profile [40]. In each of the above cases, the decreased ligation to Zn^{2+} increases its (+) which in turn enhances the OH_2 acidity. From the example of **4**:$Zn^{2+}-OH_2$, one can see that a hydrophobic environment tends to disfavor charge localization so that $Zn^{2+}-OH^-$ is produced at a lower "pH" than would be the case in a more highly charge-stabilizing medium such as H_2O.

More evidence for the latter effect comes from studies of the activation parameters for ionization of tren (**7a**) and Me₆ tren (**7b**) [45] complexes of Co^{2+} and Zn^{2+} (Table 5).

TABLE 5. Apparent pK_a values for acid dissociation of H_2O coordinated to tren and Me₆ tren complexes of Co^{2+} and Zn^{2+} at 25°C, μ=1.0 NaClO₄ [45]

Complex	pK_a	$\Delta H°$(kcal/mol)	$\Delta S°$(cal/°K/mol)
(Co tren H_2O)$^{2+}$	10.22±0.01	0.22±0.65	−46.06±2.20
(Zn tren H_2O)$^{2+}$	10.26±0.02	2.84±0.45	−37.40±1.50
(CoMe₆ tren H_2O)$^{2+}$	8.80±0.04	8.10±0.45	−13.03±1.50
(ZnMe₆ tren H_2O)$^{2+}$	9.00±0.01	11.76±0.45	− 1.72±1.50

The Me_6 tren complexes of Co^{2+} and Zn^{2+} each have pK_a values for the $M^{2+}-OH_2$ (in a 5-coordinate environment) ~ 1.2-1.4 units lower than the corresponding tren complexes. Molecular models suggest that the Me_6 groups tend to form a hydrophobic ring around the $M^{2+}-OH_2$ while in the tren complexes, the $M^{2+}-OH_2$ is near the hydrophilic NH_2 groups. From the data in Table 5, the entropies of ionization for the Me_6 tren complexes are far less negative than is the case for the tren complexes which indicates that less reordering of the solvent accompanies ionization of the former. This probably stems from the fact that the more hydrophobic $(Me_6$ tren $M-OH_2)^{2+}$ face leads to a situation generally conducive to the ordered, ice-like structure for the contacting H_2O. On the other hand, the adjacent solvent to (tren $M-OH_2)^{2+}$ should be less highly ordered, and its entropy decrease on protolysis greater, than will be the case for the corresponding $(Me_6$ tren $M-OH_2)^{2+}$ complexes. It is of note that an ordered, ice-like structure for H_2O in the active site of CA has been observed by x-ray crystallography [2,3].

Can a low pK_a for a $M^{2+}OH_2$ system generate an anion retaining sufficient nucleophilicity to attack CO_2? The nucleophilicity of MOH^- complexes toward CO_2 hydration can be expressed as a Brønsted relationship (Fig. 14) of pK_a vs log k_2^{max} [46].

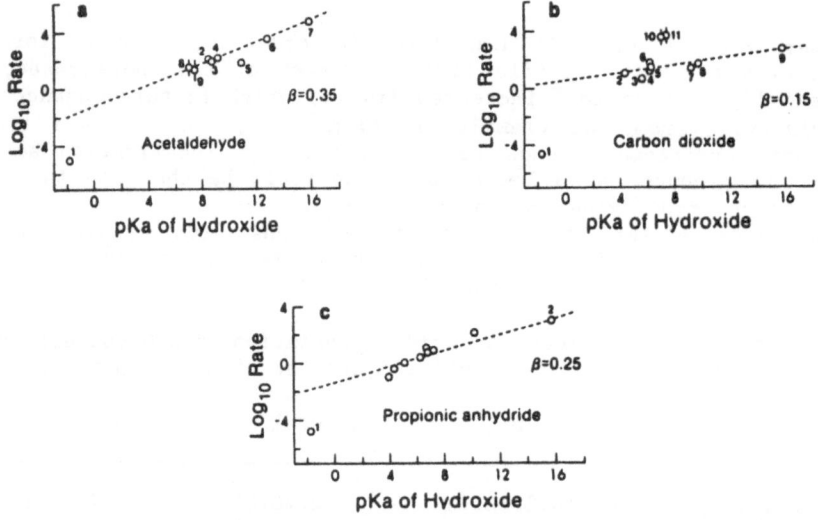

FIGURE 14. Graphic representation of the Brønsted relationship between pK_a of conjugate acids of various $M-OH^-$ and log k_2^{max} for attack on various carbonyl species. (Redrawn from [46] which also defines identity of species corresponding to #'s)

For CO_2 hydration, some representative values are given in Table 6 [46]. The slope of the Brønsted plot (β) in Fig. 14 is 0.15 [43a] – 0.35 [46] showing a general insensitivity of k_2^{max} to pK_a. This has an obvious advantage to the enzymic system since a low β and pK_a allows a large local concentration of Zn^{2+}-OH^- in the active site without much concomitant loss of nucleophilicity.

TABLE 6. Second order rate constants (k_2^{max}) for CO_2 attack and pK_a values for hydroxide nucleophiles at 25°C [46]

Nucleophile	pK_a (conjugate acid)	k_2 ($M^{-1}s^{-1}$)
OH^-	15.7	6000
$Zn^{2+}(OH_2)_xOH^-$	9–10	~1000[a]
glycylglycine $Cu^{2+}OH^-$	9.4	580
$(NH_3)_5Rh^{3+}OH^-$	6.8	470
$(NH_3)_5Ir^{3+}OH^-$	6.7	590
$(NH_3)_5Co^{3+}OH^-$	6.4	220–310
$(NH_3)_5Cr^{3+}OH^-$	5.3	> 11
H_2O	-1.57	0.0007[b]

a. Calculated from data of Tabushi and Kuroda [47], at pH=7. $k_{Zn^{2+}(OH_2)_x} \simeq 8\ M^{-1}s^{-1}$; given a pK_a of 9–10 for aquozinc, then ~ 1% of the metal ion is present as hydroxide.

b. $0.037\ s^{-1}/55\ M$

Although several model systems (including those in Table 6) have been investigated for their CO_2 hydration ability, the great bulk [43,47,48] do not approximate the known active site of the enzyme, both with respect to numbers and types of M^{2+} ligands.

The first demonstration of catalysis of interconversion of HCO_3^- and CO_2 by a complex containing three imidazoles approximating the Zn^{2+}-binding site of CA, appears to be that of Brown and coworkers who studied Zn^{2+}-complexes of phosphines **3a,b,c** [38], **8** [49], and **9** [49].

8   **9**

The catalytic viability was established by watching the approach to equilibrium from both sides of the HCO_3^- CO_2 reaction where the observed rate constant for the approach is $k_{obsd} = k_f + k_r$. Catalysis by the phosphine:Zn^{2+} chelates was evaluated by determining k_{obsd} in the absence and presence of ligand, Zn^{2+}, and ligand + Zn^{2+}. Neither Zn^{2+} nor ligand alone accelerates the attainment of equilibrium, but **3c**:Zn^{2+} does with k_2^{max} being 900 ± 60 $M^{-1}s^{-1}$ at pH 6.4-6.8 (80% EtOH/H_2O, HEPES buffer, 25°C, μ=0.2 $NaClO_4$) [38b]. Neither **3a** nor **3b** as their Zn^{2+} complexes display any activity nor do they show spectroscopic evidence of a 4(5)coordination for their Co^{2+} complexes [38b].

Since the <u>iso</u> propyl groups in **3b**:Zn^{2+} severely restrict access of ligands to the metal surface [39] (as might be required during a 5-coordinate catalytic CO_2 hydration cycle), less highly encumbered phosphines **8** and **9** were synthesized and their ability to facilitate the $CO_2 \rightleftharpoons HCO_3^-$ equilibrium investigated [49]. Apparently, removing the encumbrance to ligand access does increase the catalytic viability since the k_2^{max} of **8**:Zn^{2+} and **9**:Zn^{2+} are 1420 $M^{-1}s^{-1}$ and 2700 $M^{-1}s^{-1}$ respectively (80% EtOH/H_2O, HEPES, pH 6.4-6.6, 25°C, μ=0.2 $NaClO_4$).

Finally, it has long been known that monovalent anions inhibit the catalytic viability of CA by either binding to the M^{2+} or interupting the proton shuttle of eq. 5. For the above complexes, low concentrations (5×10^{-4} - 2×10^{-3} M) of ions like Cl^-, Br^-, I^-, and NO_3^- inhibit the ability of **3c**:Zn^{2+}, **8**:Zn^{2+} and **9**:Zn^{2+} to facilitate the $CO_2 \rightleftharpoons HCO_3^-$ interconversion [38b,49]. This stems from an association of the ion and M^{2+} in the models which blocks access of HCO_3^-, H_2O (OH^-) and CO_2 to the metal.

The models discussed in sections 4a and b all have a number of deficiencies. Perhaps the best systems are those in which the ligating groups are three imidazole units since these at least approximate the active site histidines. Understandably, development in the model area has been slow, and is dictated by what one can readily synthesize which may not be what one would optimally <u>like</u> to synthesize. Two other deficiencies attendent to any type of model are unlikely to be overcome, and this relates to their small size relative to the enzyme. The first rests the fact that the geometry of the small-molecule chelates is dictated by the actual binding of the metal. This invariably yields smaller affinities for M^{2+} than is the case for the enzyme whose active site is "preformed" and apparently very similar in geometry for the <u>holo</u>- and <u>apo</u>-enzymes. The second feature relates to the fact that the models do not contain a "binding" site for their substrates. As such their catalytic viability must be viewed in terms of a second order rate process whereas the enzyme, once having bound the substrate establishes the $CO_2 \rightleftharpoons HCO_3^-$ interconversion through a unimolecular process. Rate comparisons are therefore made using k_2^{max} for the model and k_{cat}/K_M for the enzyme.

There remains one major hurdle to overcome before the models can be considered as reasonable catalytic mimics for the enzymatic process.

This relates to the high value of k_{cat}/K_M for CAI or II, 10^7-10^8 $M^{-1}s^{-1}$ as compared with the best models ($\sim 10^3$). No M-OH$^-$ complex reported to date has a reported rate constant for CO_2 hydration which exceeds free OH (~ 6000 $M^{-1}s^{-1}$) although in the enzyme, the active site Zn^{2+}-OH$^-$ must be at least 10^4-fold more active (and perhaps more since $CO_2 \rightleftharpoons HCO_3^-$ interconversion is not rate limiting). It is tempting to involve additional roles for the active site constituents that in some way activate the CO_2 toward nucleophilic attack by Zn^{2+}-OH$^-$, perhaps by neutralization of the incipient charges in the transition state.

c. THEORETICAL MODELS

We conclude the discussion with a few studies [50-52] that have been reported calculations wherein of the attack of M-OH$^-$ on CO_2 is modelled by quantum mechanical means of different sophistication. It should be pointed out that these calculations are essentially gas phase ones which may not be completely accurate representations of the situation extant in solution or the enzyme active site.

Pullman [50] has reported the calculated energetics for the attack of **10** on CO_2 to produce **11**. For the starting point (**10**) the HO---C distance was 3.0Å with the =O---Zn^{2+} distance being 3.4Å; the three NH$_3$

10   **11**

Zn^{2+} ligands were chosen as representations of the imidazole ligands to minimize computational costs. The calculations indicate that in **10**, the stabilization energy is already appreciable (-10.3 kcal/mol) and that there is essentially no activation energy in the transformation to an energy minimized **11** (-13.3 kcal/mol). Experimentally, the activation energy for OH$^-$ attack on CO_2 is +13.2 kcal/mol [53] in solution.

Pullman's analysis [50] suggests that H_2O coordinated to Zn^{2+} readily dissociates to a Zn^{2+}-OH$^-$, and the latter retains significant nucleophilicity to attack CO_2. Also the barrierless process interconversion of **10→11** might be taken as indicative of a dual role for the metal in providing a nucleophilic delivery of OH$^-$ and stabilizing by a Lewis acid role the incipient (-) on the developing bound HCO$_3^-$. On the other hand, calculations of Jönsson et al [54] on the simple attack of OH$^-$ on CO_2 (in the absence of solvent) indicate that this process too proceeds without a barrier. Thus the experimental activation barrier of 13.2 kcal/mol [53] must be attributed to solvation effects. It seems possible in the enzyme, where much of the solvation sphere of OH^- is stripped away, that a barrierless process could result without an additional role for the Zn^{2+} other than providing the nucleophile.

Allen and coworkers [51,55] have modelled the $CO_2 \rightleftharpoons HCO_3^-$ interconversion starting with a simplified representation of the active site residues (Scheme 3) and a 5-coordinate Zn^{2+} containing both H_2O and OH^-.

Scheme 3

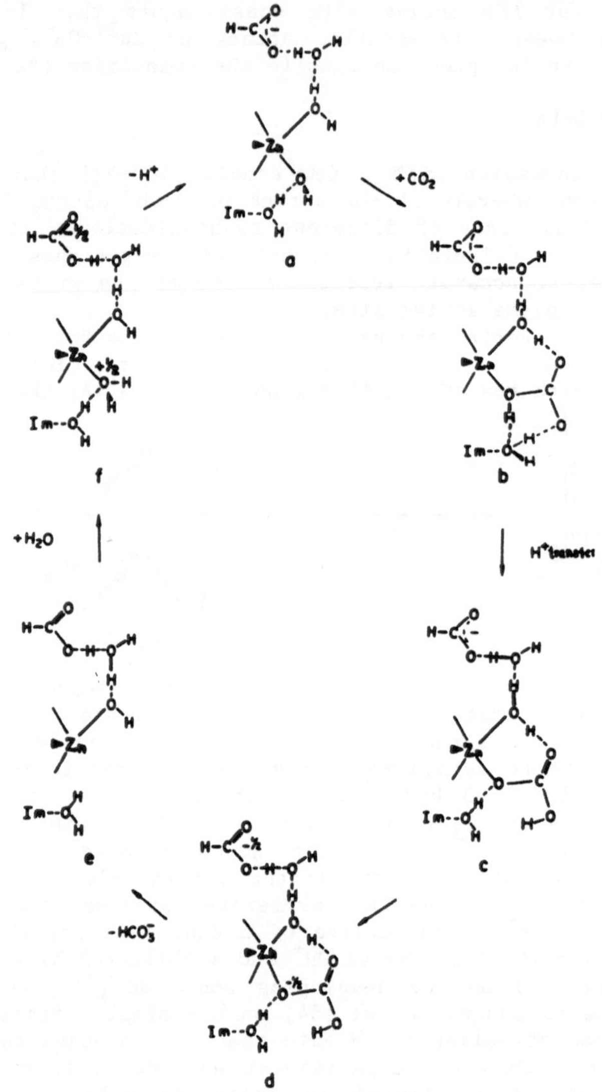

Their computed pathway involves nucleophilic attack of the $-Zn(OH_2)(OH^-)$ on CO_2 with stabilization of the coordinated HCO_3^- being provided by the fifth coordinate ligand H_2O as in b (Scheme 3). Interconversion of b and c is mediated by an intervening H_2O with the net effect being to labilize the $ZnOH_2$ to produce a 5-coordinate $-Zn-(OH^-)(OCO_2H^-)$ (d) which assists in the expulsion of HCO_3^- (e). Their computed pathway compliments a series of reports by Pocker and Diets [56] who conclude that the failure of the enzyme to catalyze the decarboxylations of alkyl carbonates (formally analogous to $HOCO_2^-$ dehydration) is because the analogous b $\rightleftharpoons$ c interconversion involving an alkyl shift is of prohibitively high energy.

Liang and Lipscomb [52] have favored a generally similar pathway for the enzyme promoted hydration of CO_2. Their computed barriers for interconversion of two configurations of $Zn^{2+}HCO_3^-$ (12$\rightleftharpoons$13) decreases

12   **13**   **14**

from 35.6 kcal/mol with no intervening H_2O bridge, to 3.5 k$_{cal}$/mol and 1.4 kcal/mol when 1 and 2 waters respectively are included [52]. In addition, the computed barrier for 12$\rightleftharpoons$13 with bridging H_2O is substantially lower than the one calculated for the cyclic $Zn^{2+}OH^-$ process involving a bidentate transition state (**14**) described by Lindskog et al [57].

5. References and Notes

1. For recent reviews see:

   a. Lindskog, S. Adv. Inorg. Biochem. 1982, **4**, 115.

   b. Lindskog, S.; Ibrahim, S.A.; Jonsson, B.-H.; Simonsson, J. in "The Coordination Chemistry of Metalloenzymes", Bertini, I.; Drago, R.S.; Luchinat, C., Eds., D. Reidel Co., Dordrecht: Holland, 1983.

   c. Silverman, D.N.; Vincent, S.H. CRC Crit. Rev. Biochem. 1983, **14**, 207.

   d. "Biophysics and Physiology of Carbon Dioxide", Bauer, C.; Gros, G.; Bartels, H., Eds., Springer-Verlag, Berlin: Heidelberg, 1980.

   e. Carter, M.; Jeffrey, S. Biochem. Soc. Trans. 1985, **13**, 531.

   f. "Zinc Enzymes", Bertini, I.; Luchinat, C.; Maret, W.; Zeppezauer, M., Eds., Birkhäuser: Boston, 1986, chapters 22-27.

   g. Silverman, D.N.; Lindskog, S. Accts. Chem. Res. 1988, **21**, 30.

   h. Lindskog, S. in "Zinc Enzymes", Bertini, I.; Luchinat, C.; Maret, W.; Zeppezauer, M., Eds., Birkhäuser: Boston, 1986, 306.

2. a. Liljas, A.; Kannan, K.K.; Bergsten, P.-C.; Waara, I.; Fridborg, K.; Strandberg, B.; Carlbom, U.; Järup, L.; Lövgren, S.; Petef, M. Nature New Biol. 1972, **235**, 131.

   b. Kannan, K.K.; Petef, M.; Cid-Dresdner, H.; Lövgren, S. FEBS Lett. 1977, **73**, 115.

   c. Notstrand, B.; Vaara, I.; Kannan, K.K. in "The Isozymes", Vol. 1, Markert, C.I., Ed., Academic Press: New York, 1975, 575.

   d. Kannan, K.K.; Notstrand, B.; Fridborg, K.; Lövgren, S.; Ohlssen, A.; Petef, M. Proc. Natl. Acad. Sci. USA 1975, **72**, 51.

   e. Kannan, K.K. in "Biophysics and Physiology of Carbon Dioxide", Bauer, C.; Gros, G.; Bartels, H., Eds., Springer-Verlag, Berlin: Heidelberg, 1980, 84.

3. a. Ericksson, E.A.; Jones, T.A.; Liljas, A. in "Zinc Enzymes", Bertini, I.; Luchinat, C.; Maret, W.; Zeppezauer, M., Eds., Birkhäuser: Boston, 1986, 317.

   b. Kannan, K.K.; Ramanadham, M.; Jones, T.A. Ann. N.Y. Acad. Sci. 1984, **429**, 49.

4. a. Tu, C.K.; Sanyal, G.; Wynns, G.C.; Silverman, D.N. J. Biol. Chem. 1983, **258**, 8807.

   b. Kararli, T.; Silverman, D.N. J. Biol. Chem. 1985, **260**, 3484.

   c. Tashian, R.E.; Hewett-Emmett, D.; Stroup, S.K.; Goodman, M.; Yu, S.L. in "Biophysics and Physiology of Carbon Dioxide", Bauer, C.; Gros, G.; Bartels, H., Eds., Springer-Verlag, Berlin: Heidelberg, 1980, 165.

5. a. Pocker, Y.; Meany, J.E. J. Am. Chem. Soc. 1965, **87**, 1809.

   b. idem. J. Phys. Chem. 1970, **74**, 1486.

   c. Pocker, Y.; Meany, J.E.; Davis, B.C. Biochemistry 1974, **13**, 1411.

6. a. Tashian, R.E.; Plato, C.C.; Shows, T.B. Science (Washington) 1963, **140**, 53.

   b. Pocker, Y.; Bjorkquist, L.C.; Bjorkquist, D.W. Biochemistry

<u>1977</u>, **16**, 3967.

c. Malmstrom, B.G.; Nyman, P.O.; Strandberg, B.; Tilander, B. in "Structure and Activity of Enzymes", <u>FEBS Symp.</u> <u>1964</u>, No. 1, 121.

d. Lo, K.W.; Kaiser, E.T. <u>J. Am. Chem. Soc.</u> <u>1969</u>, **91**, 4912.

7. Knowles, J.R.; Albery, W.J. <u>Accts. Chem. Res.</u> <u>1977</u>, **10**, 105.

8. a. Lindskog, S.; Malmstrom, B.G. <u>J. Biol. Chem.</u> <u>1962</u>, **237**, 1129.

   b. Hunt, J.B.; Rhee, M.-J.; Storm, C.B. <u>Anal. Biochem.</u> <u>1977</u>, **79**, 614.

9. a. Lindskog, S. <u>J. Biol. Chem.</u> <u>1963</u>, **238**, 945.

   b. Lindskog, S.; Nyman, P.O. <u>Biochim. Biophys. Acta</u> <u>1964</u>, **85**, 462.

   c. For a recent account of the properties of $M^{2+}CA$ derivatives see Bertini, I.; Luchinat, C.; Monnanni, R. in "Carbon Dioxide as a Source of Carbon: Biochemical and Chemical Uses", Aresta, M.; Forti, G., Eds., D. Reidel Publishing Co., Dordrecht, Holland, <u>1987</u>, 139.

10. a. Bertini, I.; Canti, G.; Luchinat, C.; Scozzafava, A. <u>J. Am. Chem. Soc.</u> <u>1978</u>, **100**, 4873.

    b. Bertini, I.; Luchinat, C.; Scozzafava, A. <u>Inorg. Chim. Acta</u> <u>1980</u>, **46**, 85.

    c. Jacob, G.S.; Brown, R.D.; Koenig, S.H. <u>Biochemistry</u> <u>1980</u>, **19**, 3574.

    d. Bertini, I.; Luchinat, C. <u>Accts. Chem. Res.</u> <u>1983</u>, **16**, 272.

    e. Simonsson, I.; Lindskog, S. <u>Eur. J. Biochem.</u> <u>1982</u>, **123**, 29.

    f. Bertini, I.; Luchinat, C.; Scozzafava, A. <u>Struct. Bonding (Berlin)</u> <u>1982</u>, **48**, 45.

    g. Engberg, P.; Lindskog, S. <u>FEBS Lett.</u> <u>1984</u>, **170**, 326.

11. Bertini, I.; Dei, A.; Luchinat, C.; Monnanni, R. <u>Inorg. Chem.</u> <u>1985</u>, **24**, 301.

12. a. Bertini, I.; Borghi, E.; Canti, G.; Luchinat, C. <u>J. Inorg. Biochem.</u> <u>1983</u>, **18**, 221.

    b. Led, J.J.; Neesgaard, E.; Johansen, J.T. <u>FEBS Lett.</u> <u>1982</u>, **147**, 74.

    c. Williams, T.; Henkins, R.W. <u>Biochemistry</u> <u>1985</u>, **24**, 2459.

    d. Stein, P.J.; Merrill, S.P.; Henkins, R.W. <u>J. Am. Chem. Soc.</u> <u>1977</u>, **99**, 3194.

13. Led, J.J.; Neesgaard, E. <u>Biochemistry</u> <u>1987</u>, **26**, 183.

14. a. Kalifah, R.G. <u>Proc. Natl. Acad. Sci. USA</u> <u>1973</u>, **70**, 1986.

    b. Lindskog, S.; Coleman, J.E. <u>ibid.</u> <u>1973</u>, **70**, 2025.

    c. Prince, R.H.; Wooley, P.R. <u>Bioorg. Chem.</u> <u>1973</u>, **2**, 337.

15. a. Jonsson, B.H.; Steiner, H.; Lindskog, S. <u>FEBS Lett.</u> <u>1976</u>, **64**, 310.

    b. Silverman, D.N.; Tu, C.K. <u>J. Am. Chem. Soc.</u> <u>1975</u>, **97**, 2263.

16. Eigen, M. <u>Angew. Chem., Int. Ed. Engl.</u> <u>1964</u>, **3**, 1.

17. a. Rowlett, R.S.; Silverman, D.N. <u>J. Am. Chem. Soc.</u> <u>1982</u>, **104**, 6737.

    b. Pocker, Y.; Jamjic, N.; Miao, C.H. in "Zinc Enzymes", Bertini, I.; Luchinat, C.; Maret, W.; Zeppezauer, M., Eds., Birkhäuser:

Boston, 1986, 341.
18. a. Steiner, H.; Jonsson, B.H.; Lindskog, S. Eur. J. Biochem. 1975,
    59, 253 (human CAII).
    b. Pocker, Y.; Bjorkquist, D.W. Biochemistry 1977, 16, 5698
    (bovine CAII).
19. Venkatasubban, K.S.; Silverman, D.N. Biochemistry 1980, 19, 4984.
20. Sirs, J.A. Trans. Faraday Soc. 1957, 54, 201.
21. Pocker, Y.; Bjorkquist, D.W. J. Am. Chem. Soc. 1977, 99, 6537.
22. a. Pocker, Y.; Stein, J.E. quoted in ref. 21.
    b. Kresge, A.J.; Preto, R.J. J. Am. Chem. Soc. 1965, 87, 4593.

23. For a process involving specific acid catalysis such as

$$L_3O^+ + LOCO^- \rightleftharpoons L_2O + L_2\overset{+}{O}C\overset{O}{\underset{O^-}{}} \quad (L=H,D)$$

Øa   Øb   Øc   Ød

the fractionation factors (Ø) [24] for L_3O^+ (0.69), $LOCO_2^-$ (1.0),
L_2O (1.0), $L_2\overset{+}{O}-CO_2^-$ (0.69) lead to a calculated $k_D/k_H = (0.69)^2/(0.69)^3 = 1.45$. If the process is a more complicated general acid one:

$$L_3O^+ + LOCO_2^- \rightleftharpoons \left[\cdots \right]^{\ddagger} \rightarrow \cdots$$

Øa   Øb Øc Ød

the calculated $k_D/k_H = \emptyset^{\ddagger}/\emptyset_R = (\emptyset b)^2 (\emptyset c)(\emptyset d)/(\emptyset a)^3$ could assume
values $2 > k_D/k_H > .3$ [24].

24. For a discussion of fractionation factors see:
    a. Venkatasubban, K.S.; Schowen, R.L. CRC Crit. Rev. Biochem.
    1984, 17, 1.
    b. Alvarez, J.; Schowen, R.L. in "Isotopes in Organic Chemistry",
    Buncel, E.; Lee, C.C., Eds., Elsevier: Amsterdam, Vol. 7, 1987, 1.
    c. Kresge, J.; More O'Ferrall, R.A.; Powell, M.F. ibid. 1987, 177.
25. a. Paneth, P.; O'Leary, M.H. J. Am. Chem. Soc. 1985, 107, 7381.
    b. Marlier, J.F.; O'Leary, M.H. ibid. 1984, 106, 5054.
26. a. Paneth, P.; O'Leary, M.H. Biochemistry 1987, 26, 1728.
    b. Paneth, P.; O'Leary, M.H. ibid. 1985, 24, 5143.
27. Simonsson, F.; Jonsson, B.-H.; Lindskog, S. Eur. J. Biochem. 1979,
    93, 409.
28. a. Silverman, D.N.; Tu, C.K. J. Am. Chem. Soc. 1975, 97, 2263.
    b. Tu, C.K.; Silverman, D.N. ibid. 1975, 97, 5935.
29. Silverman, D.N.; Tu, C.K.; Lindskog, S.; Wynns, G.C. ibid. 1979,
    101, 6734.

30. Kararli, T.; Silverman, D.N. Ann. N.Y. Acad. Sci. 1984, 429, 129.

31. Silverman, D.N.; Tu, C.K. Biochemistry 1986, 25, 8402.

32. Hassinoff, B.B. Arch. Biochem. Biophys. 1984, 233, 676 has shown that in the presence of a glycerol visogen, the activity of CA is reduced which suggests that diffusional processes contribute, at least in part, to determining the rate. This is contrary to the conclusions reached from the Carbon KIE [26a] studies which employed sucrose (0.5 M HEPES, pH 8.2, 25°C). However, the Hassinoff study employed buffers consisting of 1-2 mM phosphate (pH 7.70, 6.64 for CO_2 hydration and HCO_3^- dehydration respectively) which is lower than the [buffer] required to ensure that deprotonation of the enzyme is not rate limiting [17,18]. Thus, it seems likely that the results of the Hassinoff experiment was attributable to a viscosity dependence on the $HE-ZnOH^- + B: \rightarrow EZnOH^- + B-H^+$ as in section 3a. Similar viscosity effects on CO_2 hydration and HCO_3^- dehydration (at 20 mM buffer) were observed by Pocker et al and attributed to the k_{cat} term [17b].

33. a. Ghannam, A.F.; Tsen, W.; Rowlett, R.S. J. Biol. Chem. 1986, 261, 1164.
    b. Kogut, K.A.; Rowlett, R.S. ibid. 1987, 262, 16417.

34. a. Lindskog, S. J. Mol. Catal. 1984, 23, 357.
    b. Rowlett, R.S. J. Prot. Chem. 1984, 3, 369.

35. a. Brown, R.S.; Huguet, J.; Curtis, N.J. in "Metal Ions in Biological Systems", Sigel, H., Ed., Marcel Dekker, Inc.: New York, Vol. 15, 1983, 55.
    b. Brown, R.S. in "Carbon Dioxide as a Source of Carbon", Aresta, M.; Forti, G., Eds., D. Reidel: Dordrecht, NATO ASI Series C, 1987, 169.

36. Lindskog, S. Struct. Bonding (Berlin) 1970, 8, 153.

37. a. Bertini, I.; Canti, G.; Luchinat, C.; Mani, F. Inorg. Chem. Acta 1980, 46, L91.
    b. Benelli, C.; Bertini, I.; DiVaira, M.; Mani, F. Inorg. Chem. 1984, 23, 1422.

38. a. Huguet, J.; Brown, R.S. J. Am. Chem. Soc. 1980, 102, 7571.
    b. Brown, R.S.; Curtis, N.J.; Huguet, J. ibid. 1981, 103, 6953.
    c. Curtis, N.J.; Brown, R.S. J. Org. Chem. 1980, 45, 4038.

39. Read, R.J.; James, M.N.G. J. Am. Chem. Soc. 1981, 103, 6947.

40. Brown, R.S.; Salmon, D.; Curtis, N.J.; Kusuma, S. J. Am. Chem. Soc. 1982, 104, 3188.

41. Jacob, G.S.; Brown, R.D.; Koenig, S.H. Biochemistry 1980, 19, 3754.

42. a. Martin, R.B. Proc. Natl. Acad. Sci. USA 1974, 71, 4146.
    b. Sillen, L.G.; Martell, A.E. in "Stability Constants of Metal Ion Complexes", Chem. Society Special Publications #17, 1964 and supplement No. 1, 1971.

43. a. Wooley, P. J. Chem. Soc. Perkin II 1977, 318.
    b. Wooley, P. Nature (London) 1975, 258, 677.

44. Groves, J.T.; Olson, J.R. Inorg. Chem. 1985, 24, 2715.

45. Coates, J.H.; Gentle, G.J.; Lincoln, S.F. Nature (London) 1974, 249, 773.
46. Martin, R.B. J. Inorg. Nucl. Chem. 1976, 38, 511.
47. Tabushi, I.; Kuroda, Y. J. Am. Chem. Soc. 1984, 106, 4580.
48. a. Breslow, E. in "The Biochemistry of Copper", Peisach, J.; Aisen, D.; Blumberg, W.E., Eds., Academic Press: New York, 1966, 149.
    b. Tabushi, I.; Kuroda, Y.; Mochizuki, A. J. Am. Chem. Soc. 1980, 102, 1152.
49. Slebocka-Tilk, H.; Cocho, J.L.; Frakman, Z.; Brown, R.S. J. Am. Chem. Soc. 1984, 106, 2421.
50. Pullman, A. Ann. N.Y. Acad. Sci. 1981, 367, 340 and references therein.
51. a. Cook, C.M.; Allen, L.C. Ann. N.Y. Acad. Sci. 1984, 429, 84 and references therein.
    b. Allen, L.C. Ann. N.Y. Acad. Sci. 1981, 367, 383.
52. Liang, J.-Y.; Lipscomb, W.N. Biochemistry 1987, 26, 5293.
53. Pinsent, B.R.W.; Pearson, L.; Roughton, F.J.W. Trans. Faraday Soc. 1956, 52, 1512.
54. Jönsson, B.; Karlström, G.; Wennerström, H. J. Am. Chem. Soc. 1978, 100, 1658.
55. Cook, C.M.; Haycock, K.; Lee, R.H.; Allen, L.C. J. Phys. Chem. 1984, 88, 4875.
56. a. Pocker, Y.; Deits, T.L. J. Am. Chem. Soc. 1982, 104, 2424.
    b. Pocker, Y.; Davison, B.L.; Deits, T.L. ibid. 1978, 100, 3564.
    c. Pocker, Y.; Deits, T.L.. ibid. 1981, 103, 3949; 1983, 105, 980.
    d. idem. Ann. N.Y. Acad. Sci. 1983, 429, 76.
57. Lindskog, S.; Engberg, P.; Forsman, C.; Ibrahim, S.A.; Jonsson, B.-H.; Simonsson, I.; Tibell, L. Ann. N.Y. Acad. Sci. 1984, 429, 61.

THE MECHANISM OF ACTION OF CARBONIC ANHYDRASE

L. BANCI,[1] I. BERTINI,[1*] C. LUCHINAT[2] and J.M. MORATAL[3]
[1]Department of Chemistry, University of Florence, Italy
[2]Institute of Agricultural Chemistry, University of Bologna, Italy
[3]Department of Inorganic Chemistry, University of Valencia, Spain.

ABSTRACT. A survey of the structure—function relationship in the enzyme carbonic anhydrase is presented. The coordination number around the metal ion in derivatives inhibited with some anions is stressed as well as the role of the cavity in determining the coordination number. Inferences on the enzyme—substrate interactions are presented.

1. Introduction

A simple reaction in which CO_2 can participate is that with H_2O:

$$CO_2 + H_2O \rightleftharpoons HCO_3^- + H^+$$

The reaction transforms a gas into hydrated ions and viceversa. The reaction occurs in most living organisms and is fundamental to animal breathing and plant photosynthesis [1]. By itself, the reaction at physiological pH is kinetically slow, well below the needed rates ($k = 3.5 \times 10^{-2}$ s^{-1}) [2]. Even if it were not so, Nature would have had control of the reaction through a catalyst. So, the enzyme carbonic anhydrase (CA) is used which in some cases is capable of increasing the reaction rate well above the H^+ diffusion limit, which at pH 7 is 10^4 s^{-1} [3,4].

The isoenzymes of CA are numerous depending on the source [3,5–8]. They are distinguished by using roman numerals. CA II isoenzymes display a k_{cat} of 1×10^6 s^{-1}, CA I of 1.5×10^5 s^{-1}, and CA III of 3×10^3 s^{-1} at pH = 7 [9,10]. They differ also in the profile of the activity versus pH [11] (Figure 1). CA II has a pK_a around 6, CA I around 7 and CA III around 5. It was shown that such profiles depend on more than one acidic group and therefore that there are more than one pK_a's [12,13].

The function of the enzyme is that of providing $Zn-OH$ moieties at physiological pH values. The coordinated OH group performs the nucleophilic attack. Indeed, the uncatalyzed reaction

$$CO_2 + OH^- \rightleftharpoons HCO_3^-$$

is fast ($k = 8.5 \times 10^3$ M^{-1} s^{-1}) [14]. Above pH 9, this reaction is faster than that of hydration.

The X—ray structure has been recently refined for HCA II and HCA I [15,18], where H stays for human. The 260— or 261—aminoacids (for the human I and the human II isoenzymes respectively) single chain protein has a rugby ball shape

181

M. Aresta and J. V. Schloss (eds.),
Enzymatic and Model Carboxylation and Reduction Reactions for Carbon Dioxide Utilization, 181–197.
© 1990 *Kluwer Academic Publishers.*

(Figure 2) with a conical cavity, about 15 Å deep, at the bottom of which the zinc ion sits. The cavity is half hydrophilic and half hydrophobic (Figure 3). The zinc ion is coordinated to three histidines as shown in Figure 4. The fourth ligand to zinc is

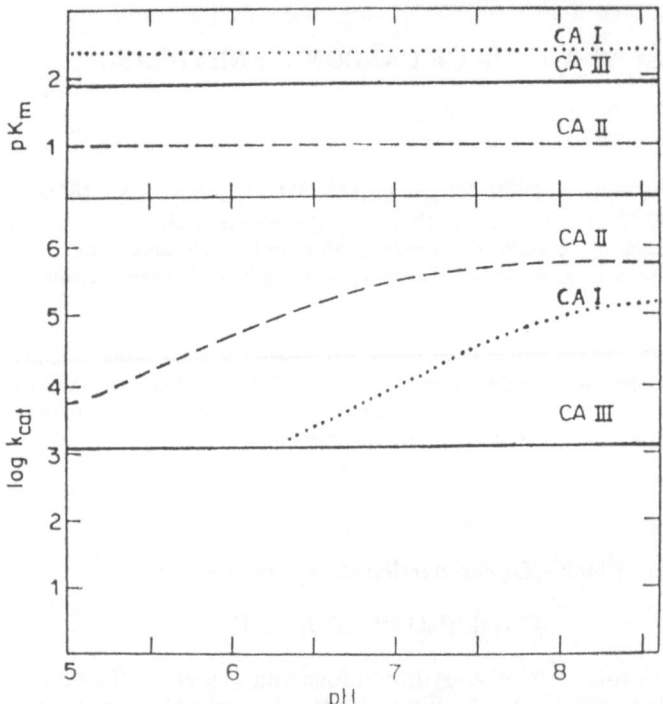

Figure 1. pH dependence of k_{cat} and K_m values for CO_2 hydration catalyzed by carbonic anhydrase I, II, III isoenzymes [Redrawn from refs. 3,10].

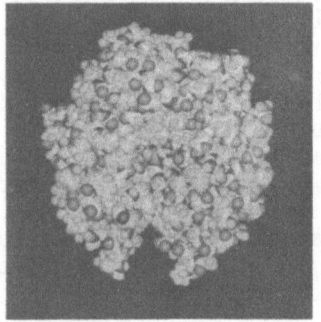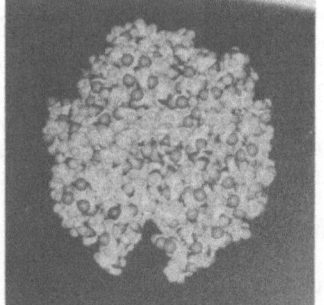

Figure 2. Stereo view of human carbonic anhydrase II, represented as CPK model. The white spheres are carbon atoms; the light grey spheres are hydrogen atoms and the dark grey spheres are either nitrogen or oxygen atoms. The entrance of the active cavity is oriented toward the bottom of the picture. The large dark sphere at the center of the molecule is the zinc ion [16].

either a water molecule or an OH group, depending on pH. They form a hydrogen bond with OH of Thr–199. It is possible that at low pH values two water molecules are coordinated to zinc.

2. About the Zn–OH Moiety

When water coordinates to a metal ion, its acidity increases. The pK_a of the reaction

$$\text{\textbackslash}M-OH_2 \rightleftharpoons \text{\textbackslash}M-OH^- + H^+$$

varies with the charge of the metal ion, the coordination number and the nature of the donor atoms. It can be as low as 2 in hexaaqua metal(III) complexes or 7–9 as in

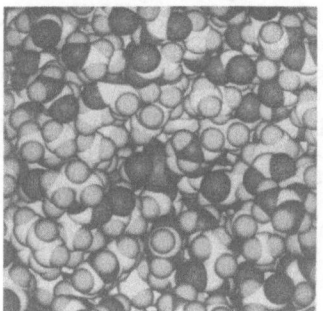 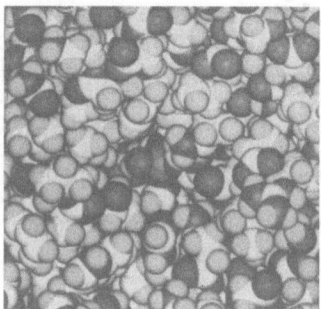

Figure 3.　　Stereo view of the active cavity of human carbonic anhydrase II represented as CPK model. The large sphere at the bottom of the central cavity is the zinc ion with the OH ligand on the left. The upper wall of the cavity is hydrophilic. The right and lower walls of the cavity are hydrophobic [16].
either a water molecule or an OH group, depending on pH. They form a hydrogen bond with OH of Thr–199. It is possible that at low pH values two water molecules are coordinated to zinc.

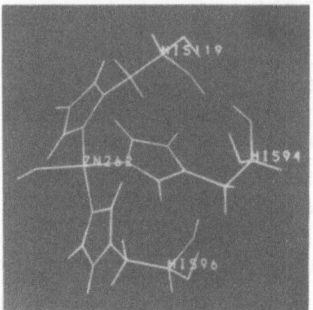 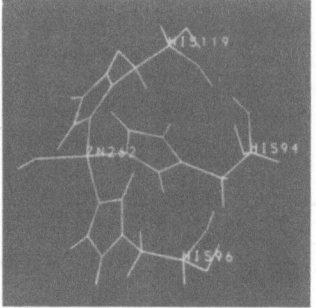

Figure 4.　　Stereo view of the donor groups bound to the zinc ion in the active site of carbonic anhydrase (The numbering is that for HCA II) [16].

184

many zinc and cobalt(II) complexes containing a water molecule. This point has largely been addressed by inorganic chemists who have synthesized many models for this acid–base equilibrium [19–29]. A theoretical study has shown that the number and charge of the donor atoms is capable of affecting the pK_a of the coordinated water by a few units [30]. The lowest pK_a is expected for a tetrahedral complex having three neutral donor groups. However, it is now becoming apparent that the driving force for the water dissociation is the presence of positively charged groups inside the cavity. For example CA III has Arg and Lys groups in the cavity and therefore the Zn–OH form is present at relatively low pH. CA I and II have at least one His in the cavity which becomes positive with a pK_a around 6.5.

Once the existence of the Zn–OH group at physiological pH is accepted, some considerations are needed about the nucleophilicity of the Zn–OH group. The nucleophilicity is a kinetic parameter relative to the reaction of the nucleophilic attack. It is generally considered that the nucleophilicity of a group is associated to its negative charge and to the energy of the highest occupied molecular orbital (HOMO). Indeed the larger the negative charge, the larger is the rate of the attack. On the other hand, the higher the energy of the HOMO, the closer to the energy of the low unoccupied molecular orbital (LUMO) of the substrate and the larger the rate of the attack [30]. In Figure 5 the charge on oxygen is plotted against the energy of the HOMO of the Zn–OH group. It appears that OH⁻ is the best nucleophile followed by the Zn–OH moiety independently of the charge and coordination number of the chromophore. In particular the charge on oxygen does

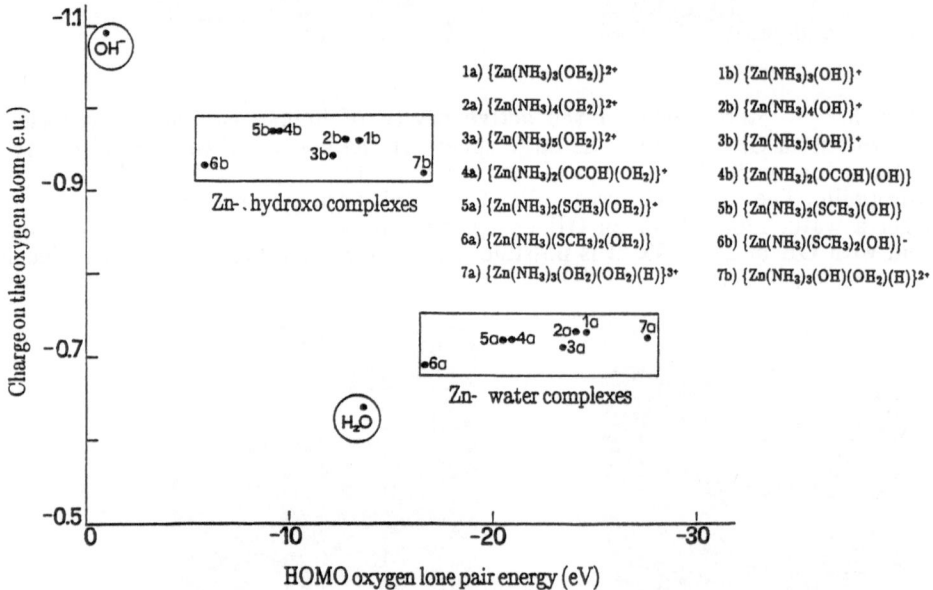

Figure 5.   Charges on the oxygen atom against the HOMO oxygen lone pair energy for a series of Zn–OH$_2$ and Zn–OH complexes. The nucleophilicity of the oxygen is expected to increase from right to left (increasing HOMO energy) as well as from bottom to top (increasing negative charge) [Redrawn from ref. 30].

not vary appreciably among the many OH containing moieties.

3. What Attracts CO_2 Inside the Cavity

Hydrogencarbonate is attracted by the zinc(II) ion, possibly helped by the hydrophilic part of the cavity. The dissociation constant for HCO_3^- is weaker than the steady state K_m value, indicating that the latter is not a true dissociation constant. CO_2 is possibly attracted by the hydrophobic part of the cavity. Indeed CO_2 is more soluble in organic solvents than in water.

^{13}C NMR T_1 measurements have been performed on the $CO_2 - HCO_3^-$ system in the presence of copper(II)–substituted CA [31–33]. Copper is needed because it provides an inactive derivative so that CO_2 and HCO_3^- exchange slowly on the NMR time scale and because the unpaired electron of copper(II) may allow the investigation of copper–carbon interaction.

In Figure 6 the T_1^{-1} values as a function of CO_2 concentration are shown. It appears that the carbon nucleus of CO_2 senses the unpaired electron of copper, for example, much more than in the presence of SOD, which also contains copper. However no saturation of the binding site(s) is observed up to a CO_2 concentration of 1 M. If saturation occurred, a sigmoidal decrease of T_1^{-1} would have been expected. The data could be analyzed in terms either of a single CO_2 molecule directly coordinated to the metal ion with an upper limit for the affinity constant of 0.1 M^{-1} or of several molecules concentrated within a few Ångstroms from the metal

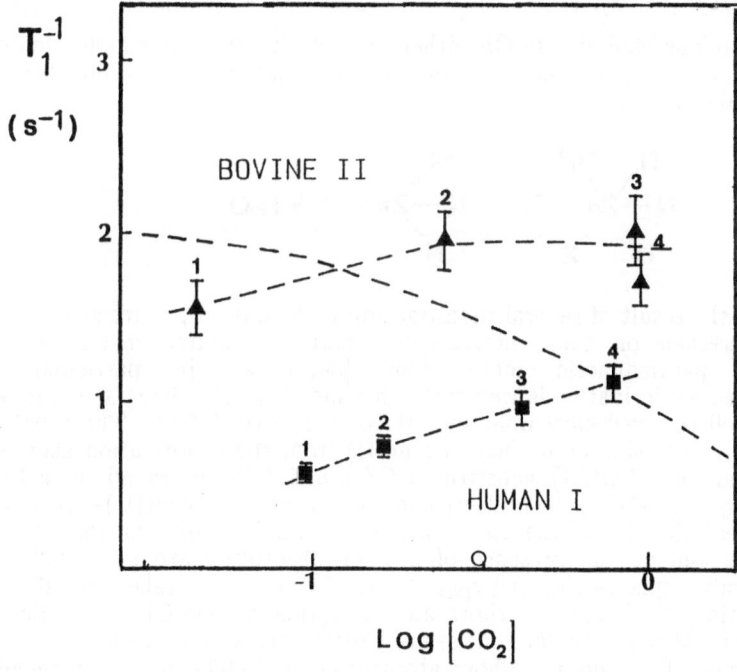

Figure 6.   20 MHz, 20⁰ C T_1^{-1} values of the ^{13}C nucleus of CO_2 in the presence of CuBCA II (▲) and CuHCA I (■) as a function of CO_2 concentration.

[33]. In contrast with CO_2, the nuclear relaxation rate values of hydrogencarbonate are consistent with a direct binding to the metal ion with an affinity constant of 3.5×10^{-3} M^{-1} [31,33]. Actually the linewidth is determined by the exchange rate of free and bound HCO_3^-, which for HCA II and BCA II at pH 7 and 25^0 is 4.3×10^2 and 5.0×10^3 s^{-1}, respectively [33].

Since $K_m = 0.012$ M [34], the inability to saturate the binding site(s) of CO_2 indicates that K_m for CO_2 hydration is not a thermodynamic dissociation constant. These data point at an increase in solubility of CO_2 in the hydrophobic region of the cavity.

The question remains of whether a direct CO_2 oxygen–zinc bond takes place. The idea would be tempting, because it would provide another role for the Zn ion, i.e. to activate CO_2 for the nucleophilic attack by OH^- and to allow a concerted attack. The model complexes described during this symposium do not provide any evidence for this kind of interaction [35].

Ab initio calculation [36] and molecular dynamics studies [37] show that in the gas phase CO_2 does indeed interact with a ZnN_3O moiety giving rise to a five coordinated derivative. If such derivative exists in this enzyme, it is a transition state of the reaction pathway. As observed in the analysis of the ^{13}C T_1^{-1} data on CO_2 interaction with the copper(II) ion, this possible intermediate would have an affinity constant $< 10^{-1}$ M^{-1}, i.e. it would be quite unstable. However this step is not essential, as Zn–OH might bind CO_2 in the gas phase without energy barrier.

4. Are There Other Evidences About The Role of the Cavity?

It is known that anions bind zinc in CA either by replacing the water molecule or by expanding the coordination number [38–40]. In many cases an equilibrium between the two species takes place.

$$
\begin{array}{c}
\text{N}\diagdown\;\;\text{OH}_2 \\
\text{N}-\text{Zn} \\
\diagup\;\;\diagdown \\
\text{N}\;\;\;\;\text{X}
\end{array}
\rightleftharpoons
\begin{array}{c}
\text{N}\diagdown \\
\text{N}-\text{Zn}-\text{X}\;+\;\text{H}_2\text{O} \\
\diagup \\
\text{N}
\end{array}
$$

This finding was the result of several investigation techniques. Spectroscopic studies have been undertaken on some derivatives where the native metal ion was substituted by paramagnetic metal ions [38,41–52]. In particular the cobalt(II)–substituted derivative is well suited for such investigations, as it provides indicative and well resolved electronic and ^1H NMR spectra [53,54]. The cobalt(II) ion has a behavior very similar to that of zinc(II) from the coordination chemistry point of view, and the cobalt(II)–substituted CA's maintain almost all the activity of the native enzymes [55–57]. The electronic spectra of the cobalt(II)–substituted CA were taken as indicative of the coordination number depending on the intensity of the absorption and on the presence of a weak absorption around 14,000 cm^{-1} [40,58]. In particular, spectra showing $\epsilon_{max} < 150$ M^{-1} cm^{-1} were taken as indicative of five coordination. Such spectra show an absorption at $\simeq 14,000$ cm^{-1}. Spectra showing $\epsilon_{max} > 300$ $M^{-1}cm^{-1}$ (as for the case of CoBCA II) were taken as due to four coordinated species. They do not show absorption at 14,000 cm^{-1}. Intermediate spectra are indicative of equilibria between four– and five–coordinated species [38–40, 60]. Of course the intensity of the spectra by itself cannot be an absolute diagnostic criterion, except in the extreme cases, because the intensity depends also

187

on the presence of charge transfer transitions close in energy. It was observed that five coordinated species give rise to relatively sharp ^{1}H NMR spectra of the cobalt(II) derivatives [54]. Five coordinated cobalt(II) derivatives have several low lying energy levels which allow electron relaxation to be fast [61]. Under such conditions, the ^{1}H NMR lines are sharp. A measure of this effect is provided by the nuclear T_1 values. As a consequence of the presence of low lying energy levels the magnetic anisotropy is large, and this causes unequivalence among similar protons in the ^{1}H NMR spectra as well as further shifts of the protons which sense the magnetic susceptibility anisotropy of the cobalt(II) ion. In other words, protons within 7–8 Å from the metal ion, even if belonging to groups not directly coordinated to the metal ion, experience some shift outside the diamagnetic envelope of the spectrum. So, some criteria are available for guessing the coordination number, which can be combined with others, based on magnetic circular dichroism (MCD) measurements and on the rates of electronic relaxation [60].

Despite many information can be obtained on the coordination number, the factors governing the type of coordination were not known. With the help of a recent X–ray report [62], we try now to account for the resulting coordination number and the geometry on the basis of the properties of the active site cavity. The X–ray structure of the NCS⁻ derivative is today available [62]. As we had proposed [41], the compound is five coordinated. With respect to the uninhibited structure, the water molecule has moved slightly from its original position (termed A site) to accommodate the extra ligand, though it seems to maintain the H–bond with the OH group of Thr–199. This new binding site for the OH group is called C site. NCS⁻ binds in the hydrophobic pocket (B site). The terminal sulphur atom gives rise to contacts with Val–143, Leu–198, and Trp–209 (Figure 7).

We believe that these hydrophobic interactions are those which determine the type of coordination chemistry. Indeed, we have proposed that NCO⁻ gives rise to a tetrahedral derivative. We would like to substantiate this important point. The electronic spectra (Figure 8) are quite different in shape and intensity [60]. It would be unexpected if changing the terminal atom did not alter the coordination geometry around cobalt. The ^{1}H NMR spectra are also different (Figure 9) in that the nuclear

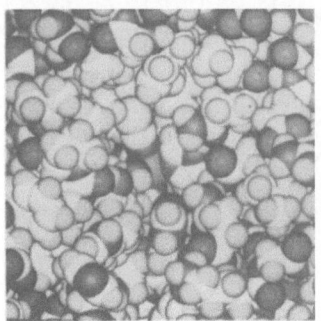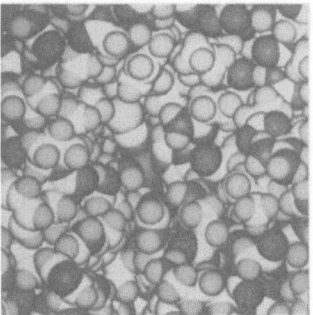

Figure 7. Stereo view of the active cavity of the NCS⁻ adduct of human carbonic anhydrase II viewed as CPK model [62]. The NCS⁻ ion binds at the B binding site; from comparison with Figure 3, it appears that this site is more buried than the binding site of the OH⁻ ion in the active form (A site). The water molecule in the NCS⁻ adduct occupies the C site, which is pointing more towards the entrance of the cavity.

T_1's are larger for the NCS⁻ derivative (Table I), the signals are more spread, and the signals close to the diamagnetic positions are spread over a larger range (Figure 10). So, there is no doubt that the NCO⁻ derivative is different from that with NCS⁻. If we try to accommodate NCO⁻ in the site where OH is present in the uninhibited derivative (site A) we may see that NCO⁻ can interact through H bond with the NH of Thr–199, the NH of Thr–200 and possibly with the OH of Thr–199 (Figure 11).

The coordination number in model compounds is a fine balance among many factors. In proteins the interactions with the cavity have to be taken into account. The change from the soft sulphur of NCS⁻ to the hard oxygen of NCO⁻ is possibly shifting the coordination number from four to five. Sulphur provides energy by interaction with the hydrophobic pocket of the cavity. In the case of NCO⁻, the hydrophilic interactions prevail and it binds in the C site. Interestingly, N_3^- has an intermediate behavior, which brings to some sort of equilibria between the two species. In fact the electronic spectrum (Figure 8) has just an intermediate absorption and the ¹H NMR spectra show a spreading of the signals intermediate between that of the NCO⁻ and NCS⁻ derivatives. In addition, the T_1 values are a

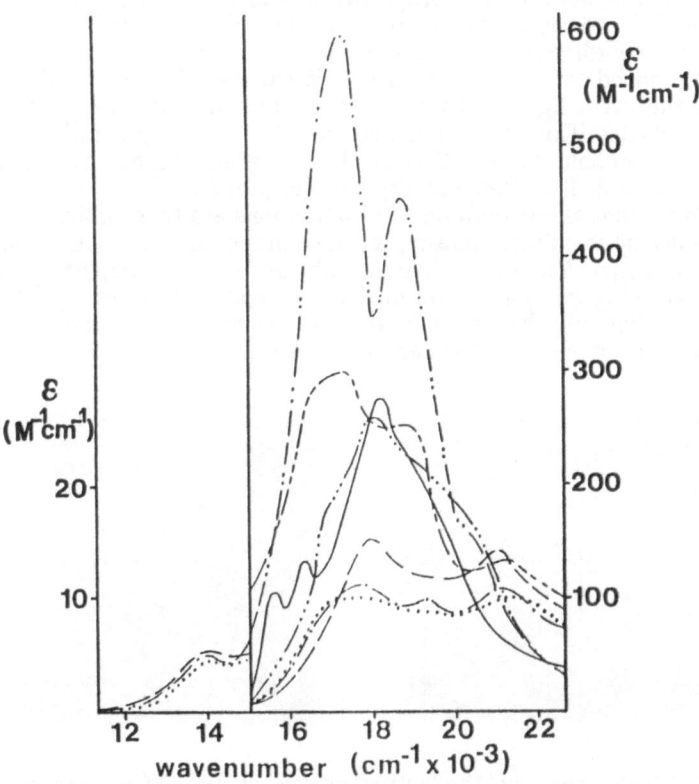

Figure 8.   Electronic spectra of unbuffered solutions of some anion adducts of cobalt(II)–substituted BCA II at pH 6.2–6.4 (——— native, ···· + NCS⁻, —·—·— + CH₃COO⁻, – – – + NO₃⁻, — ··· — + ClO₄⁻, – – – – – + N₃⁻, –– · · –– + NCO⁻) [Redrawn from ref. 60].

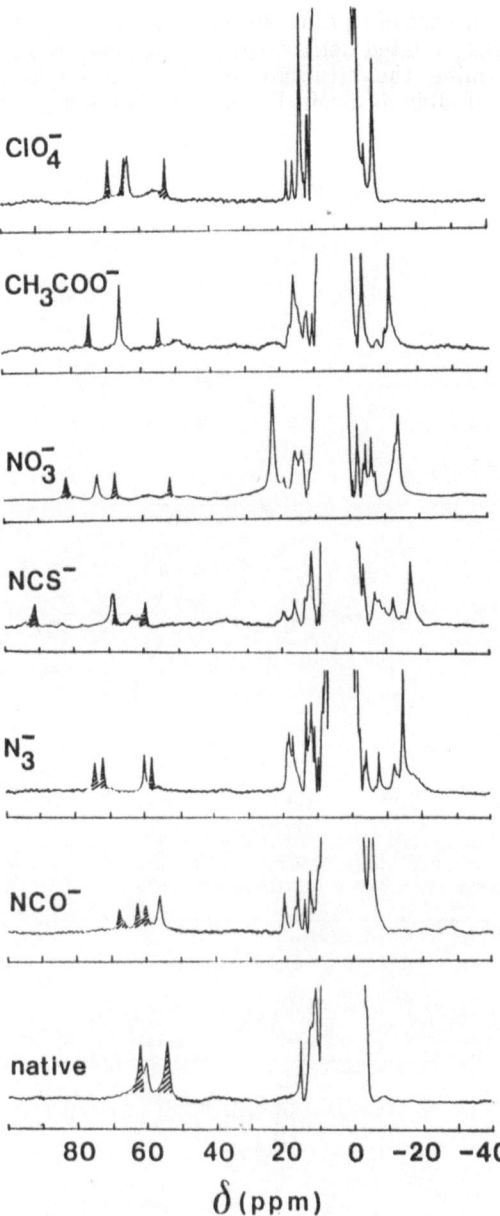

Figure 9.   200 MHz ^{1}H NMR spectra at 300 K of some anion adducts of cobalt(II)-substituted BCA II in HEPES (2[4–(2-hydroxyethyl)–1-piperazinyl] ethanesulfonic acid) buffer at pH 6.2–6.4. The dashed signals disappear when the spectra are recorded in D_2O [Redrawn from ref. 60].

further support for the presence of equilibrium between four and five coordination.

This series of closely related derivatives is a clear example of the potentiality of the cavity to determine the structure of the coordination environment of cobalt(II) and zinc(II). Iodide is a soft ligand. One would therefore expect five

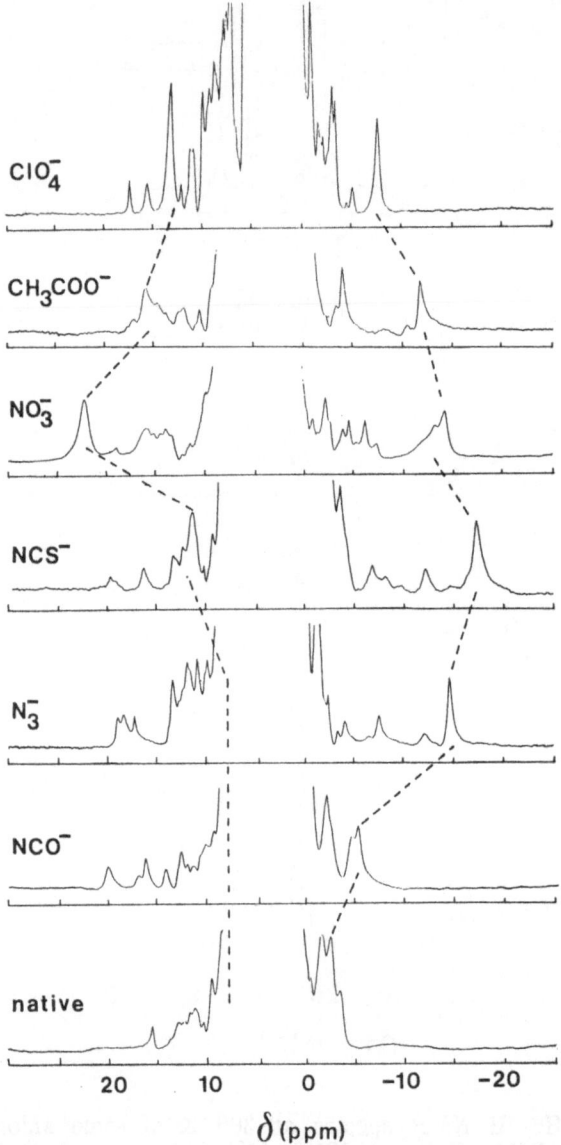

Figure 10. Expanded "quasi–diamagnetic" region of the 200 MHz ^{1}H NMR spectra of some anions derivatives of cobalt(II)–substituted BCA II. Experimental conditions are the same as in Figure 9. [Redrawn from ref. 60].

coordination as indeed happens. Iodide binds in the B binding site as shown in Figure 12 [63].

On the other hand, cyanide, OH⁻ and SH⁻ give rise to four coordinated derivatives. They all interact with the hydrophilic part of the cavity (Figure 13). Indeed, the terminal N atom of the CN⁻ group can be accommodated at 2.8 A from the nitrogen of the NH of Thr—199, thus providing the possibility of a strong H—bond. The electronic spectra are very high in intensity and are similar to those of

Table 1. 200 MHz ^{1}H NMR Shifts (ppm) and T_1 Values (ms) for Some Anion Derivatives of Bovine Cobalt(II)—Substituted Carbonic Anhydrase[a]

Signal	no anion	+ CH₃COO⁻	+ NO₃⁻	+ ClO₄⁻	+ NCS⁻	+ NCO⁻	+ N₃⁻
His NH	62.1 (7.3)	66.0 (20.3)	6.2 (17)	69.0 (9.5)	68.7 (18)	62.1 (3.1)	57.8 (6.2)
His—119 m—H	60.3 (7.3)	66.0 (20.3)	70.1 (12)	63.4 (5.7)	69.0 (18)	55.5 (3.6)	59.9 (7.8)
His—119 NH	54.2 (8.1)	75.0 (19.5)	78.7 (12)	64.3 (5.4)	91.2 (14)	59.8 (2.5)	74.1 (5.5)
His NH	54.2 (8.1)	55.0 (22.5)	51.1 (27)	52.4 (14.6)	59.5 (25)	67.3 (2.2)	71.7 (5.4)
downfield CH₃	<6	15.7 (24)	20.1 (19)	13.3 (10)	11.4	<10	<10
upfield CH₃	≈ −3	−11.8 (40)	−13.1 (28)	−7.7 (20)	−17.3 (44)	−5.6 (7)	−14.9 (17)

[a]Only the meta—like signals for the histidine rings are reported.

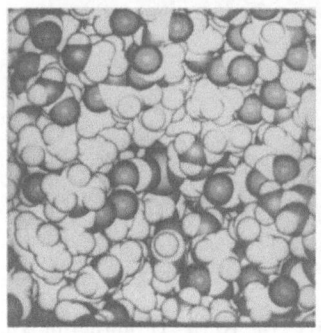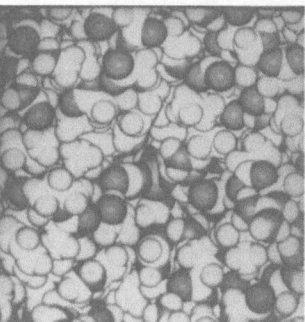

Figure 11. Stereoview of the active cavity of HCA II showing the proposed metal binding mode of NCO⁻. The anion is bound to the central zinc ion and is interacting, through the terminal oxygen atom, with the OH of Thr—199 and, behind this, with the NH of the same residue.

the NCO⁻ adduct (Figure 8). All of these ligands are very strong and it is reasonable to consider that strong ligands favor lower coordination numbers.

Finally sulfonamidates and acetazolamidates bind as anions and the NH donor group behaves like OH⁻ and SH⁻ (Figure 14, [62]), giving rise to H–bond with the OH of Thr–199. By looking at the structures of sulfonamidates derivatives which have been recently made available it appears that the SO_2 oxygens do not bind the metal ion [62]. The closest Zn–O distance is 3.1 Å. In an early report, sulfonamidates were proposed to bind in a bidentate fashion. Our laboratory kept proposing the monodentate behavior of these ligands based on spectroscopic criteria [40,41]. We know now that one of the oxygen atoms of SO_2 is H–bonded with the NH of Thr–199, while the other oxygen atom is interacting with Val–121 and

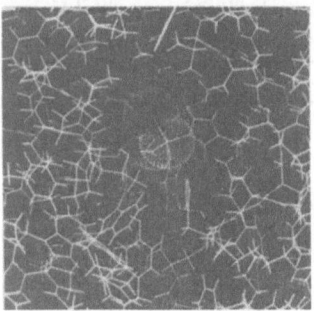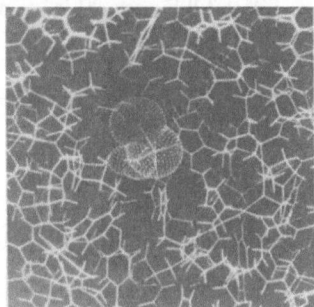

Figure 12.   Stereoview of the active cavity of HCA II showing the proposed metal binding mode of I⁻. The central zinc ion is surrounded by a dotted spherical surface at the Van der Waals radius. The volume occupied by the I⁻ and the OH_2 ligands are also shown. The I⁻ ligand is located in the hydrophobic pocket formed by Val–121, Val–143, Leu–198 and Trp–209. The OH_2 ligand is H–bonded to the OH group of Thr–199.

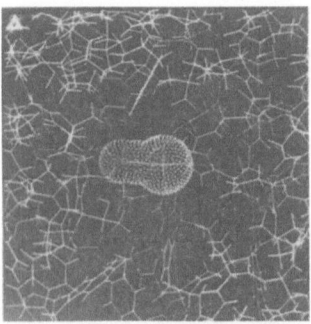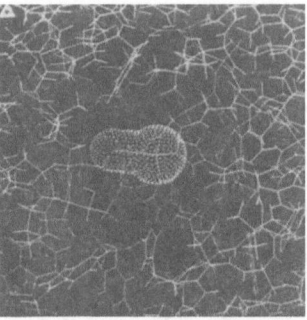

Figure 13.   Stereoview of the active cavity of HCA II showing the proposed binding mode of CN⁻. The central zinc ion is surrounded by a dotted spherical surface at the Van der Waals radius. The volume occupied by the CN⁻ ligand is also shown according to the carbon and nitrogen radii. The CN⁻ nitrogen is H–bonded to the NH proton of Thr–199, whose OH group (behind the CN⁻ ion) could also be H–bonded to the CN⁻ nitrogen.

Val—143, in the hydrophobic cavity. Even hard atoms may contribute to the overall stability of the adduct through hydrophobic interactions when they are prevented to form hydrophilic interactions.

5. The Case of Nitrate, Acetate and Perchlorate

Nitrate, perchlorate and acetate bind the metal ion in CA with an affinity higher than they do with free zinc ion. In model complexes they may bind in a monodentate or bidentate fashion. The spectroscopic criteria indicate that nitrate and acetate give rise to essentially pure five coordinated species. The electronic spectra are similar to that of NCS$^-$ (Figure 8). The NMR T_1 values (Table 1) and the spreading of the NMR signals around the diamagnetic region are indicative of five coordination (Figure 10).

Five coordination can be reached as NCS$^-$ does, i.e. with a water molecule as fifth ligand, or through a bidentate behavior centered at the A site. This matter is important in order to understand the way of interaction of the natural substrate HCO$_3^-$.

We have measured the water ^{1}H T_1^{-1} values as a function of magnetic field

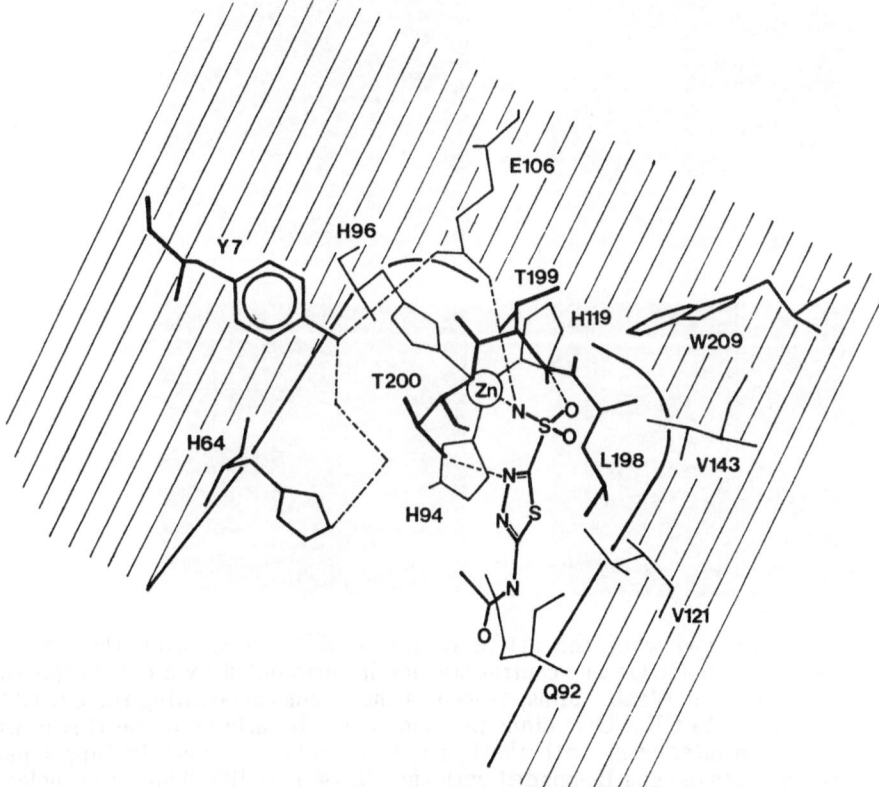

Figure 14.   Schematic drawing of the Diamox (acetazolamide) adduct of HCA II. Diamox interacts with Val—121, Val—143, Leu—198 and Gln—92.

(Nuclear Magnetic Relaxation Dispersion, NMRD). We have interpreted the data in terms of a water molecule bound to cobalt simultaneously with NO_3^- [53]. Furthermore the similarity the electronic spectra with those of the CH_3COO^- and NCS^- adducts (Figure 8) favors the idea of the same kind of coordination geometry.

Let us now analyze the acetate derivative with computer graphics pictures in order to understand the problem better. Acetate fits in the B site with the CH_3 group pointing towards Val–121 and an oxygen atom of the carboxylic group hydrogen bonded with the NH of Thr–199. (Figure 15). Water would be in the C site. A bidentate behavior of acetate (Figure 15) would probably cause the CH_3 group pointing towards the hydrophilic side of the cavity. It is reasonable to propose that this is not the case. Nitrate adduct has similar NMR spectra (Figures 9 and 10). This is a good indication of the similarity of the adducts. On the basis of the spectroscopic data, that indicate five coordination, and on the fact that NMRD measurements indicate the presence of a bound water molecule, we can propose a monodentate behavior for nitrate binding. Perchlorate has an intermediate behavior: both electronic and NMR spectra indicate a coordination equilibrium between four and five coordination.

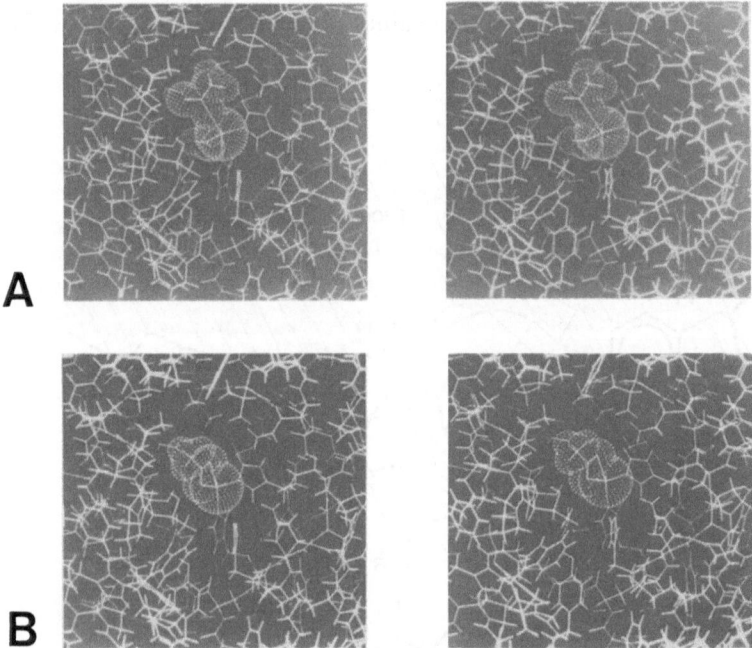

A

B

Figure 15.   Stereoview of the active cavity of HCA II showing the proposed binding modes of CH_3COO^-. The central zinc ion is surrounded by a dotted spherical surface at the Van der Waals radius, as well as the atoms constituting the CH_3COO^- and OH_2 ligands. (A) CH_3COO^- binds in a monodentate fashion at the B site, with the methyl group interacting with the hydrophobic side of the cavity (upper part) and the carboxylate oxygen H–bonded with the NH of Thr–199. The water molecule is in the C site, H–bonded with the OH of Thr–199 and possibly of Thr–200. (B) CH_3COO^- binds in a bidentate fashion. The CH_3 group is pointing towards the hydrophilic side of the cavity.

6. About the Catalytic Mechanism

CO_2 is attracted into the cavity by hydrophobic interactions, and the nucleophilic attack is performed by the Zn–bound OH moiety, providing a transition state with a bidentate bicarbonate anion. As already discussed, no evidences are reported for a specific binding site for CO_2, and indeed it is not necessary to the catalytic cycle. After the attack, the transition state evolves toward the detachment of the Zn–bound OH moiety of bicarbonate.

We propose that $HCO_3{}^-$ undergoes an equilibrium between four and five coordination.

When it is five coordinated it may behave like nitrate and acetate. The five coordination is achieved by the binding of a water molecule that helps in the release of the reaction product.

7. Perspectives

Alteration of the hydrophilic or hydrophobic properties of the cavity is expected to dramatically affect the catalytic mechanism of the enzyme. Two laboratories (Dept. of Biochemistry, Umea University, Sweden and Dept. of Chemistry, Duke University, NC, USA) have expressed the enzyme in E. Coli and started to modify specific residues by SDM. The changes in the structure and, therefore, in the interatomic distances, induced by the residues substitution, could be detected through X–ray investigations and ¹H NMR. Attempts are pursued in this laboratory to obtain interproton distances through NOE experiments on the ¹H NMR spectra.

Ab initio calculations at various degrees of sophistication should, in principle, answer the question of whether CO_2 interacts with the zinc ion. Molecular dynamics calculations, once the charge distribution in the active site is calculated and the potential parameters for zinc are evaluated, should provide a more quantitative picture of the interactions between inhibitors and the cavity. This is just a beautiful example of enzymology applied to a simple reaction !

196

8. References

1) *'Carbon Dioxide as a Source of Carbon'*, Aresta, M. and Forti, G., Eds., D. Reidel Publishing Company, Dordrecht, (1987).
2) Faurholt, C., *J. Chem. Phys.*, **21**, 400 (1924).
3) *'Biology and Chemistry of the Carbonic Anhydrases'*, Ann. N.Y. Acad. Sci. USA, Vol.429, Tashian, R.E , Hewett–Emmett, D., Eds., New York, (1984).
4) Silverman, D. N. and Lindskog, S., *Acc. Chem. Res.*, **21**, 30 (1988).
5) Lindskog, S., in *Zinc Enzymes, Metal ions in Biology*, Vol. 5, Spiro, T. G., Ed., Wiley–Interscience, New York, p.78, (1983).
6) Tu, C. K., Sanyal, G., Wynns, G. C., and Silverman , D. N., *J. Biol. Chem.*, **258**, 8867, (1983).
7) Kararli, T., and Silverman, D. N., *J. Biol. Chem.*, **260**, 3484 (1985).
8) Engberg, P., Millqvist, E., Pohl, G., and Lindskog, S., *Arch. Biochem. Biophys.*, **241**, 628 (1985).
9) Khalifah, R. G., *J. Biol. Chem.*, **246**, 2561 (1971).
10) Lindskog, S., in *Zinc Enzymes*, Bertini, I., Luchinat, C., Maret, W., and Zeppezauer, M., Eds., Birkhauser, Boston, p.307 (1986).
11) Lindskog, S., Henderson, L.E., Kannan, K.K., Liljas, A., Nyman, P.O., and Strandberg, B., in *The Enzymes*, Vol. 5, 3rd ed., Boyer, P.D., Ed., Academic, New York, p.587 (1971).
12) Simonsson, I., and Lindskog, S., *Eur. J. Biochem.*, **123**, 29 (1982).
13) Pocker, Y., and Miao, C.H., *Biochem.*, **26**, 8481 (1987).
14) Edsall, J.T. and Wyman, J., *Biophys. Chem.*, Acad. Press, New York, 1, 550 (1958).
15) Eriksson, E. A., Jones, T. A. and Liljas, A., in *Zinc Enzymes*, Bertini, I., Luchinat, C., Maret, W., and Zeppezauer, M., Eds., Birkhauser, Boston, p.317 (1986).
16) Eriksson, E. A., Jones, T.A., and Liljas, A., *Proteins*, **4**, 274, (1989).
17) Liljas, A., Kannan, K.K., Bergsten, P.–C., Waara, I., Fridborg, K., Strandberg, B., Carlbom, U., Jarup, L., Lovgren, S., Petef, M., *Nature*, **235**, 131 (1972) .
18) Kannan, K.K.,and Ramanadham, M., in *Biology and Chemistry of the Carbonic Anhydrases*, Ann. N. Y. Acad. Sci. USA, Vol. 429, Tashian, R.E , Hewett–Emmett, D., Eds., New York, p. 49 (1984).
19) Sillen, L.G., and Martell, A.E., *Stability Constants of Metal–Ion Complexes*, Special Public. No 25, The Chemical Society, London, 1971.
20) Billo, E. J., *Inorg. Nucl. Chem. Lett.*, **11**, 491 (1975).
21) Dei, A., Paoletti, P., and Vacca, A., *Inorg. Chem.*, **7**, 865 (1968).
22) Bertini, I., Canti, G., Luchinat, C., and Mani, F., *Inorg. Chem.*, **20**, 1670 (1981).
23) Burki, S., Thesis, University of Basel, Basel, Switzerland (1977).
24) Wolley, P., *Nature*, **258**, 677 (1975).
25) Groves, T.J., and Olson, J.R., *Inorg. Chem.*, **24**, 2715 (1985).
26) Groves, J.T., Chambers, R.R., Jr., *J. Am. Chem. Soc.*, **106**, 630 (1984).
27) Bertini, I., Luchinat, C., and Messori, L., *Inorg. Chem.*, **21**, 3426 (1982).
28) Zompa, L.J., *Inorg. Chem.*, **17**, 2531 (1978).
29) Kimura, E., Koike,T., Toriumi, K., *Inorg. Chem.*, **27**, 3687 (1988)
30) Bertini, I., Luchinat, C., Rosi, M., Sgamellotti, A., and Tarantelli, F., submitted for publication.
31) Bertini, I, Borghi, E., and Luchinat, C., *J. Am. Chem. Soc.*, **101**, 7069 (1979).
32) Bertini, I., Borghi, E., Canti, G., and Luchinat, C., *J. Inorg. Biochem.*, **18**,

221 (1983).

33) Bertini, I., Luchinat, C., Monnanni, R, Roelens, S., and Moratal, J.M., *J. Am. Chem. Soc.*, **109**, 7855 (1987).

34) Khalifah, R.J., *J. Biol. Chem.*, **8**, 2561 (1971).

35) Brown, R.S., (this volume).

36) Merz, K.M., Jr., Hoffmann, R., and Dewar, M.J.S., *J. Am. Chem. Soc.*, **111**, 5636 (1989).

37) Merz, K.M., Jr., and Kollman, P.A., *J. Am. Chem. Soc.*, **111**, 5649 (1989).

38) Bertini, I., Luchinat, C., and Scozzafava, A., *Struct. Bonding*, **48**, 45 (1982).

39) Lindskog, S., *Adv. Inorg. Biochem.*, **4**, 115 (1982)

40) Bertini, I.,and Luchinat, C., *Acc. Chem. Res.*, **16**, 272 (1983).

41) Bertini, I., Canti, G., Luchinat, C., Scozzafava, A., *J. Am. Chem. Soc.* **100**, 4873 (1978).

42) Bertini, I., Luchinat, C., and Monnanni, R., in Ref. 1.

43) Bertini, I., Luchinat, C., and Viezzoli, M.S., in *Zinc Enzymes*, Bertini, I., Luchinat, C., Maret, W., and Zeppezauer, M., Eds., Birkhauser, Boston, p. 27 (1986).

44) Bertini, I., and Luchinat, C., in *Metal Ions in Biological Systems*, Sigel H., Ed., M. Dekker Inc., New York, Vol 15, p.101 (1983).

45) Lanir, A., Navon, G., *Biochemistry*, **11**, 3536 (1972).

46) Bertini, I., Borghi, E., Luchinat, C., and Monnanni, R., *Inorg. Chim. Acta*, **67**, 99 (1982).

47) Led, J.J., Nesgaard, E., and Johansen, J.T., *FEBS Lett.*, **147**, 74 (1982).

48) Lindskog, S., Nyman, P.O., *Biochim. Biophys. Acta*, **85**, 462 (1964).

49) Lindskog, S., Malmstrom, B.G., *J. Biol. Chem.*, **237**, 1129 (1962).

50) Fitzgerald, J.J., Chasteen, N.D., *Biochemistry*, **13**, 4338 (1974).

51) Shinar, H., Navon, G., *Biochim. Biophys. Acta* **334**, 471 (1974).

52) Lindskog, S., *Structure and Bonding*, **8**, 153 (1970).

53) Bertini, I., and Luchinat, C., *Adv. Inorg. Biochem.*, **6**, 71 (1984).

54) Bertini, I., and Luchinat, C., *NMR of Paramagnetic Molecules in Biological Systems*, Benjamin/Cummings, Menlo Park, CA, 1986.

55) Coleman, J.E., *Biochemistry*, **4**, 2644 (1979).

56) Duff, T.A., and Coleman, J.E., *Biochemistry*, **5**, 2009 (1966).

57) Lindskog, S., *J. Biol. Chem.*, **238**, 945 (1963).

58) Bertini, I., Luchinat, C., Scozzafava, A., *Inorg. Chim. Acta*, **23**, L23 (1977).

59) Rosenberg, R.C., Root, C.A., Wang, R.–H., Cerdonio, M., and Gray, H.B., *Proc. Natl. Acad. Sci. U.S.A.*, **70**, 161 (1973).

60) Banci, L., Bertini, I., Luchinat, C., Donaire, A., Martinez, Ma.J., and Moratal, J.M., *Comments in Inorg. Chem.*, (in press).

61) Banci, L., Bertini, I., Luchinat, C., *Magn. Reson.Rev.*, **11**, 1 (1986).

62) Eriksson, A.E., Kylsten, P.M., Jones, T.A., and Liljas, A., *Proteins*, **4**, 283 (1989).

63) Bergsten, P.–C., Waara, I.; Lovgren, S.; Liljas, A.; Kannan, K.K.; Bengtsson, U., in *Proc. of the Alfred Benzon Symp IV*, Rorth, M., Astrp, P., Eds., Munskgaard, Copenhagen, p. 363 (1972)

Computer Simulation of Zinc-Containing Enzymes

Kenneth M. Merz Jr.

Department of Pharmaceutical Chemistry
University of California, San Francisco
San Francisco, California 94143

Rational drug design (RDD) or the design of bioactive molecules using the computer is one of the holy grails of computational chemistry. To date, however, the application of RDD has yet to fulfill all its promise. The thermodynamic cycle - free energy perturbation method (TC-FEPM)[1], which has just appeared on the scene has, in principal, the potential to allow us to acheive the goal of true RDD. The TC-FEPM allow us for the first time to routinely determine the free energy of an associative or activated process on a macromolecular system. Germane to our present discussion, this method allows us to obtain the relative free energy of binding of a pair of inhibitors (drugs) to a target protein. In our continuing effort to define the range of applicability of this method we have undertaken studies on the inhibition of the zinc metalloenzymes thermolysin[1] and human carbonic anhydrase II (HCAII).[2]

In this method the Gibbs free energy can be calculated according to eqn 1[3]

$$G(\lambda+\Delta\lambda) - G(\lambda) = -k_B T \ln \langle \exp\{-(H(\lambda+\Delta\lambda) - H(\lambda)) / k_B T\}\rangle_\lambda \quad (1)$$

where k_B is the Boltzmann's constant, T is the absolute temperature, $H(\lambda+\Delta\lambda)$ and $H(\lambda)$ are the hamiltonians at the states λ and $\lambda+\Delta\lambda$, and the $\langle\rangle_\lambda$ indicates the ensemble average at the intermediate points along the conversion pathway defined by the coupling parameter λ. A simulation is then run between the states $\lambda=1$ to $\lambda=0$ (or the reverse) where the free energy is evaluated at each of the intermediate λ points. The total free energy for this change is then a sum of the free energy at all of the intermediate points along the $\lambda=1$ to $\lambda=0$ pathway according to eqn 2

$$\Delta G = \Sigma G(\lambda+\Delta\lambda) - G(\lambda) \quad (2)$$

In order to determine the $\Delta\Delta G_{bind}$ between one inhibitor and another we

M. Aresta and J. V. Schloss (eds.),
Enzymatic and Model Carboxylation and Reduction Reactions for Carbon Dioxide Utilization, 199–209.
© 1990 Kluwer Academic Publishers.

make use of the following thermodynamic cycle

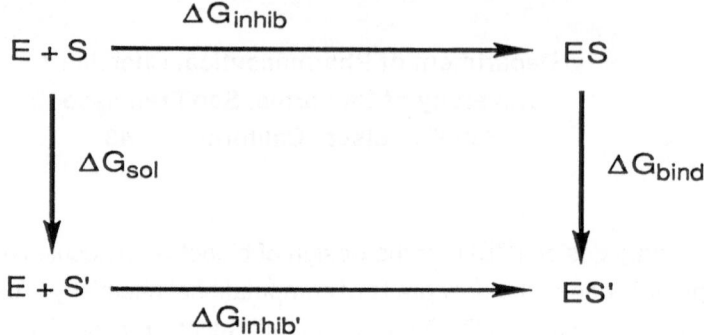

$$\Delta\Delta G_{bind} = \Delta G_{inhib'} - \Delta G_{inhib} = \Delta G_{bind} - \Delta G_{sol}$$

where ΔG_{inhib} and $\Delta G_{inhib'}$ represent the free energy of binding of the inhibitor S and S', respectively. ΔG_{sol} represents the free energy of solvation difference between S and S', and , lastly ΔG_{bind} represents the free energy of binding difference between S and S' in the enzyme active site. Using simulations we are currently unable to determine the ΔG_{inhib} terms; however, it is possible to "mutate" one inhibitor into another and thereby determine ΔG_{sol} and ΔG_{bind}. Since we are dealing with a state function (ΔG) the following relationship, based on the thermodynamic cycle, holds

$$\Delta\Delta G_{bind} = \Delta G_{inhib'} - \Delta G_{inhib} = \Delta G_{bind} - \Delta G_{sol} \quad (3)$$

Thus, we have at our disposal all that is necessary to determine the $\Delta\Delta G_{bind}$ between two inhibitors.

Computations of this sort will allow industrial and academic users to design novel inhibitors of targeted enzymes. In order for this approach to be useful in the design of drugs the following will be required: if the disease can be corrected by the disruption of a specific enzymatic pathway the structure of this enzyme will have to be known in order to carry out simulations. A target compound, which is a known inhibitor of the enzyme, is the next requirement for a sucessful simulation because, by the very nature of the TC-FEPM, only small changes to the inhibitor can be made at any one time.[1] In other words, one should not expect this method to make predictions of what a useful lead compound might be. The investigator will then have to decide what could be a potential change to this inhibitor that will make it a more powerful drug. Once a plan has been formulated for the simulations a force field describing the the enzyme and inhibitor will have to be decided upon and subsequent to this the necessary simulations can be executed to determine the $\Delta\Delta G_{bind}$.

In the present discussion we will be focusing on inhibitors of thermolysin[1] and HCAII[2] both of which have a catalytically essential zinc ion. Inhibition of the the former enzyme is not therapeutically useful; however, a closely related enzyme, angiotensin converting enzyme, has been the target of extensive drug design efforts.[4] Inhibition of CA is of therapeutic importance, being useful in the control of glaucoma.[5] It is the goal of the present contibution to test the limitations of the TC-FEPM as well as attempting to further understand the inhibition of these enzymes.

The series of thermolysin inhibitors that were studied are the Cbz-Gly^P-(X)-Leu-Leu, where X=NH,O,CH$_2$. The NH and O inhibitors have been studied both crystallographically[6] and experimentally[7], while the CH$_2$ compound has not be the subject of scrutiny. The interesting aspect of the NH and O compounds is the 4.1 kcal/mol value observed for their $\Delta\Delta G_{bind}$. The cause of this difference is the presence of a hydrogen bonding interaction (see stereo plots in Figure 1 below) between the main chain carbonyl of Ala 113 with the hydrogen from the NH group, which is necessarily absent in the O compound.[6,7,8]

We used the AMBER[9] force field for our simulations, and since it has a coulombic term (q_iq_j/r_{ij}) we have to determine atomic point charges for the inhibitor. We determined the charges for these inhibitors by using a partially optimized 4-31G[10] geometry in a STO-3G*[11] electrostatic potential calculation.[12] The enzyme simulations were carried out using the GIBBS module of the AMBER[13] suite of programs using 20 windows with 500 steps (0.002 psec time step) of equilibration and 300 steps (0.001 psec time step) of sampling being done at each window. A nonbonded cutoff of 8Å (updated every 100 MD steps) was used along with a 15Å belly (*i.e.* all residues 15Å from the active site were not included in the MD simulation) centered at the zinc ion. The active site was capped with 200 TIP3P[14] water molecules, which were harmonically constrained to be at most 18Å from the zinc ion. SHAKE[15] was used to constrain all bond distances to their equilibrium values, and the temperature was maintained at 300 K by coupling to a temperature bath.[16] We assumed that the zinc ion has a +2.0 charge and as a result we had to restrain several atoms bound to the zinc because of the propensity of the zinc ion to be attracted to other charged groups in active site vicinity. The atoms that were restrained are the two oxygens bound to zinc from the inhibitor, the two oxygens from Glu 166, and the nitrogens from His 142,146. These latter two sets of atoms form the coordination sphere around the zinc ion. The solution phase simulations were identical to the enzyme ones except that we used the NPT (P = 1 Atm, T = 300K) ensemble[16] with periodic boundary conditions and 800 TIP3P[14] water molecules to simulate the aqueous enviroment of the inhibitor only. Since all simulations employed explicit solvent we used a constant dielectric of one.

The results are given in Table 1 and 2.

Table 1. Calculated Free Energies for NH -> O in kcal/mol.

		NH -> O		
Run		A -> B	B -> A	Ave.
ΔG_{sol}		F[a] -0.617	F +0.282	-0.464±0.27
		R +0.765	R -0.190	
ΔG_{bind}		F +3.83	F -1.881	+2.86±0.87
		R -3.32	R +2.43	

$\Delta\Delta G_{bind} = \Delta G_{bind} - \Delta G_{sol} = 3.3\pm1.1$ kcal/mol (experimental 4.1 kcal/mol)

a) In this table and Table 2 F and R represent the free energies computed in the ($\lambda+\Delta\lambda$) and ($\lambda-\Delta\lambda$) directions, respectively. See equation 1.

Table 2. Calculated Free Energies for $CH_2 \rightarrow NH$ in kcal/mol.

Run		$A \rightarrow B$		$B \rightarrow A$		Ave.
ΔG_{sol}		F -2.217		F $+2.367$		-2.395 ± 0.28
		R $+2.099$		R -2.627		
ΔG_{bind}		F -2.72		F $+2.14$		$+2.40 \pm .52$
		R $+2.94$		R -1.798		

$\Delta\Delta G_{bind} = \Delta G_{bind} - \Delta G_{sol} = -0.005 \pm .80$ kcal/mol (experimental -0.9 kcal/mol)

We find that the NH -> O simulations reasonably reproduce ($\Delta\Delta G_{bind}$ (calc) = 3.3 kcal/mol) the experimental $\Delta\Delta G_{bind}$ of 4.1 kcal/mol. In our previous simulations we obtained much better agreement with experiment ($\Delta\Delta G_{bind}$ (calc) = 4.21 kcal/mol), and we trace this difference to our new choice of charges and to a modified set of parameters. Thus, not unexpectedly, we find the TC-FEPM to sensitive to the choice of charge model and parameters. What might be the best approach for the determination of atomic charges is not yet clear, but, obviously, determination of electrostatic potential charges at the basis set that best reproduce the electrostatic properties of the molecule is desireable.[12] From the $CH_2 \rightarrow$ NH simulations we predict that the CH_2 compound is nearly as good an inhibitor as is the NH compound. From the NH -> O and $CH_2 \rightarrow$ NH simulations we can determine that the $\Delta\Delta G_{bind}$ for $CH_2 \rightarrow$ O is 5.64 kcal/mol. Hence, we arrive at the inhibitor ordering of NH $\approx CH_2 >$ O.

What are the differences between the compounds that give this spread in the relative $\Delta\Delta G_{bind}$? The difference between the NH and O compounds can be directly traced to the Ala 113 C=O⋯NH hydrogen bond interaction, which cannot be present in the O compound. In the CH_2 compound this interaction is also not possible, but since the Ala 113 C=O⋯CH_2 interaction is not unfavorable or favorable we would expect it to bind better than the

O compound because for the latter compound it is certainly an unfavorable repulsive interaction. Furthermore, the CH_2 compound, relative to the NH and O compounds, has an unfavorable ΔG_{sol} because of the hydrophobic effect. This effect counterbalances the unfavorable ΔG_{bind} (relative to the NH compound), which ultimately leads to the observed $\Delta\Delta G_{bind}$.

Paul Bartlett of Berkeley has recently sythesized and determined the free energy of binding (K_i) of the CH_2 compound. This value was unknown to us at the time we carried out this research, thus our calculated $\Delta\Delta G_{bind}$ is a true *prediction*. Our calculated $\Delta\Delta G_{bind}$ for CH_2 -> NH is -0.005 ± 0.80 kcal/mol, while the experimental value is -0.09 kcal/mol. The agreement is encouraging.

For HCAII the following inhibitors were studied: p-hexylbenzenesulfonamide (**1**), benzenesulfonamide (**2**), and p-hexylbenzenesulfonate (**3**). Sulfonamides are the classic HCAII inhibitors and all commercially available compounds that are HCAII inhibitors are sulfonamides. We have carried out our HCAII study in collaboration with scientists at Merck, Sharp and Dohme, with an eye on helping them further understand HCAII inhibition and to potentially design new HCAII inhibitors.

1: X = NH, R = hexyl
2: X = NH, R = H
3: X= O, R = hexyl

The charges for these inhibitors were determined from STO-3G*[11] electrostatic potential calculations.[12] The geometries used for these calculations were determined by first doing a 6-31G* geometry optimization on a CH_3-SO_2NH^- (or CH_3-SO_3^-) fragment. The resulting structure was then transferred onto a benzene ring or a p-hexylbenzene moiety. This was then subjected to an AM1[17] geometry optimization where the C-SO_2NH (or C-SO_3) portion of the geometry was fixed at the 6-31G* coordinates and the remaining geometric variables were optimized at the AM1 level. The enzyme simulations were again carried out using the GIBBS module of the AMBER[13] suite of programs using 40 windows with 250 steps (0.001 psec time step) of equilibration and 150 steps (0.001 psec time step) of sampling being done at each window. A nonbonded cutoff of 10Å (updated every 50 MD steps) was used along with a 15Å belly centered at the zinc ion. The active site was capped with 300 TIP3P[14] water molecules, which were harmonically constrained to be at most 22Å from the zinc ion, and, finally, the temperature was maintained at 300K by coupling to a temperature bath.[16] The solution phase simulations were identical to the enzyme ones except that we used the NPT (P = 1 Atm, T = 300K) ensemble[16] with periodic boundary conditions and 800 TIP3P water molecules to simulate the aqueous enviroment. Instead of using a 10Å nonbonded cutoff we had to use an 8Å cutoff because this distance was roughly one-half of the smallest dimension of the box. Since all simulations employed explicit solvent we used a constant dielectric of one.

Initially, we attempted to again describe to zinc ion as a dipositive point charge as we had done for thermolysin, but we found that this would not work in the present case due to the proximity of Glu 106. With our active site set up in this way the Glu 106 would collapse into the zinc ion forming a hexa-coordinated zinc and a rather distorted HCAII active site. We found that it was necessary to reduce the zinc ion charge to +0.8[18] and to explicitly incorporate zinc to histidine nitrogen bonds with a force constant of 100 kcal/Å. The excess positive charge from the zinc ion was then displaced onto the imidazole rings of the histidines that are bound to zinc. We also included bond angles between N-Zn-N and Zn-N-C, which have force constants of 50 kcal/degree and 10 kcal/degree, respectively. We then found that we needed to reduce the sulfonamide oxygen charges from about -0.6 to -0.5 in order for the correct zinc-oxygen bond distance to be predicted from an AMBER minimization run. The displaced charge was then placed on the sulphur atom reducing its charge from 1.0 to about 0.8. Armed with this description of the active site, which is able to reproduce the experimentally observed HCAII·sulphonamide active site structure[2], we carried out our simulations.

The results are given in Table 3.

Table 3. Calculated ΔG_{sol}, ΔG_{bind} and $\Delta\Delta G_{bind}$ for the processes 1 --> 3 (I and II) and 1 --> 2 (III). The experimental $\Delta\Delta G_{bind}$ are given for comparison. All values are in kcal/mol

Run		ΔG_{sol}	ΔG_{bind}	$\Delta\Delta G_{bind}$	$\Delta\Delta G_{bind}$
I		+3.02±1.26	+5.65±0.15	+2.63±1.35	+2.73
II		-1.24±1.02	+1.21±0.96	+2.45±0.74	+2.73
III		+7.73±0.24	+12.42±0.35	+4.69±0.14	+1.57

We first carried out simulations in which we converted **1** to **2**, which is a change that involves the conversion of a hexyl group into a hydrogen. We wanted to study this because we had had initial sucess in converting alanine to phenylalanine[19] using the united atom approximation, but recently Brooks etal[20], have had difficulty in reproducing the ΔG_{sol} between propane and ethane and suggested that this was due to the use of the united atom approximation. Thus, we have decided to tackle this problem by using the all atom AMBER force field[9], with an eye on determining how this level of sophistication will preform. We first carried out this simulation using the "shrink" protocol; in this approximation we slowly shrink, for example, the carbon carbon bond distances from 1.53Å down to 0.4Å (the distance for the resulting dummy atom dummy atom bond) during the free energy calculation. This was the approach used our previous effort.[19] The value we calculate for ΔG_{sol} is 3.02 kcal/mol in favor of **1**. This is clearly at variance with chemical intuition, which would predict that the opposite would be true. However, we find that when we complete the thermodynamic cycle that the calculated $\Delta\Delta G_{bind}$ of 2.63 kcal/mol agrees well with the experimental value of 2.73 kcal/mol. This suggests that the error in the calculated ΔG_{sol} and ΔG_{bind} is similar and cancels when $\Delta\Delta G_{bind}$ is determined. We next carried out a simulation in which we did not shrink the bonds, which we call a "no-shrink" simulation. The results in this case are much more encouraging: we find that the ΔG_{sol} value is more in line with chemical intuition. The $\Delta\Delta G_{bind}$ value of 2.45 kcal/mol is in good agreement with the experimental value. In conclusion, we find that the all-atom approximation using a no-shrink simulation is capable of reproducing the ΔG_{sol} for hydrophobic changes.

Next we determine the $\Delta\Delta G_{bind}$ for the conversion of a **1** to **3**. This change is, contrary to our previous simulations, a purely electrostatic change. Our calculated $\Delta\Delta G_{bind}$ of 4.69 kcal/mol is in fair agreement with the experimental value of 1.57 kcal/mol. In these simulation we find that the hysteresis is much smaller than that we observed in the previous simulations. This is because we are making much smaller perturbations in the present set of simulations and as a result we are probably adequately sampling configurational space. The major source of error between the calculated and observed $\Delta\Delta G_{bind}$ is our electrostatic model. It is our opinion that in order to improve our results we will have to carry out ESP calculations on the zinc ion and its associated ligands. In this way we have a set of charges that have been determined in not an *ad hoc*, but a consistent manner.

The TC-FEPM method has again been relatively successful in studying drug enzyme interactions, but some new problems have come to light. We first note that the charge model can be a very important factor in determining if the simulation will be sucessful. Care should be taken in the determination of the charges to be used, and the best suggestion is that the charges should be obtained using an electrostatic potential calculation at the level of sophistication that best reproduces the electrostatic properties of the molecule.[12] In the simulation of purely hydrophobic changes we suggest that the all-atom "no-shrink" method should be used. We also suggest that simulations in which one can break down a large change (*i.e.* hexyl to hydrogen) into a series of smaller ones this should be done. We have observed very significant hysteresis in our simulations when we have attempted to make very large changes and suspect that it can be minimized by using smaller perturbations or by carrying out longer simulations. As more extensive simulations are done using the TC-FEPM a clearer picture of the "best" approach to doing these simulations will emerge. Once this has been accomplished the applicability of this approach to the rational design of drugs is manifest.

References

(1) For a review on thermolysin see: Matthews, B. W. *Acc. Chem. Res.* in press.

(2) For reviews on carbonic anhydrase see: Linskog, S. In *Zinc Enzymes* ; Spiro, T. G. , Ed.; John Wiley & Sons: New York 1983; pg 77; Silverman, D. N.; Lindskog, S. *Acc. Chem. Res.* **1988**, *21*, 30.

(3) For a good review on methods to determine free energies from macromolecular simulations see: Mezei, M.; Beveridge, D. L. *Ann. N. Y. Acad. Sci.* **1986**, *482*, 1; for a good critical review of the TC-FEPM see: van Gunsteren, W. F. *Protein Eng.* **1988**, *2*, 5.

(4) Ondetti, M. A.; Rubin, B.; Cushman, D. W. *Science,* **1977**, *196*, 441; Cushman, D. W.; Cheung, H. S.; Sabo, E. F.; Ondetti, M. A. *Biochemistry,* **1977**, *16*, 5484; Petrillo, E. W. Jr.; Ondetti, M. A. *Med. Res. Rev.* **1982**, *2*, 1; Patchett, A. A.; Harris, E.; Tristham, E. W.; Wyvratt, M. J.; Wu, M. T.; Taub, D.; Peterson, E.R.; Ikelar, T. J.; ten Broeke, J.; Payne, L. G.; Ondeyka, D. L.; Thorsett, E. D.; Greenlee, W. J.; Lohr, N. S.; Hoffsommer, R. D.; Joshua, H.; Ruyle, W. V.; Rothrock, J. W.; Aster, S. D.; Maycock, A. L.; Robinson, F. M.; Hirschmann, R. *Nature* **1980**, *288*, 280; Patchett, A. A.; Cordes, E. H. *Advan. Enzymol.* **1983**, *57*, 1.

(5) Zimmerman, J. J. *Ann. Ophthalmol.* **1978**, *10*, 509; Becker, B. *Am J. Ophthalmol* **1954**, *37*, 13; Friedenwald, J. C. *Am. J. Ophthalmol* **1949**, *32*, 9.

(6) Tronrud, D. E.; Holden, H. M.; Matthews, B. W. *Science* **1987**, *235*, 571.

(7) Bartlett, P. A.; Marlowe, C. K. *Science* **1987**, *235*, 569.

(8) Bash, P. A.; Singh, U. C.; Brown, F. K.; Langridge, R.; Kollman, P. A. *Science* **1987**, *235*, 574; Merz, Jr., K. M.; Kollman, P. A. accepted for publication in *J. Am. Chem. Soc.*.

(9) All-atom force field: Weiner, S. J.; Kollman, P. A.; Nguyen, D. T.; Case, D. A. *J. Comput. Chem.* **1986**, *7*, 230; for the united-atom force field: Weiner, S. J.; Kollman, P. A.; Case, D. A.; Singh, U. C.; Ghio, C.; Alagona, G.; Profeta, S.; Weiner, P. *J. Am. Chem. Soc.* **1984**, *106*, 765.

(10) Ditchfield, R.; Hehre, W. J.; Pople, J. A.; *J. Chem. Phys.* **1971**, *54*, 724; Hehre, W. J.; Lathan, W. A. *J. Chem. Phys.* **1972**, *56*, 5255.

(11) Collins, J. B.; Schleyer, P. v. R.; Binkley, J. S.; Pople, J. A. *J. Chem. Phys.* **1976**, *64*, 5142.

(12) Singh, U. C.; Kollman, P. A. *J. Comput. Chem.* **1984**, *108*, 129.

(13) Singh, U. C.; Weiner, P. K.; Caldwell, J. W.; Kollman, P. A.. AMBER (UCSF), version 3.0.

(14) Jorgensen, W. L.; Chandrasekhar, J.; Madurea, J.; Impey, R. W.; Klein, M

L. *J. Chem. Phys.* **1983**, *79*, 926.

(15) van Gunsteren, W. F.; Berendsen, H. J. C. *Mol. Phys.* **1977**, *34*, 1311.

(16) Berendsen, H. J. C.; Potsma, J. P. M.; van Gunsteren, W. F.; DiNola, A.; Haak, J. R. *J. Chem. Phys.* **1984**, *81*, 3684.

(17) Dewar, M. J. S.; Zoebisch, E. G.; Healy, E. F.; Stewart, J. J. P.; *J. Am. Chem. Soc.* **1985**, *107*, 3902.

(18) The charge of 0.8 was arrived at in the following way: we first carried out a number of AM1[17] calculations on (Im)$_3$ZnX compounds, where Im is imidazole and X was hydroxide, H_2O, NH_3, NH_2^-. Drom these calculations we found that the zinc charge was usually 0.6; however, a charge of this magnitude was too small to allow for an effective electrostatic interaction between the zinc and the sulfonamide inhibitor (*i.e.* bond formation). We found, though, that if we increased the zinc charge to 0.8 that the zinc sulfonamide interactions were optimal and gave the correct bond lengths between the zinc and the sulfonamide.

(19) Singh, U. C.; Brown, F. K.; Bash, P. A.; Kollman, P. A. *J. Am. Chem. Soc.* **1987**, *109*, 1607.

(20) Fleischman, S. H.; Brooks, C. L. *J. Chem. Phys.* **1987**, *87*, 3029.

MECHANISM OF CARBOXYLATION BY BIOTIN ENZYMES

ALBERT S. MILDVAN, DAVID C. FRY, AND ENGIN H. SERPERSU
Department of Biological Chemistry
The Johns Hopkins University School of Medicine
725 N. Wolfe Street
Baltimore, MD 21205 USA

ABSTRACT. Biotin enzymes catalyze the carboxylation of substrates (RH) in two steps:

$$MgATP + HCO_3^- + E\text{-biotin} \rightleftharpoons MgADP + P_i + E\text{-biotin-}CO_2^- + H^+ \qquad (1)$$

$$E\text{-biotin-}CO_2^- + RH \rightleftharpoons R\text{-}CO_2^- + E\text{-biotin} \qquad (2)$$

Mechanistic studies suggest that step (1) occurs by the phosphorylation of bicarbonate to form carboxyphosphate, which carboxylates the enolate of biotin possibly via bound CO_2 as an intermediate. In step (2) carboxybiotin decomposes into bound CO_2 and the biotin enolate which deprotonates S to form a carbanion, facilitated in some cases by a bound metal. The carbanion is then carboxylated by bound CO_2.

Introduction

Six enzymes use the prosthetic group D–Biotin as a carboxylate carrier in carboxylation reactions. The substrates which are thereby carboxylated are pyruvate, urea, or esters of Coenzyme A. Several comprehensive reviews of these enzymes have been written (1-5). This review will focus on the mechanisms of these enzymes. All biotin enzymes are large (> 200,000) consisting of subunits of three types, a carboxybiotin synthetase, a carboxybiotin carrier and a carboxytransferase (Fig. 1). Each overall carboxylation reaction catalyzed by a biotin enzyme is reversible and may be broken into two reversible steps:

$$MgATP + HCO_3^- + E\text{-biotin} \overset{Mg^{2+}}{\rightleftharpoons} MgADP + P_i + E\text{-biotin-}CO_2^- + H^+ \qquad (1)$$

$$E\text{-biotin-}CO_2^- + RH \rightleftharpoons R\text{-}CO_2^- + E\text{-biotin} \qquad (2)$$

Step (1), a synthetase reaction, utilizes MgATP to activate bicarbonate and forms carboxybiotin. Step (2) is a transcarboxylation in which the carboxybiotin intermediate carboxylates the substrate RH. Between these steps, carboxybiotin must migrate or translocate between the two active sites. The biotin prosthetic group, a derivative of urea, is covalently attached to the biotin carrier subunit via an amide

211

M. Aresta and J. V. Schloss (eds.),
Enzymatic and Model Carboxylation and Reduction Reactions for Carbon Dioxide Utilization, 211–219.
© 1990 *Kluwer Academic Publishers.*

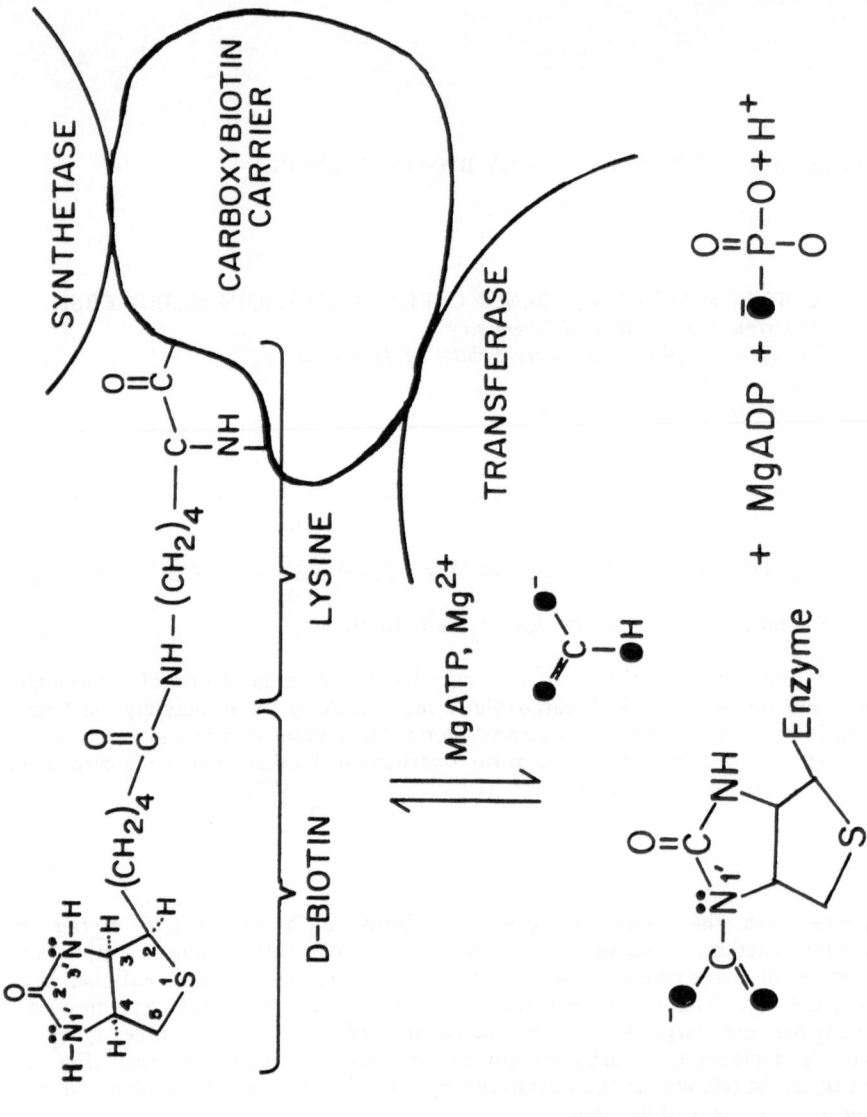

Figure 1. Structure of enzyme-bound biotin and site of biotin carboxylation.

213

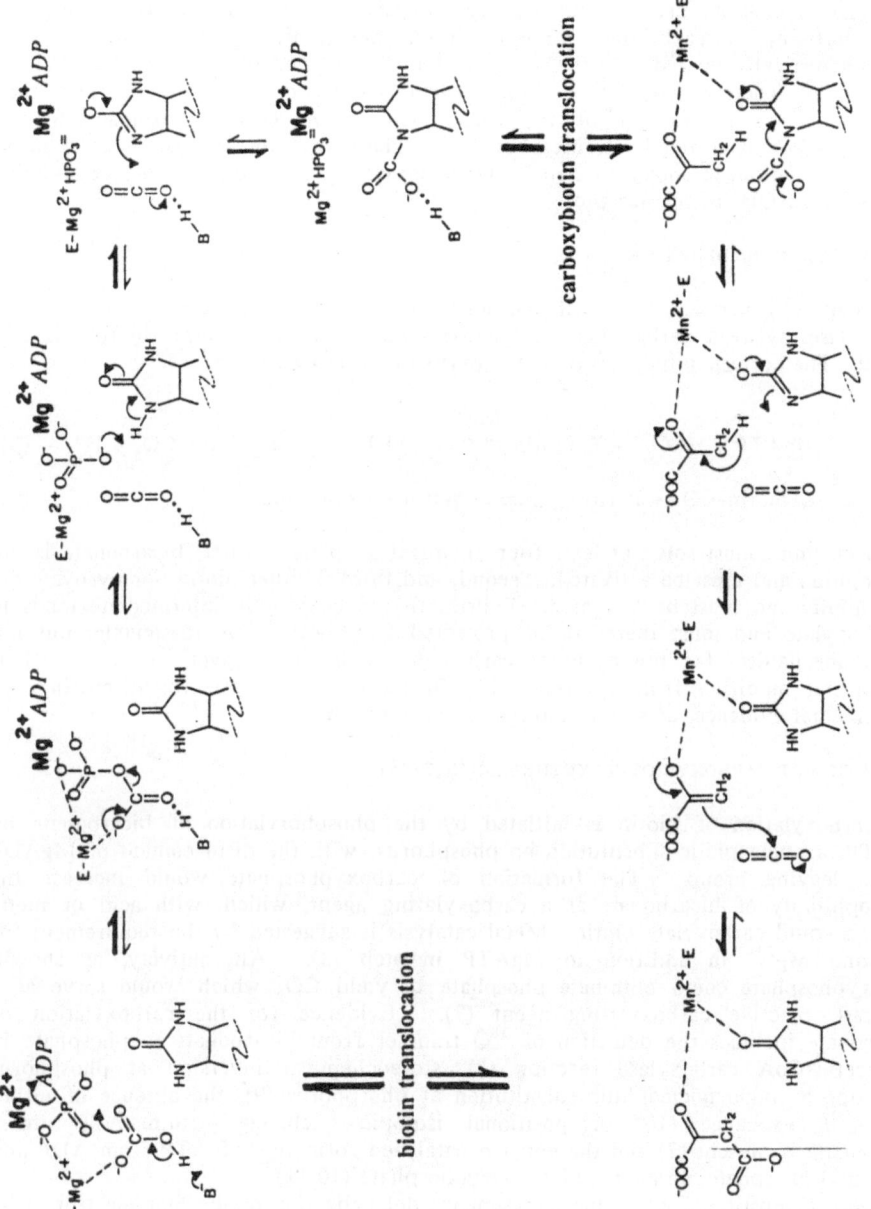

Figure 2. Possible mechanism of pyruvate carboxylase, modified from reference 2.

bond to an ϵ amino group of lysine (Fig. 1). Its long (14 Å) arm would permit it to translocate over large distances but this may not be necessary since the two active sites appear to be near to each other on one biotin enzyme, transcarboxylase (6).

Mechanistically related reactions catalyzed by biotin enzymes are the transport of Na^+ out of bacteria coupled to the decarboxylation of β-keto acids and their thioesters, and the transfer of the carboxylate group between oxaloacetate and propionyl-CoA catalyzed by transcarboxylase. These reactions consist of consecutive transcarboxylations as shown in step 2. Decarboxylation is simply a transcarboxylation with water or OH^- as the acceptor.

MECHANISTIC CONSIDERATIONS

For greatest generality, we will consider in detail the mechanism of a biotin carboxylase, pyruvate carboxylase, making use of experiments on other biotin enzymes as well. The corresponding reactions for pyruvate carboxylase are:

$$MgATP + HCO_3^- + \text{E-biotin} \overset{Mg^{2+}}{\rightleftharpoons} MgADP + P_i + \text{E-biotin-}CO_2 + H^+ \quad (3)$$

$$\text{E-biotin-}CO_2 + \text{Pyruvate} \rightleftharpoons \text{E-biotin} + \text{oxaloacetate} \quad (4)$$

Biotin enzymes must solve at least four chemical problems. First, bicarbonate is not electrophilic and must be activated. Second, and third, neither biotin nor pyruvate are nucleophilic and must be activated. Fourth, the carboxybiotin intermediate tends to decarboxylate and must therefore be protected from water. A reasonable, but not unique mechanism for the pyruvate carboxylase reaction is given in Fig. 2. This mechanism, modified from reference 2, will serve as the basis for discussing the experimental evidence, as well as alternative mechanisms.

MECHANISM OF CARBOXYBIOTIN SYNTHETASE (STEP 3)

The carboxylation of biotin is initiated by the phosphorylation of bicarbonate by MgATP, a nucleophilic substitution on phosphorus, with the displacement of MgADP as the leaving group. The formation of carboxyphosphate would increase the electrophilicity of bicarbonate as a carboxylating agent, which, with acid or metal catalysis could carboxylate biotin. Metal catalysis is suggested by the requirement for a second Mg^{2+}, in addition to MgATP in step (3). Alternatively, as shown, carboxyphosphate could eliminate phosphate to yield CO_2 which would serve as a localized, reactive carboxylating agent (7). Evidence for the carboxylation of bicarbonate includes the detection of ^{18}O transfer from bicarbonate to phosphate in the acetyl-CoA carboxylase reaction (8), stereochemical inversion at phosphorus indicating a single nucleophilic substitution at phosphorus (9), the absence of either ATP-ADP exchange (10) or positional isotopic exchange within ATP unless bicarbonate is present (2) and the enzyme-catalyzed formation of ATP from ADP and carbamoyl phosphate, an analog of carboxyphosphate (10,11).

Biotin, a substituted urea, due to resonance delocalization of the lone electron pairs is not a strongly basic or nucleophilic species, and is therefore difficult to carboxylate. A solution to this problem was proposed in 1965 by Retey and Lynen (12) who suggested that biotin is tautomerized or "enolized" by the enzyme, to form a more nucleophilic species (Fig. 3), and Bruice and coworkers obtained evidence that isourea analogs were 10^{10}-fold more nucleophilic than biotin (13,14). No direct evidence for

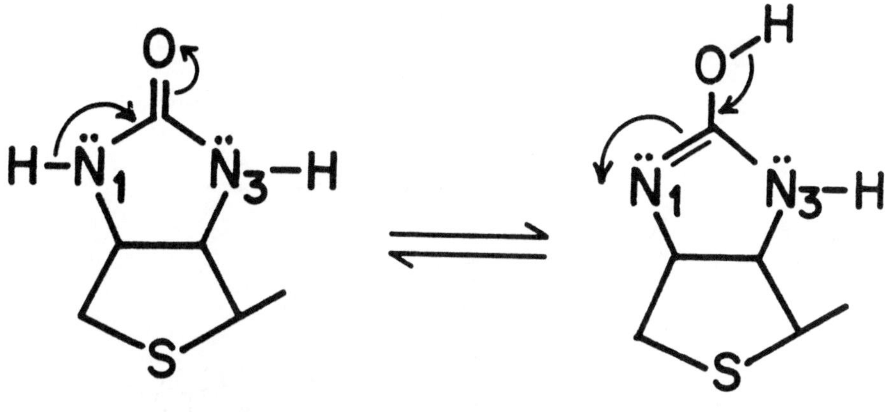

Figure 3. Tautomerization of biotin.

the enolization of biotin existed until an NMR study of the exchange of the NH protons of biotin itself and of several biotin derivatives with water protons revealed that under mild conditions (25°, pH 7.4) the enzymatically reactive N-1 proton exchanges with water protons at a rapid rate, comparable to the rates of carboxylation of enzyme-bound biotin (1,15-17). Hence, in principle, no additional catalysis by the enzyme over that provided by 10^{-7} M [OH⁻] at neutral pH is necessary to deprotonate the 1-NH group in the enolization of biotin (1,2). The immediate product of this deprotonation, the imidate or "enolate" of biotin is even more nucleophilic than the enol or isourea tautomer and would readily attack the localized CO_2 or carboxyphosphate to form carboxybiotin. A question as yet unanswered is whether the enol or the enolate of biotin is the nucleophilic species which is actually carboxylated. While a seemingly minor issue at this point in the reaction, it is relevant to a later fundamental question, namely which atom of carboxybiotin, N-1 or the carbonyl oxygen, deprotonates the substrate pyruvate prior to its carboxylation?

CARBOXYBIOTIN AND BIOTIN TRANSLOCATION

The carboxybiotin must next translocate to the transcarboxylation site of the enzyme to participate in step (4). The translocation of carboxybiotin must be highly efficient since decarboxylation must be avoided and, in the case of the enzyme transcarboxylase even the proton exchange on biotin is incomplete (18). On transcarboxylase, the two sites between which carboxybiotin migrates are, at most 7 Å apart (6), requiring only facile rotations about two single bonds of the biotin side chain for the translocation to take place (19). As shown by model reactions, the acid catalyzed decarboxylation of carboxybiotin is slowed 30 fold by the presence of the lower ring of biotin, ascribed to steric hindrance of formation of the tetrahedral intermediate at N-1 in the acid catalyzed decarboxylation (20).

CARBOXYLATION OF PYRUVATE BY CARBOXYBIOTIN (STEP 4)

In this electrophilic substitution, carboxybiotin trades a carboxylate group for a proton at C-3 of pyruvate with retention of configuration at C-3 (21). For this to occur, pyruvate must be activated to form a nucleophilic species, such as the enolate of

216

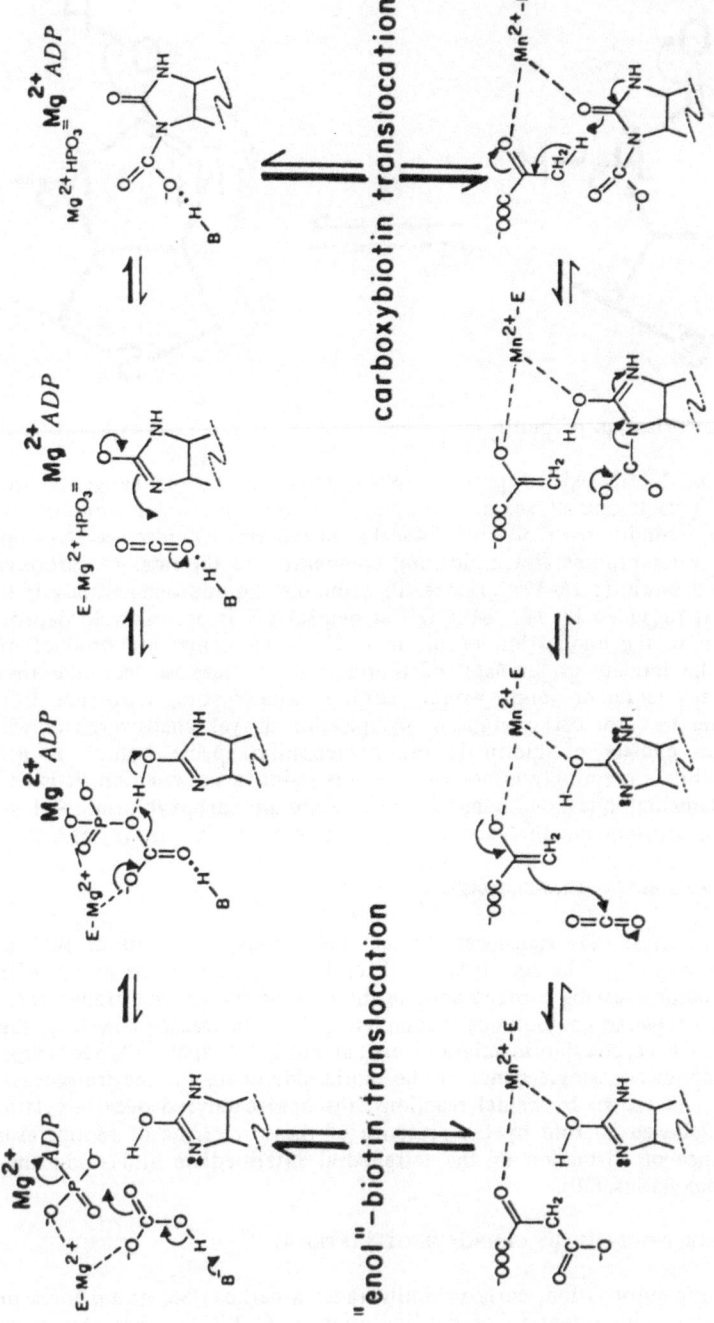

Figure 4. Alternative mechanism of pyruvate carboxylase.

pyruvate. For the enzymes pyruvate carboxylase and transcarboxylase, both of which catalyze step (4), this problem may have been solved by the discovery that these are both metalloenzymes, with the divalent cation bound nearest to the carbonyl oxygen of bound pyruvate (22-25). Such proximity would promote the enolization of pyruvate.

The simplest possible mechanism, a concerted electrocyclic process for the transcarboxylation of pyruvate, was initially proposed because of the failure of pyruvate carboxylase to detritiate 3-tritiopyruvate unless carboxybiotin had formed (12,26). However this mechanism has subsequently been ruled out by three independent experiments. First, propionyl-CoA carboxylase not only carboxylates β-fluoropropionyl CoA but also eliminates HF from this substrate indicating a carbanion intermediate which disproportionates (27). Second, both pyruvate carboxylase and transcarboxylase catalyze the stereospecific protonation of pre-formed enol pyruvate (28). Third, double-isotope effect experiments with both transcarboxylase (29) and pyruvate carboxylase (30) indicate separate, partially rate-determining carboxyl and proton-transferring steps. Hence the enolization of pyruvate precedes its carboxylation.

In the mechanism shown, pyruvate is deprotonated by N-1 of the imidate of biotin generated by the decarboxylation of carboxybiotin. This decarboxylation, which might be facilitated on the enzyme by distortion of the carboxylate plane out of the urea plane of carboxybiotin (31), yields localized CO_2 which then completes the reaction. In the case of the enzyme transcarboxylase, 5% of the ^3H removed from 3-tritiopyruvate survives biotin translocation and is conserved in the propionyl CoA product (18). Hence the most likely ^3H-carrier is biotin itself which translocates over the very short distance of ≤ 7 Å (6).

Alternatively, pyruvate could be deprotonated by the carbonyl oxygen of carboxybiotin, thereby promoting the decarboxylation of carboxybiotin to yield localized CO_2, which, in turn, carboxylates the enolate of pyruvate (Fig. 4). If this were the case, then the enol of biotin, the immediate product of this alternative process must, in the transcarboxylase reaction, translocate to the propionyl CoA site with conservation of 5% of the tritium removed from 3-tritiopyruvate. While the N-1 of biotin would lose tritium more slowly than the -OH of enol-biotin, the latter alternative cannot be excluded on kinetic grounds. This alternative mechanism is summarized in Fig. 4.

Acknowledgements

This work was supported by National Institutes of Health Grant DK 28616 to A.S.M.

References

1. Mildvan, A.S., Fry, D.C., and Serpersu, E.H. (1988) 'Mechanistic role of biotin in enzymatic carboxylation reactions', in S. A. Kuby (ed.), A Study of Enzymes, Vol. 2, CRC Press, Boca Raton (in press).
2. Knowles, J.R. (1989) 'The mechanism of biotin-dependent enzymes', Ann. Rev. Biochem. (in press).
3. Dakshinamurti, K. and Bhagaran, H.N. (eds.) 'Biotin', Ann. N.Y. Acad. Sci. 447, N.Y. Acad. Sci., p.441.
4. Wood, H.G. and Barden, R.E. (1977) 'Biotin enzymes', Ann. Rev. Biochem. 46, 385-413.

5. Attwood, P.V. and Keech, D.B. (1984) 'Pyruvate carboxylase', Current Topics in Cellular Regulation 23, Academic Press, N.Y., pp. 1-55

6. Fung, C.H., Gupta, R.K., and Mildvan, A.S. (1976) 'Magnetic resonance studies of the proximity and spatial arrangement of propionyl CoA and pyruvate on a biotin metalloenzyme, transcarboxylase', Biochemistry 15, 85-92.

7. Sauers, C.K., Jencks, W.P., and Groh, S. (1975) 'The alcohol-bicarbonate-water system', J. Am. Chem. Soc. 97, 5546-5553.

8. Kaziro, Y., Hass, L.F., Boyer, P.D., and Ochoa, S. (1962) 'Mechanism of the propionyl CoA carboxylase reaction II. Isotopic exchange and tracer experiments', J. Biol. Chem. 237, 1460-1468.

9. Hansen, D.E. and Knowles, J.R. (1985) 'N-Carboxybiotin formation by pyruvate carboxylase: The stereochemical consequences at phosphorus', J. Am. Chem. Soc. 107, 8304-8305.

10. Ashman, L.K. and Keech, D.B. (1975) 'Sheep kidney pyruvate carboxylase', J. Biol. Chem. 250, 14-21.

11. Polakis, S.E., Guchait, R.B., Zwergel, E.E., Lane, M.D., and Cooper, T.G. (1974) 'Acetyl CoA carboxylase system of *E. coli*', J. Biol. Chem. 249, 6657-6667.

12. Retey, J. and Lynen, F. (1965) 'Zur biokemischen funktion des biotins', Biochem. Z. 342, 256-271.

13. Hegarty, A.F., Bruice, T.C., and Benkovic, S.J. (1969) "Biotin and nucleophilicity of 2-methoxy-2-imidazoline toward sp^2 carbonyl carbon', J. Chem. Soc. Chem. Commun., 1173-1174.

14. Bruice, T.C. (1976) 'Some pertinent aspects of mechanism as determined with small molecules', Ann. Rev. Biochem. 45, 331-373.

15. Fry, D.C., Fox, T.L., Lane, M.D., and Mildvan, A.S. (1985) 'Exchange characteristics of the amide protons of d-biotin and derivatives', J. Am. Chem. Soc. 107, 7659-7665.

16. Serpersu, E.H., Mildvan, A.S., Fox, T., Fry, D.C., and Lane, M.D. (1987) 'A re-examination of the acid catalyzed exchange rates of the amide protons of D-biotin and its methyl ester', Biochemistry 26, 4160.

17. Perrin, C.L. and Dwyer, T.J. (1987) 'Proton exchange in biotin: a reinvestigation with implications for the mechanism of CO_2 transfer', J. Am. Chem. Soc. 109, 5163-5167.

18. Rose, I.A., O'Connell, E.L., and Solomon, F. (1976) 'Intermolecular tritium transfer in the transcarboxylase reaction', J. Biol. Chem. 251, 902-904.

19. DeTitta, G.T., Parthasarathy, R., Blessing, R.H., and Stallings, W. (1980) 'Carboxybiotin translocation mechanisms suggested by deffraction studies of biotin and its vitamers', Proc. Natl. Acad. Sci. U.S.A. 77, 333-337.

20. Tipton, P.W. and Cleland, W.W. (1988) 'Mechanisms of decarboxylation of carboxybiotin', J. Am. Chem. Soc. 110, 5866-5869.

21. Rose, I.A. (1970) 'Stereochemistry of pyruvate kinase, pyruvate carboxylase, and malate enzyme reactions', J. Biol. Chem. 245, 6052-6056.

22. Scrutton, M.C., Utter, M.F., and Mildvan, A.S. (1966) 'Pyruvate carboxylase VI. The presence of tightly bound manganese', J. Biol. Chem. 241, 3480-3487.

23. Northrop, D. and Wood, H.G. (1969) 'Transcarboxylase V. The presence of bound zinc and cobalt', J. Biol. Chem. 244, 5801-5807.

24. Fung, C.H., Mildvan, A.S., Allerhand, A., Komoroski, R., and Scrutton, M.C. (1973) 'Interaction of pyruvate with pyruvate carboxylase and pyruvate kinase as studied by paramagnetic effects on ^{13}C relaxation rates', Biochemistry 12, 620-629.

25. Fung, C.H., Mildvan, A.S., and Leigh, J.S. (1974) 'Electron and nuclear magnetic resonance studies of the interaction of pyruvate with transcarboxylase', Biochemistry 13, 1160-1169.

26. Mildvan, A.S., Scrutton, M.C., and Utter, M.F. (1966) 'Pyruvate carboxylase VII. A possible role for tightly bound manganese', J. Biol. Chem. 241, 3488-3498.

27. Stubbe, J., Fish, S., and Abeles, R.H. (1980) 'Are carboxylations involving biotin concerted or nonconcerted?', J. Biol. Chem. 255, 236-242.

28. Kuo, D.J. and Rose, I.A. (1982) 'Utilization of enolpyruvate by the carboxybiotin form of transcarboxylase: evidence for a nonconcerted mechanism', J. Am. Chem. Soc. 104, 3235-3236.

29. O'Keefe, S.J. and Knowles, J.R. (1986) 'Enzymatic biotin-mediated carboxylation is not a concerted process', J. Am. Chem. Soc. 108, 328-329.

30. Attwood, P.V., Tipton, P.A., and Cleland, W.W. (1986) 'Carbon-13 and deuterium isotope effects on oxaloacetate decarboxylation by pyruvate carboxylase', Biochemistry 25, 8197-8205.

31. Thatcher, G.R.J., Poirier, R., and Kluger, R. (1986) 'Enzymic carboxyl transfer from N-carboxybiotin. A molecular orbital evaluation of conformational effects in promoting reactivity', J. Am. Chem. Soc. 108, 2699-2704.

CARBAMATE SYNTHASES AND KINASES

VICENTE RUBIO, HUBERT G. BRITTON[1], LEANDRO B.
RODRIGUEZ-APARICIO[2] AND ISABEL CLIMENT
Instituto de Investigaciones Citológicas de la Caja de
Ahorros de Valencia (Centro Asociado del CSIC)
Amadeo de Saboya, 4
Valencia, E-46010, Spain

ABSTRACT. In the synthesis of carbamoyl phosphate from HCO_3^- and NH_3 by carbamoyl phosphate synthetases HCO_3^- is first activated by ATP (ATP_A), with formation of enzyme-bound carboxyphosphate and ADP. Mono and divalent cations are required in this reaction. Carboxyphosphate appears also to be formed with biotin-dependent carboxylases and with phosphoenolpyruvate carboxylase. With the carbamoyl phosphate synthetases carboxyphosphate is believed to react with ammonia in a second step, to give enzyme-bound carbamate, which would then be phosphorylated by another molecule of ATP (ATP_B). However, a concerted mechanism for the conversion of carboxyphosphate to carbamoyl phosphate has not been ruled out.
Ureotelic carbamoyl phosphate synthetase, a polypeptide of 160 kDa composed of 4 domains of about 40, 40, 60 and 20 kDa (from the N-terminus), is activated allosterically by N-acetyl-L-glutamate (AG^3). AG binds with the highest affinity to the central complex and favors its formation. The activation is associated with a dramatic increase in affinity for ATP_A, K and Mg. We have located, using photoaffinity labeling, the site for AG in the 20 kDa COOH-terminal domain of the enzyme. We propose that this domain interacts with the site for ATP_A, and that the binding of AG exposes this site. A model for the site for AG is presented, based on a study using analogs of AG. The site is composed of two subsites connected by a mobile bridge, and relative movements of the two subsites appear crucial for activation.

---

[1] Permanent address, Department of Physiology, St. Mary's Hospital Medical School, Norfolk Place, Paddington, London W2 1PG, England.
[2] Permanent address, Department of Biochemistry, Facultad de Veterinaria, Universidad de León, León, Spain.

[3] AG and acetylglutamate, N-acetyl-L-glutamate; ATP_A and ATP_B, the molecules of ATP that, in the reaction of carbamoyl phosphate synthetase, yield P_i and carbamoyl phosphate, respectively.

M. Aresta and J. V. Schloss (eds.),
Enzymatic and Model Carboxylation and Reduction Reactions for Carbon Dioxide Utilization, 221–238.
© 1990 Kluwer Academic Publishers.

1. INTRODUCTION

In the cytoplasm, "CO_2" exists predominantly as the hydrated and deprotonated form, HCO_3^-, and, in smaller proportion, as the unhydrated, dissolved gas form, CO_2. Some of the enzymes involved in "CO_2" fixation use CO_2 and others use HCO_3^- (Table 1).

Table 1. Form of "CO_2" used by various enzymes[a]

CO_2
  -Ribulose-1,5-bisphosphate carboxylase oxygenase [1]
    -Vitamin K-dependent carboxylase [2]
    -Phosphoenolpyruvate carboxykinase [3]
    -Phosphoenolpyruvate carboxytransphosphorylase [3]

HCO_3^-
    -Carbamoyl phosphate synthetases [4,5]
    -Biotin-dependent carboxylases [6,7]
    -Phosphoenolpyruvate carboxylase [8]

---

[a]Enzymes involved in "CO_2" fixation in anaerobic microorganisms [9] are not included because of lack of information on the chemical form of "CO_2" used.

Although HCO_3^- is more abundant (about 20-fold at pH 7.4) than CO_2 and its charge may facilitate binding to catalytic sites in enzymes, CO_2 crosses membranes readily and is more susceptible than HCO_3^- to nucleophilic attack, and thus is more suitable for carboxylation. Mechanistically, the enzymes that use CO_2 have a simpler task than the enzymes that use HCO_3^-. We will deal here only with enzymes that use HCO_3^- as the substrate for carboxylation.

A common feature of the reactions catalyzed by these enzymes is that P_i is one of the products (Fig.1). This P_i originates from the splitting of a high-energy phosphate bond of ATP (carbamoyl phosphate synthetase and biotin-dependent carboxylases) or phosphoenolpyruvate (phosphoenolpyruvate carboxylase). Carbamoyl phosphate synthetases use, in addition, another molecule of ATP to provide the phosphoryl group of carbamoyl phosphate.

Another common feature is that one of the three oxygen atoms of HCO_3^- is incorporated in P_i (Fig.1) [4,6,8], indicating that there is direct interaction between the phosphoryl group that is split and the HCO_3^-. This may be taken as evidence for a sequential mechanism, with an initial step of HCO_3^- activation by the high-energy phosphate compound, and a second step of attack of the resulting intermediate by the acceptor (a N atom in carbamoyl phosphate synthetase and biotin enzymes; the double bond of phosphoenolpyruvate in phosphoenolpyruvate carboxylase). Therefore, it was puzzling that neither carbamoyl phosphate synthetase [10-12] nor the biotin carboxylase subcomponent of acetyl CoA carboxylase from E. coli [7,13] catalyze substantial [^{14}C]ADP-ATP or ^{32}P$_i$-ATP exchanges in the absence of the acceptor (NH$_3$ or biotin, respectively). One of these exchanges would be expected if there is an initial activation of HCO$_3^-$ with release of ADP or P$_i$. Although for many years this was considered a serious difficulty in accepting a sequential mechanism for these enzymes, we now know that the activation of HCO$_3^-$ takes place without release of products (see below). Our understanding of the process of activation derives in great part from studies with carbamoyl phosphate synthetase, and we will discuss first the findings with this enzyme.

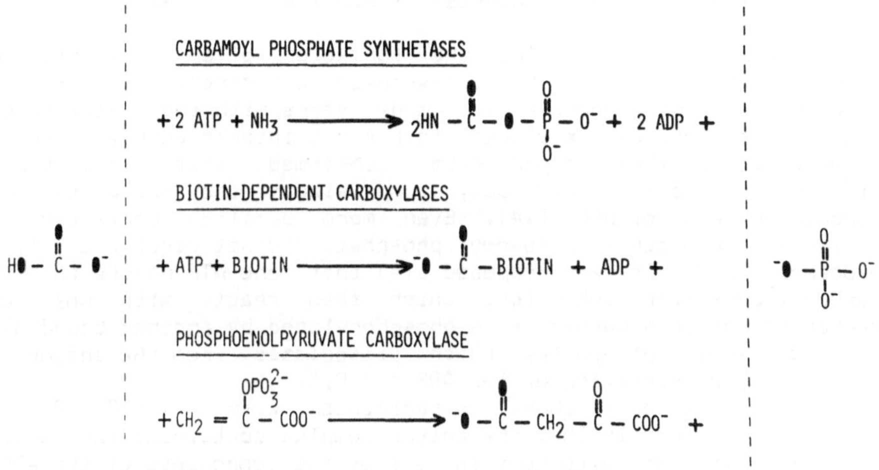

Figure 1. Reactions catalyzed by enzymes that fix HCO$_3^-$ illustrating common features. Although many carbamoyl phosphate synthetases use glutamine, the latter is split to yield ammonia, which is the true substrate. For biotin-dependent carboxylases only the reaction of carboxybiotin synthesis is shown. The oxygens in HCO$_3^-$ are shown in black to illustrate the fact that two appear in the carboxylated product and the third in P$_i$.

2. ACTIVATION OF HCO$_3^-$ BY CARBAMOYL PHOSPHATE SYNTHETASE

Carbamoyl phosphate synthetase catalyzes the reaction:

$$2ATP + HCO_3^- + NH_3 \longrightarrow 2ADP + H_2NCO_2PO_3H^- + P_i \qquad (1)$$

NH$_3$ rather than NH$_4^+$ is the true substrate [14,15]. In many carbamoyl phosphate synthetases it is generated from glutamine by a glutaminase component, but the ureotelic enzyme cannot use glutamine and has a low K$_m$ for ammonia [16]. The two ATP molecules have different roles. The molecule that yields P$_i$ is here designated ATP$_A$ and the molecule that yields the phosphoryl group of carbamoyl phosphate is designated ATP$_B$.

Metzenberg et al. [17] found that the frog enzyme has a slow HCO$_3^-$-dependent ATPase activity in the absence of ammonia and that it catalyzes also a partial reaction in which a molecule of ATP is synthesized from a molecule of ADP and of carbamoyl phosphate. These findings were taken as evidence for a sequential mechanism with an initial step of activation of HCO_3^- by ATP. Subsequently, the intermediate was formulated as enzyme-bound carboxyphosphate, since bicarbonate contributes one oxygen atom to phosphate [4], and the following mechanism was proposed (reactions 2-4; see [18])

$$E + HCO_3^- + ATP \longrightarrow E.CO_3PO_3^{3-} + ADP \qquad (2)$$
$$E.CO_3PO_3^{3-} + NH_3 \longrightarrow E.CO_2NH_2^- + P_i \qquad (3)$$
$$E.CO_2NH_2^- + ATP \longrightarrow NH_2CO_2PO_3^{2-} + ADP + E \qquad (4)$$

More direct evidence for the formation of an active bicarbonate intermediate was provided by pulse-chase experiments [14,19]. These experiments were negative for many years with the ureotelic enzyme until a procedure was developed to prevent interference from traces of ammonia [14]. These experiments confirmed that activation of bicarbonate requires ATP but, surprisingly, they failed to show the expected release of ADP [14]. Even more puzzling, conversion of the active intermediate to carbamoyl phosphate did not require addition of exogenous ATP. It was proposed [14] that "one ATP reacts to give the carbonyl phosphate anhydride, which then reacts with NH_3 to give carbamate and then carbamate is phosphorylated by another bound ATP to give carbamoyl phosphate, which dissociates from the enzyme first, followed by dissociation of two ADP and P_i".

Subsequent pulse-chase experiments with $[\tau^{-32}P]ATP$ [20,21] demonstrated directly that the enzyme complex containing the activated form of bicarbonate also contains all of the components of the ATP_A and ATP_B molecules. Addition of ammonia in the chase released bound ATP_A as ADP and P_i and bound ATP_B as ADP and carbamoyl phosphate. When the pulse period with $[\tau^{-32}P]ATP$ was ended by addition of $HClO_4$, ATP_B was recovered as ATP but ATP_A was released as ADP and P_i. Thus, the terminal phosphoryl group of the ATP_A molecule had become labile in the central complex containing the activated bicarbonate, as expected if carboxyphosphate was formed. The fact that products are not released unless ammonia is present explains the lack of substantial isotope exchanges between labeled ADP or P_i and ATP in the absence of ammonia.

Carbamoyl phosphate synthetase requires a monovalent metal cation such as K^+ and free divalent metal cations in addition to those required in the formation of the Me-ATP complexes [22-24]. The bicarbonate-dependent ATPase reaction, which reflects the activation of bicarbonate, also requires mono and divalent metal cations [17,23,24]. It appears that these metals are involved in the step of activation of bicarbonate.

3. CARBOXYPHOSPHATE IS THE ACTIVATED FORM OF BICARBONATE IN THE REACTION OF CARBAMOYL PHOSPHATE SYNTHESIS

The formation of carboxyphosphate was demonstrated using two approaches. In one of them this intermediate was trapped in the E. coli enzyme as formate using potassium borohydride, or as the expected trimethylated derivative using diazomethane [25]. This approach has failed, in our hands (Ebner, K. and Rubio, V., unpublished results), with the rat liver enzyme.

In a second approach positional isotope exchange techniques were used [5,12,26]. Fig.2 schematizes our experiments with the mammalian enzyme [12] in which we used ATP labelled with ^{18}O in the non-bridging positions of the ß-phosphoryl group; the ß-γ bridging oxygen became labelled upon incubation with the enzyme and bicarbonate, as expected if there is reversible carboxyphosphate formation without dissociation of the products.

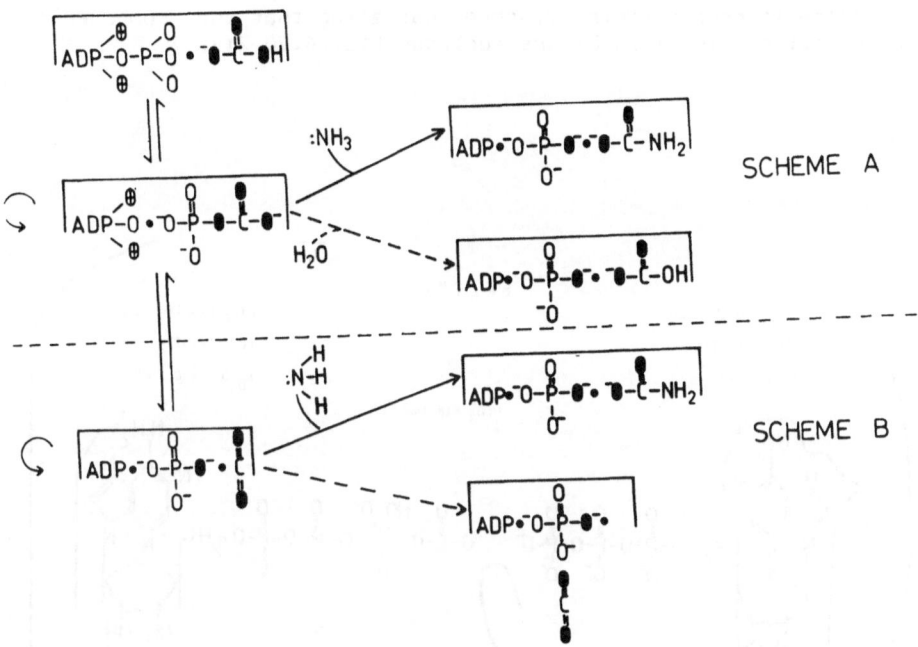

Figure 2. Diagram illustrating the formation of enzyme-bound ADP and carboxyphosphate in the reaction catalyzed by carbamoyl phosphate synthetase. In Scheme A the carboxyphosphate is the carboxylating species, whereas in Scheme B it decomposes to enzyme-bound P_i and CO_2, and the latter is the carboxylating species. Broken arrows indicate the ATPase activity. Curved arrows indicate the rotation of the ß-phosphoryl group of ADP and of enzyme-bound P_i. Oxygens labelled with crosses represent ^{18}O initially present in the non-bridging positions of the ß-phosphoryl group of ATP. Oxygens shown in black represent ^{18}O initially present in HCO_3^-.

Sauers et al. [27], on the basis of extrapolation from results with esters of carbonic acid with various alcohols, concluded that the $t_{\frac{1}{2}}$ of the carbonic-phosphoric anhydride in aqueous solution at 25°C should be <0.069 secs. We studied with the frog liver enzyme the decay of the enzyme-bound intermediate [14] and found a $t_{\frac{1}{2}}$ at this temperature of about 0.8 secs. Clearly, the enzyme protects the intermediate, possibly by excluding water from the environment of the carboxyphosphate.

Carboxyphosphate might be the carboxylating agent or it might decompose to enzyme-bound P_i and CO_2 and the latter would be the true carboxylating species. The evidence with carbamoyl phosphate synthetase [5,12,28] does not favor the latter possibility. For example, we did not find [12], as expected if CO_2 were generated reversibly from carboxyphosphate at the catalytic center, any incorporation of ^{18}O into ATP from $HC^{18}O_3^-$ (Fig. 2).

There is considerable evidence indicating that the components of the central complex (Fig.3) are secluded [12,14,20,21].

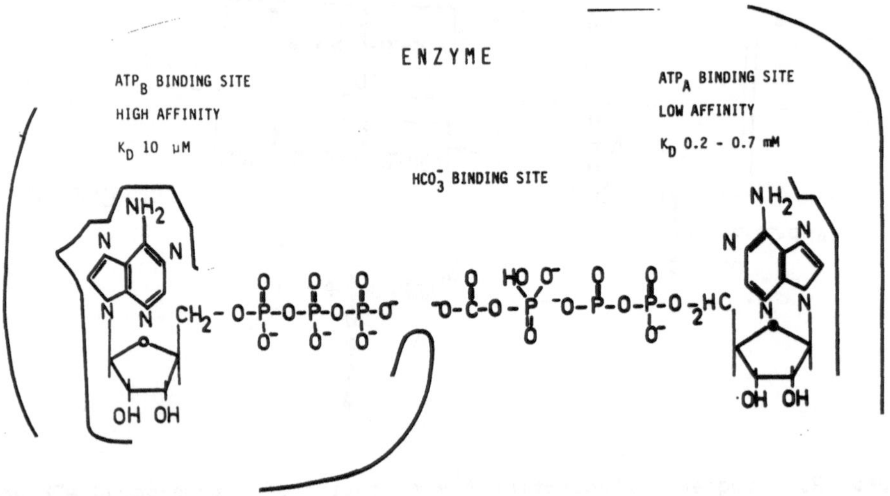

ENZYME

ATP$_B$ BINDING SITE
HIGH AFFINITY
K_D 10 μM

HCO$_3^-$ BINDING SITE

ATP$_A$ BINDING SITE
LOW AFFINITY
K_D 0.2 - 0.7 mM

Figure 3. Diagram to illustrate a structure for the enzyme-ATP$_B$-carboxyphosphate-ADP$_A$ complex (central complex). The deep cleft for ATP$_B$ illustrates the very slow dissociation of this ATP molecule from the enzyme (see text).

None of the components dissociate from this complex. ATP$_A$ and HCO$_3^-$ dissociate from the complex [enzyme-ATP$_A$-HCO$_3^-$-ATP$_B$], which is in equilibrium with the central complex (although the equilibrium is largely in favor of the carboxyphosphate), but the dissociation is not instantaneous (t$_\frac{1}{2}$ at 37°C of about 0.25-0.35 secs)[14,20,21]. The dissociation from the enzyme of the molecule of ATP$_B$ is extremely slow (t$_\frac{1}{2}$ >50 secs at 25°C, with the rat liver enzyme [21]; nearly 2 min with the E. coli enzyme, P. Llorente and V. Rubio, unpublished experiments).

In fact, ATP$_B$ can only leave if HCO$_3^-$ or ATP$_A$ dissociate from the enzyme [21], indicating that it is bound in a pocket which becomes inaccessible to the solution when HCO$_3^-$ and ATP$_A$ are also bound. Pulse-chase results demonstrate that the active centre in the rat liver enzyme exists in accessible and inaccessible configurations for the molecule of ATP$_B$ [21]: the nucleotide binds and the products dissociate from the accessible configuration, but the formation of the central complex requires the transition to the inaccessible form (Fig.4). These changes may limit the rate of the overall reaction.

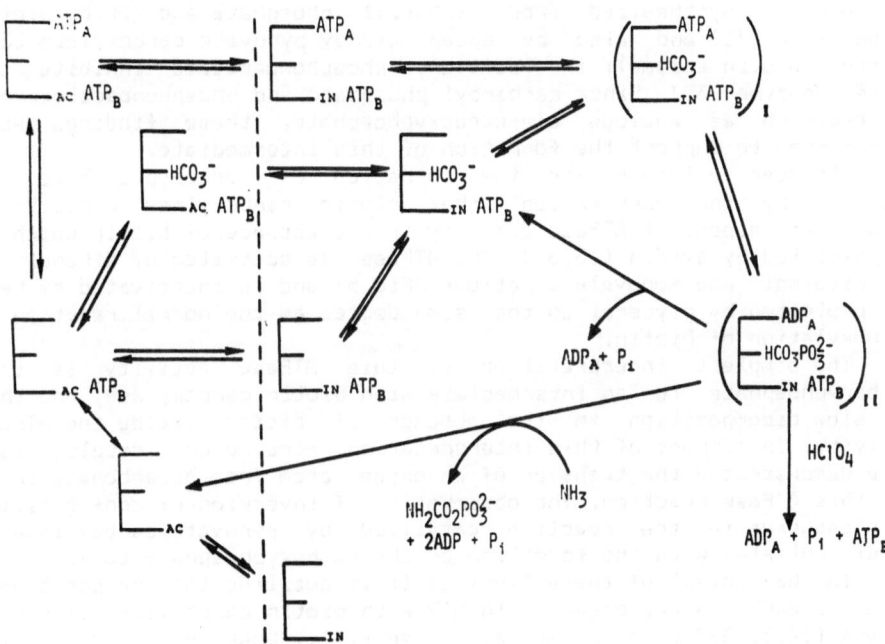

Figure 4. Reaction intermediates and reaction steps for which evidence has been obtained. E$_{AC}$ and E$_{IN}$ represent enzyme in which the ATP$_B$ binding site is in the accessible and inaccessible conformations, respectively. All the accessible conformations are to the left of the broken line. The double-headed arrow indicates that the ATP$_B$ site in this form is in reversible equilibrium with ATP in solution.

4. CARBOXYPHOSPHATE SYNTHESIS WITH OTHER ENZYMES THAT USE HCO₃⁻.

4.1. Biotin enzymes. Although carboxyphosphate was proposed as an intermediate for biotin enzymes many years ago [29], there was no clear-cut evidence for the formation of this compound with these enzymes, and concerted mechanisms were favored by many. Generally the mechanistic analysis of the carboxylation of biotin is hampered by the covalent linkage of this prosthetic group to the enzyme, which prevents the study of the activation of bicarbonate in the absence of the acceptor. However, the biotin carboxylase subcomponent of E. coli acetyl CoA carboxylase contains no biotin and catalyzes the carboxylation of free biotin [30]. The lack of ADP-ATP and P_i-ATP exchanges with this subcomponent [7,13], as well as kinetic evidence provided by steady-state and product inhibition studies [13] indicate that the products are not released prior to the addition of biotin. Although this may be interpreted to favor a concerted mechanism, the findings with carbamoyl phosphate synthetase indicate that even if carboxyphosphate is formed with biotin carboxylase, ADP release and ADP-ATP exchange need not be expected.

ATP is synthesized from carbamoyl phosphate and ADP by biotin carboxylase [31] and also by sheep kidney pyruvate carboxylase [32] (another biotin enzyme); in addition, phosphonoacetate inhibits the latter enzyme [32]. Since carbamoyl phosphate and phosphonoacetate may be regarded as analogs of carboxyphosphate, these findings were interpreted to support the formation of this intermediate.

Stronger evidence for the formation of carboxyphosphate was provided by the observation that biotin carboxylase exhibits a bicarbonate-dependent ATPase activity in the absence of biotin which is not affected by avidin [13,33]. The ATPase was activated by ethanol or by divalent and monovalent cations (Fig.5) and is inactivated by heat and protected by glycerol to the same degree as the normal reaction of carboxylation of biotin.

The simplest interpretation of this ATPase activity is that carboxyphosphate is an intermediate with biotin carboxylase, and that its slow decomposition in the absence of biotin yields the ATPase activity. In support of this interpretation, more recent results [34] have demonstrated the transfer of an oxygen atom from bicarbonate to P_i in this ATPase reaction. The observation of inversion of configuration at phosphorus in the reaction catalyzed by pyruvate carboxylase is consistent also with the formation of the carboxyphosphate [35].

In the context of these findings it is puzzling that no positional isotopic exchange was observed in ATP with biotin carboxylase either by others [34,36,37] or by ourselves (Climent, I., Lowe, G. Rubio, V., unpublished results). A large forward commitment of the carboxyphosphate or the lack of torsional freedom of the β phosphoryl group of enzyme-bound ADP are possible reasons for the failure to observe this exchange.

Overall, there is good indirect evidence for the formation of carboxyphosphate with biotin enzymes but it would be highly desirable to demonstrate the presence of this intermediate more directly. This task may prove difficult in view of the instability of the

intermediate and the apparently small proportion of the enzyme charged
with it in the absence of biotin.

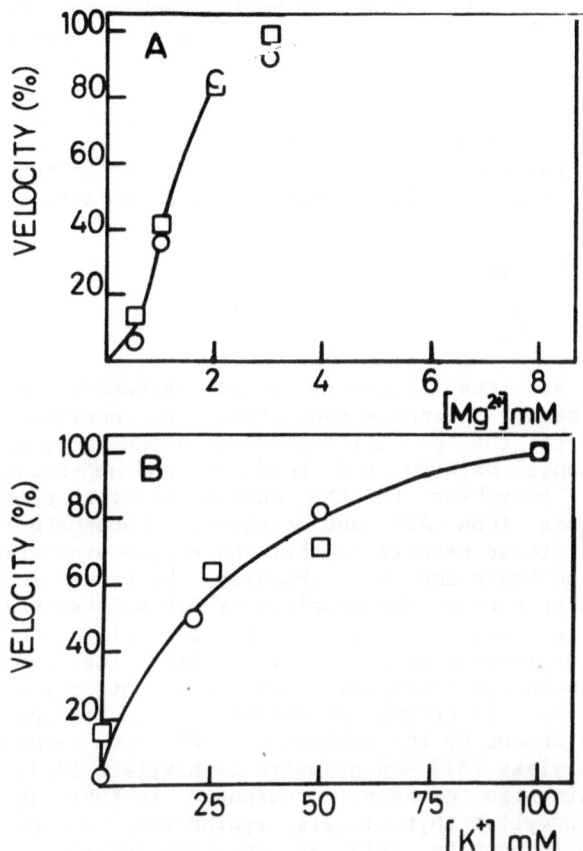

Figure 5.
Influence of the
concentration of
Mg²⁺ and K⁺ on the
ATPase (□) and
carboxylase (○)
activities of
biotin carboxylase.
In (A) the concen-
tration of ATP was
1 mM and the con-
centrations
indicated on the
x-axis are total
Mg²⁺.

4.2 Phosphoenolpyruvate carboxylase. Although a concerted mechanism was
favored earlier [8], more recent evidence based on a) the
quantification of carbon isotope effects for the carbon atom arising
from bicarbonate [38], b) the stereochemical course at phosphorus [39],
and c) the bicarbonate-dependent dephosphorylation of phosphoenol-α-
ketobutyrate without concomitant bicarbonate fixation [40] are
consistent with a stepwise mechanism involving an enzyme-bound
carboxyphosphate intermediate.

**5. SYNTHESIS OF CARBAMOYL PHOSPHATE FROM CARBOXYPHOSPHATE BY CARBAMOYL
PHOSPHATE SYNTHETASE.**

We showed [14] in pulse chase experiments with the frog liver enzyme
that NH₃ rather than NH₄⁺ is the true substrate for this reaction, and
that the reaction is relatively fast, with a second-order rate constant

for NH₃ at 25°C of 8.10⁴ M⁻¹s⁻¹. The rapidity of the reaction reflects the fact that ATP$_B$ is already bound to the enzyme and may react immediately to give carbamoyl phosphate. If, as formulated in reactions 2-4, carboxyphosphate reacts first with NH₃ to yield carbamate, which is then phosphorylated by ATP$_B$, it appears reasonable to have ATP$_B$ ready for reaction within the enzyme, since carbamate is unstable, particularly if water is present (M. Aresta, personal communication).

Nevertheless, the fact that the reaction with NH₃ is fast and that carboxyphosphate, ATP$_B$ and NH₃ are bound simultaneously to the enzyme would also render attractive a concerted mechanism of reaction between these three reagents with no intermediate formation of carbamate.

The best evidence for the intermediacy of carbamate in the reaction catalyzed by carbamoyl phosphate synthetase was provided by experiments [26] in which the E. coli enzyme was shown to catalyze positional isotopic exchange between the bridging and non-bridging oxygens of [¹⁸O]carbamoyl phosphate in the course of the partial reaction of ATP synthesis from ADP and carbamoyl phosphate. The simplest interpretation of these results is that the enzyme synthesizes reversibly enzyme-bound carbamate and ATP. However, the value of this finding depends very much on whether the reaction of ATP synthesis from ADP and carbamoyl phosphate truly reflects the reversal of a normal step of carbamate phosphorylation, or whether it reflects the reversal of the step of carboxy-phosphate synthesis. This latter interpretation is supported by the fact that carbamoyl phosphate is an analog of carboxyphosphate, as evidenced by the synthesis of ATP from carbamoyl phosphate by biotin carboxylase [31] and pyruvate carboxylase [32], two enzymes that do not synthesize carbamoyl phosphate in their normal reaction. Furthermore, formyltetrahydrofolate synthetase, an enzyme that synthesizes formylphosphate [41] as an intermediate, also catalyzes the synthesis of ATP from ADP and carbamoyl phosphate, and in this case the latter presumably acts as an analog of formylphosphate. Therefore, until the role of the partial reaction of ATP synthesis by carbamoyl phosphate synthetase is clarified, the potential intermediacy of carbamate in the reaction of carbamoyl phosphate synthesis by carbamoyl phosphate synthetase will remain uncertain.

6. ACTIVATION OF CARBAMOYL PHOSPHATE SYNTHETASE BY ACETYLGLUTAMATE.

Carbamoyl phosphate synthetases catalyze the initial step in the pathways of urea, arginine and pyrimidine biosynthesis. These enzymes are subjected to allosteric modulation in many organisms. The ureotelic enzyme is activated by N-acetyl-L-glutamate. The activation is dramatic (activity in the absence of acetylglutamate, <2 % of that at saturation of the activator, in the usual assay [42]). Acetylglutamate is partially replaced in activating the enzyme by a number of chemically unrelated cryoprotecting agents [42]. These cryoprotectants increase the affinity for acetylglutamate, as expected if they trigger the same conformational change in the enzyme as acetylglutamate. Although there is considerable evidence for the induction of conformational changes in the enzyme by acetylglutamate (see [43]), these changes are characterized only partly. Activation by acetylglutamate is associated with the generation of a relatively "high affinity" (K_m, about 0.2 mM) site for the molecule of ATP_A, and also increases considerably the apparent affinity of the enzyme for K^+ and Mg^{2+} [24]. It is probable that these two metals bind at or close to the ATP_A binding site, as expected if they are involved in the formation of the carboxyphosphate. The affinity for acetylglutamate is much higher in the central complex (Fig.3) than in the free enzyme [44] as a consequence of changes in the rate of its association [45]. Thus, binding of acetylglutamate favors the formation of the central complex, and this may be a key factor in the activation.

The ureotelic carbamoyl phosphate synthetase consists of a single polypeptide of about 160 kDa [46]. Binding studies with the rat liver enzyme demonstrated a single site for acetylglutamate per polypeptide chain [44]. Limited proteolysis revealed that the enzyme is composed, from the N-terminus, of 4 domains of about 40, 40, 60 and 20 kDa [47-50] (Fig. 6). To understand the mechanism of activation of the synthetase it was essential to determine the location of the binding site for acetylglutamate in the enzyme. This was also relevant because other carbamoyl phosphate synthetases are susceptible to a number of positive modulators although not to acetylglutamate; in view of the homologies between the various synthetases for which the sequences are known (see [46]), it is probable that modulatory sites in different synthetases are derived from a single ancestral site.

We have demonstrated [43], using a combination of limited proteolysis with elastase and trypsin and photoaffinity labeling with N-chloroacetyl-[^{14}C]-L-glutamate, that acetylglutamate binds to the 20-kDa COOH-terminal domain of the rat liver enzyme (Fig. 6). This excludes the possibility that the acetylglutamate binding site evolved from an ancestral substrate site for glutamine (glutamine is not a substrate of the ureotelic enzyme, but is a substrate for other types of carbamoyl phosphate synthetase): this substrate binds to the small subunit of the E. coli enzyme, which is homologous to the NH_2-terminal domain of the rat liver enzyme (see [46]).

The site for acetylglutamate is within the moiety of ureotelic carbamoyl phosphate synthetase that is homologous to the large subunit of the E. coli enzyme [46]. In this microorganism, the large subunit is responsible for carbamoyl phosphate synthesis from NH_3 and it also contains the sites for the positive effectors ornithine and IMP [51]. Ornithine in the E. coli enzyme [51] and acetylglutamate in the mammalian enzyme [24] increase the apparent affinity for ATP, and glycerol partly replaces these activators in both enzymes [42,52]. With carbamoyl phosphate synthetase II (the enzyme that catalyzes the first step of pyrimidine synthesis in mammals) phosphoribosyl pyrophosphate, another positive effector, also increases the apparent affinity for ATP and again is partly replaced by glycerol [53]. Thus, activation of the synthetases by these effectors exhibits similarities that may reflect a common mechanism. Therefore, there may be a topographically equivalent site that would have evolved to yield the particular specificity for each enzyme.

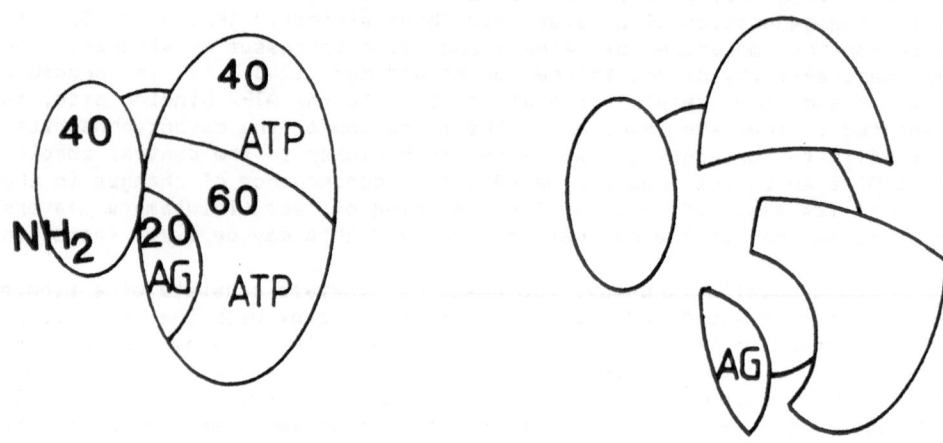

Figure 6. Diagram to show the domain structure of rat liver carbamoyl phosphate synthetase, as revealed by limited proteolysis. The figures given are approximate masses in kDa. The NH₂ denotes the amino-terminal end of the polypeptide. Putative ATP binding sequences have been identified tentatively in the 2nd and 3rd domains of 40 and 60 kDa. The binding site for acetylglutamate (AG) is located in the COOH-terminal domain of 20 kDa. In the presence of AG and ATP there is a single point of preferential proteolytic cleavage in the bridge between th N-terminal domain and the rest of the molecule, and thus the three domains in the carboxy-terminal moiety are assumed to be tightly associated. In the presence of acetylglutamate alone, two bridges between these domains become accessible, and this is represented in the figure by illustrating the enzyme in a more relaxed configuration with exposed bridges between the domains.

Binding of acetylglutamate to the 20-kDa COOH-terminal domain greatly increases the susceptibility of rat liver carbamoyl phosphate synthetase to proteolytic attack at the region that joins the 20 kDa COOH-terminal domain to the rest of the molecule [47,50]; thus, a region about 20 kDa from the COOH-terminus is exposed. ATP protects this region from proteolysis. Binding of acetylglutamate is also known to allow access of ATP to the site for the molecule of ATP_A [24]. Thus, this point of proteolytic cleavage is likely to be at or close to the binding site for ATP_A. Indeed, one of two highly conserved putative ATP binding sequences tentatively identified by Nyunoya et al. [46] is located at 22-kDa from the COOH-terminus. The site for ATP$_A$ is involved in making the mixed carbonic-phosphoric anhydride, which must be excluded from water. The COOH-terminal domain of 20 kDa may provide protection by shielding the site for ATP$_A$. We propose that acetyl-glutamate binding relaxes the hinge region between the 20-kDa COOH-terminal domain and the rest of the molecule, allowing movement to occur, thus exposing the site for ATP$_A$ (Fig. 7). We also propose that an equivalent domain in other carbamoyl phosphate synthetases is involved in binding of positive effectors.

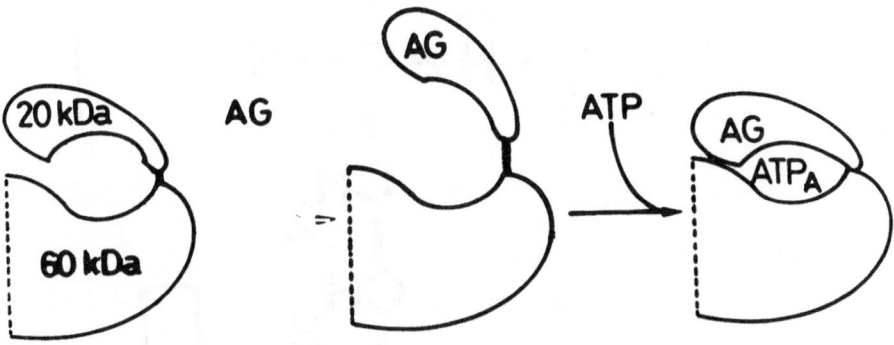

Figure 7. Proposed model for the action of acetylglutamate and ATP$_A$. A part of the COOH-terminal moiety of the enzyme including the 20 kDa and the adjacent 60 kDa domains is presented. Acetylglutamate binding is proposed to relax this region and to expose the site for ATP$_A$. On binding of ATP$_A$ the enzyme adopts a tight conformation to minimize hydrolysis of carboxyphosphate.

In a different approach, we have studied the acetylglutamate binding site of the rat liver enzyme [54], using a competitive binding assay to examine the affinity of the enzyme for analogs of acetyl-glutamate in which the groups attached to the chiral α-carbon were systematically modified. The results of this study showed that the conformation of bound acetylglutamate may be similar to that of the minimum energy conformer and that there is a strong "chelate" effect between the groups in the molecule of acetylglutamate that are involved in the binding. The two carboxyls of acetylglutamate are essential and the acetamido group enhances binding 5000-fold. The relative positions of the α-COO$^-$ and the acetamido group are critical, and the acetamido group is bound within a cavity of limited dimensions; it appears to form hydrogen bonds with the enzyme that may be important in activation. The α and γ-carbons and the δ-carboxyl are in close contact with the enzyme, and the subsite for the γ-carbon and the δ-carboxyl is mobile; this mobility is probably involved in the activation.

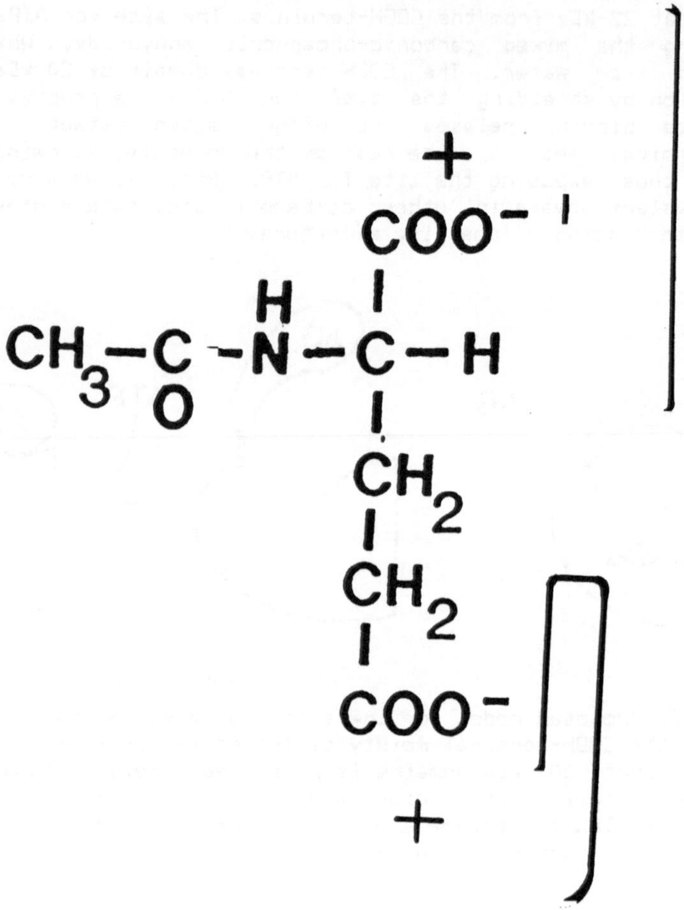

234

Figure 8. A model for the binding site of carbamoyl phosphate synthetase I for acetylglutamate, based on work using analogs of this activator.

The ß-carbon is at the surface of the enzyme and the hydrogens attached to it face the solution; phenyl substituents at this carbon have virtually no influence on the binding and activation. From these results, a model for the site of acetylglutamate was constructed, and is presented in Fig. 8. This model stresses the participation of salt bridges with the carboxyls, of hydrogen bonds with the acetamido group, and the existence of physical restrictions in the cavities surrounding the various carbons except the β-C. In addition, a flexible connection between the subsites for the γ-COO⁻ and the α-COO⁻ and acetamido groups is shown. We propose as a working hypothesis that the replacement of intramolecular hydrogen bonds in the protein by intermolecular hydrogen bonds with the acetamido group triggers conformational changes that result in the relocation of the subsite for the γ-COO⁻, with concomitant activation.

7. ACKNOWLEDGMENTS. Supported by grants from the Spanish Government 0294/81 and PB85-0198 (Comisión Asesora de Investigación Científica y Técnica) and PB87-0189 (Dirección General de Investigación Científica y Técnica) and by an Acción Integrada from the Ministerio de Educación y Ciencia (Spain) and the British Council. V. Rubio is a member of the IIC-KUMC International Molecular Cytology Programme. Figs. 1 and 2 are reproduced by permission from Rubio, V., **Bioscience Reports** vol.6, pgs. 337 and 340, copyright (c) 1986, The Biochemical Society, London. Fig. 5 is reproduced with permission from Academic Press from reference [33].

8. REFERENCES

1. Cooper, T. G., Filmer, D., Wishnick, M., and Lane, M.D. (1969) 'The Active Species of CO_2 Utilized by Ribulose Diphosphate Carboxylase' J. Biol. Chem. **244**, 1081-1083.
2. Jones, J. P., Gardner, E. J., Cooper, T. G., and Olson, R. E. (1977) 'Vitamin K-dependent Carboxylation of Peptide-bound Glutamate' J. Biol. Chem. **252**, 7738-7742.
3. Utter. M. F., and Kolenbrander, H. M. (1972) 'Formation of Oxalacetate by CO_2. Fixation on Phosphoenolpyruvate', in P. D. Boyer (ed), **The Enzymes** (3ʳᵈ Edition), Academic Press, New York, pp. 117-168.
4. Jones, M. E., and Spector, L. (1960) 'The Pathway of Carbonate in the Biosynthesis of Carbamyl Phosphate', J. Biol. Chem. **235**, 2897-2901.
5. Wimmer, M. J., Rose, I. A., Powers, S. G., and Meister, A. (1979) 'Evidence that Carboxyphosphate Is a Kinetically Competent Intermediate in the Carbamyl Phosphate Synthetase Reaction', J. Biol. Chem. **254**, 1854-1859.
6. Kaziro, Y., Hass, L. F., Boyer, P. D., and Ochoa, S. (1962) 'Mechanism of the Propionyl Carboxylase Reaction', J. Biol. Chem. **237**, 1460-1468.
7. Polakis, S. E., Guchhait, R. B., Zwergel, E. E., Lane, M. D., and Cooper, T. G. (1974) 'Acetyl Coenzyme A Carboxylase System of Escherichia Coli', J. Biol. Chem. **249**, 6657-6667.
8. Maruyama, H., Easterday, R. L., Chang, H. C., and Lane, M. D.(1966) 'The Enzymatic Carboxylation of Phosphoenolpyruvate' J. Biol. Chem. **241**, 2405-2412.
9. Fuchs, G. 'CO_2 Reduction by Anaerobic Microorganisms: Challenge for Chemistry', this symposium.
10. Marshall, M., Metzenberg, R. L., and Cohen, P. P. (1958) 'Purification of Carbamyl Phosphate Synthethase from Frog Liver', J.

Biol. Chem. 233, 102-105.

11. Guthöhrlein, G., and Knappe, J. (1969) 'Structure and Function of Carbamoylphosphate Synthetase. On the Mechanism of Bicarbonate Activation', Eur. J. Biochem. 8, 207-214.

12. Rubio, V., Britton, H. G., Grisolia, S., Sproat, B. S., and Lowe, G. (1981) 'Mechanism of Activation of Bicarbonate Ion by Mitochondrial Carbamoyl-Phosphate Synthetase: Formation of Enzyme-bound Adenosine Diphosphate from the Adenosine Triphosphate that Yields Inorganic Phosphate', Biochemistry 20, 1969-1974.

13. Climent, I., (1986) 'Fijación Enzimática de HCO_3^-: Estudio con la Biotin Carboxilasa de E. Coli', Ph.D. Dissertation. Universidad de Valencia, Valencia, Spain.

14. Rubio, V., and Grisolia, S. (1977) 'Mechanism of Mitochondrial Carbamoyl-Phosphate Synthetase. Synthesis and Properties of Active CO_2, Precursor of Carbamoyl Phosphate', Biochemistry 16, 321-329.

15. Cohen, N.S., Kyan, F. S., Kyan, S. S., Cheung C. W. and Raijman L. (1985) 'The apparent K_m of ammonia for carbamoyl phosphate synthetase (ammonia) in situ', Biochem. J. 229, 205-211.

16. Marshall, M. (1976) 'Carbamyl Phosphate Synthetase I from Frog Liver', in S. Grisolia, R. Báguena and F. Mayor (Eds.), The Urea Cycle, John Wiley and Sons, New York, pp. 133-156.

17. Metzenberg, R. L., Marshall, M., and Cohen, P. P. (1958) 'Carbamyl Phosphate Synthetase: Studies on the Mechanism of Action', J. Biol. Chem. 233, 1560-1564.

18. Jones, M. E. (1976) 'Partial Reactions of Carbamyl-Phosphate Synthetase: A Review and an Inquiry into the Role of Carbamate', in S. Grisolia, R. Báguena and F. Mayor (Eds.), The Urea Cycle, John Wiley and Sons, New York, pp. 107-122.

19. Anderson, P. M., and Meister, A. (1965) 'Evidence for an Activated Form of Carbon Dioxide in the Reaction Catalyzed by Escherichia Coli Carbamyl Phosphate Synthetase', Biochemistry 4, 2803-2809.

20. Rubio, V., Britton, H. G., and Grisolia, S. (1979) 'Mechanism of Carbamoyl-Phosphate Synthetase. Binding of ATP by the Rat-Liver Mitochondrial Enzyme', Eur. J. Biochem. 93, 245-256.

21. Britton, H. G., Rubio, V., and Grisolia, S. (1979) 'Mechanism of Carbamoyl Phosphate Synthetase. Properties of the Two Binding Sites for ATP', Eur. J. Biochem. 102, 521-530.

22. Marshall, M., Metzenberg, R. L., and Cohen, P. P. (1961) 'Physical and Kinetic Properties of Carbamyl Phosphate Synthetase from Frog Liver', J. Biol. Chem. 236, 2229-2237.

23. Lusty, C. J. (1978) 'Carbamyl Phosphate Synthetase. Bicarbonate-Dependent Hydrolysis of ATP and Potassium Activation', J. Biol. Chem. 253, 4270-4278.

24. Rubio. V., Britton, H. G., and Grisolia, S. (1983) 'Mitochondrial Carbamoyl Phosphate Synthetase Activity in the Absence of N-Acetyl-L-Glutamate. Mechanism of Activation by this Cofactor', Eur. J. Biochem. 134, 337-343.

25. Powers, S. G., and Meister, A. (1978) 'Carbonic-Phosphoric Anhydride (Carboxy Phosphate). Significance in Catalysis and Regulation of Glutamine-Dependent Carbamyl Phosphate Synthetase', J. Biol. Chem. 253, 1258-1265.

26. Raushel, F. M., and Villafranca, J. J. (1980) 'Phosphorus-31 Nuclear Magnetic Resonance Application to Positional Isotope Exchange Reactions Catalyzed by Escherichia Coli Carbamoyl-Phosphate Synthetase: Analysis of Forward and Reverse Enzymatic Reactions', Biochemistry 19, 3170-3174.

27. Sauers, C. K., Jenks, W. P., and Groh, S. (1975) 'The Alcohol-Bicarbonate-Water System. Structure-Reactivity Studies on the Equilibria for Formation of Alkyl Monocarbonates and on the Rates of Their Decomposition in Aqueous Alkali', J. Am. Chem. Soc. 97, 5546-5553.

28. Raushel, F. M., and Villafranca, J. J. (1979) 'Determination of Rate-Limiting Steps of Escherichia Coli Carbamoyl-Phosphate Synthase. Rapid Quench and Isotope Partitioning Experiments', Biochemistry 18, 3424-3429.

29. Tietz, A., and Ochoa, S. (1958) 'Fluorokinase and Pyruvic Kinase', Arch. Biochem. Biophys. 78, 477-493.

30. Lane, M. D., Moss, J., and Polakis, S. E. (1974) 'Acetyl Coenzyme A Carboxylase', Curr. Top. Cell. Regul. 8, 139-195.

31. Polakis, S. E., Guchhait, R. B., and Lane, M. D. (1972) 'On the Possible Involvement of a Carbonyl Phosphate Intermediate in the Adenosine Triphosphate-dependent Carboxylation of Biotin', J. Biol. Chem. 247, 1335-1337.

32. Ashman, L. K., and Keech, D. B. (1975) 'Sheep Kidney Pyruvate Carboxylase. Studies on the Coupling of Adenosine Triphosphate Hydrolysis and CO_2 Fixation', J. Biol. Chem. 250, 14-21.

33. Climent, I., and Rubio, V. (1986) 'ATPase Activity of Biotin Carboxylase Provides Evidence for Initial Activation of HCO_3^- by ATP in the Carboxylation of Biotin', Arch. Biochem. and Biophys. 251, 465-470.

34. Ogita, T., and Knowles, J. R. (1988) 'On the Intermediacy of Carboxyphosphate in Biotin-Dependent Carboxylations', Biochemistry 27, 8028-8033.

35. Jansen, D. E., and Knowles, J. R. (1985) 'N-Carboxybiotin Formation by Pyruvate Carboxylase: The Stereochemical Consequence at Phosphorus', J. Am. Chem. Soc. 107, 8304-8305.

36. Wimmer, M. Unpublished experiments quoted in [34].

37. Tipton, P. A., and Cleland, W. W. (1988) 'Carbon-13 and Deuterium Isotope Effects on the Catalytic Reactions of Biotin Carboxylase', Biochemistry 27 4325-4331.

38. O'Leary, M. H., Rife, J. E., and Slater, J. D. (1981) 'Kinetic and Isotope Effect Studies of Maize Phosphoenolpyruvate Carboxylase', Biochemistry 20, 7308-7314.

39. Hansen, D. E., and Knowles, J. R. (1982) 'The Stereochemical Course at Phosphorus of the Reaction Catalyzed by Phosphoenolpyruvate Carboxylase', J. Biol. Chem. 257, 14795-14798.

40. Fujita, N., Izni, K., Nishino, T., and Katsuki, H. (1984) 'Reaction Mechanism of Phosphoenolpyruvate Carboxylase. Bicarbonate-Dependent Dephosphorylation of Phosphoenol-α-ketobutyrate', Biochemistry 23, 1774-1779.

41. Buttlaire, D. H., Balfe, C. A., Wendland, M. F., and Himes, R. H. (1979) 'Carbamyl Phosphate-Dependent ATP Synthesis Catalyzed by

238

Formyltetrahydrofolate Synthetase', **Biochim. Biophys Acta. 567**, 453-463.

42. Rubio, V., Britton, H. G. and Grisolia, S. (1983) 'Activation of Carbamoyl Phosphate Synthetase by Cryoprotectants', **Mol. Cell. Biochem. 53/54**, 279-298.

43. Rodríguez-Aparicio, L. B., Guadalajara, A.M. and Rubio, V. (1989) 'Physical Location of the Site for N-acetyl-L-glutamate, the Allosteric Activator of Carbamoyl Phosphate Synthetase, in the 20-Kilodalton COOH-Terminal Domain', **Biochemistry 28**, 3070-3074.

44. Alonso, E. and Rubio, V. (1983) 'Binding of N-acetyl-L-glutamate to Rat Liver Carbamoyl Phosphate Synthetase (Ammonia)', **Eur. J. Biochem. 135**, 331-337.

45. Britton, H. G. and Rubio, V. (1988) 'Carbamoyl-phosphate synthetase I. Kinetics of binding and dissociation of acetylglutamate and of activation and deactivation', **Eur. J. Biochem. 171**, 615-622.

46. Nyunoya, H., Broglie, K. E., Widgren, E. E. and Lusty, C. J. (1985) 'Characterisation and Derivation of the Gene Coding for Mitochondrial Carbamyl Phosphate Synthetase I of Rat', **J. Biol. Chem. 260**, 9346-9356.

47. Guadalajara, A.M. (1987) 'Estudio, Mediante Proteolisis Limitada, de la Estructura de Dominios, Unión de Ligandos y Cambios Asociados de Conformación de la Carbamil Fosfato Sintetasa de Hígado de Rata' Ph.D. Dissertation, University of Valencia, Valencia, Spain.

48. Powers-Lee, S. G. and Corina K. (1986) 'Domain Structure of Rat Liver Carbamoyl Phosphate Synthetase I', **J. Biol. Chem. 261**, 15349-15352.

49. Evans, D. R. and Balon, M. A. (1988) 'Controlled proteolysis of ammonia-dependent carbamoyl-phosphate synthetase I from syrian hamster liver', **Biochim. Biophys. Acta 953**, 185-196.

50. Marshall, M. and Fahien, L. A. (1985) 'Proximate Sulfhydryl Groups in the Acetylglutamate Complex of Rat Carbamylphosphate Synthetase I: Their Reaction with the Affinity Reagent 5'-p-Fluorosulfonylbenzoyl-adenosine', **Arch. Biochem. Biophys. 241**, 200-214.

51. Meister, A. and Powers, S. G. (1978) 'Glutamine-dependent Carbamyl Phosphate Synthetase: Catalysis and Regulation' **Adv. Enz. Reg. 16**, 289-315.

52. Rubio, V. and Llorente, P. (1982) 'Activation of Carbamoyl Phosphate Synthetase from Escherichia Coli by Glycerol', **Biochem. Biophys. Res. Commun. 107**, 1400-1405.

53. Ishida, H., Mori, M. and Tatibana, M. (1977) 'Effects of Dimethyl Sulfoxide and Glycerol on Catalytic and Regulatory Properties of Glutamine-Dependent Carbamoyl Phosphate Synthase from Rat Liver and Dual Effects of Uridine Triphosphate', **Arch. Biochem. Biophys. 182**, 258-265.

54. Britton, H., G., García-España, A., Goya, P., Rozas, I. and Rubio, V. (1989) 'Structure-Reactivity Study of the Binding of Acetylglutamate to Carbamoyl Phosphate Synthetase I', Submitted for publication.

CARBOXYPHOSPHATE; PREDICTED CHEMICAL PROPERTIES, SYNTHESIS AND ROLE AS AN INTERMEDIATE IN ENZYMIC REACTIONS.

W.B. Knight
Merck Sharp and Dohme Research Labs
Department of Enzymology
P.O. Box 2000
Rahway, NJ, 07065, USA

ABSTRACT. The evidence for the existence of carboxyphosphate as a distinct chemical entity is presented. The properties predicted for this unstable mixed phosphoanhydride are discussed. The evidence for the involvement of carboxyphosphate in the reactions catalyzed by carbamoyl-phosphate synthetases, biotin carboxylases and phosphoenolpyruvate carboxylases is reviewed. The role(s) of this species in these reactions is discussed.

1. INTRODUCTION

The interest in carboxyphosphate ($\underline{I}$), the mixed anhydride between carbonic and phosphoric acids, stems from it's putative role as an intermediate in some carboxylase reactions that utilize bicarbonate rather than CO_2 as the carbon source. These reactions include carbamoyl phosphate synthetases ($\underline{CPS}$), biotin dependent carboxylases and phosphoenolpyruvate carboxylase. Thus, carboxyphosphate has been implicated in such diverse processes as CO_2 fixation in C-4 plants, the biosynthesis of fatty acids and nitrogen metabolism including the biosynthesis of pyrimidines and arginine. From the scope of these reactions the relative importance of this high energy phosphate is evident. But, the involvement of $\underline{I}$ in enzymatic reactions is not without controversy. In fact, the existence of carboxyphosphate as a chemical entity has not been conclusively demonstrated. The biochemical and chemical evidence for the existence of $\underline{I}$ is reviewed below. The predicted properties of this unstable phosphate as well as it's potential role in enzymatic reactions are also discussed.

239

M. Aresta and J. V. Schloss (eds.),
Enzymatic and Model Carboxylation and Reduction Reactions for Carbon Dioxide Utilization, 239–258.
© 1990 *Kluwer Academic Publishers.*

2. PREDICTED CHEMICAL PROPERTIES OF CARBOXYPHOSPHATE

2.1 Properties of carboxyphosphate.

While carboxyphosphate has not been isolated and fully characterized, alkyl triesters [1] and other triesters of I [2] are known stable compounds (Figure 1). Griffith and Stiles [2] synthesized two dibenzyl carbobenzoxy phosphate derivatives via pyridine catalyzed reaction of silver dibenzyl phosphate with either benzyl or p-nitrobenzyl chloroformate. They observed that metal ions increased the rate of hydrolysis of the p-nitro analog by a factor of 2. This presumably proceeds through a complex of the metal with the carbonyl oxygens (structure V) [2]. Koshland [3] proposed a similar complex in the

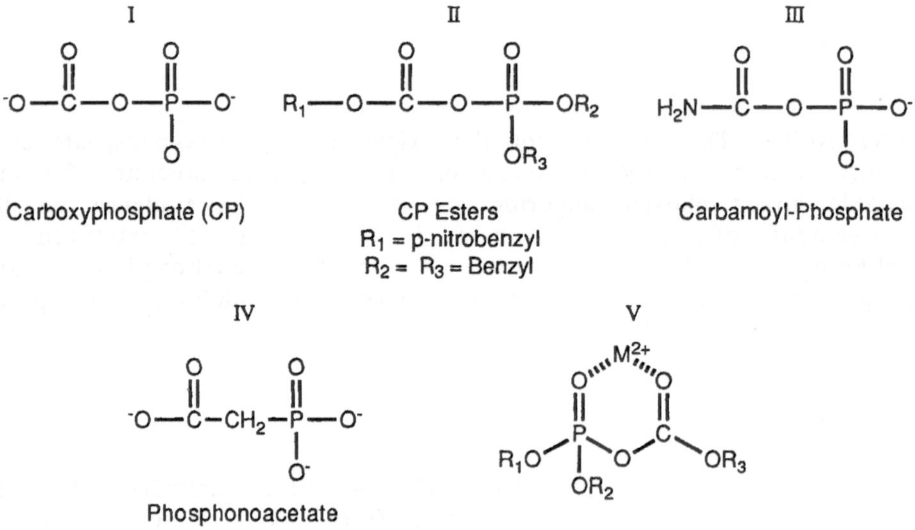

Figure 1. Structures of carboxyphosphate and analogs.

metal catalyzed hydrolysis of acetyl phosphate. But, he observed metal catalysis with only the dianion of this high energy phosphate and not the free acid. These results suggest that I should coordinate metal ions and that they may catalyze it's rate of hydrolysis in aqueous solvents. Griffith and Stiles [2] failed to observe incorporation of O-18 into the phosphate during the hydrolysis of II in $H_2^{18}O$. Therefore, the hydrolysis of esters of I proceeds via cleavage of the C-O bond and not the P-O bond. This reflects the attack of water on the carbonyl oxygen and not the phosphate. This indicates that the carbonyl is most susceptible to nucleophilic attack in the case of triesters of $\underline{I}$. Both carbamyol phosphate ($\underline{III}$) and phosphonacetate ($\underline{IV}$) can be considered analogs of $\underline{I}$, although Kluger et. al. have argued that the former could also be considered analogous to O-phosphobiotin [4]. These compounds are relatively stable and their reactivity in enzymatic reactions either as inhibitors or a phosphoryl donor in the case of $\underline{III}$ provide additional evidence for the involvement of $\underline{I}$ in the reaction.

Sauers et. al. [5] have predicted the thermodynamic properties of $\underline{I}$. They examined the stability of a series of alkyl carbonates according to the following equilibrium;

$$\text{eqn. 1} \quad \text{ROH} + \text{HCO}_3^{-} \rightleftharpoons \overset{\displaystyle O}{\overset{\|}{\text{R-OCO}^{-}}} + \text{H}_2\text{O}$$

where ROH was a series of aliphatic alcohols with different pKa's. They determined the equilibrium constants for the reaction in eqn. 1 as a function of the alkyl alcohol. The known pK_a's of the alcohols then allowed the construction of a Bronsted plot for the attack of the alkoxide on CO_2. They could then predict the equilibrium constant for the attack of phosphate on CO_2 (eqn. 2) from the pK_a of the phosphate dianion (about 12). This extrapolation suggested that the equilibrium constant for eqn. 2 should favor formation of $\underline{I}$ (with $K_{eq}=10^3 M^{-1}$) in the absence of water. This corresponds to a dissociation constant of $\underline{I}$ to CO_2 and the phosphate trianion of 0.001M.

$$\text{eqn. 2} \quad \overset{\displaystyle O}{\underset{\displaystyle O}{\overset{\|}{\underset{|}{O-P-O}}}}^{-3} + CO_2 \rightleftharpoons \overset{\displaystyle O \quad\quad O}{\underset{\displaystyle O_-}{\overset{\| \quad\quad \|}{\underset{|}{.O-P-O-C-O^-}}}}$$

They predicted that the equilibrium constant for eqn. 2 at pH 8 (dissociation to produce monoprotonated phosphate) would be $0.1M^{-1}$. This indicates that $\underline{I}$ could exist in the environment of an active site in the absence of water. The equilibrium constant for the reaction in eqn. 1 where R= PO_4H would be 0.13. Therefore, there will be little $\underline{I}$ present in aqueous solution. This equilibrium constant yields a ΔG of -3.6 kcal/mol for the hydrolysis of $\underline{I}$ (reversal of eqn. 1, R=PO_4H). This led Sauers et al. to conclude that $\underline{I}$ is not a particularly high energy phosphoester at higher pH although at lower pH hydrolysis of I to produce the dianion of phosphate and bicarbonate which would result in a more negative ΔG_{hyd}.

The life-time of $\underline{I}$ has been estimated in both enzyme active sites and free in solution. Sauers and co-workers determined the rate of decomposition of the alkyl carbonates. A Bronsted plot of these rate constants versus the pKa for the alcohols allowed these workers to estimate the rate constant for the hydrolysis of $\underline{I}$ to be ≥ 10 sec^{-1} to produce the phosphate dianion and carbonate. Powers and Meister [6,7a] suggest that the half-life of enzyme bound $\underline{I}$ in 67% DMSO is 2.5 min while the data of Sauers et. al. [5] predict a half-life of less than 0.1s in aqueous solution. Rubio and Grisiola report a half-life of enzyme bound $\underline{I}$ of 2.8s in a H$_2$O/acetone mixture from their studies with frog liver CPS [8].These results suggest that organic solvents or the exclusion of water from an enzyme active site might stabilize $\underline{I}$.

2.2 Efforts to synthesize $\underline{I}$.

Griffith and Stiles [2] reported in 1965 that the initial attempts to synthesize $\underline{I}$ from phosphate and CO$_2$ via metal catalysis were unsuccessful although they did not report any experimental conditions. In their studies with CPS, Powers and Meister reported the trapping of products derived from $\underline{I}$. in control reactions in the absence of enzyme. They isolated trimethyl carboxyphosphate from their control reactions upon the addition of diazomethane. They attribute this to the metal catalyzed phosphorylation of bicarbonate by ATP prior to trapping with diazomethane (eqn 3). Ogita and

eqn. 3 $\text{MgATP} + \text{HCO}_3^- \;\rightleftharpoons\; \text{MgADP} + \;^-\text{O}-\overset{\overset{\displaystyle O}{\|}}{\underset{\underset{\displaystyle O^-}{|}}{P}}-\text{O}-\overset{\overset{\displaystyle O^-}{\|}}{C}-\text{O}^-$

$$^-\text{O}-\overset{\overset{\displaystyle O}{\|}}{\underset{\underset{\displaystyle O^-}{|}}{P}}-\text{O}-\overset{\overset{\displaystyle O}{\|}}{C}-\text{O}^- + \text{CH}_3\text{N}_2 \longrightarrow \text{CH}_3\text{O}-\overset{\overset{\displaystyle O}{\|}}{\underset{\underset{\displaystyle OCH_3}{|}}{P}}-\text{O}-\overset{\overset{\displaystyle O}{\|}}{C}-\text{OCH}_3$$

Knowles [27] made a similar observation in their control reactions when they attempted to trap $\underline{I}$ in the biotin carboxylase reaction. They suggest that the trimethyl carboxyphosphate is produced either via eqn. 3 or more likely by attack of bicarbonate on the methyl esters of ATP that are produced upon addition of diazomethane. What ever the route for the nonenzymatic production of the triesters it indicates that bicarbonate can attack phosphate esters or phosphoanhydrides.

   Knight et al. [9] attempted to exploit the predicted properties of $\underline{I}$ in synthesis. They reasoned that $\underline{I}$ might be relatively stable in the absence of water. In addition, if the role of $\underline{I}$ is to provide a molecule of CO$_2$ to enzyme active sites then the formation of $\underline{I}$ from CO_2 and phosphate in an enzyme active site must be reversible since the reactions catalyzed by biotin carboxylases are reversible. In addition, Powers and Meister [6,7a] reported the exchange of phosphate into $\underline{I}$ was catalyzed by CPS indicating reversible formation of $\underline{I}$ from P_i and CO_2. Finally the reversible formation of CO_2 at an enzyme active site is suggested by the multiple incorporation of O-18 into phosphate during the PEPC reaction with phosphoenol-α-ketobutyrate discussed below (vida infra). These observations suggest that a high mole fraction of carbon dioxide and phosphate might lead to the production of $\underline{I}$.

Knight and co-workers [9] distilled CO_2 into capillary tubes containing

eqn. 4 $((butyl)_4N)_3\, PO_4 + CO_2 \rightleftharpoons ((butyl)_4N)_3^{+3}$

$$\left(O-\overset{\displaystyle O}{\underset{\displaystyle O}{\overset{\displaystyle \|}{\underset{\displaystyle |}{P}}}}-O-\overset{\displaystyle O}{\overset{\displaystyle \|}{C}}-O \right)^{-3}$$

tri-tetrabutyl ammonium phosphate in either DMF or DMSO under vacuum. The tubes were sealed and allowed to warm to 25 deg. examined by P-31 NMR. This produced a high mole fraction of liquid CO_2 in solution under approximately 60 atms of pressure. They observed the formation of a new species at -5.0 ppm upfield of an external phosphoric acid standard. This chemical shift is in the range predicted for $\underline{I}$ from model compounds. For example the P-31 chemical shift of pyrophosphate is -7 ppm upfield from phosphoric acid. A similar species was produced in both DMF and DMSO.

When the pressure was released from the reaction the unknown product slowly decomposed to phosphate. The decomposition was instantaneous upon the addition of water. In addition, they report the observation of C-13 coupling ($J_{c-p} = 1.2$ hz) when the reaction was conducted in the presence of $^{13}CO_2$. These workers report that the signal to noise ratio in that particular experiment was such that they need to confirm this result before publishing the data [9a]. These observations are consistent with carboxyphosphate as the identity of the unknown. But, Knight et. al. were unable to trap this intermediate with methoxide or by addition to a solution of pyruvate carboxylase and pyruvate. This is not surprising considering the lability of the unknown in water. Under some conditions (high concentrations of phosphate and CO_2) pyrophosphate was produced. They accredit this to the attack of phosphate on $\underline{I}$ (see scheme 1).

Scheme 1

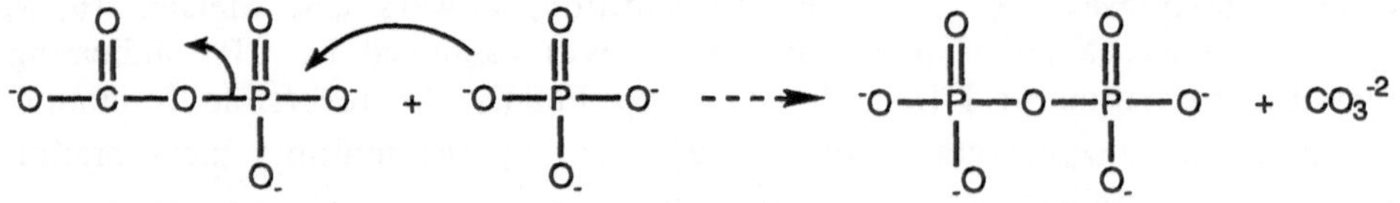

Knight et al. suggest that the unknown could either be I or possibly one of the two other structures shown below. VI is inconsistent with the observed

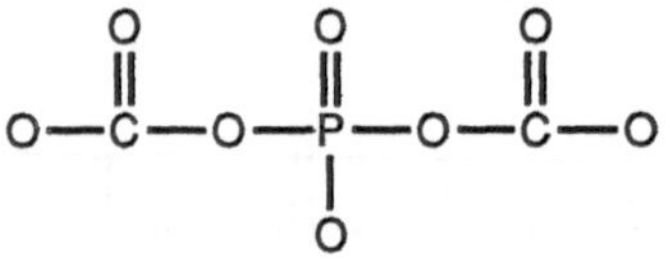

| VI | VII |

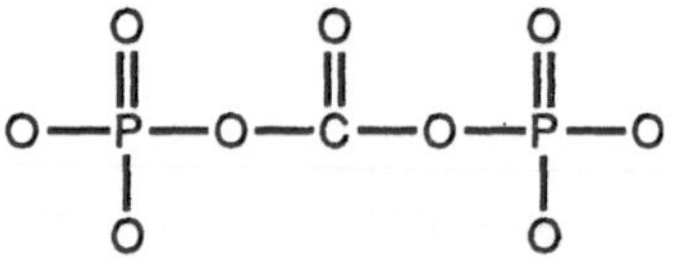

C-13 coupling pattern as well as the chemical P-31 shift. The P-31 chemical shift of VI should be closer to that of the β−phosphorus of tripolyphosphates such as ATP (-21 ppm). VII could result from the nucleophilic attack of the carbonyl oxygen of carboxyphosphate on the phosphate of a second molecule of carboxyphosphate. This seems unlikely but can not be ruled out based on their preliminary data.

3. THE PARTICIPATION OF I IN ENZYMATIC REACTIONS

3.1 General evidence.

The enzymes that are thought to utilize I have several common features. They carry out their respective carboxylations using bicarbonate as the carbon source. They require a high energy phosphate either ATP or PEP. In addition, they require a divalent cation for maximal activity. The divalent cation is typically magnesium.

Establishing the identity of short-lived, relatively unstable enzyme-bound intermediates is a difficult task to say the least. Carboxyphosphate is no exception to this tenet. When faced with this dilemma, one analyzes the results of a series of different experiments to deduce the identity of the intermediate. While any single piece of evidence on it's own is insufficient to establish the chemical mechanism, the entire body of data will often suggest the identity of an intermediate(s) along the reaction pathway. This approach has led workers to suggest that I is an intermediate in the reactions discussed below.

The lines of evidence that indicate that $\underline{I}$ acts as an enzyme intermediate are discussed below. While not all of these experiments have met with success for each enzyme thought to utilize $\underline{I}$, fulfilling several of these criteria is reasonable evidence that $\underline{I}$ is an intermediate in that reaction.

3.1.1. The incorporation of O-18 into the phosphate product from $HC^{18}O_3$ indicates that the substrate is bicarbonate and not CO_2 [36]. The other two O-18's are incorporated into the carboxylated product. This observation can be explained either by a concerted mechanism involving the direct attack of bicarbonate on the phosphoryl donor concomitant with carboxylation of the acceptor or by a stepwise mechanism involving $\underline{I}$.

3.1.2. In the absence of a carboxyl acceptor the enzymes catalyze the hydrolysis of the high energy phosphate in the presence of bicarbonate but not in it's absence.

$$\text{eqn. 5} \quad X \sim P + H_2O \xrightleftharpoons{HCO_3} X + P_i$$

3.1.3. Analogs of $\underline{I}$ can act as either a phosphoryl donor or in some cases potent inhibitors.

3.1.4. The trapping of carboxyphosphate by either reduction or alkylation is direct evidence for it's participation in the reaction. In the ideal situation the amount of trapped material would be stoichiometric with enzyme.

3.1.5. The stereochemistry of the phosphate transfer from the donor should be consistent with the formation of an intermediate. A single enzymatic phosphoryl transfer to bicarbonate should proceed with inversion of stereochemistry at phosphorus [10]. But, Kluger et al. [4] have argued that pseudorotation of the phosphate ester could result in the formation of $\underline{I}$ from O-phosphobiotin with adjacent attack resulting in the retention of stereochemistry. This process was originally suggested by Westheimer from the results of the hydrolysis of cyclic phosphoesters [11]. To date there has been no evidence for pseudorotation of phosphoryl groups during enzymatic phosphoryl transfer reactions.

The methods for determining the stereochemistry of phosphoryl transfer reactions have been recently reviewed [12]. Currently the prevalent method involves the use of a thiophosphoryl substrate that has been labeled with the isotopes of oxygen and was developed by Webb and Trentham [13] and Tsai independently [14]. The chiral thiophosphate product is then converted to ATPβS via a series of enzymatic reactions of known stereochemistry and analyzed by P-31 NMR.

3.1.6. In reactions that utilize ATP as the phosphoryl donor the enzyme catalyzed exchange of the (β-γ) bridging oxygen with a β–phosphoryl nonbridge oxygen indicates the reversible formation of an intermediate. This technique was recently reviewed by Rose [15]. These experiments are conducted with O-18 labeled ATP. The process is monitored either by mass spectroscopy of the products or more recently by P-31 NMR. In the former the product P_i is isolated methylated and analyzed for O-18 content via mass spectroscopy [16]. The second technique relies on the difference in the phosphorus NMR chemical shift induced upon substitution of an O-16 with an O-18. This results in a .02 ppm upfield shift of the phosphorus resonance [17]. Therefore, observation of the P-31 NMR of the phosphoryl group under study can determine the degree of O-18 substitution directly. Raushel and Villafranca were the first to apply this technique to monitor a positional isotope exchange reaction [18a]. This procedure has the advantage of allowing one to monitor the exchange process continuously and thereby determine the rate for the process. The rate of this positional isotope exchange reaction must be kinetically competent or in other words must be comparable to the rate of the reverse reaction. In the most favorable cases these experiments have allowed the determination of rate constants for individual steps in the both the PIX and overall reactions.

3.1.7. Recently in several cases heavy atom isotope effects observed on the carboxylation reactions have been consistent with a stepwise mechanism involving I. These are discussed in more detail below.

In addition, the transient formation of similar phosphorylated carboxyl groups has been demonstrated in other enzymatic processes. For example γ-glutamyl phosphate (VIII) is thought to be an intermediate in the glutamine synthetase reaction (for reviews see refs. 19 and 20). Therefore, high energy phosphates are capable of phosphorylating weakly nucleophilic carboxylates in the environment of an enzyme active site.

$$
\begin{array}{c}
\text{O} \\
\parallel \\
\text{C} - \text{OPO}_3 \\
\mid \\
\text{CH}_2 \\
\mid \\
\text{CH}_2 \\
\mid \\
\text{H}_3\text{N} - \text{C} - \text{COO}^- \\
\mid \\
\text{H}
\end{array}
$$

VIII

3.2 The intermediacy of $\underline{I}$ in individual enzyme reactions.

3.2.1 *Carbamoyl phosphate synthetases (CPS)* catalyze the production of carbamoyl phosphate from bicarbonate, 2 moles of ATP and either glutamine or ammonia as a source of nitrogen (eqn. 6). There are at least three different CPS's. CPS-I is primarily found in the liver and it uses ammonia

eqn. 6 $2MgATP + HCO_3^- +$ Amine (NH_3 or glutamine)

$$\rightleftharpoons \quad 2MgADP + 2P_i + \text{carbamoyl-P}$$

as a source of nitrogen. It is primarily involved in the urea cycle of mammals. On the other hand, CPS-II is localized within the cytosol of mammals and utilizes either glutamine or ammonia. It produces carbamoyl-P as a precursor for the de novo synthesis of pyrimidine nucleotides. The bacterial enzyme utilizes either glutamine or ammonia as a source of nitrogen and provides carbamoyl-P for the biosynthesis of both pyrimidine and arginine. Jones suggested as early as 1963 that $\underline{I}$ was an intermediate in the reactions catalyzed by carbamoyl phosphate synthetases [21]. Meister reviewed the evidence for the involvement of $\underline{I}$ in this reaction in 1983 [6]. However, recent work has shed more light on the role of $\underline{I}$ in this reaction. CPS displays a bicarbonate dependent ATPase activity [22]. When the CPS reaction is conducted in the presence of $HC^{18}O_3$ a single O-18 is incorporated into the P_i product in accord with the intermediacy of $\underline{I}$ in this reaction [23]. These two observations are also consistent with a concerted transfer of the $(\gamma)-P$ of ATP to bicarbonate. Powers and Meister [7] report the trapping of $\underline{I}$ by both reduction and alkylation during the synthetase catalyzed ATPase reaction. In the former they reduced enzyme bound $\underline{I}$ with KBH_4 producing formate and phosphate. In the latter they alkylate $\underline{I}$ with diazomethane to produce the trimethyl ester of carboxyphosphate. Meister reports that the amount of $\underline{I}$ trapped approached the stoichiometry of the enzyme [6]. While the presence of glutamine and ammonia prevent trapping of the intermediate (as expected), filling the amine binding site with analogs that are unable to react enhance the amount of $\underline{I}$ trapped.

Exchange of the $(\beta-\gamma)$ bridge oxygen of ATP into the $\beta-$nonbridge position during the bicarbonate dependent ATPase reaction catalyzed by CPS's was demonstrated by a number of workers [18a,24,25]. From their positional isotope exchange experiments, Wimmer et al. [24] and Meek et al. [25] demonstrated that $\underline{I}$ was a kinetically competent intermediate in the CPS (from E. coli and hamster kidney cells respectively) catalyzed ATPase reactions. Raushel et al. [18] coupled their results from positional isotope exchange experiments and data obtained from rapid quench studies to calculate rate constants for the individual steps of the CPS ATPase reaction according to scheme 2. In their kinetic model EA is the enzyme-MgATP-HCO_3

Scheme 2
$$E + A \underset{k_2}{\overset{k_1}{\rightleftharpoons}} EA \underset{k_4}{\overset{k_3}{\rightleftharpoons}} EP \overset{k_5}{\longrightarrow} E + P$$

complex, and EP is the enzyme-MgADP-$\underline{I}$ complex. The rate constant for the formation of enzyme bound $\underline{I}$ is k_3 and the reported value of 4.2 s^{-1} is close to the k_{cat} (3.2 s^{-1}) reported for the overall reaction in the presence of glutamine. This would indicate that the formation of $\underline{I}$ is rate limiting in the overall reaction, while their value for k_5 suggested that the release of products is rate limiting for the ATPase reaction. This is not a surprising result since $\underline{I}$ must be sequestered from solvent to protect it from rapid hydrolysis.

3.2.2 *Biotin Carboxylases*. In Figure 2 the reactions catalyzed by some of the carboxylases that utilize biotin as a cofactor are presented. These enzymes are involved in diverse processes including gluconeogenesis, fatty acid biosynthesis and amino acid catabolism. These carboxylases carry out two half reactions (Figure 3). The first involves an ATP dependent carboxylation of biotin by bicarbonate which may involve $\underline{I}$ as an intermediate. In the second half reaction the carboxyl is transferred to an acceptor. The biotin carboxylase activity of acetyl-CoA carboxylase resides on a distinct subunit of the multienzyme complex and can be isolated. This protein will catalyze the carboxylation of free biotin and has greatly facilitated the study of this class of enzymes.

Acetyl-CoA Carboxylase

MgATP + HCO_3 + acetyl-CoA $\rightleftharpoons$ MgADP + P_i + malonyl-CoA

Pyruvate Carboxylase

MgATP + HCO_3 + pyruvate $\rightleftharpoons$ MgADP + P_i + oxaloacetate

Propionyl-CoA Carboxylase

MgATP + propionyl-CoA + HCO_3 $\rightleftharpoons$ MgADP + P_i + methylmalonyl-CoA

B-Methylcrotonyl-CoA Carboxylase

MgATP + B-methylcrotonyl-CoA $\longrightarrow$ MgADP + P_i + methylglutaconyl-CoA

Figure 2. Carboxylase reactions that utilize biotin.

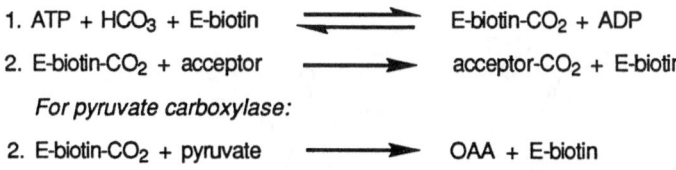

1. ATP + HCO$_3$ + E-biotin $\rightleftharpoons$ E-biotin-CO$_2$ + ADP

2. E-biotin-CO$_2$ + acceptor $\longrightarrow$ acceptor-CO$_2$ + E-biotin

For pyruvate carboxylase:

2. E-biotin-CO$_2$ + pyruvate $\longrightarrow$ OAA + E-biotin

Figure 3 The half-reactions of carboxylases that utilize biotin as a cofactor.

Four possible mechanisms for the enzymic carboxylation of biotin have been considered (Figure 4) [4,26,27]. Mechanisms 1 and 2 involve $\underline{I}$ as an intermediate. The intermediacy of $\underline{I}$ in the biotin dependent carboxylases is supported by a number of lines of evidence. Several groups have reported the incorporation of O-18 into P$_i$ from O-18 bicarbonate during the course of these reactions [27,28b,29] and a bicarbonate dependent ATPase activity [27,28b,30]. Climent and Rubio [30] and more recently Tipton and Cleland [28b] and Ogita and Knowles [27] have demonstrated a biotin carboxylase catalyzed ATPase activity in the absence of biotin. While the rate of this reaction was only a 0.5% the rate of the ATPase reaction in the presence of biotin, both reactions displayed similar dependence on metal ions, the concentration of ATP and activation by ethanol. In addition, Ogita and Knowles demonstrated that the fourth oxygen in the phosphate produced during the biotin independent ATPase reaction is derived from bicarbonate rather than water. This observation rules out a nonspecific hydrolysis of ATP by the enzyme preparation. Therefore, all three groups conclude that this activity is inherent to biotin carboxylase. These results would rule out mechanism 3 of Figure 4. Ashman and Keech report that phosphonacetate an analog of $\underline{I}$ is a potent inhibitor of pyruvate carboxylase [31] although the reported dissociation constant is rather high. Carbamoyl phosphate another analog of $\underline{I}$ can serve as a phosphoryl donor to produce ATP from ADP in these reactions both in the presence and absence of biotin [28,31,32]. But, Kluger et al. [4] have argued that carbamoyl-P is an analog of O-phosphobiotin rather than of $\underline{I}$ in these reactions. A requirement for the phosphorylation of biotin by ATP is not consistent with the bicarbonate dependent ATPase activity in the absence of biotin. In addition, the intermediacy of O-phosphobiotin necessitates a second phosphoryl transfer that could involve $\underline{I}$ (Figure 4,#2) unless the second transfer goes via a concerted mechanism. The second transfer would result in a net retention of stereochemistry at phosphorus. But, Hansen and Knowles [33] observed inversion of stereochemistry at phosphorus during the pyruvate carboxylase reaction which is also consistent with the intermediacy of $\underline{I}$. Although, Kluger et al. [4] would argue that if one of the phosphorylation steps in mechanism 2 or 3 occurred via a pseudorotatory route the net result would be inversion of stereochemistry.

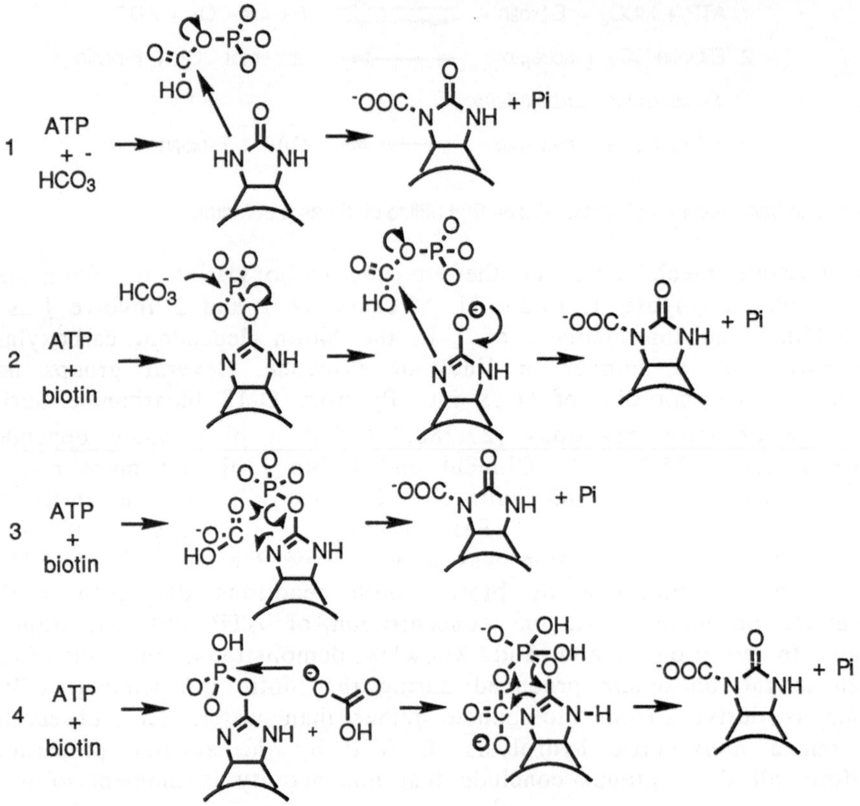

Figure 4. Four possible mechanisms to the enzymic carboxylation of biotin (adapted from references 4, 26, 27)

While the evidence presented above suggests a role for I in biotin carboxylase reactions attempts to confirm the presence of I in these reactions by either direct trapping or positional isotope exchange studies have been unsuccessful [27,28]. The success of intermediate trapping experiments rely both on the life time of the intermediate and the accessibility of the intermediate to the trapping reagent as well as the reactivity of the reagent. The instability of I in aqueous solutions requires that it be tightly sequestered within an enzyme's active site and inaccessible to solvent. Therefore, failure to trap an intermediate can not be taken as solid evidence against it's intermediacy in the reaction. With these caveats only a positive result from a trapping experiment can be used to interpret a mechanism.

The failure to observe a PIX in the biotin carboxylase reaction also does not preclude the intermediacy of I in these reactions. These experiments require that the β-phosphate of the ADP produced be torsionally symmetrical or in layman's terms free to rotate in the enzyme active site. In fact, Tipton and Cleland [28b] present evidence that the nucleotide may be coordinated to two divalent metal ions in the active site of biotin carboxylase. This coordination scheme was suggested due to the lack of reactivity of thionucleotides with the enzyme when magnesium fulfilled the requirement for metal ions. Their study indicated that at least one of the nonbridge oxygens of the α-phosphate and both of the nonbridge oxygens of the β-phosphate of ATP are coordinated to the two metal ions. This coordination would hardly allow the β-phosphate of ADP to rotate freely. The failure to observe a PIX in these reactions could also be due to a short life-time for the enzyme bound intermediate. Finally, Tipton and Cleland have suggested that a large forward commitment would also result in the failure to observe PIX in the biotin carboxylase reaction. That is, once $\underline{I}$ is formed the reaction is committed to the production of products. This conclusion was suggested by their observation of a small C-13 isotope effect on V/K in this reaction. In general a large forward commitment will decrease the observed isotope effect on a bond breaking step in an enzymatic reaction from the intrinsic isotope effect on this step. While these are all excellent arguments to explain the failure to observe $\underline{I}$ via these techniques, one can not ignore the possibility that $\underline{I}$ is not an intermediate in these reactions.

3.2.3 *Phosphoenolpyruvate carboxylase* (PEPC) is the key carboxylating enzyme in plants that utilize the C-4 [34] and CAM pathways [35]. Therefore, this enzyme and the CPS discussed above are perhaps the most pertinent systems to the mission of this conference that have been reported to utilize $\underline{I}$ as an intermediate.

PEPC catalyzes the reaction in eqn. 7. Three mechanisms can be suggested for the PEPC reaction (Figure 5) [33]. Maruyama et al. [36] found that O-18

eqn. 7 $PEP + HCO_3 \longrightarrow OAA + P_i$

was incorporated into phosphate and oxaloacetate when O-18 bicarbonate was used as the substrate. This suggests that the substrate is bicarbonate and not carbon dioxide thus ruling out mechanism 3. They suggested a concerted cyclic mechanism for the carboxylation reaction involving a single transition state to explain their results (mechanism 1 in Figure 5). But, both mechanisms 1 and 2 in Figure 5 are consistent with these observations. In addition, Diaz et al. found that O-18 was incorporated into phosphate during the maize PEPC catalyzed hydrolysis of bromo-PEP to produce bromopyruvate and P_i [37]. O'Leary and Rife [38] observed synergistic inhibition of PEPC by oxalate and carbamoyl-P. The ternary complex formed between these inhibitors and PEPC could be considered analogous to the E·I·enolpyruvate complex. Hansen and Knowles [40] found that the carboxylase reaction proceeds with inversion of stereochemistry at phosphorus. These results are consistent with the intermediacy of I in the reaction.

The results from the isotope effect studies of O'Leary et. al. [39] suggested the stepwise mechanism of Figure 5 with I was an intermediate in this reaction. In this mechanism bicarbonate attacks the PEP phosphoryl producing I and the enolate of pyruvate. In the second step the carboxyl of I is transferred to the enolate to produce oxaloacetate and phosphate. O'Leary et al. observed only a small C-13 isotope effect on the V/K for this reaction. They predicted that a concerted mechanism would produce significantly larger C-13 effects than those observed. In addition, if the second step of mechanism 2 was rate limiting, they predicted a larger isotope effect. On the other hand, the observed isotope effect was consistent with mechanism 2 in which the rate limiting step is the formation of I form PEP and bicarbonate.

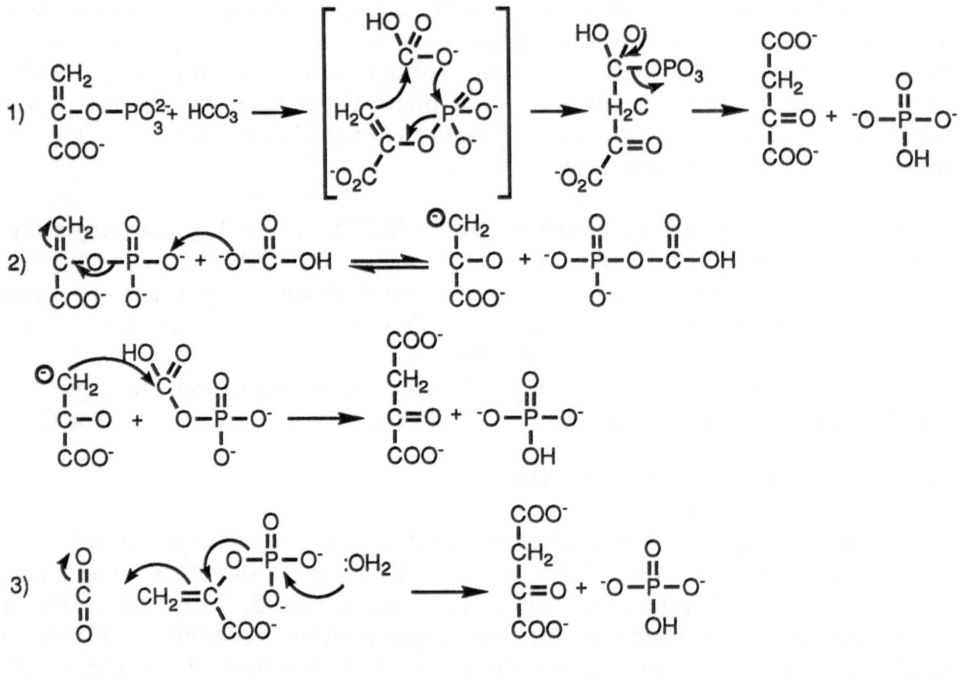

Figure 5. Three mechanisms proposed for the PEP carboxlylase reaction. (Adapted from refs 39 and 40)

Perhaps the best evidence for the intermediacy of I was provided by Fujita et al. [41] and more recently by O'Leary and O'Laughlin [42]. These workers demonstrated that E. coli PEPC and the PEPC from Zea Mays respectively catalyze the bicarbonate dependent cleavage of phosphoenol-α-ketobutyrate to phosphate and α-ketobutyrate thus uncoupling the phosphatase reaction from the carboxylation reaction. This is not consistent with a concerted mechanism for PEPC. In addition, these workers observed more than a single equivalent of O-18 incorporated into P_i from $HC^{18}O_3$. This could result from isotopic exchange via $\underline{I}$ (see Figure 6). This observation is consistent with the reversible decarboxylation of $\underline{I}$ to form phosphate and CO_2 in the enzyme active site. Rotation of the phosphate followed by a reversal of the steps to produce bicarbonate and phosphoenol-α-ketobutyrate CO_2 would result in multiple incorporation of O-18. This mechanism requires a reasonable lifetime for enzyme bound $\underline{I}$. However, they were unable to trap $\underline{I}$ with diazomethane. Their trapping experiments suffer from the same caveats discussed above for biotin carboxylases.

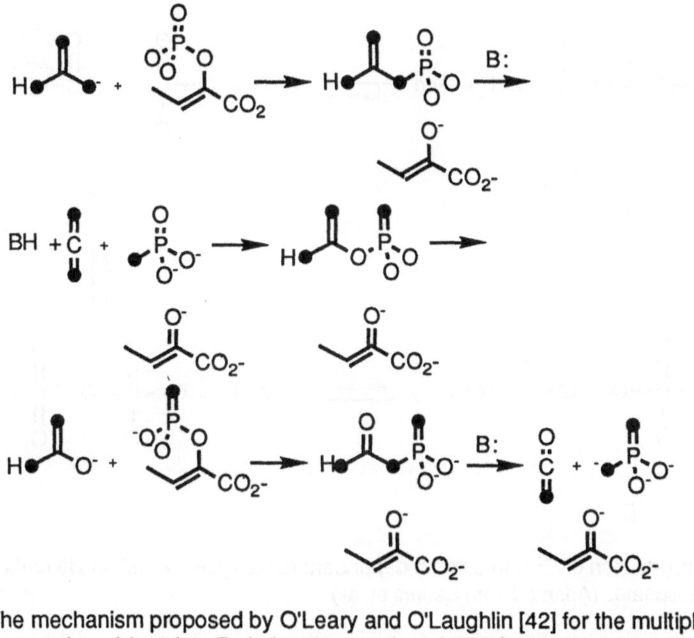

Figure 6. The mechanism proposed by O'Leary and O'Laughlin [42] for the multiple incorporation of 0-18 into P_i during the reaction of PEPC with phosphoenol-2-ketobutyrate. (● = 0-18)

4. WHY CARBOXYPHOSPHATE?

The binding of the small neutral planar CO_2 molecule by an enzyme will produce very little of the binding energy necessary to overcome the loss in entropy required to position it in the proper orientation for nucleophilic attack. The binding of bicarbonate would be expected to yield substantial binding energy from both the donation and acceptance of hydrogen bonds as well as electrostatic interactions of the negative charge with positively charged active site residues. In addition, the most predominant form of CO_2 in solution at physiological pH is bicarbonate, so it is not surprising that some enzymes have evolved to utilize it rather than free CO_2. On the other hand bicarbonate is a rather poor electrophile.

Sauers et al. [5] have suggested that the role of this intermediate in ATP dependent biotin carboxylases and the carbamoyl synthetase reactions is to produce an electrophilic enzyme bound CO_2 according to Figure 7. This overcomes the necessity of providing a good nucleophile at the active site. They proposed that I decarboxylates in the enzyme active site to produce free CO_2 which is the actual carboxylating agent.

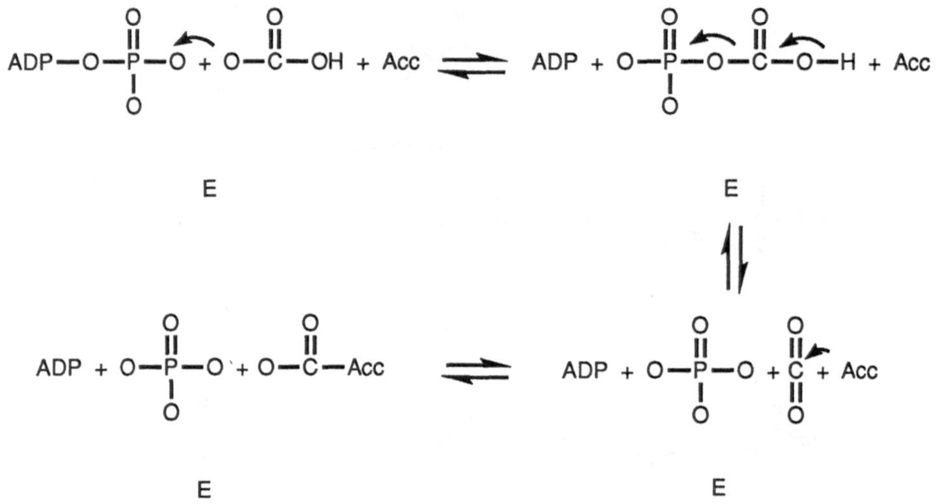

$$ADP-O-\overset{\overset{O}{\parallel}}{\underset{\underset{O}{|}}{P}}-O + O-\overset{\overset{O}{\parallel}}{C}-OH + Acc \rightleftharpoons ADP + O-\overset{\overset{O}{\parallel}}{\underset{\underset{O}{|}}{P}}-O-\overset{\overset{O}{\parallel}}{C}-O-H + Acc$$

E   E

$$ADP + O-\overset{\overset{O}{\parallel}}{\underset{\underset{O}{|}}{P}}-O + O-\overset{\overset{O}{\parallel}}{C}-Acc \rightleftharpoons ADP + O-\overset{\overset{O}{\parallel}}{\underset{\underset{O}{|}}{P}}-O + \overset{\overset{O}{\parallel}}{\underset{\underset{O}{\parallel}}{C}} + Acc$$

E   E

Figure 7. The production of CO_2 in a biotin dependent carboxylase reaction via carboxy phosphate. (Adapted from Sauers et. al.)

The intermediacy of CO_2 is supported by the report of Powers and Meister [7a] of a CPS catalyzed exchange of phosphate into I, but they presented no data in support of this observation. This exchange would likely proceed via decarboxylation of I in the active site to produce free CO_2. If this were the case for the CPS reaction one could expect to observe washout of O-18 from the γ-phosphate during the ATPase reaction (see Figure 8). Neither Meek et al. [25] nor Raushel and Villafranca [18a] observed such a process. In addition, Rubio et al. [43] failed to observe the incorporation of O-18 from $HC^{18}O_3$ into ATP during the CPS catalyzed ATPase reaction. Tipton and Cleland [28b] also failed to observe washout of O-18 from the γ-phosphate of ATP in the biotin carboxylase reaction. While these results argue against reversible formation of free carbon dioxide at the active site, they are not conclusive. The success of these experiments require not only that CO_2 be formed reversibly but that the P_i be free to rotate in the active site. If this is not the case, then I would reform with the same orientation of the oxygens and there would be no washout of O-18. Tipton and Cleland suggest that metal coordination limits the torsional symmetry of the phosphate in the active site of biotin carboxylase. Metal coordination of the phosphate in the active site of CPS's may also limit free rotation as they to possess metal ions at the active site. Raushel and Villafranca [18a] argue that the failure to observe O-18 washout in the CPS reaction is because the rate of reformation of I from P_i and CO_2 is slow relative to release of I.

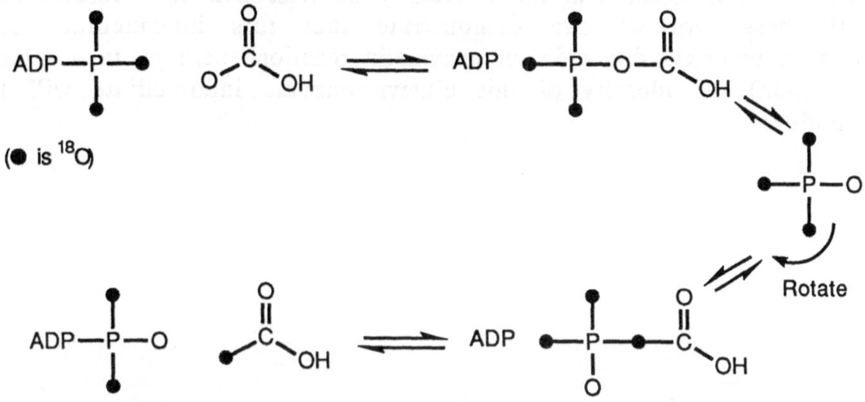

Figure 8. General mechanism for 0-18 washout from ATP during a reversible ATPase reaction. (Taken from reference 18a)

According to the hypothesis put forth by Sauers et. al. [5], the energy from the hydrolysis of ATP is coupled to the production of a molecule of CO_2 in the correct orientation for reaction. A highly reactive molecule is thus produced due to it's low entropy and resulting high free energy. The localization in the active site also produces a high local concentration which is used to drive the reaction. O'Leary et al. [39] used similar reasoning in rationalizing the role of I in the PEPC reaction. They argue rather persuasively that the high energy of the enolphosphate can be coupled to carbon-carbon bond formation via the intermediacy of I. In addition, I functions to activate bicarbonate for nucleophilic attack by the enol carbon. In this case the enol is a rather potent nucleophile and may attack the carbon atom of I directly without a requirement for prior decarboxylation of $\underline{I}$ to produce enzyme bound CO_2. But, perhaps the best evidence for the reversible formation of CO_2 from $\underline{I}$ during an enzymatic reaction comes from the PEPC system (vida supra). It would appear that in this case the enzyme has evolved to take advantage of both the inherent nucleophilicity of the enol but also the electrophilicity of enzyme bound CO_2.

5. CONCLUSIONS

There is substantial evidence that carboxyphosphate is an intermediate in enzymic carboxylation reactions that utilize bicarbonate as the carbon source. Additional work is required to demonstrate that I is a discrete chemical entity. The high pressure NMR experiments underway in the laboratory of W.W. Cleland at the University of Wisconsin may resolve this issue. If these workers can demonstrate that this intermediate can function as a carboxyl donor in an enzymatic reaction (perhaps in a mixed solvent system) the identity of this elusive enzyme intermediate will be established.

REFERENCES

1. D.S. Tarbell and M.A. Insalaco. Proc. Nat. Acad. Sci. USA 57, 233 (1967).
2. D.L. Griffith and M. Stiles. J. Amer. Chem. Soc. 87, 3710 (1965).
3. D.E. Koshland. J. Amer. Chem. Soc. 74, 2286 (1952).
4. R. Kluger, P.P. Davis and P.D. Adawadkar. J. Amer. Chem. Soc. 101, 5995 (1979).
5. C. K. Sauers, W.P. Jencks and S. Groh. J. Amer. Chem. Soc. 97, 5546 (1975).
6. A. Meister. Trans. New York Academy of Sciences, 117-128 (1983).
7.(a) S.G. Powers and A. Meister. Proc. Natl. Acad. Sci. USA 73, 3020 (1976); (b) S.G. Powers and A. Meister. J. Biol. Chem. 253, 1258 (1978); (c) A. Meister and S.G. Powers. Adv. Enz. Reg. 16, 289 (1978).
8. V. Rubio and S. Grisolia. Biochemistry 16, 321
9.a.W.B. Knight, P.V. Atwood and W.W. Cleland. Fed. Proc., Fed. Amer. Soc. Exp. Biol. 45(6), 1607 (1986). b. W.B. Knight, P.V. Atwood, and W.W. Cleland. unpublished results. 10. Knowles, J.R. Annu. Rev. Biochem. 49, 877 (1980).
11. F.H. Westheimer. Acc. Chem. Res. 1, 70 (1968).
12. J.A. Gerlt, J.A. Coderre and S. Mehdi. Adv. Enz.
13. M.R. Webb and D.R. Trentham. J. Biol. Chem. 255, 1775 (1980).
14. M.D. Tsai. Biochemistry 19, 5310 (1980).
15. I.A. Rose. Adv. Enz. 50, 361 (1979).
16. C.F. Middlefort and I.A. Rose. J. Biol. Chem. 251, 5881 (1976).
17. M. Cohn and A. Hu. Proc. Natl. Acad. Sci. USA, 75, 3133 (1978).
18. a) F.M. Raushel, P.M. Anderson and J.J. Villafranca Biochemistry 17, 558 (1978); F.M. Raushel and J.J. Villafranca. ibid. 18., 3424 (1979).
19. A. Meister. The Enzymes 10, 699 (1974).
20. M.J. Wimmer and I.A. Rose. Ann. Rev. Biochem. 47, 1031 (1978).
21. M.E. Jones. Science 140, 1373 (1963).
22. P.M. Anderson and A. Meister. Biochemistry. 5., 3157 (1966).
23. M.E. Jones and L. Spector. J. Biol. Chem. 235, 2897 (1960).
24. M.J. Wimmer, I.A. Rose, A.G. Powers and A. Meister. J. Biol. Chem. 254, 1854 (1979).
25. T.D. Meek W.E. Harsten and C.W. Brosse. Biochemistry 26, 2584 (1987).
26. J.C. Wallace, N.B. Phillips, M.A. Snoswell, G.J. Goodall P.V. Atwood and D.B. Keech. Ann. N. York. Acad. Sci. 447, 169 (1985).
27. T. Ogita and J.R. Knowles. Biochemistry 27, 8028 (1988).
28. (a) P.A. Tipton and W.W. Cleland. Biochemistry 27, 4317 (1988) (b) ibid., 4325 (1988).

258

29. Y. Kaziro, L.F. Hass, P.D. Boyer and S. Ochoa. J. Biol. Chem. 237, 1460·(1962).
30. I. Climent and V. Rubio. Arch. Biochem. Biophys. 251, 465 (1986).
31. L.K. Ashman and D.B. Keech. J. Biol. Chem. 250, 14 (1975).
32. (a) S.E. Polakis, R.B. Guchhait and M.D. Lane. J. Biol. Chem. 247, 1335 (1972); (b) S.E. Polakis, R.B. Guchhait, E.E. Zwergel, M.D. Lane and T.G. Cooper. ibid. 249, 6657 (1974).
33. D.E. Hansen and J.R. Knowles. J. Amer. Chem. Soc. 107, 8304 (1985).
34. M.D. Hatch and C.R. Slack. Ann. Rev. Plant Physiol. 21, 141 (1970).
35. C.B. Osmond. Ann. Rev. Plant Physiol. 29, 379 (1978).
36. H. Maruyama, R.L. Easterday, H.C. Chang and M.D. Lane.
J. Biol. Chem. 241, 2405 (1966).
37. E. Diaz, J.T. O'Laughlin and M.H. O'Leary. Biochemistry 27, 1336 (1988).
38. M.H. O'Leary and J.E. Rife. unpublished results.
39. (a) M.H. O'Leary, J.E. Rife and J.D. Slater. Biochemistry 20, 7308 (1981); (b) M.H. O'Leary and P.Paneth.
Biophosphates and Their Analogues - Synthesis. Structure. Metabolism and Activity, p303. Elsevier, Amsterdam (1987).
40. D.E. Hansen and J.R. Knowles. J. Biol. Chem. 257, 14795 (1982).
41. N. Fujita, K. Izui, T. Nishino and H. Katsuki. Biochemistry 23, 1774 (1984).
42. J.T. O'Laughlin and M.H. O'Leary. unpublished results.
43. V. Ribio, H.G. Britton, S. Grisiola, B.S. Sproat and G. Lowe. Biochemistry 20, 1969 (1981).

BIOLOGICAL UTILIZATION OF CARBON DIOXIDE. ENZYMIC CATALYSIS PATTERNS INVOLVING BIOTIN, ATP, AND BICARBONATE

RONALD KLUGER
Lash Miller Chemical Laboratories
Department of Chemistry
University of Toronto
Toronto, Canada M5S 1A1

ABSTRACT. The interaction of a carboxylase enzyme with bicarbonate, ATP, biotin, and a substrate results in the production of the carboxylated substrate and the hydrolysis of ATP. This very efficient utilization of carbon dioxide (from bicarbonate) contains many mechanistic questions related to the nature of reactive intermediates. The first part of the reaction involves the formation of N-carboxybiotin from ATP, bicarbonate, and biotin. An array of possible mechanisms can be constructed in terms of the order of reaction between the various components. It is proposed that only by first constructing a set of all possible classes of mechanism and then eliminating those that are disproven by experiment can the mechanism be deduced. The analysis leads to consideration of mechanisms which have not previously been presented involving the direct attack of biotin upon bicarbonate. The reaction in which the carboxyl group of N-carboxybiotin is transferred to the substrate involves a process in which N-carboxybiotin becomes more reactive in the presence of substrate. It is proposed that this is the result of conformational properties of the cofactor which are controlled by the interaction of enzyme and substrate. Model reactions which bear on the various processes give good chemical analogies that provide insight into the reasonableness of various mechanisms.

1. Background

1.1 Nonphotosynthetic Biochemical Utilization of Carbon Dioxide

Most organisms which are incapable of photosynthesis nonetheless can efficiently utilize the elements of carbon dioxide for biosynthesis, deriving these elements from dissolved bicarbonate. The energy for this must come from other chemical sources and typically involves the coupled cleavage of ATP. While the understanding that carbon dioxide incorporation and ATP hydrolysis are somehow related gives a thermodynamic plausibility to the process, the means by which the energy is transferred and utilized is a complex question involving some of the most fundamental problems in enzymology and bioenergetics.

In this article I will deal specifically with reaction patterns of the group of enzymes which utilize biotin as a cofactor to promote ATP-dependent

259

M. Aresta and J. V. Schloss (eds.),
Enzymatic and Model Carboxylation and Reduction Reactions for Carbon Dioxide Utilization, 259–271.
© 1990 *Kluwer Academic Publishers.*

carboxylation processes. These carboxylases catalyze the substitution of the CO_2 group in place of a proton at a position in a molecule at which a carbanion will form, typically at a carbon α to a carbonyl group. The overall reaction is shown in Figure 1 and the structure of biotin is in Figure 2.

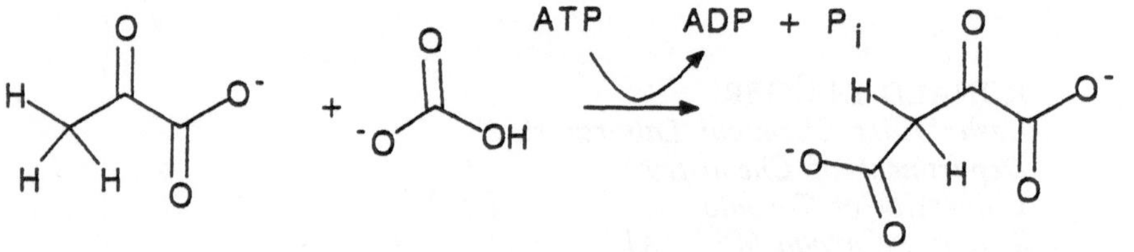

Figure 1. The reaction catalyzed by the biotin-dependent enzyme, pyruvate carboxylase.

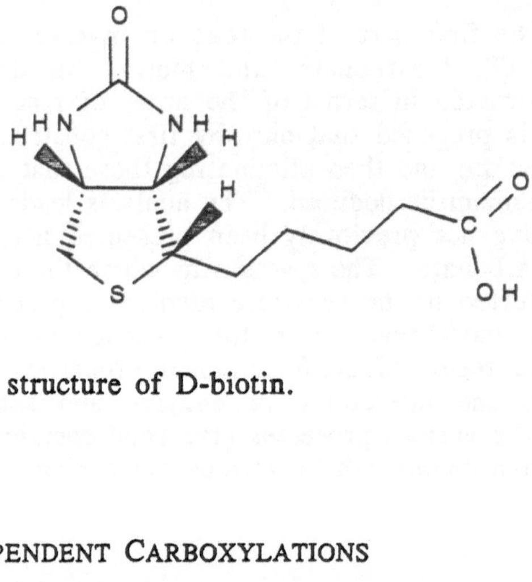

Figure 2. The structure of D-biotin.

1.2. BIOTIN-DEPENDENT CARBOXYLATIONS

Biotin-dependent carboxylase reactions involve two steps. First, biotin itself is a reactant in a process involving stoichiometric amounts of bicarbonate and ATP. This results in the formation of N-carboxybiotin, a process first proposed by Lynen [1] who trapped this species with diazomethane. Lane later showed that synthetic N-carboxybiotinol is a substrate in acetyl CoA carboxylase [2].

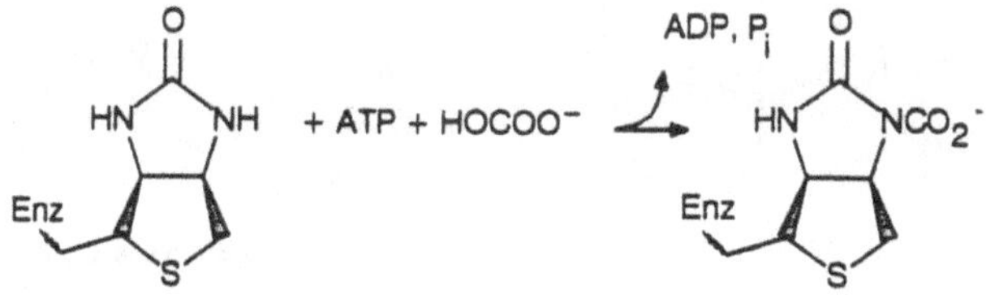

Figure 3. Formation of N-carboxybiotin

The mechanism of formation of N-carboxybiotin and the transfer to the substrate are important in relation to carbon dioxide utilization. The system is very efficient and the energy for the fixation is provided by the net hydrolysis of ATP. The hydrolysis of ATP is a net exergonic process, while the formation of N-carboxybiotin from bicarbonate and biotin is significantly endergonic due to the stability of bicarbonate relative to carbon dioxide in water (the large degree of dissociation of carbonic acid in neutral solution affects the equilibrium as well). Understanding of the mechanism of the biochemical system may have far-reaching general applicability for developing methodology for carbon dioxide utilization.

The relationship of ATP hydrolysis to the utilization of bicarbonate by biotin-dependent carboxylases was the subject of an important study by Kaziro, Boyer and Ochoa [3]. They reacted ^{18}O-bicarbonate and ATP with the biotin-dependent enzyme propionyl-CoA carboxylase. They found that the [18]O from bicarbonate appears in the inorganic phosphate produced from ATP as well as in the carboxyl group of the product, methylmalonyl-CoA. Therefore, at some point an oxygen atom derived from bicarbonate must add to the phosphate group derived from the γ-phosphate residue of ATP. These authors proposed that this is consistent with a reaction in which biotin, bicarbonate, and ATP react in a concerted process. More generally, the result shows that ATP, or a species derived from ATP, functions in the mechanism by phosphorylating bicarbonate or an intermediate derived from bicarbonate.

While this information does not specify a mechanism it places limits on the possibilities. If the interaction occurs directly as the initial reaction step, ATP and bicarbonate would react to form the anhydride of carbonic acid and phosphoric acid, carboxyphosphate (Figure 4).

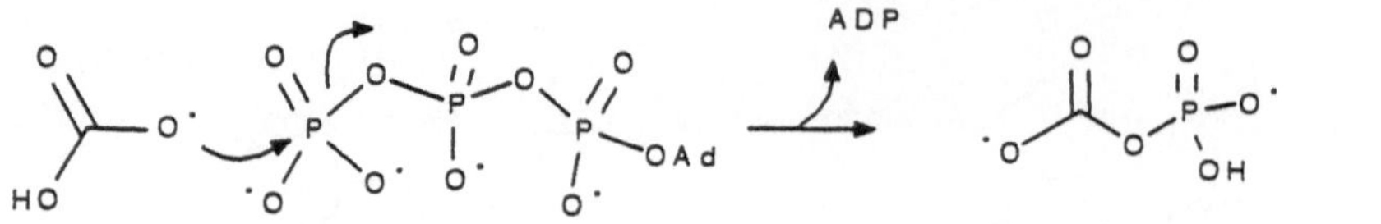

Carboxyphosphate

Figure 4. The interaction of ATP and bicarbonate may produce carboxyphosphate.

This material might then go on to react with biotin to form N-carboxybiotin. In this case, the transfer of oxygen occurs directly as carboxyphosphate expels phosphate in the process of carboxylating biotin. This would be expected to occur with net inversion of relative configuration of the ligands about phosphorus (which is detected by using oxygen isotopes to make the γ-phosphorus stereogenic) [4]. The elegant work of Hansen and Knowles showed that the reaction at phosphorus does occur with inversion [5].

However, carboxyphosphate has not been observed or trapped in this or any other enzymic or nonenzymic reaction. Based on estimates from linear free energy relationships of the reactivity of alkyl monocarbonates, Sauers, Jencks and Groh estimated that the half life of carboxyphosphate is on the order of 70 milliseconds [6]. Alkyl monocarbonates decompose by a unimolecular mechanism in which carbon dioxide is generated and other carbonate derivatives should react by a similar mechanism. This means that unless carboxyphosphate reacts very rapidly with a nucleophile it will certainly decompose. This led Sauers et al to the suggestion that in the enzymic carboxylation of biotin, carboxyphosphate is produced but it spontaneously decomposes to carbon dioxide and inorganic phosphate [6]. The nascent carbon dioxide localized adjacent to biotin has a much lower entropic barrier to reaction than does dissolved carbon dioxide which is dispersed through a large volume. This is an intriguing suggestion which has been difficult to test.

1.3. ISOTOPE EXCHANGE PATTERNS

While isotopic exchange reactions with tracers are often useful probes of mechanism, in the case of biotin mechanisms this has not been the case. A partial reaction in which ATP cleavage occurs at a kinetically competent rate independent of a second process in a good indication that an intermediate phosphorylation process occurs. However, in biotin systems, exchange only occurs when all components are present [7]. Such a result is consistent with exchange occurring by the reaction going to completion and reversing, thus not implicating any intermediate. One might be tempted to argue that this implicates a concerted reaction mechanism but enzymes often have complex binding requirements (an ordered binding mechanism, for example, with a central ternary complex) so that the lack of exchange cannot confirm any mechanism.

1.4. CATALYSIS OF ALTERNATIVE REACTIONS BY BIOTIN

Another approach to testing a mechanism is finding if the enzyme can catalyze a kinetically competent process in a related reaction. This approach has been used to support a mechanism involving carboxyphosphate. Lane and co-workers found that biotin carboxylase will transfer phosphate from carbamyl phosphate to ADP [8]. However, we have pointed out that it is impossible to know if carbamyl phosphate is recognized by the enzyme as an analogue of carboxyphosphate or as analogue of some other phosphorylated intermediate, such as O-phosphobiotin [9] (Figure 5).

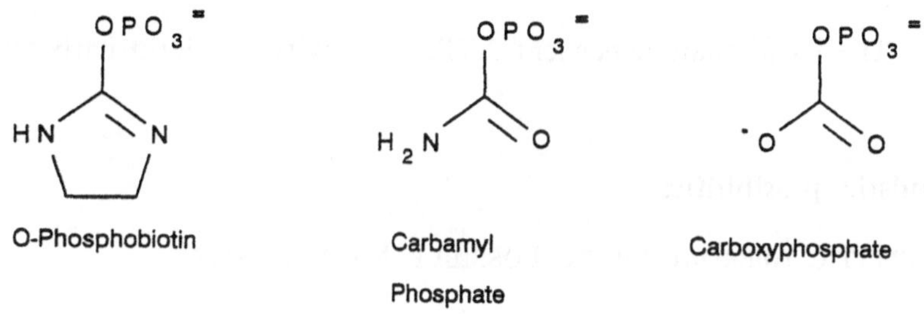

O-Phosphobiotin   Carbamyl   Carboxyphosphate

                         Phosphate

Figure 5. Is carbamyl phosphate an analogue of carboxyphosphate or of O-phosphobiotin?

Another alternative reaction cited in support of carboxyphosphate as an intermediate is the very slow (.005 times the net rate of the normal reaction) bicarbonate-dependent ATPase activity found in acetyl CoA carboxylase [10,11]. While this was reported as evidence that the biotin-dependent reaction involves carboxyphosphate, such a slow rate may indicate of that water is entering a site normally occupied by biotin which attacks bicarbonate which in turn reacts with ATP (Figure 6). Ogita and Knowles studied the same reaction and report that ^{18}O from bicarbonate is transferred to the phosphate derived from ATP in this activity [12]. Again, this is consistent with either interpretation.

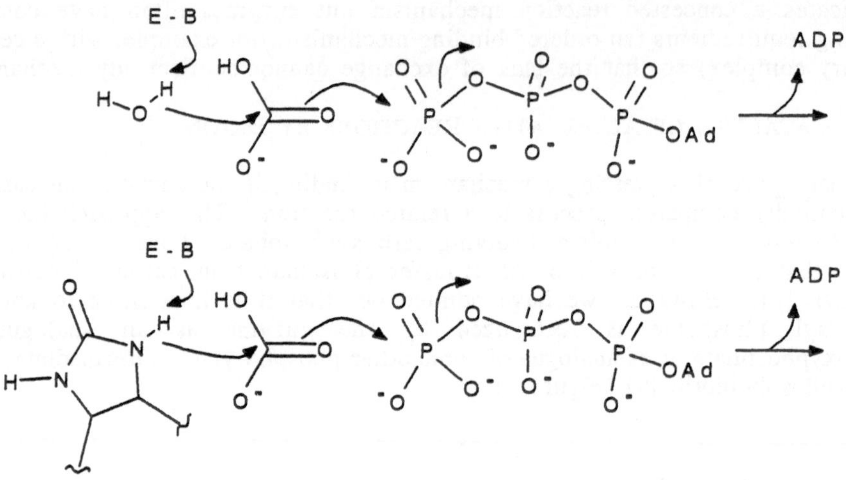

Figure 6. The bicarbonate-dependent ATPase activity of biotin carboxylase.

2. Mechanistic possibilities

2.1. SYSTEMATIC APPROACHES TO POSSIBLE MECHANISMS

In order to evaluate mechanisms for the ATP-dependent carboxylation of biotin we sought a more systematic approach to the problem. What mechanisms are possible and which of these are eliminated by direct experimental evidence and which by their chemical unreasonableness? How can we set up a full collection of possible mechanisms? The deductive mechanistic approach is an analogy to the deductive structural method used by Fischer to determine the structure of D-glucose. In that case, all possible structures consistent with all nonstereochemical criteria were considered and experiments were conducted which systematically ruled out structures until there was a unique solution.

We propose a similar deductive approach with mechanisms. If we can consider all possibilities according to some formal set of patterns, then we can use experiments to rule out members of the set. This is in contrast to the inductive approach in which a mechanism is proposed which is consistent with the experimental data at the time but the uniqueness of the mechanism is not a criterion. The finding of experimental data which does not fit the favored mechanism leads to a new ad hoc mechanism or to revisions of the original.

2.2. A SYSTEMATIC SET OF MECHANISMS

The mechanisms we shall consider all involve the potential interaction of ATP, bicarbonate and biotin on an enzyme to produce N-carboxybiotin. The systematic approach requires a basis for recognizing a mechanism as belonging to a particular class and all mechanisms must be classified uniquely [13]. We propose that the molecularity of the key step, formation of the nitrogen-carbon bond in N-carboxybiotin is the primary mechanistic descriptor. The secondary descriptor is the number of intermediates (ignoring proton transfers) involved in processes subsequent to binding of reactants to the enzyme and prior to reaction of N-carboxybiotin with the substrate. The process is either stepwise or concerted. This makes no assumption about which step in the mechanism is rate-determining. Thus, the carboxyphosphate mechanism involves an initial binary process (ATP + bicarbonate) with the C-N bond formation step also being binary (biotin + carboxyphosphate). There is one essential intermediate, carboxyphosphate. The totally concerted mechanism, which is discussed in the next section is ternary with no intermediate and therefore is designated as a ternary concerted mechanism.

2.2.1. *Ternary Mechanisms.*

A mechanism with no intermediate necessarily involves a termolecular transition state with all the reactants (biotin, ATP and bicarbonate) leading to all the products in the key step. The mechanism was proposed in response to the discovery that oxygen is transferred from bicarbonate to ATP [3] and in regard to the fact that carboxylases do not catalyze partial exchange reactions [7]. Recently Perrin and Dwyer [14] have shown that it is reasonable to expect that the conjugate base of biotin will be the likely reactant rather than biotin itself in any mechanism. Therefore, the ternary mechanism is a direct displacement by the urea nitrogen of biotin (as its conjugate base) at the carbonyl center of bicarbonate with hydroxide expelled directly as it simultaneously attacks ATP. The substitution mechanism at phosphorus is unspecified for the purposes of the present analysis although one would expect an intermediate to form.

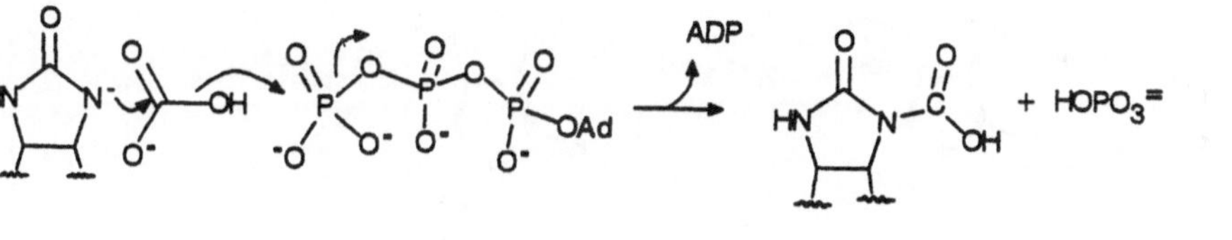

Figure 7. The termolecular concerted mechanism for carboxylation of biotin.

Concerted displacements at carbonyl centers must compete with stepwise mechanisms which involve the formation of tetrahedral intermediates. The latter pattern has been confirmed in most cases which have been examined and only in the rare case where there is a weak nucleophile and a very easily displaced leaving group has there been evidence for a mechanism without an intermediate (presumably, the intermediate is too unstable to exist in these cases) [15]. This does not describe the attack of biotin on bicarbonate, where the leaving group, hydroxide, is a strong base and the nucleophile the conjugate base of biotin, also a strong base. The function of ATP in such a mechanism is also unclear since it does not promote the addition or formation process but merely accepts the departing hydroxide, a function that could better be filled by a proton. So while the termolecular concerted mechanism is consistent with all available data, it is chemically unreasonable and is ruled out on that basis.

2.2.2. *Ternary Mechanisms with Intermediates.* The next class of mechanism involves the same three reaction components in the characteristic transition state but a tetrahedral addition intermediate is formed. This is a ternary stepwise mechanism. The intermediate would necessarily involve the central reactant, bicarbonate. The products result upon decomposition of the intermediate. The conjugate base of the urea group adds to bicarbonate while the γ-phosphate of ATP traps the intermediate. This produces a phosphate ester of the tetrahedral intermediate derived from bicarbonate and the conjugate base of biotin, with ADP a co-product. The phosphorylated intermediate then expels phosphate to give the product.

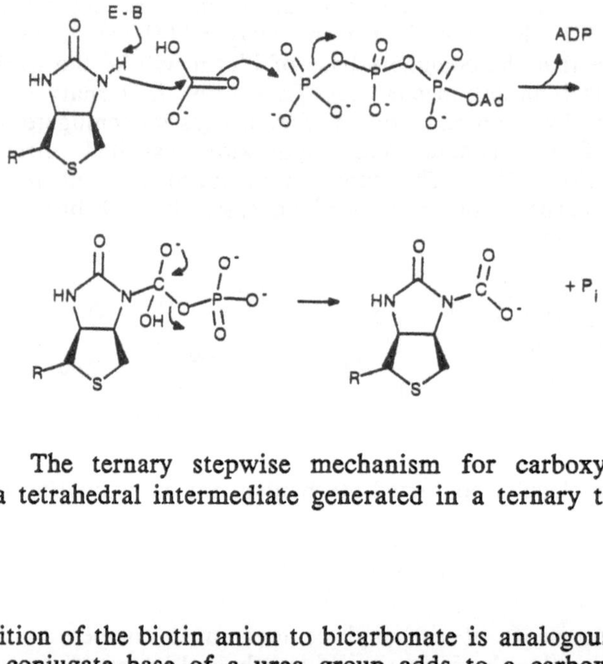

Figure 8. The ternary stepwise mechanism for carboxylation of biotin involving a tetrahedral intermediate generated in a ternary transition state.

The addition of the biotin anion to bicarbonate is analogous to the reaction in which the conjugate base of a urea group adds to a carboxylate anion. This reaction was observed by Blagoeva et al and was cited by those authors in support of carboxyphosphate mechanisms [16]. We feel that it is has wider implications in that it shows that the conjugate base of a urea is nucleophilic toward carboxyl groups in general. Thus, it is also a model for mechanisms in which biotin attacks bicarbonate directly.

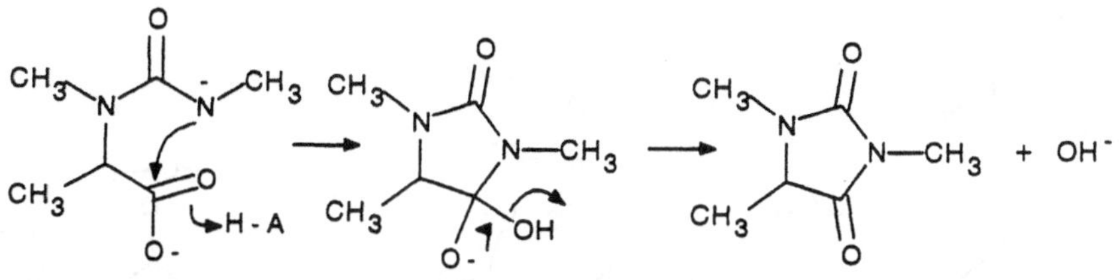

Figure 9. A model reaction for the attack of the conjugate base of biotin on bicarbonate.

The mechanism also requires that the oxygen anion generated in the incipient tetrahedral intermediate attack ATP. It is not clear whether such a process is likely to be concerted with the addition step but there is a good chemical analogy for the overall process. It was found by Ramirez that the conjugate base of a carbonyl hydrate very readily attacks an adjacent phosphate ester [17]. Recent work by Scott Taylor in our laboratory has confirmed and extended the work of Ramirez.

2.2.3. *Binary mechanisms.* In the next class of mechanisms, two of the three components form an intermediate which reacts with the third component. One of the products may be generated prior to reaction with the third component. If the product is released, then the enzyme should be capable of catalyzing an exchange reaction involving the component that is released. If either biotin or bicarbonate reacts with ATP, then ADP would potentially be released. This exchange has not been observed but enzymes do not necessarily release products as they are formed.

All binary mechanisms must involve at least one intermediate. The initial intermediate can result from reaction of any two of the three components. The combination of ATP and bicarbonate gives carboxyphosphate as an intermediate which then reacts with biotin (See Figure 4). This binary mechanism has been considered in detail and is widely advocated [7]. There are also two other general classes of binary mechanism. The combination of biotin and ATP might lead to the formation of a phosphorylated biotin derivative, O-phosphobiotin [9,18,19]. We have proposed that O-phosphobiotin can be readily generated by the reaction of ATP with biotin (Figure 10) and it should react with bicarbonate to give N-carboxybiotin [9,19].

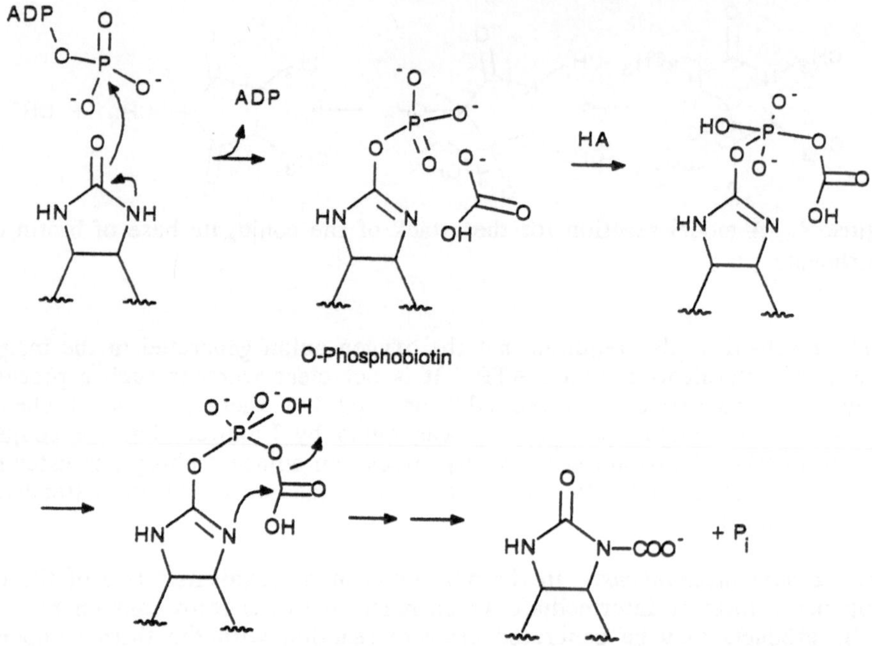

Figure 10. The reaction of ATP and biotin could produce O-phosphobiotin which then reacts with bicarbonate.

Perič and co-workers have investigated models for the reaction of ATP and biotin to give N-phosphobiotin and this can also lead to formation of N-carboxybiotin by reaction with bicarbonate [20].

The third possible general binary mechanism involves the reaction of the conjugate base of biotin with bicarbonate to produce an addition intermediate which then reacts with ATP [13]. This is like the stepwise ternary mechanism in that the conjugate base of biotin attacks bicarbonate to give a tetrahedral intermediate. In the binary case the attack on ATP occurs subsequent to formation of the intermediate (Figure 11).

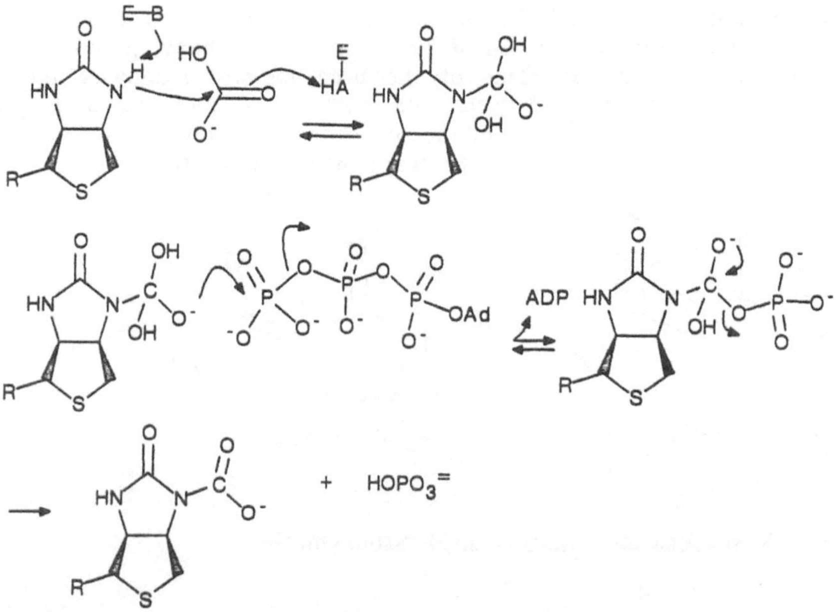

Figure 11. A binary mechanism in which bicarbonate and the conjugate base of biotin react initially.

As mentioned earlier, the tetrahedral intermediate should be a reactive nucleophile toward ATP based on the reactivity of the conjugate base of a carbonyl hydrate toward an internal phosphate ester [17]. This mechanism as well as the ternary mechanisms predict inversion of configuration at phosphorus, in agreement with experimental observations. The direct use of bicarbonate is efficient in that trapping by biotin of a secondary product, such as carboxyphosphate or carbon dioxide, is not required. The attack of biotin on bicarbonate is no worse than the attack on carboxyphosphate would be and is analogous to the reverse of the reaction in which the carboxyl group of N-carboxybiotin is transferred to the conjugate base of the substrate.

3. Carboxyl transfer from N-carboxybiotin

After biotin is converted to N-carboxybiotin, carbon dioxide is not yet utilized. The elements of carbon dioxide must be transferred from biotin to the acceptor substrate. N-carboxybiotin must be capable of transferring the carboxyl function to the acceptor yet it must be resistant to transfer of the group to unproductive acceptors, such as water. Wallace and co-workers showed that the binding of a substrate or an analogue of the substrate to pyruvate carboxylase causes the reactivity of N-carboxybiotin toward carboxyl transfer to increase [21]. The change in reactivity might easily be accomplished by a change in conformation about the bond between the nitrogen of biotin and the carboxyl group. When the carboxyl group is in the plane of the urea moiety, resonance overlap stabilizes the carbon-nitrogen between the carboxylate and the urea [22]. Rotation of the carboxyl group out of the plane destroys this stabilization and enhances the reactivity of the carboxyl toward nucleophilic attack [7,22].

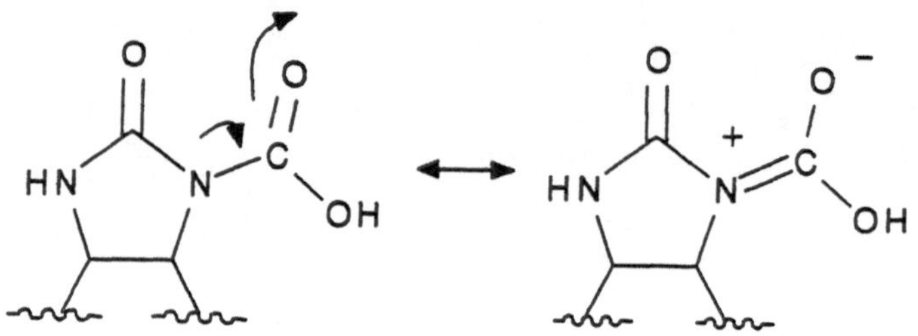

Figure 12. Resonance and rotation in N-carboxybiotin.

The transfer of the carboxyl group to the carbanion derived from the substrate should occur readily. We have made a number of substituted cyclic ureas and find that the coplanar conformation is likely to predominate, even when bulky groups are in adjacent positions. An alternative, which we are currently investigating involves binding of the N-carboxyurea to a receptor which forces the carboxyl group out of the plane of the urea.

4. Concluding Remarks

The reaction patterns we have seen as possibilities for carbon dioxide utilization by biotin-dependent systems only represent the broadest categories of reactions. The detailed mechanism of any of these is worthy of further detailed analysis. The deductive approach assures us that we have presented ourselves with many possibilities before we find the necessary limiting evidence to eliminate incorrect mechanisms. It is still very early in the development of our understanding of these systems. They continue to present a remarkably complex puzzle for what appears to be a simple system.

5. Ackowledgements

Continued support through an Operating Grant from the Natural Sciences and Engineering Council of Canada is acknowledged.

REFERENCES

1. Lynen, F., Knappe, J., Lorch, E., Juetting, G., Ringelmann, E., (1959) Angew. Chem. **71**, 485-86.
2. Guchait, R. B., Polakis, S.E., Hollis, D. Fenselau, C., and Lane, M.D. (1974) J. Biol. Chem. **249**, 6646-6656.
3. Kaziro, Y., Hase, L.F., Boyer, P.D., and Ochoa, S. (1962) J. Biol. Chem. **237**, 1460-1468.
4. Knowles, J.R. (1980) Annu. Rev. Biochem. **49**, 877-1919.
5. Hansen, D. E. and Knowles, J. R. (1985) J. Am. Chem. Soc. **107**, 8304-8305.
6. Sauers, C.K., Jencks, W.P., and Groh, S. (1975) J. Am. Chem. Soc. **97**, 5546-53.
7. Knowles, J.R. (1989) Ann. Rev. Biochem. **58**, in press,
8. Polakis, S.E., Guchait, R.B., Lane, M.D. (1972) J. Biol. Chem. **247**, 1335-1337.
9. Kluger, R., Davis, P.P. and Adawadkar, P.D. (1979) J. Am. Chem. Soc. **101**, 5995-6000.
10. Tipton, P.A., Cleland W.W. (1988) Biochemistry **27**, 4325-4331.
11. Climent, I. Rubio, V. (1986) Arch. Biochem. Biophys. **251**, 465-70.
12. Ogita, T., Knowles, J.R. (1988) Biochemistry **27**, 8028-33.
13. Kluger, R. (1989) Bioorg. Chem. 17, in press.
14. Perrin, C. A., and Dwyer, T.J. (1987) J. Am. Chem. Soc. **109**, 5163-5167.
15. Ba-Saif, S., Luthra, A., Williams, A. (1987) J. Am. Chem. Soc. **109**, 6362-6368.
16. Blagoeva, I.B., Pojarlieff, I.B., and Kirby, A.J. (1984) J. Chem. Soc. Perkin Trans. 2, 745-751.
17. Ramirez, F., Hànsen, B., Desai, N.B. (1962) J. Am. Chem. Soc. **84**, 4588.
18. Lynen, F. (1967) Biochem. J. **102**, 381-400.
19. Kluger, R. and Adawadkar, P.D. (1976) J. Am. Chem. Soc. **98**, 3741-3742.
20. Blonski, C., Gasc, M.B., Hegarty, A.F., Klaebe, A., Periè, J.J. (1984) J. Am. Chem. Soc. **106**, 7523-7529.
21. Goodall, G.J., Prager, R., Wallace, J.C., Keech, D.B. (1983) FEBS Lett. **163**, 6-9.
22. Thatcher, G.R.J., Poirier, R., and Kluger, R. (1986). J. Am. Chem. Soc. **108**, 2699-2704.

CHEMICAL ROLES OF B_{12}-DERIVATIVES IN THE BACTERIAL C_1-METABOLISM

Bernhard Kräutler
Laboratory of Organic Chemistry
Swiss Federal Institute of Technology
Universitätstrasse 16
CH-8092 Zürich, Switzerland

ABSTRACT. Vitamin B_{12} derivatives are central cofactors in a series of organo-metallic and redox-processes, essential in bacterial metabolism and which depend on the unique coordination chemistry of the cobalt-corrins. They owe their particular reactivity at cobalt mainly to the binding by the corrin ligand, but, in the complete corrins, also to the influence by the nucleotide function. The biological roles focus around the unusual Co-C-bond, that is stable towards physiological acids in the coenzyme-forms. In bacteria, the enzymatic reactions catalyzed by cobalt corrins have a central role in the C_1-metabolism in particular, where methylcorrinoids and reduced, nonalkylated corrinoids have been identified as B_{12}-coenzyme forms.

1. INTRODUCTION

The corrinoids, derivatives of vitamin B_{12} (1,cyanocobalamin, see Figure 1), play several central roles in particular in the metabolism of microorganisms [1-3], which appear to uniquely possess the capacity in nature to build up the B_{12} structure also [4]. The corrin ligand, first discovered in vitamin B_{12} [5], is related to the natural porphyrin ligands, both by structure [6] and due to its biosynthesis [7]. The nucleotide loop is a structural property, unique to the "complete" corrinoids among the natural cofactors known today.

273

M. Aresta and J. V. Schloss (eds.),
Enzymatic and Model Carboxylation and Reduction Reactions for Carbon Dioxide Utilization, 273–292.
© 1990 *Kluwer Academic Publishers.*

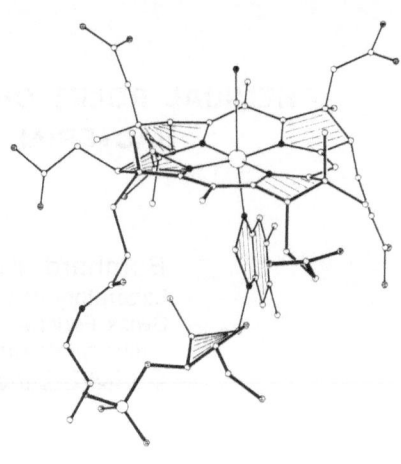

Rickes,Brink,Koniusky,Wood,Folkers
Smith,Parker
(1948)

D.C.Hodgkin et al.(1955)

Figure 1. Left: structural formula of vitamin B$_{12}$ (**1**); right: three dimensional structure of **1** according to the x-ray analysis.

The corrin-bound cobalt center has been characterized in three oxidation levels: as a hexacoordinate, diamagnetic and pseudo-octahedrally ligated Co(III)-ion (two axial ligands) [5], as a pentacoordinate, paramagnetic, low spin Co(II)-ion (one axial ligand) [8] and as a diamagnetic Co(I)-ion, which, according to the available evidence [9], is coordinated by the corrin nitrogens only. The reduction and oxidation steps interconverting the Co(III)-, Co(II)- and Co(I)-forms of vitamin B$_{12}$-derivatives have been characterized electrochemically [10]. The structure of Co(III)corrins has been studied by x-ray analysis [5,11] (e.g. methylcobalamin **2** [12]) and nmr-spectroscopy [13]. Information on the structure of Co(II)corrins stems from esr- [14] and x-ray-analysis [5b,8,15], one of which concerns the "coenzyme" cob(II)-alamin (**3**, "B$_{12}$") [15].

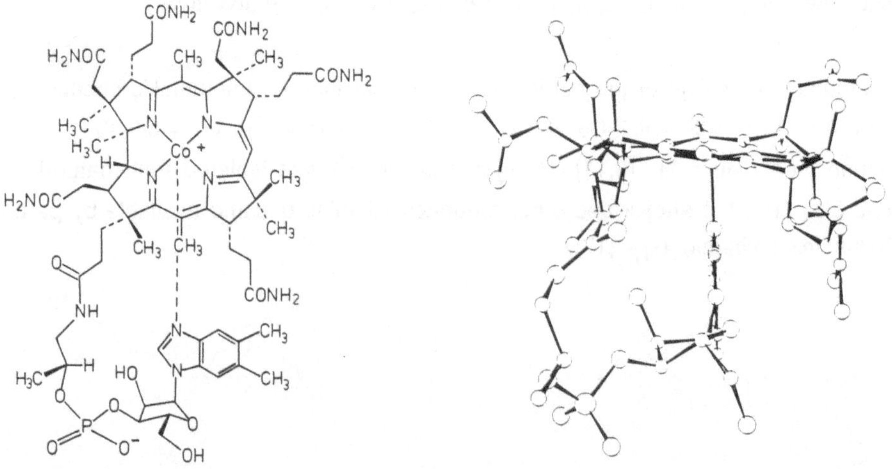

Figure 2a. Left: structural formula of cob(II)alamin (**3**, "B$_{12r}$"); right: three dimensional structure of **3**.

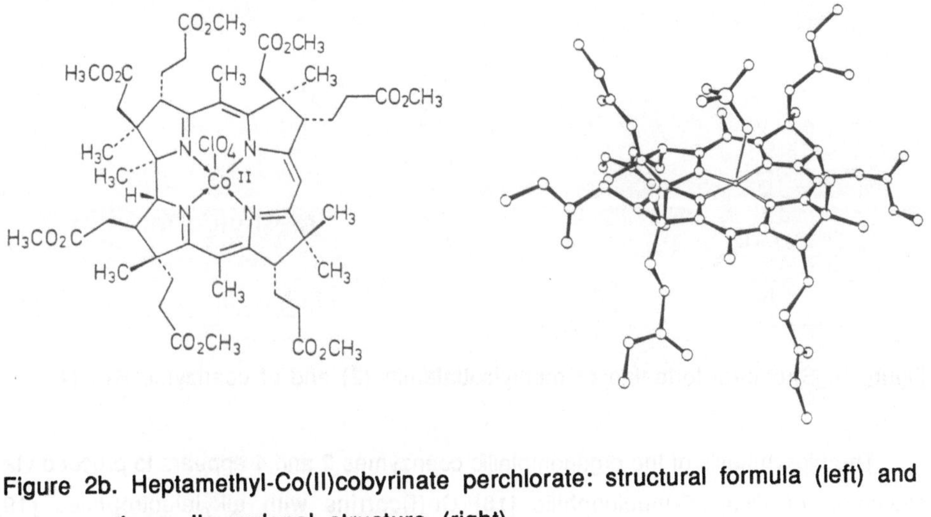

Figure 2b. Heptamethyl-Co(II)cobyrinate perchlorate: structural formula (left) and three dimensional structure (right)

<u>Figure 2</u>. The structures of paramagnetic Co(II)corrins

2. ORGANOMETALLIC CHEMISTRY OF COBALT-CORRINOIDS

Most metabolic functions of the corrinoids depend on their organometallic chemistry, uncovered in 1961 with the x-ray analysis of the coenzyme B_{12} (**4**, adenosyl-cobalamin, see Figure 3) [5,11]. Methylcobalamin (**2**) was isolated from natural sources also [16], but after it and other organocobalamins became available by partial synthesis from vitamin B_{12} [17].

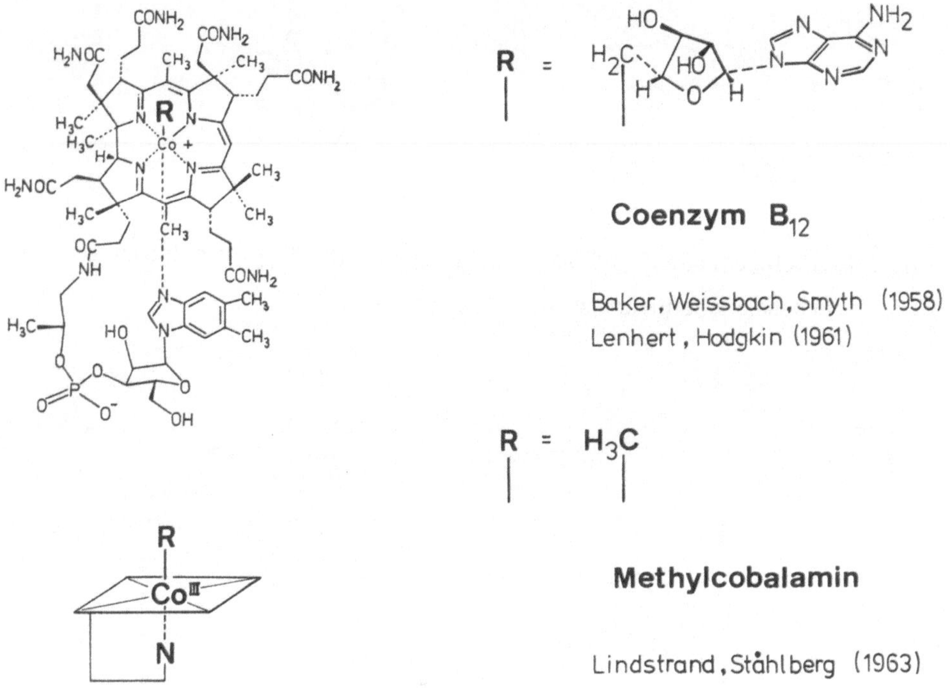

Coenzym B_{12}

Baker, Weissbach, Smyth (1958)
Lenhert, Hodgkin (1961)

Methylcobalamin

Lindstrand, Ståhlberg (1963)

Figure 3. Structural formulae of methylcobalamin (**2**) and of coenzyme B_{12} (**4**)

The biosynthesis of the organometallic coenzymes **2** and **4** appears to proceed via alkylation of ("super"-)nucleophilic [18] Co(I)corrins with alkylelectrophiles [19]. In solution, alkylation of Co(I)corrins occurs by nucleophilic displacement at carbon [20] or via electron transfer induced radical reactions [21]. With complete corrinoids the alkylation takes place with high facial selectivity, but in nucleotide free corrins alkylation can occur on either side of the Co-center [22]. Also the radicaloid Co(II)-corrins or the electrophilic Co(III)corrins can be cobalt-alkylated [19].

The coenzyme **4** has a role as a cofactor in some ten chemically complex enzymatic rearrangement reactions [1b] and in the cobamide dependent ribonucleotide reduction [23], which appear to be based on the tendency of **4** to easily cleave its Co-C-bond homolytically : its homolytic bond dissociation energy amounts to about 30 kcal/mol only(see Figure 4) [24].

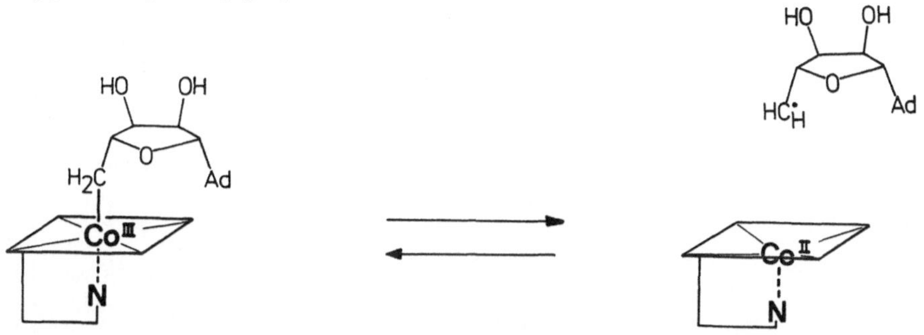

in Lösung : $\Delta H_{diss}(298)$ = 28,6 kcal/mol   (J.Halpern et al. 1984)
28,6 kcal/mol 31,5 kcal/mol   (R.G.Finke et al. 1984)
——→ $\overrightarrow{k}$ (RT) ~ 10^{-10} sek^{-1}

Figure 4. Coenzyme B$_{12}$, a "reversible source of organic radicals" (J.Halpern [24])

The natural processes involving methylcorrinoids, such as methylcobalamin (**2**), are quite diverse (see Figure 5): (a) the methyl group transfer from methyl-tetra-hydrofolate (and related methyl-tetrahydropterins) to homocysteine, to give methion-ine [3a]; (b) a related function in the methylation step of coenzyme M to methyl-co-enzyme M in the bacterial methanogenesis [25]; (c) a mechanistically less understood role in the assembly of the acetyl group during the acetyl-CoA pathway of bacterial fix-ation of carbon dioxide [26] and (d) methylcorrinoids are the source of the methyl group of the highly toxic methylmercuric ions [27]. While the methyl group transfer in the thiol alkylations presumably proceeds via nucleophilic displacement reactions (i.e. formally as a methyl cation) [28], the methyl group abstraction by the mercuric ions formally concerns a methyl anion [27], and a radicaloid abstraction of the cobalt-bound methyl group might also be relevant in the bacterial C1-metabolism [3b,29].
Accordingly, three modes of Co-C-bond cleavage are important for the biological roles of the methylcorrinoids (see Figure 6) [30].

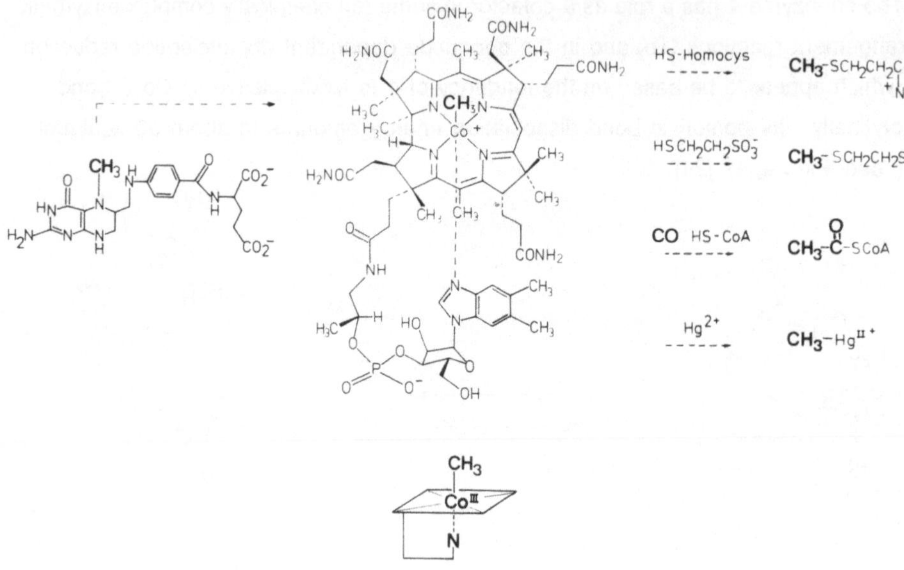

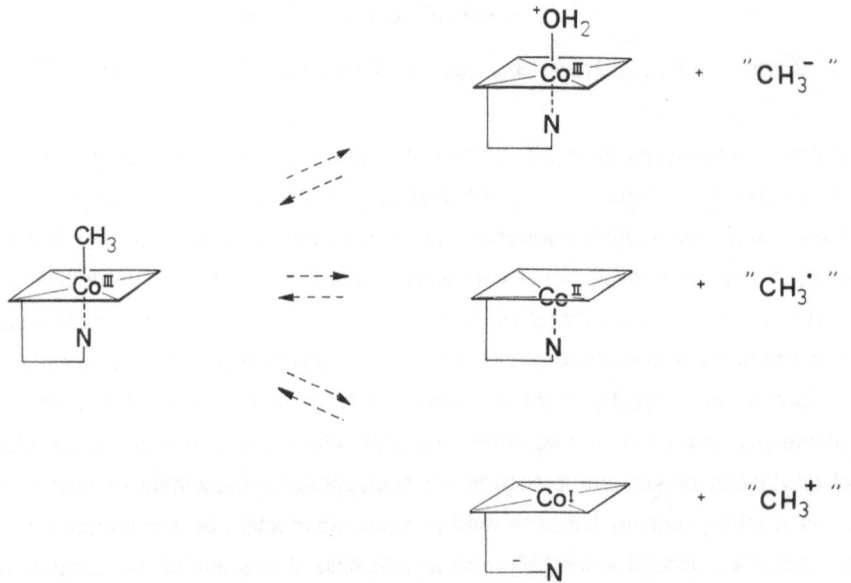

Figure 5. Roles of methylcorrinoids in biological methyl transfer reactions.

Figure 6. Important modes of cleavage of the Co-C-bond of methylcobalamin and related methylcorrinoids.

The unique nucleotide function of the complete corrinoids modifies the strength of the Co-C bond [31,32] and also controls the facial selectivity at the cobalt center [22a]. For methyl-corrinoids the effect of the nucleotide coordination on the strength of the Co-CH$_3$-bond (i.e. its thermodynamic trans effect) can be studied in methyl group equilibration experiments [31] between complete corrinoids (such as the cobalamins) and incomplete corrinoids (such as cobinamides, see Figure 7). The equilibria inform on the difference in the Co-C bond strengths in the methylated species.

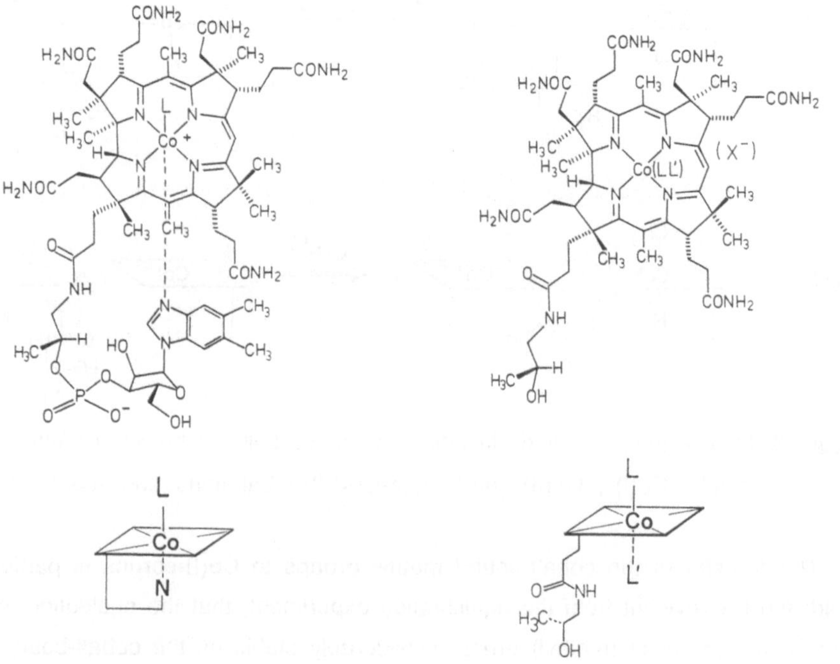

Figure 7. Structural formulae (and corresponding symbols) of cobalamins (left) and of cobinamides (right).

Such methyl group transfer and equilibration experiments can be carried out between methyl-Co(III)corrins and Co(II)corrins, to inform on the differences of the homolytic Co-C bond strengths [31]. Methyl group transfer between methyl-Co(III)-corrins and Co(I)corrins [31] or between methyl-Co(III)corrins and nonalkylated Co(III)corrins [33] lead to equilibria, that inform on the differences of the two types of heterolytic Co-C bond strengths (see Figure 8).

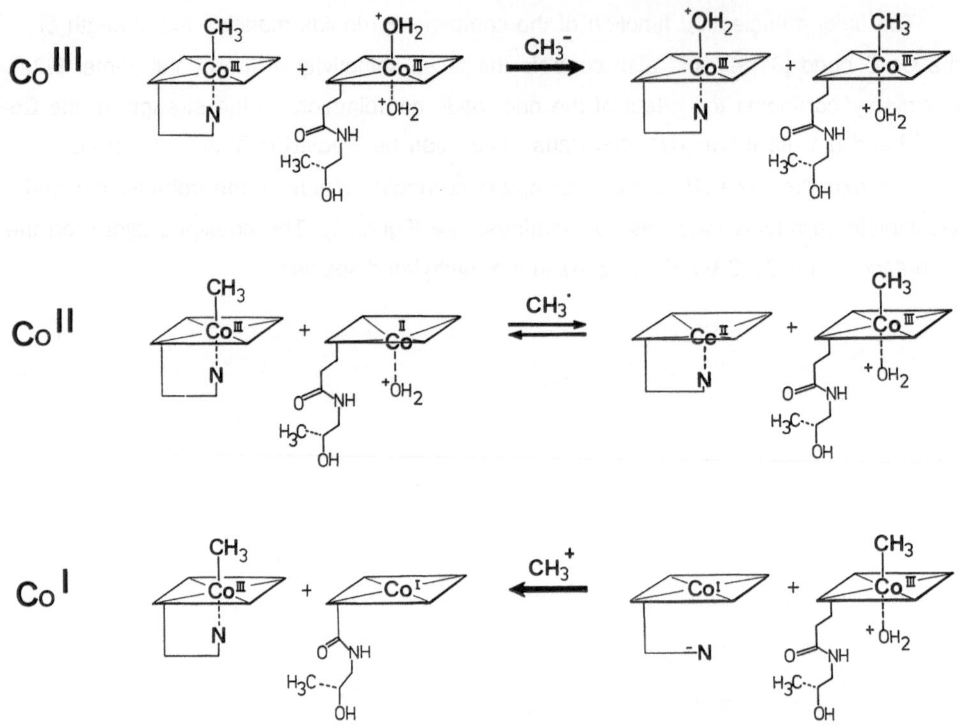

Figure 8. Methyl group equilibria in neutral aqueous solution between methyl-
Co(III)-, Co(I)-, Co(II)- and aquo-Co(III)-cobalamins and -cobinamides.

The transfer of the cobalt-bound methyl groups to Co(I)-corrins is particularly rapid and it is evident from the equilibration experiment, that the nucleotide coordination in the complete (methyl)corrins considerably stabilizes the cobalt-bound methyl group against abstraction by nucleophiles (as Co(I)corrins) [31]. On the other hand, the effect of the nucleotide base on the kinetics of the methylation reaction is less investigated and the sequence of events in the alkylation of Co(I)corrins (an "oxidative trans-addition") less well understood (see Figure 9): In this reaction, a transition occurs from a tetracoordinate Co(I)center to a hexacoordinate Co(III)center, i.e. with formation of two axial bonds. In solution, evidence has been provided for the absence of axial ligation to cobalt in Co(I)corrins, so that a pathway involving pentacoordinate alkyl-Co(III)corrin intermediates is indicated there [34]. Based on nmr-studies, such coordinatively unsaturated methylation intermediates would be less stable by ca. 12 kcal/mol than the methylation product **2** [35].

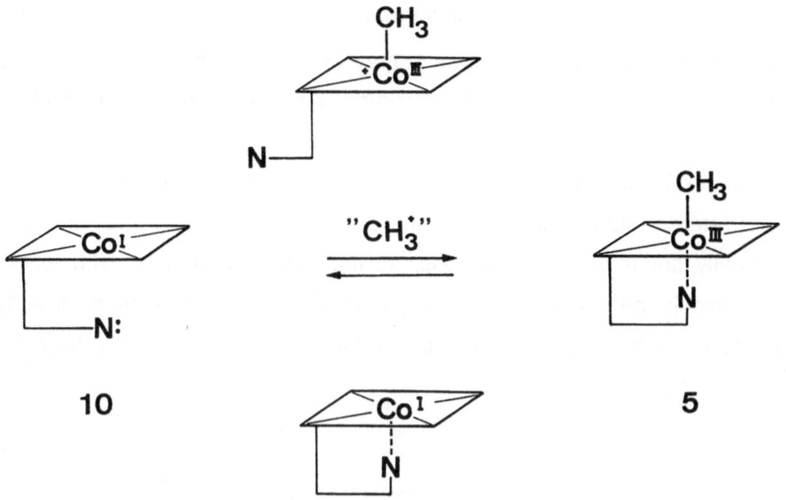

Co^I-corrin: **– supernucleophilic Co^I center**
(Schrauzer u. Mit. 1968)
(nearly) quadr. planar coord. d^8-ion cf.:

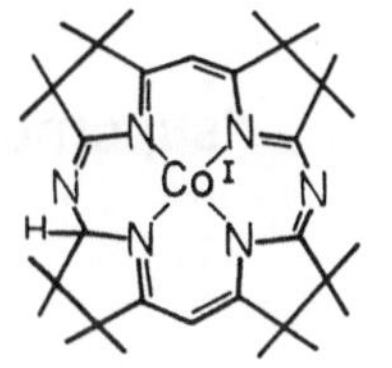

R. Scheffold
(1979, 1984)

Figure 9. Reactivity of Co(I)corrins in methylation via nucleophilic substitution reactions: concerning the sequence of bond forming events in the oxidative trans addition.

Clearly then, the nucleotide group allows the particular modification of the reactivity in alkylation reactions at the corrin-bound Co(I)-center, presumably the type of reaction most relevant to cobamide dependent biological methyl group transfer.

The cobalt-bound methyl group is transferred rapidly at ambient temperatures to Co(II)-corrins. The corresponding cobalamin/cobinamide methyl transfer equilibrium indicates the homolytic Co-C BDE of methylcobalamin to be slightly larger (by about 0.4 kcal/mol) than in the methyl-cobinamide [31]. On the other hand, the nucleotide trans coordination in coenzyme B_{12} was deduced to lead to a Co-C bond weakening by *ca.* 0.7 kcal/ mol in **4** (e.g. compared to the situation in the adenosyl-cobinamide) [31].

The vitamin B_{12}-derivatives correspondingly possess a series of important re-activities in organometallic reactions, to which they owe their unique position as catalysts in nature (see Figure 10): As Co(I)corrins, they are strong nucleophiles, which are selectively alkylated on the upper face, and quite strong reducing agents. Co(II)corrins are persistent and powerful radical traps, which react with organic radicals without large structural changes in the corrin parts of the molecules. Alkyl-Co(III)corrins finally are latent alkyl radicals, as they easily homolyse with formation of free organic radicals, and versatile alkylating agents as well, due to their capacity to transfer their cobalt-bound alkylgroup to nucleophilic, radicaloid and electrophilic alkyl-group acceptors.

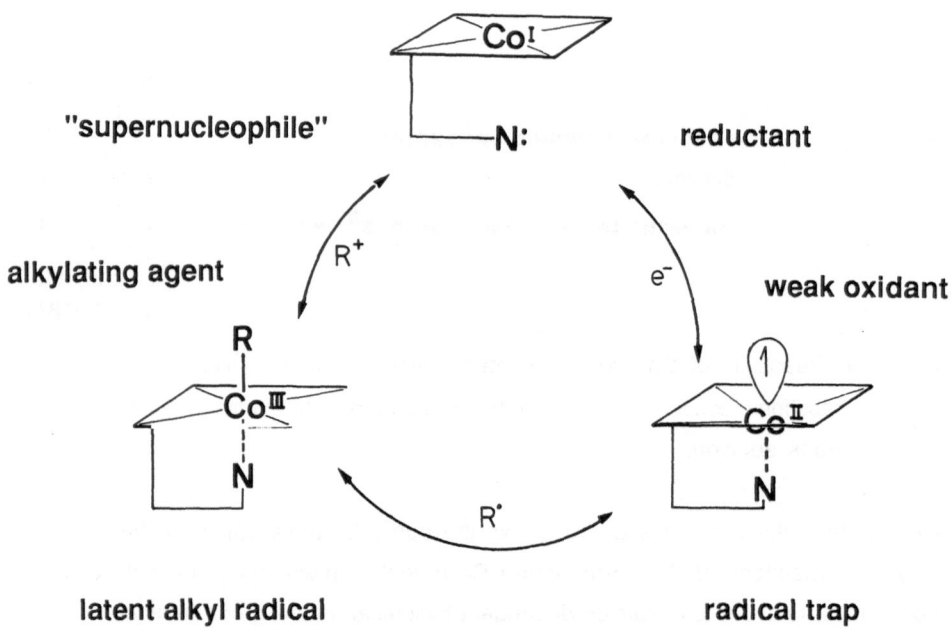

Figure 10. Important basic reactivities of vitamin B_{12}-derivatives in biological organometallic reactions.

3. ROLES OF CORRINOIDS IN BACTERIAL C$_1$-METABOLISM

Vitamin B$_{12}$-derivatives take up central roles in the metabolism of anaerobic bacteria (see Figure 11) and correspondingly are abundant in methanogenic, acetogenic and a series of sulfur metabolizing/sulfate reducing bacteria [29,36]. In some cases they are present to an extent of more than 0.1% of the bacterial dry mass. In particular methyl-corrinoids have been shown to be intermediates in a series of important enzymatic methyl group transfer/activation steps in bacterial C$_1$-metabolism [3a,25,26], but nonalkylated corrinoids might have a function as electron transfer redox-catalysts in the energy conserving formation of methane in methanogens [37].

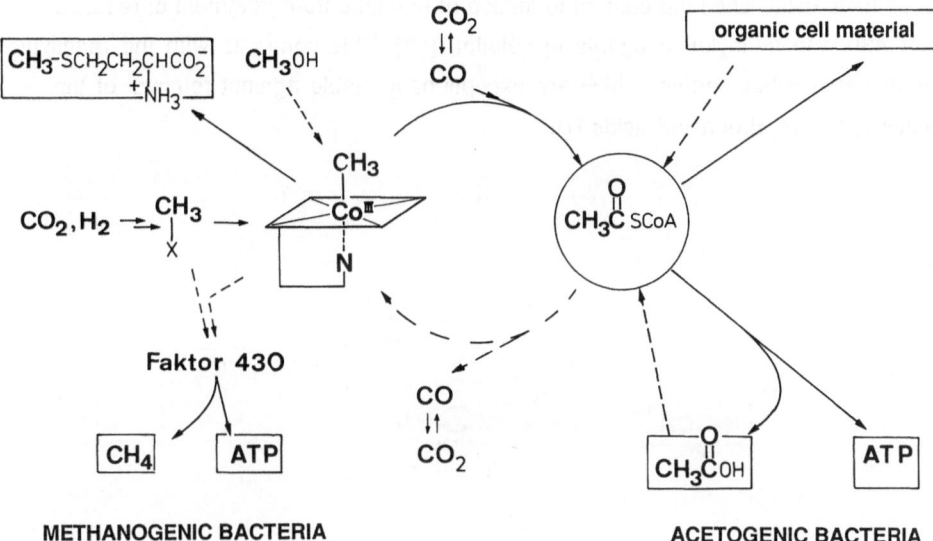

METHANOGENIC BACTERIA   **ACETOGENIC BACTERIA**

<u>Figure 11</u>. Role of methyl-corrinoids in bacterial metabolism

A role as methylating agents in the cobamide dependent bacterial biosynthesis of methionine from homocysteine was early recognized in Escherichia coli [3a,28]. As meanwhile quite clear, the transfer of an intact methylgroup from N-methyl-tetra-hydrofolate (CH$_3$-THF) to homocysteine is catalyzed via an enzyme bound methyl-corrin intermediate. By stereochemical labelling of the methylgroup of CH$_3$-THF, the transfer to the thiol with overall retention of configuration at the methyl carbon could be shown [28b].

This is in accordance with the mechanistic picture developed earlier, that the methyl group is first abstracted from the CH_3-THF by the Co(I)-corrinoid (one inversion) and then nucleophilically displaced from the methylcorrin intermediate by the thiol (second inversion = overall retention)[3a,28,38].

The cobalt-bound methyl group of methyl-corrinoids has also been found to be incorporated into methane in methanogenic bacteria [39]. A chemical mechanism for the formation of methane from methyl-corrins was searched for and found under reducing conditions [40]. Meanwhile however, the newly discovered porphinoid nickel complex, factor 430 [6,41], has been shown to be the coenzyme in the energy yielding enzymatic formation of methane by reductive demethylation of methylcoenzyme M [42]. Model studies have established the ease of formation of methane from treatment of reduced factor 430 with methylating agents in solution [43]. This contrasts with the reactivities of methylated cobalt-corrins, which are exceptionally stable against removal of the organic ligand by Broensted acids [1].

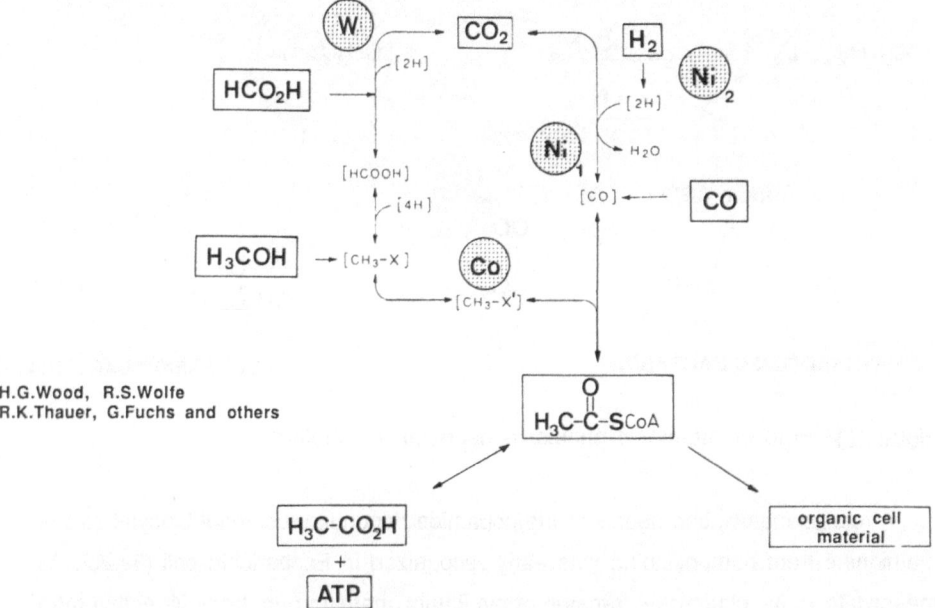

Figure 12. Flow diagramm illustrating the build-up of acetyl-coenzyme A from carbon dioxide and hydrogen in anaerobic acetogens (W = formate dehydrogenase; Ni1 = carbon monoxide dehydrogenase; Ni2 = hydrogenase; Co = corrinoid enzyme).

A third role of the methyl-corrinoids in bacterial metabolism was uncovered, when it was found that their cobalt-bound methyl group was incorporated into acetic acid in acetogens, such as Clostridium thermoaceticum [44]. This enzymatic transformation again lacked a chemical precendent. In the meantime, this step was recognized to be part of the mechanism of autotrophic fixation of carbon dioxide via the acetyl-coenzyme A pathway, a mode now recognized to be used in autotrophic methanogens and acetogens, and also in other anaerobic autotrophic bacteria [26,45].

$$?$$

$$CH_3X + CO + HSCoA \longrightarrow H_3CCSCoA + H^+ + X^-$$

$$X = B_{12} - \text{derivative} \qquad (R.K.Thauer, 1983)$$

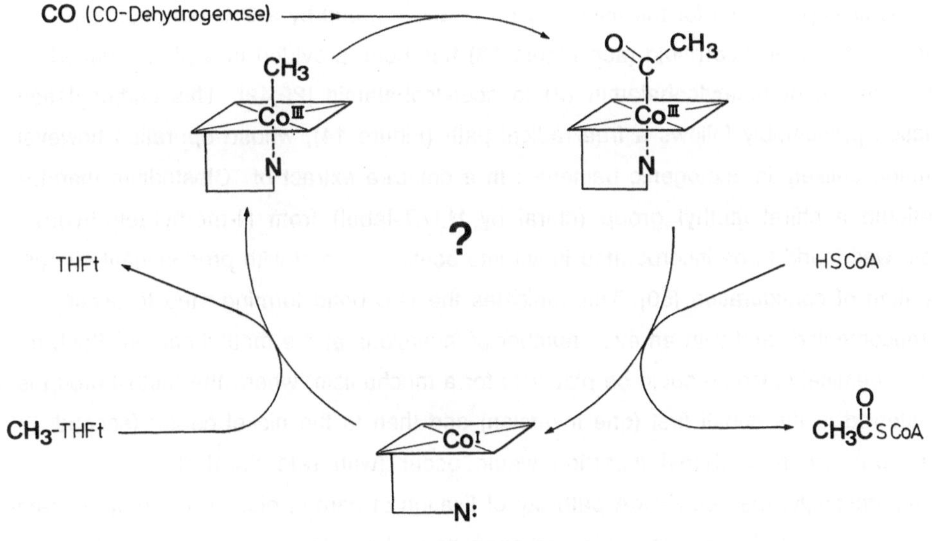

Figure 13. A hypothetical formation of the acetyl-group of acetyl-CoA by a C-C-bond formation via carbonylation of a methylcorrinoid.

Central to the acetyl-coenzyme A-path of fixation of carbon dioxide is the assembly of the acetyl group of acetyl-CoA, which involves a series of transition metalloenzymes (see Figure 12) [26,45]. In acetogenic bacteria, Clostridium thermoaceticum be ing the one investigated best, carbon dioxide is enzymatically reduced in two branches. In one of these carbon dioxide is first reduced to formate at the oxygen sensitive tungsten enzyme "formate-dehydrogenase". The formate is then incorporated as the formyl group of N-formyl-tetrahydrofolate and reduced to the methyl group of N-methyl-tetra-hydrofolate. The methyl group then is transferred to the corrinoid. In the second branch, carbon dioxide is reduced first to carbon monoxide by the oxygen sensitive "carbon-monoxide dehydrogenase" a little characterized nickel-iron-enzyme [46].

In a complex process, the intact methyl group from the methyl-corrinoid , carbon monoxide and coenzyme A combine to acetyl-coenzyme A. A currently actively invest-igated question concerns the metal-site at which the hypothetical and crucial C-C-bond forming step occurs. For such a carbonylation reaction an organometallic process is to be considered [47], but it is not clear from the microbiological studies, whether it occurs at the corrinoid cobalt center or at the nickel-iron center of the carbon-monoxide dehydrogenase [46,48].

Chemical precedent for the assembly of an acetyl group by a C-C-bond forming reaction at a methyl-corrinoid (see Figure 13) has been provided in a photoinduced carbonylation of methylcobalamin (2) to acetylcobalamin [29,49]. This carbonylation reaction presumably follows a free radical path (Figure 14), whose operation however is made unlikely in acetogenic bacteria : In a cell-free extract of Clostridium thermo-aceticum a chiral methyl group (chiral by H,D,T-label) from N-methyl-tetrahydro-folate was found to be incorporated intact into acetic acid and with predominant overall retention of configuration [50]. This indicates the C-C-bond forming step to occur stereocontrolled and with an even number of inversions at the methyl-carbon. Such a stereochemical outcome could be plausible for a mechanism, where the methyl group is transferred to the cobalt first (one inversion) and then to the nickel center (second inversion), where carbonyl insertion would occur (with retention)[50].

Interestingly, the acetyl-CoA pathway of fixation of carbon dioxide not only operates in acetogens, but also in autorophic methanogens and sulfate reducing bacteria. In addition, the acetate degradation to methane and carbon dioxide in methanogens and to carbon dioxide in sulfate reducing heterotrophs appears to be based on the same set of reactions, but in the reverse sense [26a,45].

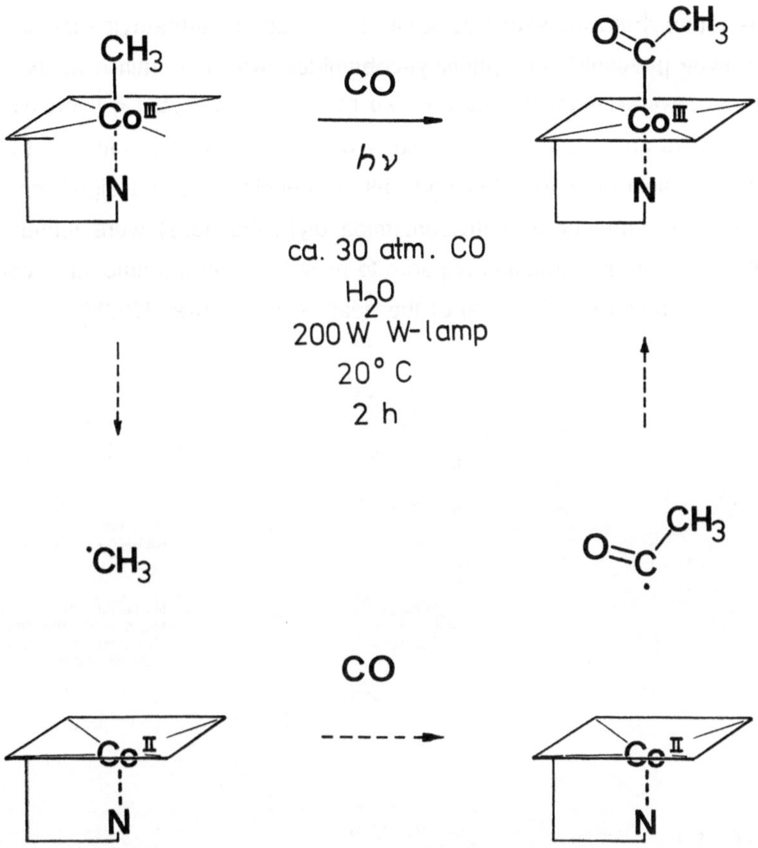

Figure 14. Photoinduced carbonylation of methylcobalamin (**2**, R = CH₃):
illustration of the hypothetical radical mechanism to acetylcobalamin
(R = COCH₃)

In a wide range of anaerobic bacteria accordingly central roles are given to the corrins and indeed in the bacterial organisms employing the acetyl-coenzyme A pathway corrinoids characteristically are abundant. The corrinoids from acetogens, when isolated in their Co-cyano-forms, were predominantly found not to be vitamin B_{12} (see Figure 15), a 5,6-dimethylbenzimidazolyl-cobamide, but rather to differ by their nucleotide

288

base: the 5-methoxybenzimidazolyl-cobamides, 5-methoxy,6-methylbenzimidazolyl-cobamides or even p-cresolyl- and phenolyl-cobamides, which contain a noncoordinating aglycon. In several methanogens likewise not vitamin B_{12}, but 5-hydroxybenzimidazolyl-cobamides and 7-adeninyl-cobamides were found. Also in some, but not all, of the investigated sulfur-metabolizing and sulfate-reducing bacteria significant amounts of corrinoids (mostly 5-methylbenzimidazolyl-cobamides) were found. In several of these bacteria a correlation appears to exist between the amount of corrins present and the operation (or absence) of the acetyl-CoA pathway [29,36].

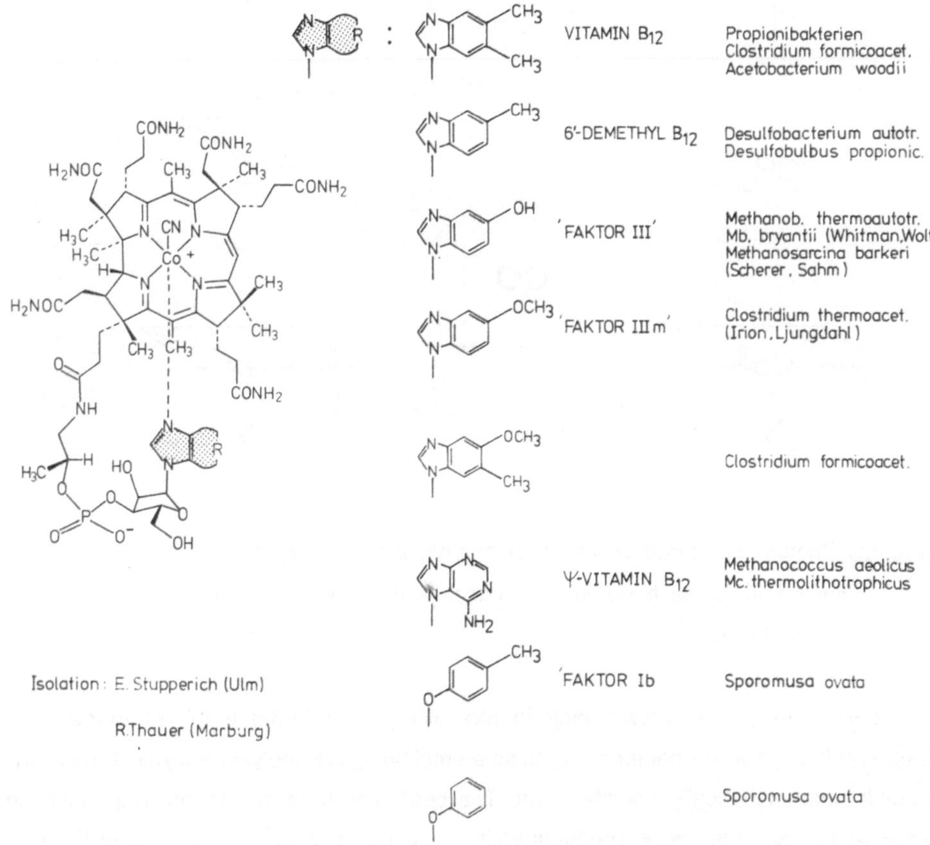

Figure 15. Corrins from methanogenic, acetogenic, sulfate-reducing and sulfur-metabolizing bacteria (Co-cyano-forms)

This structural variety appears to be a consequence, possibly, of the biosynthetic availability of the corresponding nucleotide bases [36a]. Secondly, the nucleotide structure offers a twofold potential for adjusting the reactivity of the enzyme-bound corrin: firstly, the nucleotide coordination at the corrin-bound cobalt-center directly affects the strength of the cobalt-carbon bond and the redox-properties of the corrins [31]; secondly, by way of the structure-sensitive binding between the corrin and the apoenzyme, the nucleotide base can also contribute indirectly to the reactivity of the protein-bound corrin [36c]. The structural variety among the bacterial corrins still is remarkable, since they play such central roles (e.g. in the fixation of carbon dioxide). It would represent a special case of the rule (see e.g.[6]), that important types of cofactors have remained highly conserved structurally in the course of biological evolution. Apparently this structural conservation strictly applies to the cobamide part of the complete corrins, but not to the nucleotide base.

The acetyl-CoA pathway of autotrophic fixation of carbon dioxide in anaerobic bacteria is chemically unique in several ways: It makes use of organometallic transformations at oxygen-labile metallo-enzymes, and accordingly has been found only in anaerobic organisms. Secondly, in acetogens it not only provides a means for autotrophic fixation of carbon dioxide, but also for the production of adenosine-triphosphate (ATP) [26].

The course taken by the acetyl-coenzyme A pathway demonstrates a remarkable and efficient path for the reduction of carbon dioxide. It is centered around organometallic transformations at transition metal ions, whose metal-carbon bonds typically are weak. Correspondingly a potential appears to be provided for energy rich organic functional groups in nature. Aside from their metabolic relevance these naturally occuring organometallic transformations are of interest as exemples for the economic and ecologic synthesis of basic organic molecules from carbon dioxide.

Acknowledgements: I would like to thank Prof. R.K.Thauer (University of Marburg, FRG), who activated my interest in this topic, and to Dr. E.Stupperich and Prof. G.Fuchs (University of ULM, FRG) for fruitful cooperations and helpful discussions since. I am grateful to Doz. Dr.C.Kratky and W.Keller (University of Graz, Austria), for their important x-ray crystallographic contributions, and to the Swiss National Science foundation, for financial support.

REFERENCES

[1] (a) Recent reviews of chemistry and biochemistry of B_{12}: W. Friedrich, "Vitamin B_{12} und verwandte Corrinoide" in R.Ammon & W.Dirscherl (eds.): Fermente,Hormone & Vitamine III/2, Thieme Verlag, Stuttgart (1975); (b) D.Dolphin (ed.) B_{12}, Vol. I & Vol. II, Wiley & Sons, New York, 1982.

[2] B.T.Golding &D.N.R.Rao in M.I.Page & A.Williams (eds.)"Enzyme Mechanisms" Royal Soc. Chem.,London, 1987, pp 404.

[3] (a) R.G.Matthews in R.L.Blakley & S.J.Benkovic (eds.) "Folates & Pterins", Wiley & Sons, New York 1984, Vol. I, pp 497; (b) B.Kraeutler in P.J.Craig & F.Glockling (eds.) " The Biological Alkylation of Heavy Elements", Royal Soc. Chem.London, 1988, pp 31.

[4] W.S.Beck in [1b], Vol.II, pp. 1.

[5] (a) D.C.Hodgkin, A.W.Johnson & A.R.Todd in K.Schoffield (ed.), Spec.Publ. No.3, The Chemical Society, London 1955, pp. 109; (b) J.P.Glusker in [1b], Vol.I, pp 24.

[6] A.Eschenmoser, Angew.Chem. 100, 5 (1988).

[7] F.J.Leeper, Nat.Prod.Rep. 2, pp.19 and pp. 561 (1985).

[8] (a) B.Kraeutler,W.Keller, M.Hughes, C.Caderas & C.Kratky, J.Chem.Soc., Chem.Commun 1987, 1678.(b) B.Kraeutler,W.Keller & C.Kratky in H.Fischer & H.Heimgartner "Organic Free Radicals",Springer Verlag, 1988, 99.

[9] L.Walder, G.Rytz,U.Voegeli,R.Scheffold & P.Engel, Helv.Chim.Acta 67, 1801 (1984).

[10] D.Lexa & J.-M.Saveant, Acc.Chem.Res. 16, 235 (1983).

[11] (a) P.G.Lenhert & D.C.Hodgkin, Nature(London), 192,937 (1961); (b) P.G.Lenhert, Proc.Roy.Soc.(London) A303, 45 (1961).

[12] M.Rossi,J.P.Glusker, L.Randaccio, M.F.Summers, P.J.Toscano & L.G.Marzilli, J.Am. Chem.Soc. 107,1729 (1987).

[13] A.Bax, L.G.Marzilli & M.F.Summers, J.Am.Chem.Soc. 109, 566 (1987).

[14] J.R.Pilbrow in ref [1b], Vol.I, pp. 431.

[15] B.Kraeutler,W.Keller & C.Kratky, J.Am.Chem.Soc., in press.

[16] K.Lindstrand & K.B.Stahlberg, Acta Med.Scand. 174, 665 (1963).

[17] (a) K.Bernhauer, O.Mueller & G.Mueller, Biochem.Z.,336, 102 (1962); (b) E.L. Smith, L.Marvyn, A.W.Johnson & N.Shaw, Nature(London), 194, 1175 (1962).

[18] G.N.Schrauzer, E.Deutsch & R.J.Windgassen, J.Am.Chem.Soc. 90,2441 (1968).

[19] (a) K.L.Brown, in ref. [1b], Vol.I, pp. 245; (b)H.P.C.Hogenkamp, in ref. [1b], Vol.I, pp. 295; (c) J.Halpern, in ref. [1b], Vol.I, pp. 501; (d) B.T.Golding, in ref. [1b], Vol.I, pp.543.

[20] M.Fountoulakis, J.Retey, W.E.Hull & B.Zagalak, in B.Zagalak & W.Friedrich (eds.) "Vitamin B_{12}", Proceedings of the Third European Symposium on Vitamin B_{12} and Intrinsic Factor, Zürich, 1979, W.de Gruyter, Berlin, 1979, pp.169.

[21] R.Breslow & P.L.Khanna, J.Am.Chem.Soc.,98, 1297 (1976); R.Breslow, ibid, 98, 6765 (1976).

[22] (a)B.Kraeutler & C.Caderas, Helv.Chim.Acta 67, 1891 (1984); (b) W. Friedrich & R.Messerschmidt, Z.Naturforschg, 25b, 972 (1970).

[23] R.L.Blakley, in ref. [1b], vol.II pp.381.

[24] (a) J.Halpern, Science 227, 869, 1985; (b) B.P.Hay & R.G.Finke, J.Am. Chem.Soc. 108, 4820 (1986).

[25] P.v.d.Meijden, B.W.te Brömmelstroet, C.M.Poirot, C.v.d.Drift & G.D.Vogels, J.Bacteriol. 160, 629 (1984).

[26] (a) G.Fuchs, FEMS Microbiol.Rev. 39, 181 (1986);(b) L.G. Ljungdahl, Annu. Rev.Microbiol. 40, 415 (1986); (c) H.G.Wood, S.W.Ragsdale & E. Pezacka, FEMS Microbiol.Revs.39, 345 (1986); (d) R.K. Thauer, Nachr.Chemie Techn.& Lab, 36, 993 (1988).

[27] (a) J.M.Wood, in P.J.Craig & F.Glockling (eds.) "The Biological Alkylation of Heavy Elements", Royal Soc.Chem.London, 1988, pp. 62; (b) idem, in ref. [1b], Vol.II, pp. 151.

[28] (a) R.T.Taylor, in ref. [1b], Vol.II, pp. 307; (b) T.M.Zydowsky, L.F. Courtney, V. Frasca, K.Kobayashi, H.Shimizu, L.D.Yuen, R.G.Matthews, S.J. Benkovic & H.G.Floss, J.Am.Chem.Soc. 108, 3152 (1986).

[29] B.Kraeutler, Chimia, 42, 91 (1988).

[30] B.Kraeutler, Chimia, 41, 277 (1987).

[31] B.Kraeutler, Helv.Chim.Acta 70, 1268 (1987).

[32] (a) G.N.Schrauzer & J.H.Grate, J.Am.Chem.Soc. 103, 541 (1981); (b) J.M.Pratt in ref. [1b], vol.I, pp. 326.

[33] Y.T.Fanchiang, G.T.Bratt & H.P.C.Hogenkamp, Proc.Natl.Acad.Sci.,USA 81,2698 (1984).

[34] B.Kraeutler, Oesterr.Chemie Zeitschr. 90, 2 (1989).

[35] P.A.Milton & T.L.Brown, J.Am.Chem.Soc. 99, 1390 (1977).

[36] (a) B.Kraeutler, J.Moll & R.K.Thauer, Eur.J.Biochem. 162, 275 (1987); (b) E.Stupperich, H.J.Eisinger & B.Kraeutler, Eur.J.Biochem. 172, 459 (1988); (c) E.Stupperich & B.Kraeutler, Arch. Microbiol. 149, 268 (1988); (d)B. Kraeutler, H.P.Kohler & E.Stupperich, Eur.J.Biochem. 176, 461 (1988); (e) E.Irion & L.Ljungdahl, Biochemistry 4, 2780 (1965).

[37] (a) W.Dangel, H.Schulz, G.Diekert, H.König & G.Fuchs, Arch.Microbiol. 148, 52 (1987); (b) C.M.Poirot, S.W.M.Kengen, E.Valk, J.T.Keltjens, C.v.d.Drift & G.D.Vogels, FEMS Microbiol. Lett. 40, 7 (1987) (c) J.Ellermann, R.Hedde-rich, R.Böcher & R.K. Thauer, Eur. J.Biochem. 172, 669 (1988); (d) W.B. Whitman & R.S.Wolfe, J.Bacteriol. 164, 165 (1985).

[38] H.P.C.Hogenkamp, G.T.Bratt & S.Sun, Biochemistry 24, 6428 (1985).

[39] T.C.Stadtman, Science 171, 859 (1971).

[40] G.N.Schrauzer, Pure Appl.Chem. 33, 545 (1973).

[41] (a) A.Pfaltz, B.Jaun, A.Fässler, A.Eschenmoser, R.Jaenchen, H.G.Gilles,G. Diekert & R.K.Thauer, Helv.Chim.Acta 65, 828 (1982): (b) D.A.Livingston, A.Pfaltz, J.Schreiber, A.Eschenmoser, D.Ankel-Fuchs, J.Moll,R.Jaenchen & R.K.Thauer, Helv.Chim.Acta 67, 334 (1984).

[42] (a) R.P.Gunsalus & R.S.Wolfe, FEMS Microbiol.Lett. 3, 191 (1978); (b) R. S.Wolfe, Trends.Biochem.Sci. 10, 396 (1985); (c) D.Ankel-Fuchs, R.Hüster, E.Mörschel, S.P.J.Albracht & R.K.Thauer, System.Appl.Microbiol.7, 383 (1986).

[43] B.Jaun & A.Pfaltz, J.Chem.Soc.,Chem.Commun. 1988, 293.

[44] L.Ljungdahl & H.G.Wood, in ref [1b], vol. II, pp. 166.

[45] R.K.Thauer, Eur.J.Biochem. 176, 497 (1988).

292

[46] G.Diekert, G.Fuchs & R.K.Thauer in "Microbial Gas Metabolism: Mechanistic, Metabolic & Biotechnological Aspects", R.K.Poole & C.S.Dow (eds.), Academic Press, New York, 1985, p. 115.

[47] e.g. J.P.Collman, L.S.Hegedus, J.R.Norton & R.G.Finke "Principles and Application of Organotransition Metal Chemistry", University Science Books, Mill Valley, 1987, pp. 355 & pp. 619.

[48] E.Pezacka & H.G.Wood, J.Biol.Chem. 263, 16000 (1988).

[49] B.Kraeutler, Helv.Chim.Acta 67, 1053 (1984).

[50] (a) H.Lebertz, H.Simon, L.F.Courtney, S.J.Benkovic, L.D. Zydowsky, K.Lee & H.G.Floss, J.Am.Chem.Soc. 109, 3173 (1987); (b) S.A.Raybuck, N.R.Bastian, L.D.Zydowsky, K.Kobayashi, H.G. Floss, W.H.Orme-Johnson & C.T.Walsh, J.Am.Chem. Soc. 109, 3171(1987).

ONE-CARBON METABOLISM OF ANAEROBIC BACTERIA:
CHALLENGE FOR CHEMISTRY

Georg Fuchs
Abteilung Angewandte Mikrobiologie
Universität Ulm
D-7900 Ulm
Federal Republic of Germany

ABSTRACT. CO_2 plays several roles in the metabolism of anaerobic organisms, and metals are involved in the biotic CO_2
utilizing systems in different ways. First, in the absence of
other oxidized inorganic compounds, CO_2 acts as the ultimate
electron acceptor in the anaerobic carbon cycle. Methanogenic
bacteria reduce CO_2 to CH_4 with hydrogen which is released by
microorganisms performing the anaerobic oxidation of organic
matter. This "CO_2 respiration", in analogy to "O_2 respiration", provides energy for growth of these bacteria. Second,
CO_2 can be used by many strictly anaerobic microorganisms as
the sole carbon compound from which all cellular carbon is
derived. The reduction of CO_2 to carbon monoxide was found to
play an important biological role in cell carbon fixation.
Even in those organisms, which depend on preformed organic
compounds, CO_2 is required as additional carbon source for
carboxylations and reductive carboxylations in biosynthesis.
The reactions of CO_2 fixation and reduction are mostly different from those used by the green plants. Third, an increasing number of organic compounds were found to require carboxylations, prior to further anaerobic degradation. Examples
are the utilization of acetone or phenolic compounds as
carbon and energy sources. The substitution with CO_2 yields
carboxylic acids and therefore facilitates the further metabolism of these compounds. The knowledge of the utilization
of CO_2 by anaerobes may stimulate research and development in
the chemical use of CO_2 and in catalysis of these reactions.

M. Aresta and J. V. Schloss (eds.),
Enzymatic and Model Carboxylation and Reduction Reactions for Carbon Dioxide Utilization, 293–300.
© 1990 Kluwer Academic Publishers.

INTRODUCTION

Anaerobic conditions exist in many parts of this planet, i.e.
in many parts of the soil and groundwater, in sediments of
freshwater and marine environments, or in the intestinal
tract of animals. In these environments the consumption of
oxygen by aerobic or facultative aerobic organisms exceeds
its supply. Organisms enabled to grown in the dark and in the
absence of molecular oxygen nevertheless have developed numerous types of metabolic pathways which can be coupled to biological energy conversion: In marine environments organic
material is oxidized to CO_2 with the concomitant reduction of
sulfate to sulfide by sulfate reducing bacteria. In terrestrial and freshwater ecosystems, CO_2 is formed by the oxidation of organic matter or protons can act as electron
acceptors. Proton reduction, i. e. hydrogen formation, is
energetically unfavorable. Therefore, CO_2 acts as the major
and final electron acceptor in anaerobic metabolism in non-
marine environments. For energetic aspects of CO_2 reduction
see [1, 3]; for biochemical aspects see also [4-6].

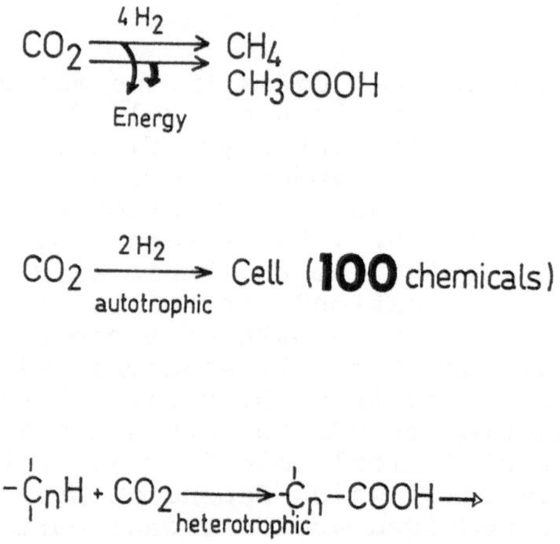

$$CO_2 \xrightarrow[\text{Energy}]{4\,H_2} \begin{array}{l} CH_4 \\ CH_3COOH \end{array}$$

$$CO_2 \xrightarrow[\text{autotrophic}]{2\,H_2} \text{Cell} \ (\mathbf{100}\,\text{chemicals})$$

$$-\overset{|}{\underset{|}{C}}_nH + CO_2 \xrightarrow[\text{heterotrophic}]{} \overset{|}{\underset{|}{C}}_n-COOH \longrightarrow$$

Fig. 1   Roles of CO_2 in anaerobic microorganisms, (a) as
electron acceptor in "CO_2 respiration" for energy
generation, (b) as carbon source in biosynthesis, and
(c) as means for facilitating further metabolism of
different organic compounds.

CO_2 can be reduced to CH_4 by methanogenic bacteria according to

$$CO_2 + 4\ H_2 \longrightarrow CH_4 + 2\ H_2O \qquad \Delta G_0' = -131\ kJ/mol\ CH_4$$

or to acetate by acetogenic bacteria according to

$$2CO_2 + 4\ H_2 \longrightarrow CH_3COOH + 2\ H_2O \quad \Delta G_0' = -95\ kJ/mol$$
$$acetate.$$

Acetate formed by fermentations will finally be cleaved to CH_4 and CO_2 by methanogenic bacteria according to

$$CH_3COOH \longrightarrow CH_4 + CO_2 \qquad \Delta G_0' = -37\ kJ/mol\ CH_4.$$

Therefore, the ultimate fate of organic matter under these conditions is disproportionation to CO_2 and CH_4 (biogas)

$$n\ \langle CH_2O \rangle \longrightarrow \frac{n}{2}\ CO_2 + \frac{n}{2}\ CH_4.$$

A variety of anaerobic bacteria perform carboxylation reactions or reductive carboxylations during metabolism of organic compounds. In addition, many anaerobic bacteria are able to synthesize all cell constituents from CO_2 plus H_2 only; CO_2, inorganic nitrogen and sulfur compounds are the sources of cellular carbon, nitrogen and sulfur (Fig. 1).

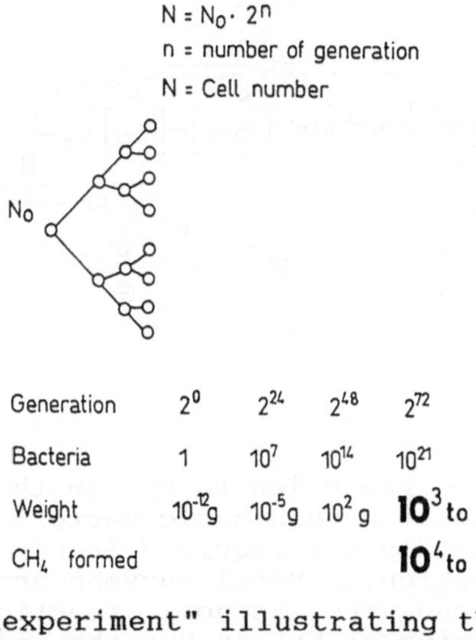

$$N = N_0 \cdot 2^n$$
n = number of generation
N = Cell number

Generation	2^0	2^{24}	2^{48}	2^{72}
Bacteria	1	10^7	10^{14}	10^{21}
Weight	$10^{-12}g$	$10^{-5}g$	$10^2\ g$	10^3 to
CH_4 formed				10^4 to

Fig. 2   "Gedankenexperiment" illustrating the potential and the role of CO_2 fixation in biosynthesis and of CO_2 reduction to CH_4 in the anaerobic carbon cycle. This graph (upperpast) shows the exponential growth of microorganisms and the growth equation. Below, the increase in cell number, weight of these cells and the amount of CH_4 formed in metabolism (if these bacteria were methanogenic bacteria) is given. Note that the generation time can be as short as 1 h.

These processes are highly efficient as illustrated by a "Gedankenexperiment" (Fig. 2). A methanogenic bacterium weighs only 10^{-12} g. When growing by dividing into two cells with a generation time of two hours it would produce, after 2 days (24 generations), $2^{24} = 10^7$ bacteria; after 6 days (72 generations) the cell mass would increase to one thousand tons, and approximately ten thousand tons of methane would have been produced provided that the supply of inorganic nutrients is unlimited.

In order to appreciate these microbial performances one has to keep in mind that these reactions are conducted at ambient temperature, neutral pH, and in aqueous systems; especially noteworthy, if the processes are sufficiently exergonic, they are even coupled to the synthesis of ATP (see Fig. 3). This last aspect makes the reactions involving CO_2 fixation and reduction in anaerobes so different from other biological or industrially used carboxylations. However, the chemical principles underlying these reactions have much in common.

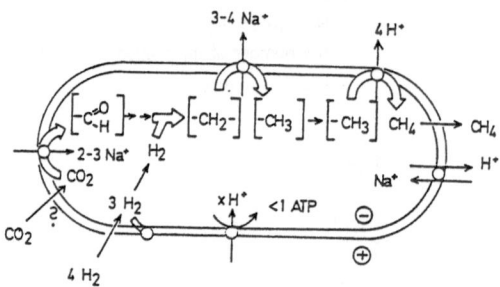

Fig. 3   Role of CO_2 reduction to CH_4 in the energy
          metabolism of methanogenic bacteria. Note that
          several metals are required in CO_2 – and C_1 –
          transformation. Several enzymes are integrated into
          the membrane. The exergonic reactions are coupled to
          the extrusion of Na^+ or H^+, the endergonic reactions
          use this electrochemical gradient. Na^+ and H^+ can be
          exchanged by an antiporter. Finally, the proton
          gradient and the membrane potential achieved drive
          the endergonic synthesis of ATP. The reducing
          equivalents from 4 H_2 are provided by three
          different types of hydrogenases which also contain
          metals for hydrogen activation.

ROLE OF CO_2 REDUCTION TO CH_4 OR CH_3COOH IN ENERGY METABOLISM OF ANAEROBIC BACTERIA

CO_2 is reduced by methanogenic bacteria to CH_4 (Fig. 3). This process which has been studied intensively in the past few years has been reviewed [1, 2, 4, 7], and therefore will not be covered here. The interested reader is referred to the competent reviews. The CO_2 reduction process is used for energy conservation. In a similar manner acetogenic bacteria perform the synthesis of acetate from CO_2 [1, 3, 5, 6]. These reactions involve various transition metals (Mn, Fe, Co, Ni, Zn, Mo, W) and trace elements (Se). For the role of nickel see [7].

CARBOXYLATIONS IN INTERMEDIARY METABOLISM AND IN DEGRADATION OF PHENOLIC COMPOUNDS BY ANAEROBIC BACTERIA

Carboxylation reactions play an important role in the intermediary metabolism of all living beings. In anaerobes, reductive carboxylations are central to their carbon metabolism when grown on C_1- or C_2-compounds. A survey of these reactions has been given elsewhere [8]. In Fig. 4 the most common CO_2 utilizing reactions and processes in anaerobic microorganisms are listed.

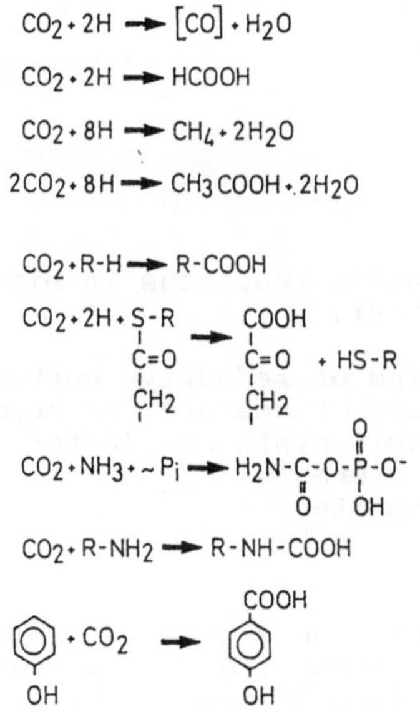

$$CO_2 + 2H \longrightarrow [CO] + H_2O$$

$$CO_2 + 2H \longrightarrow HCOOH$$

$$CO_2 + 8H \longrightarrow CH_4 + 2H_2O$$

$$2CO_2 + 8H \longrightarrow CH_3COOH + 2H_2O$$

$$CO_2 + R-H \longrightarrow R-COOH$$

$$CO_2 + 2H + \underset{\substack{| \\ C=O \\ | \\ CH_2 \\ |}}{S-R} \longrightarrow \underset{\substack{| \\ C=O \\ | \\ CH_2 \\ |}}{COOH} + HS-R$$

$$CO_2 + NH_3 + \sim P_i \longrightarrow H_2N-\underset{\underset{O}{\|}}{C}-O-\underset{\underset{OH}{|}}{\overset{\overset{O}{\|}}{P}}-O^-$$

$$CO_2 + R-NH_2 \longrightarrow R-NH-COOH$$

Fig. 4   Survey of the most common CO_2 utilizing reactions and processes in anaerobic microorganisms

Two reductive carboxylation reactions which are used in cell carbon synthesis of many anaerobes are given in Fig. 5. The electron donor for 2-oxoacid synthases is reduced ferredoxin (for details of the reaction see [9, 10]), whereas the formation of 3-hydroxyacids such as isocitrate requires NADPH.

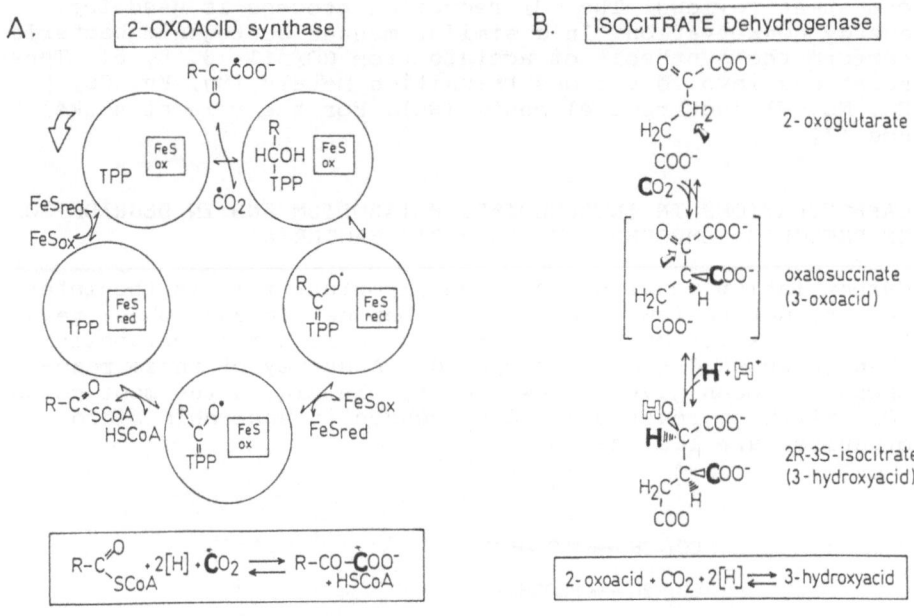

Fig. 5 Reductive carboxylation reactions in biosynthesis of many anaerobic bacteria.

A. Proposed mechanism of reductive carboxylation of coenzyme A thioesters of acetate and succinate, catalyzed by 2-oxoacid synthases (after [9]).
FeS, iron-sulfur centers
TPP, thiamin diphosphate
CoA, coenzyme A
red, reduced
ox , oxidized
The circle represents the 2-oxoacid synthase.
Note the role of a carbon radical in synthesis.
The enzyme differs from 2-oxoacid dehydrogenases.

B. Isocitrate dehydrogenase reaction. This is an example for a reversible reaction used for reductive carboxylation in several anaerobes. In aerobic organisms, the enzyme functions in oxidative decarboxylation.

One of the best studied chemical carboxylations, the carboxylation of phenol (Kolbe-Schmitt-Synthesis), has recently been found to have a biological counterpart (Fig. 6) [11]. Anaerobic bacteria have been isolated with phenol as sole source of cell carbon. Phenol oxidation to CO_2 is coupled to nitrate reduction to N_2 as the energy yielding process.

Growth of these organisms on phenol was dependent on the presence of CO_2, whereas growth on 4-hydroxybenzoate or benzoate was CO_2-independent. Whole cells grown on phenol and extracts of these cells catalyzed the exchange of $^{14}CO_2$ into the carboxyl group of 4-hydroxybenzoate; this reaction was not catalyzed by cells grown with 4-hydroxybenzoate. These and other observations suggested that phenol and possibly other phenolic compounds are metabolized in the absence of oxygen by carboxylation to 4-hydroxybenzoate. The enzyme tentatively named "phenol carboxylase" is induced by phenol. The isotope exchange reaction is a partial reaction of the overall carboxylation process.

The enzyme is inactivated by oxygen and requires Mn^{2+} for activity. The affinity for the substrates CO_2 (K_m = 1 mM) and 4-hydroxybenzoate (K_m = 0.2 mM) are reasonably high, and the specific activity can well account for growth on phenol (0.12 μmol CO_2 incorporated into 4-hydroxybenzoate $\cdot$ min$^{-1} \cdot$ mg protein^{-1}). CO_2 rather than bicarbonate is the actual substrate of the enzyme as judged from studies of the partial reaction, and the carboxylation appears not to be mediated by biotin. It will be of some interest to study this system, the role of metals, and to compare it to the non-enzymic carboxylation of phenol, especially in view of the role of metals in carboxylation and the regioselectivity of carboxylation. As described for phenol metabolism the initial reaction in anaerobic acetone metabolism is its carboxylation to 3-oxobutyric acid [12].

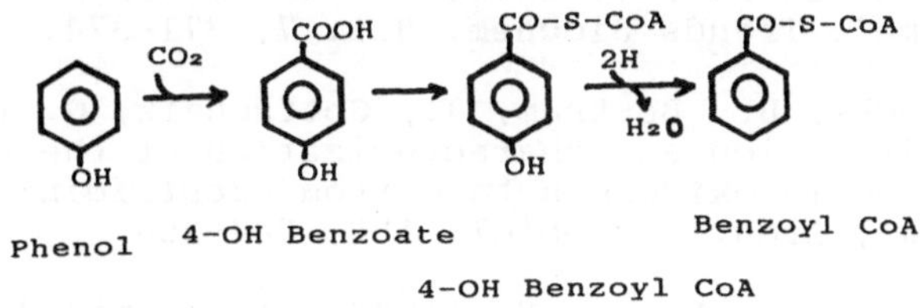

Fig.6   Role of CO_2 fixation in the anaerobic metabolism of phenolic compounds.

300

Acknowledgements: The work of the author was supported by
DFG, Fonds der Chemischen Industrie, and a Nato grant to
Prof. M. Aresta, Bari.

References

[1]   Thauer, R.K., Jungermann, K., Decker, K. (1977) 'Energy
       conservation in chemotrophic anaerobic bacteria',
       Bacteriol. Rev. 41, 100–180.

[2]   Blaut, M., Gottschalk, G. (1985) 'Evidence for a chemi-
       osmotic mechanism of ATP synthesis in methanogenic
       bacteria', Trends Biochem. Sci. 10, 486–489.

[3]   Fuchs, G. (1986) 'CO_2 fixation in acetogenic bacteria:
       Variations on a theme', FEMS Microbiol. Rev. 152, 594–
       599.

[4]   Wolfe, R.S. (1985) Unusual coenzymes of methanogene-
       sis', Trends Biochem. Sci. 10, 396–399.

[5]   Wood, H.G., Ragsdale, S.W., Pezacka, E. (1986) 'The
       acetyl-CoA pathway of autotrophic growth', Trends
       Biochem. Sci. 11, 14–18.

[6]   Ljungdahl, L.G. (1986) 'The autotrophic pathway of
       acetate synthesis in acetogenic bacteria', Annu. Rev.
       Microbiol. 40, 415–450.

[7]   Thauer, R.K. (1985) 'Nickelenzyme im Stoffwechsel von
       methanogenen Bakterien', Biol. Chem. Hoppe–Seyler 366,
       103–112.

[8]   Fuchs, G. (1989) 'Alternative pathways of autotrophic
       CO_2 fixation', in H. G. Schlegel and B. Bowien (eds.)
       Biology of autotrophic bacteria, Science Tech
       Publishers, Madison USA, pp. 365–382.

[9]   Kerscher, L., Oesterhelt, D. (1982) 'Pyruvate:
       ferredoxin oxidoreductase – new findings on an ancient
       enzyme', Trends Biochem. Sci. 7, 371–374.

[10]   Meinecke, B., Bertram, J., Gottschalk, G. (1989)
       'Purification and characterization of the pyruvate-
       ferredoxin oxidoreductase from *Clostridium acetobuty-
       licum*', Arch. Microbiol. 152, 244–250.

[11]   Tschech, A., Fuchs, G. (1989) 'Anaerobic degradation of
       phenol via carboxylation to 4-hydroxybenzoate: in vitro
       study of isotope exchange between $^{14}CO_2$ and 4-hydroxy-
       benzoate', Arch. Microbiol. 152, 594–599.

[12]   Platen, H., Schink, B. (1987) 'Methanogenic degradation
       of acetone by an enrichment culture', Arch. Microbiol.
       149, 136–141.

INCREASING PHOTOSYNTHESIS BY GENETIC MANIPULATION: DIFFICULTIES AND PROSPECTS.

Mirta N. Sivak
Robert Hill Laboratory,
University of Sheffield,
26 Taptonville Road,
Sheffield S10 5BR,
United Kingdom.

ABSTRACT   Photosynthesis by plants in terrestrial or aquatic environments is a source of biomass and a sink for CO_2 emissions. The rate of photosynthesis is limited by intrinsic thermodynamic and biological constraints, and by environmental factors. Increased $[CO_2]$ in the atmosphere and changes in climate will affect plants *via* photosynthesis and other aspects of plant metabolism, probably widening the gap between potential and actual global photosynthesis.   Although plant metabolism can be changed by genetic means like plant breeding and genetic engineering, relative ignorance in some areas of plant sciences is limiting the application of these potentially powerful resources.

1. INTRODUCTION

In photosynthesis, light energy is absorbed by chlorophyll and used to reduce CO_2 into CH_2O (this being a simple representation of carbohydrate) and O_2 is formed.   It follows that increased global photosynthesis would increase biomass and help to stabilize the CO_2 concentration in the atmosphere. Is it possible to manipulate photosynthesis in order to increase CO_2 fixation both on a global scale and in specific agricultural systems? To increase photosynthesis it is not enough to improve the mechanism of carboxylation, because the whole process proceeds at the rate of the slowest step and carboxylation is not always the limiting one. If CO_2 fixation by plants is to be manipulated to advantage, a detailed understanding of the photosynthetic process is required.

The main purpose of this paper is to outline
1) how photosynthesis is limited and regulated,
2) the likely effects of the predicted changes in climate and $[CO_2]$ on photosynthesis
3) the resources available for increasing photosynthetic CO_2 fixation, and
4) the areas of ignorance limiting the usefulness of these resources.

M. Aresta and J. V. Schloss (eds.),
Enzymatic and Model Carboxylation and Reduction Reactions for Carbon Dioxide Utilization, 301–320.
© 1990 *Kluwer Academic Publishers.*

2. WHAT LIMITS PHOTOSYNTHESIS?

2.1. The response of photosynthesis to light

If the rate of photosynthesis by a leaf is plotted against light intensity (Figure 1), the relationship is initially a straight line, and its slope a measure of the efficiency of light utilization, the quantum yield. In this part of the curve, the ultimate constraint is of a thermodynamic nature. The calorific content of sucrose, the principal end-product of photosynthesis, is about 112 Kcal per mol of CH_2O and that of a quantum mole of red light is about 42 Kcal. Although a rapid calculation gives a number of three photons for the conversion of one CO_2 into CH_2O, thermodynamic constraints raise this value to 8. This is because, according to present understanding of photosynthetic electron transport, each electron is energized twice in the so called "Z" scheme. As the electrons are transferred down a series of carriers, ATP is formed from ADP and Pi and NADP is reduced. Further considerations raise this theoretical value to about 9.

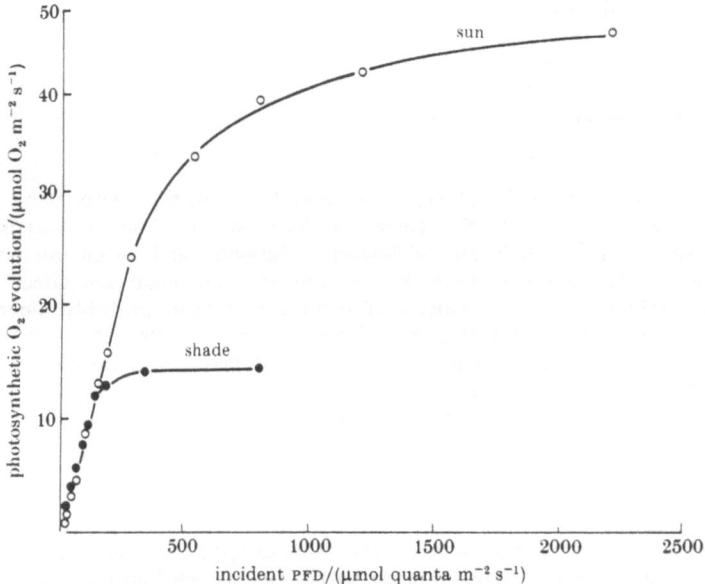

Figure 1. Relationship between the rate of photosynthesis (measured as oxygen evolution) and light intensity in shade- and sun-grown leaves. At first the rate is directly proportional to the light but, as the intensity is increased, the corresponding increase becomes smaller. The steepness in the linear part of the relationship is limited by the energy required to drive photosynthesis. The ceiling is determined by the capacity of the leaf to use the products of the "light" reactions and depends on the growth conditions and other factors (after Walker & Osmond, 1986).

Why does the curve depart from linearity? It does because light is not anymore the only limiting factor and other factors or processes become limiting.

The ATP and NADPH generated by photosynthetic electron transport are used in the reductive pentose phosphate pathway, in which the CO_2 acceptor is regenerated and triose phosphate is formed. In Figure 1, both leaves have a similar quantum yield, near the theoretical maximum. Growth conditions, however, affect other aspects of photosynthesis; shade leaves will display lower rates of dark respiration (oxygen uptake in the dark) and rate of oxygen evolution at light saturation than sun leaves. Many characteristics are affected by quality and quantity of light during development, e.g., ratio between photosystem I and II, leaf anatomy, pigment composition, and ultrastructure of the chloroplast (Melis and Harvey, 1981; Björkman, 1981). These adaptations allow the "light reactions" to proceed at or near their maximum efficiency. From the reactions that utilize the products of photosynthetic electron transport, some aspects of carboxylation will be dealt with briefly and the fate of the triose phosphate will be discussed in some detail (for a review see Sharkey, 1985).

2.2. Loss of fixed carbon

Over milenia, photosynthesis has progressively increased O_2 concentration and decreased CO_2 concentration in the atmosphere. Plants evolved different strategies to adjust to these changes. Some groups of plants developed additional metabolic pathways, associated with changes in cell or leaf anatomy, optimizing photosynthesis under these unfavourable conditions. These include photorespiration, C_4 metabolism, crassulacean acid metabolism (CAM) and mechanisms for the accumulation of inorganic carbon (CO_2 or HCO_3^-), in algae and cyanobacteria.

Like animals, plants respire, oxidising carbohydrates and other metabolites, to CO_2 and H_2O. In the light, C_3 plants also lose newly fixed carbon to the atmosphere, as CO_2, in a pathway termed photorespiration. Carbon dioxide fixation in the Benson–Calvin cycle (or reductive pentose phosphate pathway, RPPP) is catalysed by the enzyme RuBP carboxylase/oxygenase (Rubisco). In the presence of excess CO_2, only carboxylation occurs, but because Rubisco also has oxygenase activity (i.e., O_2 can compete with CO_2) in the presence of air concentrations of CO_2 and O_2 oxygenation occurs, leading to the loss of newly fixed carbon and energy. Rubisco has "improved" during evolution and much research is devoted to its further manipulation (see this volume and Ellis and Gray, 1986).

The massive concentration of Rubisco, the most abundant protein in nature, in the chloroplast stroma, may be regarded as an evolutionary response to decreasing CO_2 in the atmosphere. The photorespiratory pathway is now considered a way to bring some of the carbon "lost" by oxygenation back into the cycle, forming some useful by–products.

2.3. Mechanisms for the accumulation of inorganic carbon

The C_4 dicarboxylic acid pathway of photosynthetic carbon assimilation (the C_4 pathway) is a complex biochemical and physiological elaboration of the common photosynthetic carbon reduction cycle, the RPPP. The immediate end–product of the RPPP is a three–carbon compound. All green plants operate this sequence of carbon assimilation and most temperate species depend upon it entirely and are, therefore, called C_3 plants. Others operate an additional sequence, which yields a four–carbon compound as the first product of CO_2 fixation, and are therefore called C_4-plants. In order to provide a high, saturating CO_2 concentration to the RPPP, these plants have evolved a metabolic CO_2 pump which is energy–driven. It functions to trap atmospheric CO_2 in an outer layer of photosynthetic cells and shuttle it into an inner layer of the leaf where it is assimilated in the RPPP. Many of the worst weeds of the world are C_4 species, whilst most crops are C_3. The most economically important C_4 crops are sorghum, maize and sugar–cane, millets, and a number of pasture grasses.

The response to light is modified by the concentration of CO_2 (Figure 2). In higher $[CO_2]$, the slope is steeper (the quantum yield is higher) than in air because photorespiration, which starts by the oxygenation of RuBP, is supressed. Oxygenation of RuBP decreases the efficiency of photosynthesis.

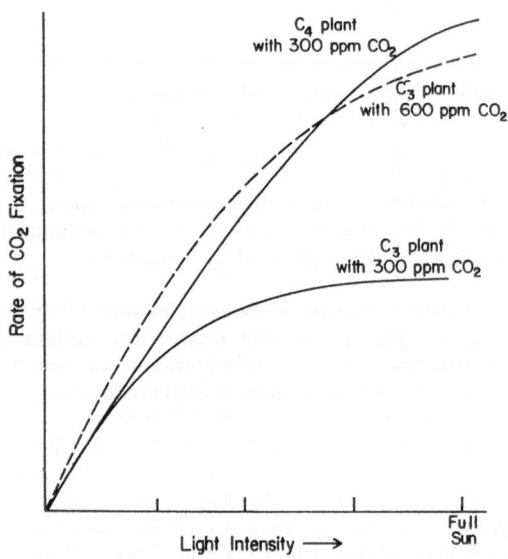

Figure 2. *Effect of CO_2 concentration on the response of photosynthesis to light intensity. Photosynthesis in air by C_3 plants saturates between 30 and 40% full sunlight intensity. For plants with the additional C_4 pathway that concentrates CO_2, photosynthesis in air increases with light intensity and at full sunlight is about twice that in a C_3 plant. The response to light of a C_3 plant in higher $[CO_2]$ approximates that of a C_4 in air.*

Mechanisms for the accumulation of inorganic carbon (IC, i.e. CO_2 or HCO_3^-), different to C_4 metabolism, have attracted the attention of plant biologists, because they also decrease photorespiration under natural conditions. There is a group of cyanobacteria (blue–green algae), which possesses the ability to concentrate IC via an energy requiring pumps. These photosynthetic microorganisms can change their relative affinities for external IC depending on the level in the external medium, and there is some active accumulation mechanism which allows

CO_2 to be concentrated inside the cell. Accumulation in excess of 10^3 that of the external medium has been demonstrated (Canvin *et al*, 1987).

3. THE FATE OF TRIOSE PHOSPHATE

The entire Benson–Calvin cycle can be divided in three phases. The initial carboxylation is followed by reduction to triose phosphate, and in the third, 5 of these C_3 molecules are rearranged to regenerate $3(C_5)$ molecules of CO_2–acceptor (Fig. 3).

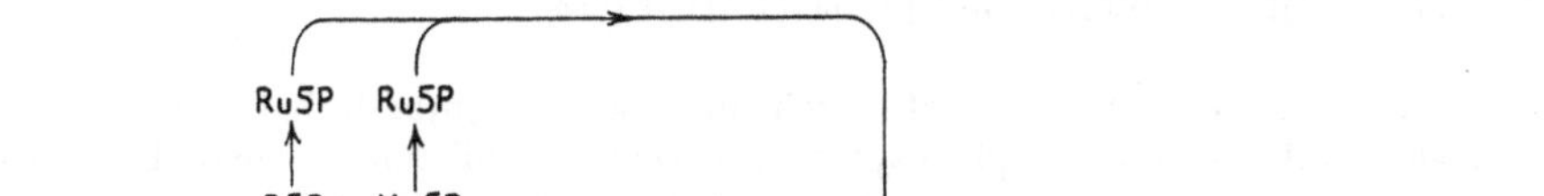

$$3CO_2 + 2H_2O + Pi \rightarrow CH_2OH \cdot CO \cdot CH_2OPO(OH)_2 + 3O_2$$

Figure 3. The reactions that lead to the regeneration of RuBP. Five out of six C3 molecules are re-arranged to give three C5 molecules (RuBP). The sixth is available for feedback or export to the cytosol via the P_i–translocator as DHAP. On the right, 3 molecules of RuBP combine with 3 molecules of CO_2 and 3 molecules of water to give 6 molecules of PGA. These are phosphorylated at the expense of ATP and the resulting DPGA is reduced by NADPH to G3P. The major part of this is converted to its isomer DHAP. Aldol condensation of these 2 triose phosphates gives a molecule of FBP which undergoes hydrolysis to F6P. This hexose phosphate is also the precursor of G6P and G1P which, after further transformation, give rise to starch. The F6P also enters the first transketolase reaction donating a 2-carbon unit to G3P to form Xu5P and E4P. The process of condensation, phosphorylation and 2-carbon transfer is repeated yielding SBP, S7P and two more molecules of pentose phosphate respectively. All 3 molecules of pentose monophosphate are finally converted to Ru5P which is phosphorylated to RuBP (after Edwards & Walker 1983).

It should be noted that the cycle consumes 9 molecules of ATP and 6 molecules of NADPH in the formation of one triose phosphate product (which can also feed back into the cycle to promote autocatalytic acceleration). In total, 5 molecules of H_2O are consumed in the cycle proper and 3 are released in the generation of assimilatory power. If the triose phosphate product were hydrolysed to give free triose in a reaction consuming 1 molecule of H_2O, there would be no net P_i consumption and the entire sequence would simplify to the classic overall equation for photosynthesis.

4. CONTROL OF CARBOHYDRATE METABOLISM

During the day, the rates of starch and sucrose synthesis have to be co-ordinated with the rate of photosynthetic carbon assimilation. There is a clear need to determine how much assimilated carbon can de diverted to sucrose and starch synthesis without unduly decreasing the amount that returns to the Benson–Calvin cycle. Conversely, when sucrose accumulates in the cytosol because the rate of export diminishes (and/or photosynthesis increases) starch begins to accumulate inside the chloroplast. During the night, the sucrose accumulated in the vacuole during the day and the starch accumulated in the chloroplast are remobilized to be used to support the metabolism of the leaf itself or to be exported as sucrose (Stitt et al., 1987). In this way, photosynthates are available "around the clock". The importance of these remobilization mechanisms is highlighted when they are disturbed. For example, mutants of the crucifer *Arabidopsis thaliana* which are unable to synthesize starch but can still synthesise sucrose will grow at the same rate that the wild type under continuous light, but this growth rate will be drastically diminished if grown in a day–night regime (Caspar et al, 1985).

4.1. The role of compartmentation and transport

The chloroplast uses light to transform CO_2, Pi and water to triose phosphate. The major metabolites transported through the phosphate translocator of the chloroplast envelope are Pi, PGA and DHAP (Figure 4). Besides the effects that these metabolites may have on sucrose and starch metabolism through their effects on the enzymes involved, levels of these metabolites could be a primary means of co-ordinating chloroplast and cytosol.

Plastids play an essential role in the biosynthetic activity of photosynthetic (as well as of non–photosynthetic) cells. For example, fatty acid biosynthesis occurs exclusively in plastids, as does the biosynthesis of the majority of amino acids. Fructan synthesis occurs in the vacuole. It seems that most of plastid's enzymes are encoded in the nuclear genome, translated in the cytosol and transported into the plastid. When similar reactions occur in the cytosol and plastids, these are catalyzed by isoenzymes, with different properties and regulatory characteristics and coded for by different genes. Concentrations of the key metabolites of photo-synthesis are central in the co-ordination of cell metabolism. Transport between compartments is as important as the activities of key enzymes in the determination of these levels. The main limitation in assessing accurately the role of transport in the control of cell metabolism is the lack of reproducible techniques for measuring metabolite concentrations in the different compartments and detailed information on the kinetic characteristics is only available for a few of the translocators.

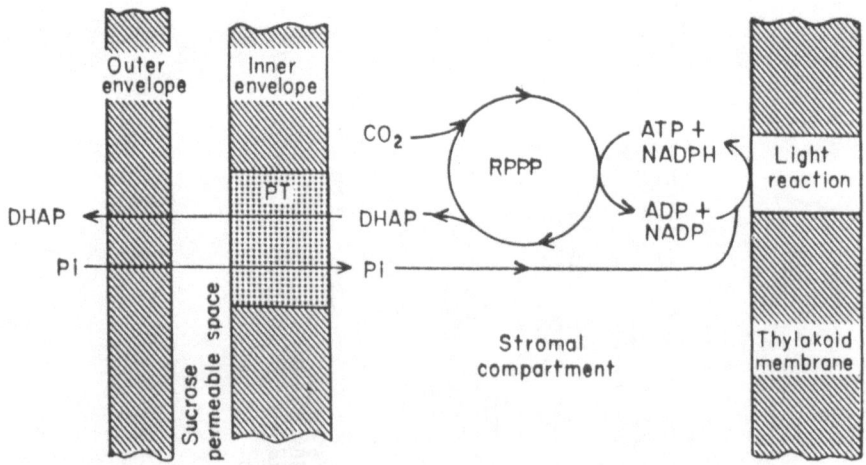

Figure 4. The Pi–translocator is located in the inner envelope and permits a strict stoichiometric exchange between external Pi and internal DHAP. Other compounds such as PGA will also exchange with PGA or DHAP. Movements may occur in either direction according to the conditions.

4.2. The fine control of sucrose and starch synthesis: the role of fructose 2,6–bisphosphate.

The role of metabolite effectors in the regulation of the key enzymatic reactions in photosynthesis has been the object of great attention during the last few years. Fructose 2,6–bisphosphate (and the enzymes that, in turn, regulate its level) plays a major role in the fine regulation of starch and sucrose synthesis as it does in the regulation of glycolysis and gluconeogenesis in animals. This effector (it is a signal metabolite rather than an intermediate in the pathways concerned) was first discovered in liver but it is also present in leaves and other plant tissues in micromolar concentrations. It modifies the activity of cytosolic fructose–1,6–bisphosphatase (FBPase) and the PP_i–Fru–6–P phosphofructophosphotransferase (PFPase). The effector is synthesised and degraded by specific enzymes, the activities of which is, in turn, affected by the concentration of other metabolites such as DHAP, PGA, P_i and Fru–6–P (Heldt and Stitt 1987; Fig. 5). According to present understanding, Fru–2,6–P_2 is involved in the fine regulation of sucrose synthesis "sensing" the availability of substrates (e.g. DHAP) and adjusting sucrose synthesis accordingly. Fru–2,6–P_2 might also be involved in the fine regulation of the partitioning of carbon between sucrose and starch.

Understanding of the intracellular transport of sucrose requires the study of the vacuolar component in the traffic of the key metabolites. This has long been a subject for speculation but the fragility of the vacuole has precluded, until relatively recently, experimental studies of the kind performed on isolated chloroplasts.

308

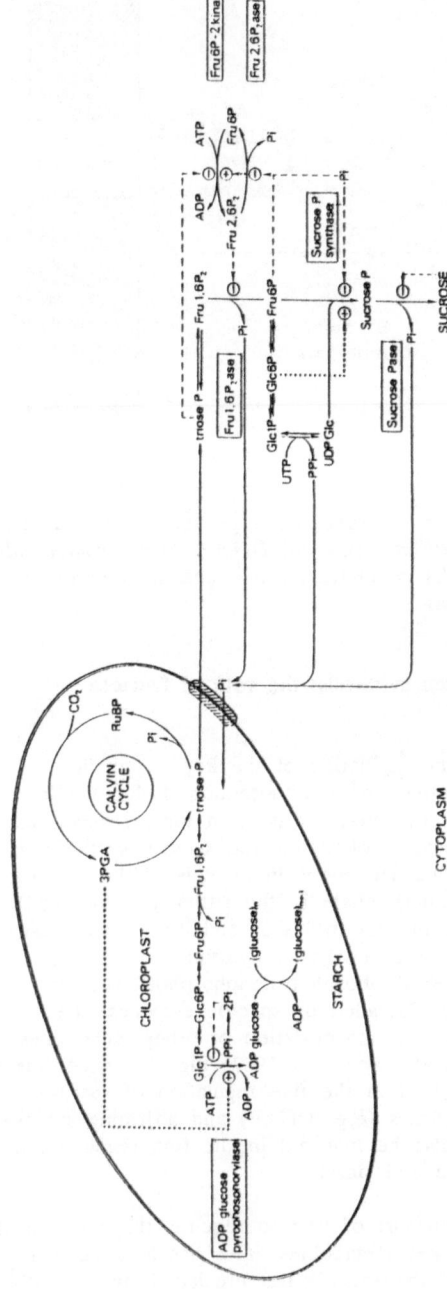

Figure 5. The fine control of sucrose and starch synthesis and the role of metabolite modulation, including that by fructose 2,6-bisphosphate. After Heldt & Stitt (1986).

If the newly–synthesised sucrose were retained by the cytosol it would decrease osmotic potential and result in swelling of the cytosol, metabolite dilution and drastic changes in cytosol metabolism.   Kinetic work involving rapid isolation of vacuoles from photosynthesising barley protoplasts showed that transport of photosynthetic products from the cytosol, across the tonoplast and into the vacuoles is a rapid process which may approach the rate of synthesis (Kaiser et al. 1982). Sucrose, malate, citrate, glutamate, glutamine and alanine are among the metabolites transported.

5. MEASUREMENT OF PHOTOSYNTHESIS *IN VIVO*

To understand a process and how it is modified by environmental or genetical factors, the starting point is to measure it. Recent advances in biochemical and biophysical methodology allow the investigation of photosynthesis in a depht that could not have been achieved a few years ago.
Measurement of different aspects of photosynthesis *in vivo* (CO_2 and O_2 exchange, light–scattering, chlorophyll fluorescence and its quenching components) is now possible, providing valuable information about how photosynthesis is limited and regulated. It also makes possible to study the effect of stress and of genetic and chemical manipulation.

5.1. Limitation of photosynthesis *in vivo* by Pi supply

The Pi requirement by the chloroplast depends on the supply of the other substrates, light and CO_2, and on the rates of the partial processes of photosynthesis in which these substrates are utilised. For example, the relative requirement of orthophosphate diminishes with photoinhibition (Walker & Osmond, 1986). At low temperatures, the [Pi] optimum for isolated spinach chloroplasts increases (Leegood & Walker, 1983), implying that an unchanged Pi supply to the chloroplast which was adequate at 20°C would be inadequate at 10°C. It is self–evident that the relative importance of the Pi–recycling processes will vary with the plant material and environmental conditions. Decreased temperature has similar effects to sequestration of cytosolic Pi (and opposite to those of Pi feeding and increased temperature) in terms of oscillatory behaviour and the effect of oxygen concentration (Sivak, 1987). Isolated chloroplasts require higher [Pi] for maximum photosynthesis at low temperatures and the lower thresholds in [CO_2] needed to initiate oscillatory behaviour at these temperatures suggest that this is also true for chloroplasts *in situ*. Alternatively (or possibly, as well as) the overall rate of Pi recycling by starch and sucrose synthesis, movement from the vacuole, is slower at these temperatures.

The plant cell can adjust its Pi supply up to a degree by increasing the rates of one or more of the processes by which it is recycled (Walker & Sivak, 1985; Sivak & Walker, 1986). Sometimes, however, Pi recycling is insufficient and symptoms resembling those shown under experimental sequestration of Pi can be observed (Walker & Sivak, 1985).   However, if the leaf can adjust the rate of the processes that re–cycle Pi to the new demands (e.g. increasing sucrose or starch synthesis following an increase in [CO_2]) the symptoms will only be transitory (Sivak & Walker, 1986; Sivak, 1987).

6. IMPROVING PHOTOSYNTHESIS. WHERE TO START?

The multiplicity of regulatory mechanisms described above suggest that plants have optimised, during milennia of evolution, photosynthetic performance in their natural environment. It does not follow, however, that photosynthesis runs at the maximum possible rate, particularly when plants are cultivated in a very different environment to that in which they evolved. Predicted changes in climate and CO_2 concentration in the atmosphere may increase that difference between potential and actual photosynthesis.

Experimental evidence indicates that RuBP regeneration cannot be separated from P_i recycling or supply (Sivak and Walker, 1986). This has implications for basic research which seeks to increase photosynthesis and plant growth by improving RuBP carboxylase and decreasing photorespiration. Decreased photorespiration should decrease the light and $[CO_2]$ thresholds at which P_i becomes limiting, firstly because less P_i will be made available through photorespiration and secondly because increased carboxylation will make higher demands on the supply of P_i, itself a substrate of photosynthesis. Symptoms of inadequate P_i may be displayed by leaves photosynthesising under conditions which impose higher P_i demands than those the plant was likely to experience under the conditions in which it was grown, e.g. higher light intensities and CO_2 concentrations. Characterisation of the syndrome, however, now makes it easier to recognise inadequate P_i supply in other circumstances which are more physiologically relevant.

7. STRESS AND ACCLIMATION TO ENVIRONMENTAL CHANGES.

As discussed above, the ultimate constraint of photosynthetic efficiency is thermodynamic and it seems that the process is already extremely efficient. Healthy leaves of C_3 species illuminated in optimum conditions require a number of photons surprisingly near the theoretical minimum, about 9, to convert a molecule of CO_2 into one of CH_2O. The catch is in the terms "healthy" and "optimum conditions". In unfavourable conditions a plant may require far more than that number. In very extreme conditions, a plant may do very badly or, worse, may not survive.

The term "healthy" implies that the leaf is fully expanded but not senescent, and that it has not been exposed to stress that reduced its efficiency in the use of light. The maximum yield seems to be rather constant (Björkman and Demmig, 1986) but the extent to which it decreases as a result of stress of any kind is extremely variable. Other variables are the rate and extent of recovery, and the perception by the plant of what are stressful conditions. Some species or varieties can acclimate rapidly to environmental changes that other plants will not tolerate. In recent years, our understanding of the mechanisms of damage, recovery and avoidance of damage has improved (Björkman, 1987). It is hoped that a detailed understanding of the molecular bases of advantageous characters will allow their introduction into other species.

Little is known about the metabolic changes that enable plants to adapt to a particular environment or to accommodate to diurnal or seasonal changes in that environment. Metabolic changes are required to respond to cold, drought or low light, to recover from stress or to acclimate to a long-term decrease in light intensity. We are familiar with the major adaptative advantages provided by C_4 or CAM metabolism but we know much less about the subtle changes that enable a C_3 plant to do well in a given environment.

One way to find an advantageous character is to look for successful species and compare them with others. Cereals, for example, are a successful group of plants, and one that is relatively tolerant to cold temperatures. It is worth noting that cereals store fructans rather than starch and it has been proposed that fructans may act as cryoprotective compounds. Another interesting characteristic is that their threshold for oscillatory behaviour is rather low in comparison with starch–synthesising species (e.g. soya). Fluorescence characteristics of cereals suggest that the violaxanthyn cycle may have a role in their rapid recovery from stress.

To evaluate the photosynthetic performance of a given species is not enough to study the plant in optimal conditions. It is important to see how the plant responds to unfavourable conditions, what is its degree of resistance to stress and how fast it recovers. To study the response to environmental changes also provides valuable information about the regulation of the photosynthetic process.

7.1. Acclimation

The better acclimation to changes in the environment observed in some species of plants (including increased [CO_2]), may be related to their capacity to adjust more eficiently the P_i supply to the chloroplast. The role of P_i re–cycling in acclimation to [CO_2] becomes evident when one of the partial processes involved is affected. When the response of photosynthesis to increased [CO_2] was studied in the starchless mutant of *Arabidopsis thaliana* refered to above (section 4), it was found that very small increments in [CO_2] elicited oscillations in photosynthesis and lead to a decline in rate (Sivak and Rowell, 1988). The response of photosynthesis to light was strongly modified (Figure 6), and feeding P_i through the petiole alleviated this and other symptoms of inadequate P_i supply brought about by the increase in [CO_2]. It was concluded that in this species the contribution of starch synthesis to P_i re–cycling is essential for efficient adaptation to changes in environmental conditions that, like increased [CO_2], pose increased demands on photosynthesis.

If better acclimation of some plant species to changes in the environment is related to their ability to adjust the rates of the P_i–recycling processes, it should be possible to improve acclimation of other species by introducing new P_i–recycling mechanisms into them or improving on the existing ones. One interesting pathway is that leading to the accumulation of fructans in leaves of temperate grasses. It has been proposed that it plays an important role in the ability of temperate species to grow in northern latitudes, by providing storage of fixed carbon in a readily accesible state under conditions of low temperatures and short daylength (Pollock *et al.* 1980). Fructans are accumulated in the vacuole, and this compartmentation may provide added flexibility. Starch is accumulated inside the chloroplast, the site of the photosynthetic process, and it has been suggested that excess starch may lead, by itself, to decreased photosynthesis, by disrupting the fine structure of the chloroplast.

312

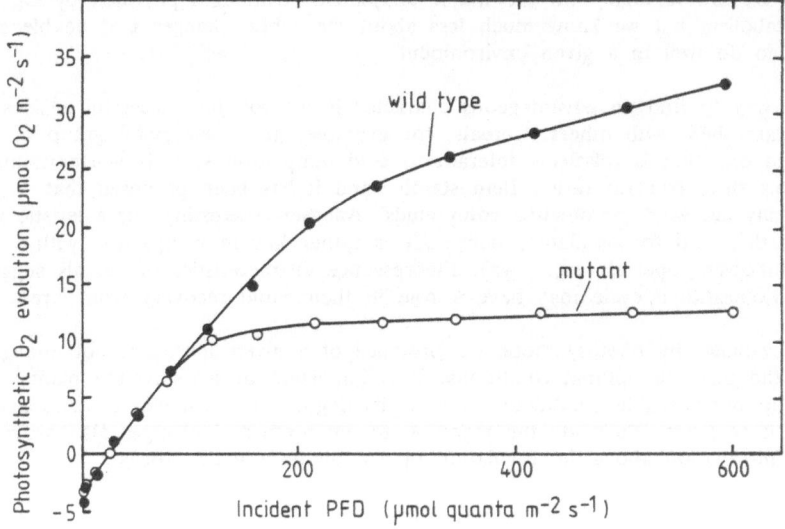

Figure 6. Limitation of photosynthesis by P_i supply to the chloroplast. Light response curves of photosynthetic oxygen evolution at CO_2 saturation in wild type and starchless mutant of Arabidopsis thaliana. *Note that the quantum yield is similar but the shapes of the curves are different, i.e. the light–saturated rate and light flux required to saturate photosynthesis in the mutant are lower (after Sivak and Rowell, 1988).*

8. PHOTOSYNTHESIS *VS.* PRODUCTIVITY

Measurement of photosynthesis by a leaf is not enough to predict the productivity of an agricultural system. Productivity from a commercial point of view depends not just on the maximum rate of photosynthesis, but also on other factors, e.g. the internal distribution of the products of photoassimilates, the growth capacity of storage tissues, the formation of new leaves. Little is known about how plants use the photosynthetic capacity of their leaves under the unfavourable conditions they encounter in the field. Resistance to diferent kinds of stress and recovery from them can be more important that the rate of photosynthesis measured in optimum conditions. Effective water utilization, early establishment of ground cover, the disposition of leaves, the rate of formation of storage organs, are just a few of the many factors, that affect photosynthesis and productivity. It is not possible to cover this kind of factor in detail but they must be taken into account when deciding how to go about genetic manipulation. How do they affect photosynthesis? Disposition or rate of formation of new leaves determine the amount of light intercepted. Formation of storage tissues affects the rate of utilization of photosynthates and, through a complex regulation, the rate of photosynthesis (for a review of this topic see Gifford, 1987).

9. INCREASED PHOTOSYNTHESIS: A MOVING TARGET

9.1. Increased CO_2 concentration

The expected changes in CO_2 concentration in the atmosphere will have important and very complex effect on plants (CO_2 being a substrate of phosynthesis). The information available on effects of CO_2 on plants has been extensively reviewed (see, for example, Lemon, 1983; Strain and Cure, 1985). The expected changes may look slow and incremental on a historical scale, but in geological terms, and in comparison with the previous decreases in $[CO_2]$ to which plants had millions of years to adapt, they are very fast, almost instantaneous. There is no reason to believe that the resulting changes in the productivity and composition of natural and man–managed ecosystems should be beneficial. Directed intervention may be required to preserve stability.

Plants can adapt to changes in the environment, in the sense of acquiring heritable characteristics enabling them and their progeny to survive and reproduce better than other plants lacking those characteristics. But we cannot expect adaptation in all species to occur over brief period. Plants with shorter growth cycles will have more generations than those with longer cycles, and a better chance to adapt to environmental changes. Acclimation, a non–heritable modification in response to the new environment, will be more important in the short term. Different species vary in their capacity to acclimate to environmental changes.

Can we accelerate these adaptive processes? Can we influence the genetic changes so they have the desired outcome? What resources are at our disposal? In some specific cases, it may be possible to manipulate the genetic composition of the plant population in order to optimise yield and increase the potential of plants to act as stabilizers of atmospheric $[CO_2]$. To achieve these aims, it may be necessary to use plant breeding and genetic engineering to compress into a few decades changes that, if left to natural evolution, would require thousands of years. Agricultural and forestry management could be adjusted to ensure that maximum yield is obtained within the environmental and economic constraints.

9.2. Plant responses to CO_2

At present, photosynthetic rates and growth of plants are limited by ambient CO_2 when other factors (e.g. light, water, nutrients, temperature) are optimal. If CO_2 limits photosynthesis for most plants in most circumstances, a simple minded extrapolation leads to the conclusion that more CO_2 in the atmosphere should mean more yield and more carbon accumulated in plants. Some greenhouse crops will respond well to the expected increase in $[CO_2]$. Indeed, CO_2 enrichment has been used for some decades now to improve yield in crops such as tomato. We also know that plants with C_3 metabolism will respond more than C_4 plants to added CO_2. But there are complications. Plant growth and yield are not controlled by photosynthesis alone. More carbohydrate produced by photosynthesis will affect metabolic and maintenance dark respiration, and $[CO_2]$ may affect the activity of specific enzymes. After the initial incorporation of CO_2 by photosynthesis, there is a hierarchy of increasingly complex processes controlling the production and allocation of proteins, starch, sucrose and lipids into the various sinks that contribute to leaf expansion, root growth, flowering and fruiting, and final yield. The timing of each part of the growth cycle is also important and CO_2 is known to affect many of the partial processes involved in growth and development in some species.

Drought resistance may become less important, because stomata are less open in higher [CO_2], thus decreasing water loss. This factor will be more important in C_4 plants, in which stomata are more sensitive to [CO_2].

Increased photosynthesis and plant growth should increase the plant demand for nutrients. The availability of nutrients already limits plant growth, and the ratio carbon/nitrogen is expected to decrease, unless more nitrogen is made available by fertlization. Increased photosynthesis will stimulate nitrogen fixation by the root nodules of the legumes and may increase migration of legumes into grassland. Photorespiration is involved in N metabolism, and the partial supression of the photorespiratory pathway by increased [CO_2] will affect N assimilation.

9.3. Climate changes

Up to a point, plants photosynthesise better as the temperature increases. A broad optimum is soon reached, however, as the advantage derived from faster carbon assimilation is offset by various deleterious factors. Photorespiration is important amongst these and therefore C_4 plants tend to have higher temperature optima. C_3 species photosaturate at about 1/5 full sunlight (i.e. photosynthesis does not increase much as the light intensity is raised above this level). Conversely, photosynthesis by C_4 species (or by C_3 species in high CO_2 or low O_2) increases with increasing light intensity up to, and beyond, full sunlight. Again, the difference is principally attributed to photorespiration, and C_4 species are likely to be at an advantage over other species under these particular conditions.

Not only a change in the global temperature is predicted as a result of the greenhouse effect, but changes in the pattern of rainfall are also forecast (and, some say, are already occuring). Some parts of the world will have an increase in rainfall whereas others will have significant decreases. Although atmospheric [CO_2] is a limiting factor, weather (particularly extremes such as frost and drought) is and will continue to be, an important source of variation in yield.

9.4. Natural ecosystems

Interactions at this level become so complex that prediction of the effects of CO_2 enrichment is extremely difficult. Competition between species will be affected by any change in climate and [CO_2]. No two species are the same, and within a given species the responses will be different according to growth conditions. These differences, added to the expected climatic changes, should result in significant changes in the composition of the plant communities, and through a complex net of ecological relationships, will bring changes in the associated animal communities.

10. RESOURCES

Increased research efforts are directed to characterise the long term response of plants to increased [CO_2] and changed climate, and to understand the biochemical mechanisms underlying these responses. The challenge is to produce the scientific information required to model the extremely complex responses of the main crop species and of natural ecosystems to such changes; and to use genetic engineering, plant breeding and other means to advantage.

10.1. *Can we circumvent the new bottlenecks?*

Increased [CO_2] will bring earlier development of leaf area, increasing early cover and the efficiency in the use of incident light in the initial stages of the crop. Increased photosynthesis will not necessarily result in higher yield. If utilization of the carbohydrates produced by photosynthesis is relatively slow, starch may accumulate in the leaves and photosynthetic rate may decrease.

The so called "inhibition of photosynthesis by sink–source inbalance" refers to decline in rate that accompanies the slow utilization of photosynthates by the "sinks", i.e. the organs that use the assimilates such as young leaves, flowers, fruits, and it can be caused by low temperature or any factor that slows down the growth rate of the sink. In some species (but not in others), increased [CO_2] leads to starch accumulation in leaves and decreased photosynthetic rates, a response that has attracted much attention because of its potential impact on yield. Some progress has been made in the understanding of this problem.

Inadequate phosphate supply was also recognised in several circumstances of physiological relevance, e.g. in leaves of plants grown in warm conditions and tranfered to relatively low temperatures, in leaves photosynthesising in saturating light and [CO_2], and in plants infected by pathogens. Symptoms of inadequate Pi supply to the chloroplast have also been described in plants grown under high [CO_2].

Decreased photorespiration in response to higher [CO_2] should decrease the light and [CO_2] thresholds at which Pi becomes limiting, firstly because less Pi will be made available through photorespiration and secondly because increased carboxylation will make higher demands on the supply of Pi, itself a substrate of photosynthesis. If limited Pi supply could be recognised as limiting photosynthesis and growth in plants growing in enriched CO_2, improved understanding of photosynthetic regulation could be used for the improvement of the crop.

10.2 *Changes in agricultural practices*

As mentioned above, if the higher yields allowed by increased [CO_2] are to be realised, increased inputs of N, P and K will be required. Increased nitrogen fixation by symbionts in leguminous crops and by free living bacteria could reduce to some extent the use of fertilizers.

For the C_3 crops (e.g. wheat, potato, rice), because they will benefit more from increased [CO_2], weeds are likely to be less of a problem. The reverse is true for the C_4 crops (the most important, corn, sorghum, sugarcane and millet).

The climate changes will result in shifts in the geographical areas more suitable for each species. Developments in herbicide technology and pest control will improve the overall situation and the same is valid for insect and infectious diseases (see below, "genetic engineering"). Increased understanding of the action of growth regulators could improve commercial yield of many crops, because it will allow the manipulation of partition of carbohydrates in the direction of increased harvest index. If, as expected, increased $[CO_2]$ leads to higher efficiency in the use of water by many crops, their cultivation could be extended to drier climates.

Forestry has different requirements and problems, because it deals with long–lived perennial species which will have too few generations to adapt to higher $[CO_2]$ and accompanying climate changes. The long life–cycles make breeding and screening for good genotypes especially difficult, but because of the importance of forests in the stability of earth climate, they must be attempted.

10.3 *Plant selection by breeding*

In spite of the complexity of plants response to increased $[CO_2]$, some generalizations can be made. For example, plants with different photosynthetic metabolism respond differently to a given change in $[CO_2]$: plants with C_3 metabolism respond more than C_4 plants to added CO_2. Thus, it should be possible to modify the productivity of a system by changing, say, from corn to wheat. This is, however, a drastic and probably impractical solution. Breeding or genetic enginering could provide a more satisfactory alternative, by introducing an advantageous character into the desired species.

The current breeding practices select those varieties best adapted to the environmental conditions of the site during the last few years of testing. Thus, the plant breeding programs incorporate the responses to changing climate and $[CO_2]$ in the part of the world where the crop is being tested. Wheat is an interesting example of this characteristic inherent to plant breeding, and the modern varieties are more resistant to a gaseous pollutant, SO_2, than varieties that were common at the beginning of this century (A. Bolan, P.J. Lea & T. Mansfield, 1989, personal communication).

Once good genotypes have been selected by conventional breeding, if the biochemical and physiological mechanisms underlying the higher yield are understood, screening in laboratory rather than in field conditions becomes possible. Better understanding should also facilitate the transfer of the beneficial characters to other varieties by breeding, and even to genetically distant species, by genetic engineering.

10.4. *Tissue culture*

Tissue culture is a relatively new methodology which allows the production of many genetically identical plants from a single individual in a relatively short time, and in a limited space. This technique has been used to obtain virus–free individuals from an infected plant. If a good genotype is obtained by plant breeding or genetic engineering, tissue culture can shorten the time necessary to obtain the large number of plants required for commercial crop production. A problem inherent to this technique is that, as genetic diversity is diminished, the likelihood of a new disease decimating the crop, increases.

10.5. *Genetic engineering*

The application of recombinant DNA technology to plants requires input information from several disciplines, especially genetics, biochemistry, biophysics and physiology. That information is scarse and inadequate and because of this limitation, molecular genetic research in plants so far has been focused on a few areas where the necessary background exists. Beneficial characters are usually controlled by several genes, but application of genetic engineering to crop improvement is at present limited to single–gene traits. A foreign gene can be introduced into what becomes a transgenic plant leading, for example, to the synthesis of an antimetabolite, trypsin inhibitor that provides protection against attack by insects (Boulter et al., 1988). This kind of character could become even more useful if higher biomass, combined with warming, increased the risk of plant diseases and attack by insects. Current research into diverse mechanisms of resistance to herbicides, insects, viruses and fungus should produce the technology needed to introduce such resistance into sensitive species. Even the introduction of resistance to some pollutants seems feasible, and synthesis of a human protein confering resistance to heavy metals has been achieved in transgenic tobacco and *Brassica napus* plants (Misra, 1988).

Much of the efforts of molecular biologists is devoted to the improvement of photosynthetic performance at low $[CO_2]$, through the modification of the carboxylating enzyme (Rubisco), introduction of IC pumps similar to those found in some cyanobacteria, or modification of the photorespiratory pathway. In the future, emphasis may have to be shifted to the partial processes likely to become limiting or (co–limiting) under increased $[CO_2]$.

The obtention of qualitative changes in plants is, however, a far more ambitious objective. We can include in this category the reduction of photorespiration in C_3 plants (making them more like C_4 plants) or the incorporation nitrogen–fixing capacity into non–legume species. These advantageous characteristics were achieved by some plants through evolution spanning thousands of years, and involve not just a single protein but a series of proteins, and a degree of compartmentation and regulation of the sort that is beyond the abilities of genetic engineering at present. From the point of view of the plant molecular biologist, the simultaneous modification of many proteins, or a major modification of leaf anatomy, is a very complicated matter. We can only guess how long it could take or whether it is at all feasible to achieve such objectives, but it is worth noting that what can be done now routinely in laboratories all over the world, was science fiction one or two decades ago. One of the essential requirements for the incorporation of new pathways, the re–targetting of a single protein into any membrane in the cell, is now possible (Ellis, 1985). And yet, for plant molecular biology to provide solutions to classical agricultural problems, a great deal more of understanding of basic plant biochemistry will be required. Plant physiologists, biochemists, molecular biologists and users will have to collaborate in the tasks of choosing the most desirable characteristics, and of evaluating the feasibility of their introduction and the effects of this genetic engineering on the overall performance of the plant.

It is generally accepted that the rate of photosynthesis is ultimately limited by the rate of electron transport and that re-oxidation of plastoquinone (a key component in the electron transport chain) is the ultimate bottleneck. It should be possible to increase the rate of electron flux through this site but, before doing that, it should be noted that many other bottlenecks are reached before this one. For example, optimal conditions of carbon fixation will be required and even then, it is more likely that a ceiling of P_i re-cycling likely to be reached first.

11. AGRICULTURE, A SOURCE OF CARBON DIOXIDE

The simplified equation for photosynthesis

$$CO_2 + H_2O \rightarrow [CH_2O] + O_2$$

describes the dual role of photosynthesis as a source of biomass and a sink for CO_2. But agriculture is, at present, a *source* of CO_2 and a *sink* for fossil fuels. This is because the high yields that produced the "grain mountains" of recent years, depend on high inputs of fertilizers, fossil fuels, herbicides and pesticides. In times of growing concern for the greenhouse effect, the turning of a process that should be a sink into a source of CO_2, is worrying enough. Other environmental issues, e.g. nitrate contamination of the ground water, and the wide use of pesticides and herbicides, add to the picture. Other more lasting effects are the loss of genetic diversity, and of forests and natural ecosystems.

12. CONCLUDING REMARKS

Environmental issues demand an improvement of the CO_2 balance of agriculture, increased re-forestation and the use of biomass as a source of energy. All of these needs demand a better agriculture that could deliver the same amount of food from a smaller area and lower inputs. It is clear that there is room for a great deal of improvement in agriculture.

Environmental changes brought about by human activities have a complex effect on plant metabolism, making the desired improvement of the photosynthetic process into a moving target.

The factors that limit yield at present are not the same factors that will be limiting in a future CO_2-enriched world. The areas of science that constrain our capacity to manage global photosynthesis at present, are not the same the will constrain it in one or two decades time. We need to keep flexibility in the way we allocate resources, human and material, to different areas of scientific research.

ACKNOWLEDGEMENTS

My work is supported by grants from the Agricultural and Food Research Council (U.K.) and by Shell Research (U.K.).

REFERENCES

Björkman, O. 1981. In: *Encyclopedia of Plant Physiology, New Series, Vol. 12 A (Physiological Plant Ecology I, Responses to Plant Environment* (O.L. Lange, P.S. Nobel, C.B. Osmond and H. Ziegler, eds.), pp 57–107. Springer, Berlin.

Björkman, O. 1987. High irradiance stress in higher plants and intercation with other stress factors. In: *Progress in photosynthesis.* (J. Biggins ed.) M. Nijhoff, vol. IV, pp. 1–8.

Björkman, O. and Demmig, B. 1986. Photon yield of O_2 evolution and chlorophyll fluorescence characteristics at 77 K among vascular plants of diverse origins. *Planta* 170; 489–504.

Boulter, D., Gatehouse, A.M.R. and Hilder, V. 1988. Trypsin inhibitor as insect protectant. Abstracts, 2nd International Congress of Plant Molecular Biology, Jerusalem, November 13–18, 1988.

Canvin, D.T., Miller, A.G. and Espie, G.S. 1987. C_3 Photosynthesis in algae: The importance of inorganic carbon concentrating mechanisms. In: *Carbon dioxide as a source of carbon.* (M. Aresta and G. Forti, eds.) pp 199–212.

Caspar, T., Somerville, C. and Huber, S. 1985. Alterations in growth, photosynthesis and rspiration in a starchless mutant of *Arabidopsis thaliana* (L.) deficient in chloroplast phosphoglucomutase activity. *Plant Physiol.* 79: 11–17.

Edwards, G. and Walker, D.A. 1983. *C3, C4: mechanisms, and cellular and environmental regulation, of photosynthesis.* Blackwell Scientific Publications, Oxford.

Ellis, J. 1985. Eucaryotic proteins retargetted among cell compartments. *Nature*, 313: 353–354.

Ellis, R.J. and Gray, J.C. (eds.) 1986. Ribulose bisphosphate carboxylase–oxygenase. *Phil. Trans. R. Soc. Lond. B.* 313: 313–469.

Forti, G. 1987. Photosynthetic electron transport: the source of electrons for CO_2 reduction in photosynthesis. In: *Carbon dioxide as a source of carbon.* (M. Aresta and G. Forti, eds.) pp 71–82.

Gifford, R.M. 1987. Barriers to increasing crop productivity by genetic improvement in photosynthesis. In: *Progress in photosynthesis Research* (J. Biggins, ed.) Vol. IV, 377–384.

Hanson, A.D., Hoffman, N.E. and Samper, C. 1986. Identifying and manipulating metabolisc stress–resistance traits. *HortScience*, 21: 1313–1317.

Heldt, H. and Stitt, M. 1986. The regulation of sucrose synthesis in leaves. In: *Progress in photosynthesis Research.* (J. Biggins ed.), pp. 675–690.

Kaiser, G., Martinoia, E. and Wiemken, A. 1982. Rapid appearance of photosynthetic products in the vacuoles isolated from barley mesophyll protoplasts by a new fast method. *Zeitschrift Pflanzenphysiol.* 94: 377–385.

Kramer, P.J. 1981. Carbon dioxide concentration, photosynthesis and dry matter production. *BioScience*, 31: 29–33.

Lange, O.L., Nobel, P.S., Osmond, C.B. and Ziegler, H. (eds.) 1981. *Encyclopedia of Plant Physiology, New series, Vol. 12A (Physiological Plant Ecology I, Responses to Plant Environment).* Springer, Berlin.

Lemon, E. (ed.) 1983. *CO_2 and plants: The response of plants to rising levels of atmospheric carbon dioxide,* AAAS Symposium Vol 84. Westview Press, Boulder, Colorado, 280 pp.

Melis, A. and Harvey, G.W. 1981. Regulation of photosystem stoichiometry, chlorophyll *a* and chlorophyll *b* content and relation to chloroplast ultrastructure. *Biochim. Biophys. Acta.* **637**: 138–145.

Misra, S. 1988. Heavy metal tolerant transgenic *Brassica napus* and tobacco plants. Abstracts, 2nd International Congress of Plant Molecular Biology, Jerusalem, November 13–18, 1988.

Pollock, C.J.; Riley, G.J.P.; Stoddart, J.L. and Thomas, H. 1980. The biochemical basis of plant response to temperature limitations. *Annual Report of Welsh Plant Breeding Station for 1979*, pp. 227–246.

Sharkey, T.D. 1985. photosynthesis in intact leaves of C_3 plants: physics, physiology and rate limitations. *Botanical Review*, **51**: 53–105.

Sivak, M.N. 1987. The effect of oxygen on photosynthetic carbon assimilation and hypotheses concerning the mechanisms involved. *photobiochem. Photobiophys.* Suppl. 145–156.

Sivak, M.N. and Walker, D.A. 1987. Oscillations and other symptoms of limitation of *in vivo* photosynthesis by inadequate phosphate supply to the chloroplast. *Plant Physiol. Biochem.* **25**: 635–648.

Sivak, M.N. and Rowell, J. 1988. Phosphate limits photosynthesis in a starchless mutant of *Arabidopsis thaliana* deficient in chloroplast phosphoglucomutase activity. *Plant Physiol. Biochem.* **26**: 493–501.

Strain, B.R. and Cure, J.D. (eds). 1985. *Direct effects of increasing carbon dioxide on vegetation*. United States Department of Energy, Office of Energy Research, Office of Basic Energy Sciences, Carbon Dioxide Research Division, 286 pp.

Stitt, M.N., Huber, S.C. and Kerr, P. 1987. Regulation of photosynthetic sucrose synthesis. In : *The Biochemistry of Plants* (M.D. Hatch and N.K. Boardman, eds.). Vol. 13, pp. 327–409. Academic Press, New York.

Walker, D.A. and Osmond, C.B. 1986. Measurement of photosynthesis *in vivo* using a leaf disc electrode: correlations between light dependence of steady–state photosynthetic oxygen evolution and chlorophyll *a* fluorescence transients. *Proc. r. Soc. L.B.* **227**: 267–280.

THE KINETIC PROPERTIES OF RIBULOSEBISPHOSPHATE CARBOXYLASE

John V. Schloss
Central Research & Development Department
E. I. du Pont de Nemours & Co.
Experimental Station, P. O. Box 80328
Wilmington, Delaware 19880-0328 USA

ABSTRACT. By a combination of isotope effects, chemical quench, and multinuclear NMR experiments, a coherent picture of the kinetic and chemical mechanism of ribulosebisphosphate carboxylase has emerged. The enzyme has a modified Theorell-Chance mechanism, in which CO_2 adds directly to the C-2 carbanion of ribulose bisphosphate. The C-2 carbanion of ribulose bisphosphate is discrete from the ene-diol intermediate, that eliminates the C-1 phosphate on chemical quench. Tautomerization of the ene-diol to the C-2 carbanion is substantially faster than the formation of this intermediate by abstraction of the C-3 proton of ribulose bisphosphate under normal conditions of saturating CO_2. The proton abstracted from C-3 of substrate is conserved and is involved in the subsequent hydration and cleavage of the 6-carbon intermediate, 2-carboxy-3-oxo-D-arabinitol bisphosphate to two molecules of 3-phosphoglycerate.

1. Difficulties of Maintaining Anhydrous CO_2 in an Aqueous Environment

For any enzyme that utilizes gaseous CO_2, physiological conditions usually conspire to make its life difficult. As a relatively good electrophile, CO_2 can be attacked directly by water to give carbonic acid (Figure 1). The mixture of CO_2 and carbonic acid will titrate with an apparent pK value of 6.34 (Umbreit et al., 1964), however the true pK for the ionization of carbonic acid is actually about 3.8 (Jencks, 1969). So, with water as nucleophile, the attack on CO_2 is not particularly favorable, and most (> 99 %) of the gaseous CO_2 dissolved in water remains unhydrated. Although carbon dioxide is relatively soluble in water, it is not a particularly abundant gas in air. As illustrated in Table 1 (Weast, 1971), CO_2 comprises only about 0.033 % of air. It is also not the only electophilic gas in air, as oxygen at 20.9 % is greater than 600 times more abundant than carbon dioxide. Oxygen is not as soluble in water as carbon dioxide. At one atmosphere of either gas (25°C) a saturated solution of carbon dioxide would be 30.9 mM vs. 1.15 mM for oxygen (Figure 2). Air equilibrated water would contain (25°C, 1 atm) 0.0102 mM carbon dioxide and 0.24 mM oxygen. The 27-fold greater solubility of carbon dioxide relative to oxygen helps to reduce the ratio of the solution concentrations of these two gasses to only 23-fold greater in favor of oxygen. However, if the solution is buffered, the two ionization constants of carbonic acid need to be considered in determining the total concentration of dissolved carbon dioxide. Although the equilibrium for hydration of CO_2 favors dissolved CO_2 over carbonic acid, at alkaline pH an increased

M. Aresta and J. V. Schloss (eds.),
Enzymatic and Model Carboxylation and Reduction Reactions for Carbon Dioxide Utilization, 321–345.
© 1990 Kluwer Academic Publishers.

322

Aqueous

$$CO_2 + H_2O \underset{}{\rightleftharpoons} H_2CO_3 \underset{}{\rightleftharpoons} H^+ + HCO_3^-$$

$$K_{eq} \approx 350 \qquad pK \approx 3.8$$

$$pK = 6.34$$

$$\frac{[H^+][HCO_3^-]}{[H_2CO_3]} = 10^{-3.8} \qquad \frac{[H^+][HCO_3^-]}{[CO_2]} = 10^{-6.34}$$

$$\frac{[CO_2]}{[H_2CO_3]} = \frac{10^{-3.8}}{10^{-6.34}} = 10^{6.34-3.8} = 350$$

Figure 1. Hydration of CO_2 to carbonic acid and the ionization of carbonic acid to bicarbonate. The equations involved that define these equilibria are also illustrated.

Table 1. Composition of Air

N_2	78.084 %
O_2	20.946
CO_2	0.033
Ar	0.934
Sum	99.997

Remaining gasses (Ne, He, CH_4, Kr, H_2, N_2O, Xe) comprise 0.003 %

fraction of the CO_2 would exist as bicarbonate and carbonate (Figure 3). At equilibrium, the concentration of CO_2 for an alkaline solution would be the same as for water, however, the total potential CO_2 in the form of bicarbonate and carbonate would increase with pH. Although the equilibrium level of CO_2 in an air-equilibrated solution would be pH independent and depend solely on temperature and the partial pressure of CO_2 in the gas phase (0.033 % of one atmosphere for air), the transient, non-equilibrium levels of CO_2 would depend on total "CO_2" (CO_2 + bicarbonate + carbonate) present in the solution, pH, rate of interconversion of different CO_2 species, and rate of equilibration

CO_2 (gas)	1 atm	0.033 % CO_2 atm (air)		
	25°C	25°C	15°C	37°C
	(mM)	(μM)		
CO_2 (aqueous)	30.9	10.2	14.2	7.51
(0.5 M KCl)	28.4	9.37	12.9	

Figure 2. The solubility of CO_2 at different partial pressures, temperatures, and in the presence of dissolved salt.

at 25°C dpK_1/dT = -0.0055 dpK_2/dT = -0.009

$$CO_2 + H_2O \rightleftharpoons H_2CO_3 \rightleftharpoons H^+ + HCO_3^- \rightleftharpoons 2H^+ + CO_3^=$$

pK_1 (apparent) = 6.343 pK_2 = 10.33

pK_1' = 6.343 - $\beta\mu^{-0.5}$ $\beta\mu^{-0.5} \approx 0.11\,\mu^{-0.5}$

at 0.1 M $NaHCO_3$ $pK_1' \approx 6.31$

Figure 3. The solution chemistry of carbon dioxide and relevant thermodynamic constants (Umbreit, 1964; Perrin and Dempsey, 1974).

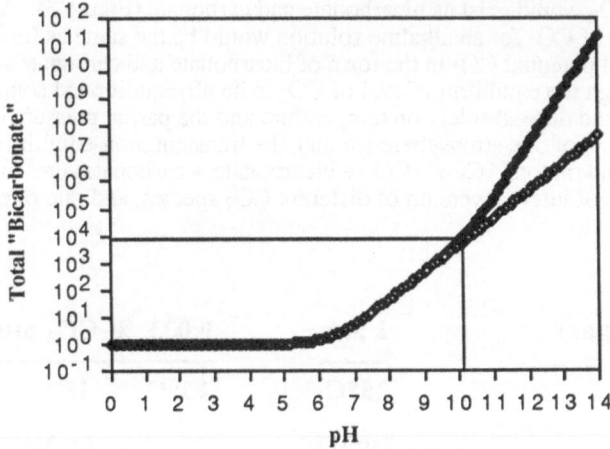

Figure 4. The relative concentration of bicarbonate required for a desired concentration of CO_2 as a function of pH. The unfilled symbols are calculated by considering only the first ionization constant of carbonic acid, while the filled symbols are calculated by considering both ionization constants.

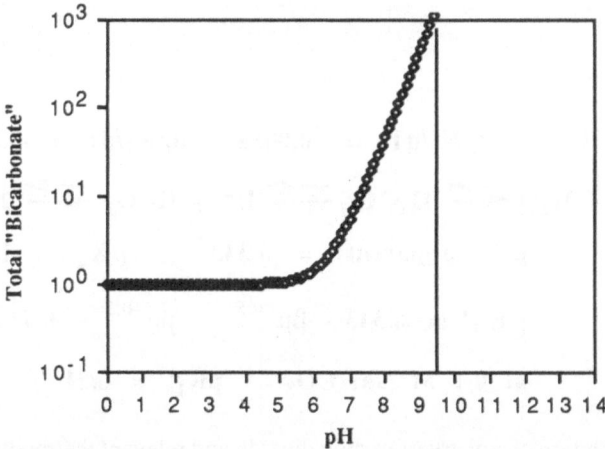

Figure 5. The relative concentration of bicarbonate required for a given concentration of CO_2 at different pH values.

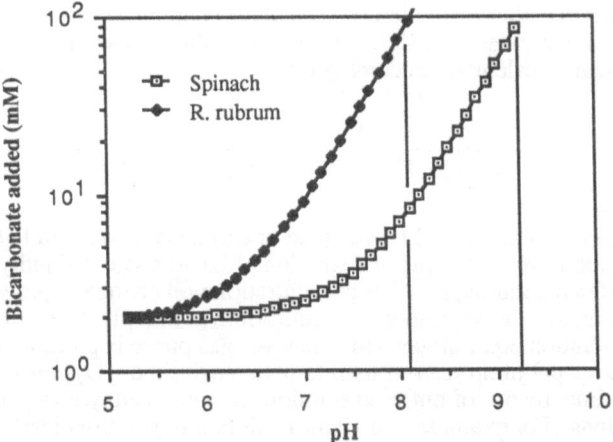

Figure 6. Concentrations of bicarbonate used in studies of the kinetic properties of the spinach and *Rhodospirillum rubrum* ribulosebisphosphate carboxylases (Van Dyk and Schloss, 1986).

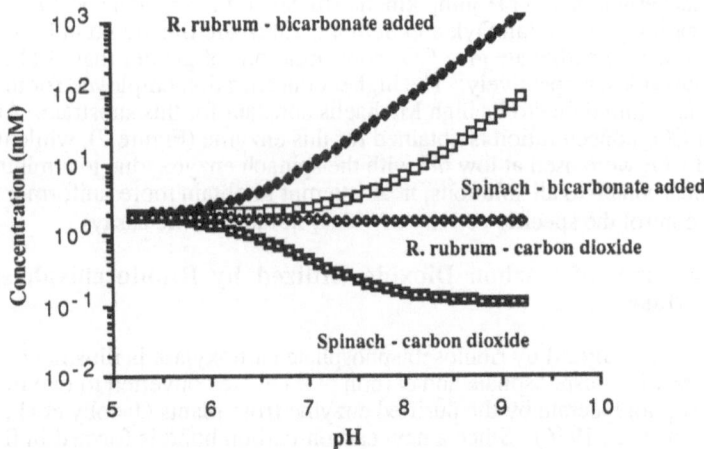

Figure 7. The bicarbonate concentrations used in pH studies of the spinach and *Rhodospirillum rubrum* ribulosebisphosphate carboxylase (as in Figure 6), and the concentrations of CO_2 that they should give.

with the gas phase. In a closed solution (no gas phase) containing a known amount of added bicarbonate or carbonate, the CO_2 concentration obtained at equilibrium in a buffered solution can be calculated according to:

$$CO_2 = \frac{C}{1 + 10^{pH - 6.343} + 10^{2pH - 16.67}}$$

where C is the concentration of total bicarbonate or carbonated added initially, and pH is the final pH at equilibrium. If the pH of a solution of bicarbonate is rapidly altered by mixing with a buffered solution, the CO_2 concentration will change, rapidly if carbonic anhydrase is present, to reflect the solution equilibrium at that pH, then slowly lose (or gain) CO_2 as the solution equilibrates with whatever gas phase is present. For biochemical studies, "pH jump" can be used to obtain a desired CO_2 concentration for a short duration of time, by use of buffered solutions, carbonic anhydrase, and stock bicarbonate solutions. For example, if a 10 mM solution of sodium bicarbonate buffered at pH 8 is prepared, initially it will have respective CO_2 and bicarbonate concentrations of 0.22 and 9.78 mM. However, that same solution after equilibrating with air would have respective CO_2 and bicarbonate concentrations of 0.01 and 0.46 mM. Reducing or eliminating the volume of air space in contact with the solution would maintain the initial concentrations. Below the second pK of carbonic acid (10.33), only the first ionization constant really needs to be considered in calculating the concentration of CO_2 (Figure 4). Above about pH 9.5, the concentration of bicarbonate required to maintain an elevated level of CO_2 (0.1 mM CO_2 would require about 0.16 M bicarbonate) becomes prohibitive for enzymic work due to the high ionic strength (Figure 5). The concentrations of bicarbonate employed in pH jump kinetic studies of the spinach and *Rhodospirillum rubrum* carboxylases (Van Dyk and Schloss, 1986) are illustrated in Figures 6 and 7. These levels of bicarbonate give CO_2 concentrations of greater than 0.11 and 1.6 mM for the two enzymes, respectively. The higher concentration employed for the *R. rubrum* enzyme is required due to its high Michaelis constant for this substrate. An almost constant CO_2 concentration is obtained for this enzyme (Figure 7), while much higher levels of CO_2 were used at low pH with the spinach enzyme due to a minimal addition of 2 mM bicarbonate to all solutions, in an attempt to obtain more uniform results, and reliably control the specific activity of $^{14}CO_2$ in radiometric assays.

2. The Form of Carbon Dioxide Utilized by Ribulosebisphosphate Carboxylase

The reaction catalyzed by ribulosebisphosphate carboxylase is illustrated in Figure 8. D–Ribulose 1,5–bisphosphate and carbon dioxide are converted to two molecules of D–3–phosphoglycerate by the purified enzyme from plants (Jacoby et al., 1956; Weissbach et al., 1956). Since a new carbon-carbon bond is formed in fixing "CO_2", and the enzyme does not require biotin, ATP, or hydrolyze any phosphate ester in catalyzing this reaction, it is reasonable to assume that the enzyme utilizes CO_2, rather than bicarbonate, due to the electrophilic character of this species. To test this point, use can be made of the rather slow rate of equilibration between CO_2 and bicarbonate in the absence of carbonic anhydrase. At temperatures below 15°C, this equilibration requires more than 60 seconds. At an enzymic reaction temperature of 10°C, Cooper et al. (1969) demonstrated that radiolabel was much more rapidly assimilated by the carboxylase when it was present as $^{14}CO_2$, than it was when it was present initially as $H^{14}CO_3^-$. In the

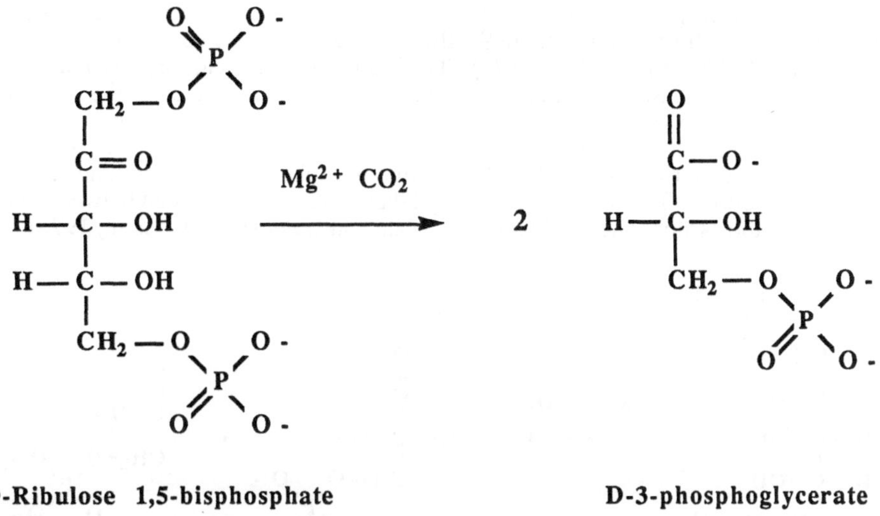

D-Ribulose 1,5-bisphosphate　　　　　　　　　　**D-3-phosphoglycerate**

Figure 8. The reaction catalyzed by ribulosebisphosphate carboxylase.

presence of carbonic anhydrase, the rate of incorporation of radiolabel from either species was identical. For synthetic purposes, use of CO_2 by an enzyme poses some real difficulties. As pointed out in the previous section, CO_2 is not the only electrophilic gas present in air-equilibrated water, and the other electrophilic gas, O_2, is present in far greater abundance (0.24 vs. 0.01 mM). Carbanions (the nucleophile in the carboxylase reaction) will readily react with either CO_2 or O_2 (Ayres, 1966; Stowell, 1979; Bates and Ogle, 1983). How does a synthetic enzyme that utilizes CO_2 discriminate effectively against the smaller electrophile O_2? There are few structural features of these two potential substrates to offer hope of a high degree of discrimination due to differential binding by the enzyme. Some difference in aqueous solubility (27–fold), noted in the previous section, would suggest that a more polar active site would favor carboxylation over oxygenation relative to a more hydrophobic site, but not by a large margin. It would seem that an enzyme that uses CO_2 synthetically (as opposed to decarboxylases that favor the reverse reaction) is doomed to waste at least a portion of the nucleophilic substrate when it runs this reaction under normal atmospheric conditions. Further, to maintain the substrate carbanion long enough for it to react with CO_2, the enzyme must shield the carbanion, as it is formed from the ene-diolate, to prevent its protonation by solvent. It would seem that the enzyme needs to provide a "dry" , polar home close to the C-2 carbanion, derived from ribulose bisphosphate, (effecting a local desolvation of both substrates) to have any hope of selectively favoring the carboxylation reaction.

3. The Oxygenase Reaction of Ribulosebisphosphate Carboxylase

Given the abundance and reactivity of oxygen, it is perhaps not surprising that ribulosebisphosphate carboxylase, an enzyme that utilizes carbon dioxide in a carbon-carbon bond forming reaction, has an oxygenase activity. It may be more surprising that this activity escaped detection for 15 years after the purification of the enzyme. Ogren and Bowes (1971) made the initial observation that O_2 was an inhibitor of the carboxylation reaction of ribulose bisphosphate carboxylase. The alternate enzymic activity responsible for this inhibition is illustrated in Figure 9 (Bowes et al., 1971; Bowes and Ogren, 1972; Andrews et al., 1973; Lorimer et al., 1973). The inhibition of the carboxylase reaction by oxygen, and the oxygenase reaction, can account for the Warburg effect, or the inhibition of photosynthesis by oxygen, and the production of glycolate during photorespiration (Tolbert, 1973). The phosphoglycerate produced in this reaction is derived from the lower three carbons of ribulose bisphosphate, while phosphoglycolate is produced from the upper two carbons. The fate of isotopically labelled oxygen is divided between one of the oxygens of the carboxylate of phosphoglycolate and water (Lorimer et al., 1973).

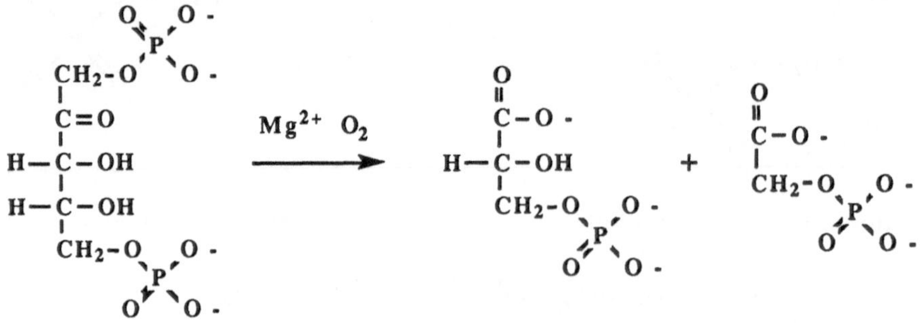

D-Ribulose 1,5-bisphosphate   D-3-phosphoglycerate   phosphoglycolate

Figure 9. The oxygenase activity of ribulosebisphosphate carboxylase.

4. Activation of Ribulosebisphosphate Carboxylase by CO^{2+} and Mg^{2+}

One of the ways that ribulosebisphosphate carboxylase seems to have developed to cope with its oxygenase activity, is an obligate activation process that is absolutely dependent on carbon dioxide and divalent metal (Lorimer et al., 1976; Badger and Lorimer, 1976; Laing and Christeller, 1976; Andrews et al., 1975). Both carboxylase and oxygenase activities require that the enzyme is "preactivated" by a carbon dioxide that is kinetically discrete from the CO_2 that becomes fixed in the carboxylation reaction (Lorimer, 1979). This activation process ensures that the enzyme does not waste ribulose bisphosphate in the oxygenase reaction, should an insufficient amount of carbon dioxide be available for the carboxylation reaction (and activation of the enzyme). Activation of the enzyme is a slow process relative to catalysis, allowing it to be distinguished kinetically from the catalytic cycle. Enzyme, that has been activated or deactivated prior to assay, will maintain its level of activity for scores of seconds following transition to conditions that favor a different extent of activation. As the enzyme from plants completes each catalytic cycle about three times each second, the carboxylase can be activated under optimal conditions and assayed briefly (15 seconds), by rapidly diluting a concentrated stock of the preactivated enzyme into an assay mixture of the desired composition. By an experimental protocol similar to this, the effects of pH, CO_2, and Mg^{2+} concentration on the activation process and catalytic activity can be determined separately. The activation process is favored by high pH and can be driven to completion at low CO_2 concentrations by high concentrations of Mg^{2+}. Similar to the carboxylation reaction, activation of the enzyme is specific for CO_2 rather than bicarbonate (Lorimer et al., 1976). Activation is an ordered process with addition of CO_2 to the ε–amino of a specific lysyl residue to form a carbamate (Lorimer, 1981; Donnelly, 1983), followed by Mg^{2+} binding. The activation scheme is summarized in Figure 10. Carbamylated enzyme, with Mg^{2+} bound (enzyme–NH–CO_2^-–Mg^{2+}), is the true form of carboxylase with any catalytic activity. Although this is an extremely labile catalyst, this is the form of the enzyme for which the kinetic properties will be reviewed.

$$\text{Enz-NH}_2 \; + \; \text{CO}_2 \; \rightleftharpoons \; \text{Enz-NH-CO}_2^- \; + \; \text{H}^+ \; \overset{\text{Mg}^{2+}}{\rightleftharpoons} \; \text{Enz-NH-CO}_2^-\text{-Mg}^{2+}$$

active

Figure 10. A scheme for the activation of ribulosebisphosphate carboxylase.

5. Chemical Quench and the Mechanism of Ribulosebisphosphate Carboxylase

Hypothetical schemes for the chemical mechanisms of the carboxylase and oxygenase activities of ribulosebisphosphate carboxylase are illustrated in Figure 11. Both reactions are presumably initiated by the abstraction of the C–3 proton of ribulose bisphosphate

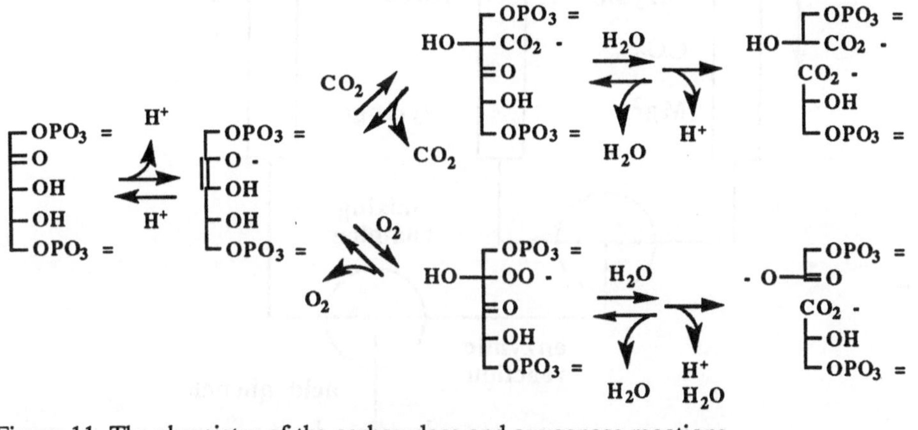

Figure 11. The chemistry of the carboxylase and oxygenase reactions.

(RuBP) to form a carbanion, that would be stabilized on the enzyme as the ene–diol. Carbon dioxide and oxygen would compete for the ene–diol of RuBP to form, respectively, 2–carboxy–3–oxo–pentitol 1,5–bisphosphate and 2–hydroperoxy–3–oxo–pentitol 1,5–bisphosphate. These intermediates would be likely to be cleaved between carbons 2 and 3 by similar mechanisms, that would involve hydration of the carbonyls at carbon 3 to form a gem–diol, and abstraction of a hydrogen ion from one of the gem–diol's hydroxyls to form the alkoxide. The initial products formed in the carboxylase and oxygenase reactions would differ, in that from the former reaction they would be 3–phosphoglycerate (free acid) and its C–2 carbanion (presumably stabilized by the adjacent carboxyl in its aci–form), while from the latter reaction they would be

3–phosphoglycerate and 2–phosphoglycolate (with concomitant formation of a carbonyl from one of the oxygens of the hydroperoxide and cleavage of the oxygen-oxygen bond to form water from the other). Stereospecific protonation of the carbanion of 3–phosphoglycerate would give the two molecules of D–3–phosphoglycerate that are obtained from the oxygenase reaction. Direct evidence for the first two intermediates of the carboxylase reaction has been obtained by chemical quench (Schloss and Lorimer, 1982; Jaworowski et al., 1984; Jaworowski and Rose, 1985; Lorimer et al., 1984; Butcher et al., 1989). Although the scheme presented for the oxygenase reaction is chemically reasonable, it is primarily inferred from the carboxylase reaction without direct evidence from trapping experiments.

Chemical quench experiments are carried out by stopping the carboxylase reaction with acid (Schloss and Lorimer, 1982; Jaworowski et al., 1984; Jaworowski and Rose, 1985; Butcher et al., 1989) or sodium borohydride (Lorimer et al., 1984) employing an apparatus similar to that illustrated in Figure 12. An apparatus is commercially available

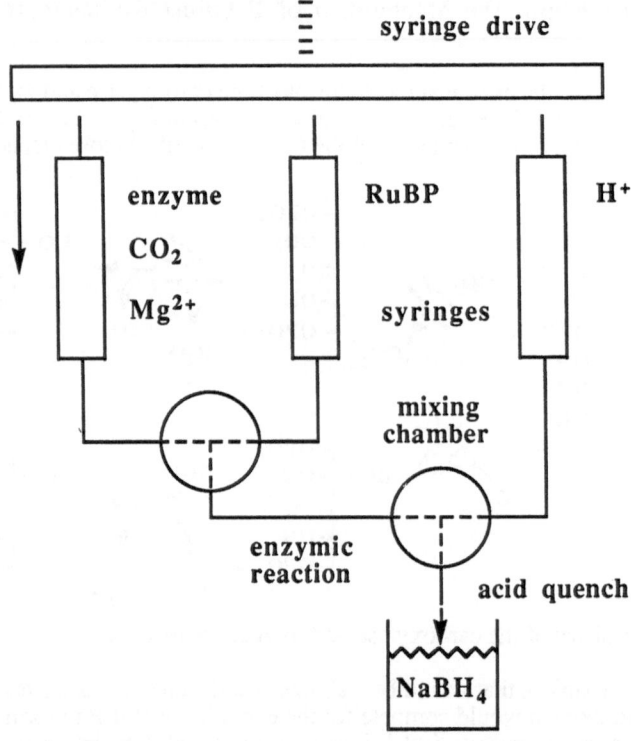

Figure 12. A schematic for a chemical quench apparatus.

from Update Instruments (Madison, Wisconsin, USA) that allows for reproducible mixing times from 6.5 ms to 2.5 s. Mixing times are controlled by driving the contents of syringes, with a sophisticated stepping motor, through an efficient mixing chamber, and varying the length of a small bore tube that connects the enzymic reaction mixture with a second chamber where it is mixed with a quench solution.

The initial enzyme–RuBP complex can be detected by quenching enzymic reaction mixtures directly with sodium borohydride (Lorimer et al., 1984). Whereas reduction of RuBP in the absence of enzyme gives a nearly equal mixture of ribitol 1,5–bisphosphate and arabinitol 1,5–bisphosphate (AraBP), in the presence of excess carboxylase, borohydride selectively attacks the *si* face of RuBP to form AraBP exclusively (Figure 13). This observation provides direct evidence for the enzyme-substrate complex, and suggests that greater than 70 % of the enzyme in steady-state reaction mixtures is in this form.

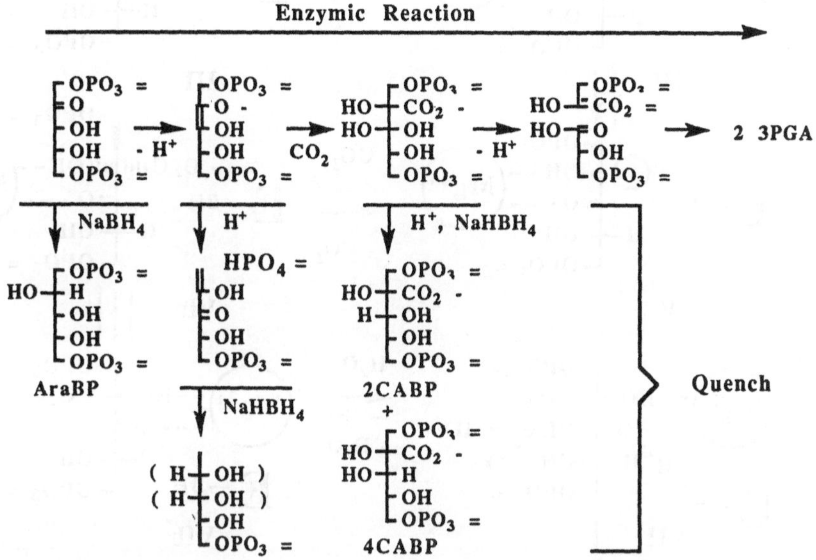

Enzymic Reaction

Figure 13. Trapping intermediates of the carboxylase reaction by chemical quench.

In the second intermediate of the reaction, 2–carboxy-3–oxo-pentitol 1,5–bisphosphate, there is a new stereocenter formed by addition of carbon dioxide to carbon 2. Evidence for this intermediate, and assignment of its stereoconfiguration to 2–carboxy-3–oxo-D–arabinitol bisphosphate (CKABP), was achieved by reduction of the intermediate to a mixture of 2–carboxy-D–arabinitol bisphosphate (2CABP) and 4–carboxy-D–arabinitol bisphosphate (4CABP) with sodium borohydride after freeing it from the enzyme by rapid acid denaturation (Schloss and Lorimer, 1982). If the enzymic reaction mixture is not quenched with acid prior to exposure to sodium borohydride, CKABP is not reduced, in contrast to the results obtained with enzyme-bound RuBP. If [3–^{18}O]RuBP is used in these experiments, the 2CABP and 4CABP obtained retain 50 % of the ^{18}O–label (Lorimer et al., 1984). Similarly, if [3–^{16}O]RuBP and H$_2$^{18}O are employed in quench experiments, the carboxypentitol bisphosphates recovered have half of the ^{18}O–labelling of solvent. These results suggest that most, if not all, of the CKABP

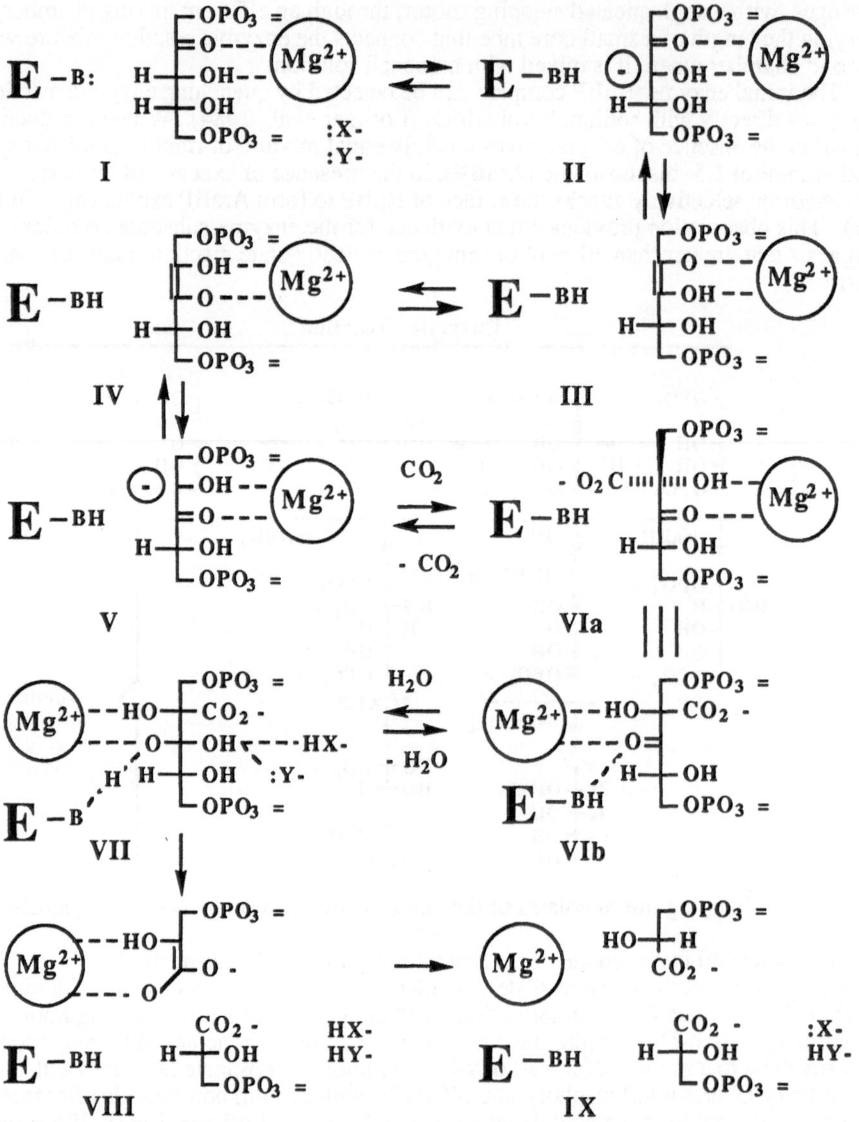

Figure 14. A detailed mechanism for the carboxylase reaction.

trapped in these experiments is present initially after acid quench as the gem-diol. The 2CABP and 4CABP obtained by reduction of the CKABP reaction intermediate can be radiolabeled by use of $^{14}CO_2$ (Schloss and Lorimer, 1982; Lorimer et al., 1984; Jaworowski and Rose, 1985), [1–^{32}P]RuBP (Jaworowski and Rose, 1985; Jaworowski et al., 1984), NaB^{3}H$_4$ (Schloss and Lorimer, 1982), and [5–^{3}H]RuBP (Butcher et al., 1989) with comparable results.

Early evidence for this intermediate had been provided by the exceptionally potent inhibition of the carboxylase by carboxypentitol bisphosphate (Siegel and Lane, 1972; 1973). Its stereochemistry had been inferred by resolution of 2–carboxyribitol bisphosphate (CRBP) and 2CABP (prepared by addition of cyanide to RuBP) and demonstration that potent inhibition of the enzyme was a property of the latter diastereomer (Pierce et al., 1980). Of the four possible stereoisomers of 2–carboxypentitol bisphosphate, with respect to carbons 2 and 3, only 2CABP and 4CABP (but not CRBP or 2–carboxyxylitol bisphosphate, CXBP) are exceptionally potent, essentially irreversible inhibitors of the carboxylase (Pierce et al., 1980; Schloss and Lorimer, 1982; Schloss, 1988). Presumably, these two inhibitors, 2CABP and 4CABP, are potent inhibitors not because of their similarity to CKABP, but to their similarity to its C–3 gem-diol (intermediate VII in Figure 14). In these two inhibitors, either of the two hydroxyl moieties of the gem-diol have been replaced by a hydrogen. Differential binding of these two analogs should, in principal, reflect the differential interaction of the enzyme with the two hydroxyls of CKABP. By use of inactivation kinetics (7.8 X 10^4 and 1.6 X 10^5 M^{-1} s^{-1}) and dual isotope exchange (^{3}H-inhibitor-enzyme complex exchange with exogenous ^{14}C–inhibitor, half-times of 1.8 and 530 days) the respective dissociation constants of 2CABP and 4CABP have been determined to be 190 fM (1.9 X 10^{-13} M) and 28 pM (2.8 X 10^{-11} M) for the spinach carboxylase, respectively (Schloss, 1988). These dissociation constants correspond to a difference in binding energy of 2.9 kcal/mol between 2CABP and 4CABP. The carboxylase seems to bind the hydroxyl of the gem–diol equivalent to that of 2CABP 2.9 kcal/mol more tightly than the hydroxyl corresponding to that of 4CABP. If the enzyme interacts fully with both hydroxyls of the gem-diol, this difference could reflect differences in hydrogen bonding between the two hydroxyls, or by consideration of the detailed reaction scheme presented in Figure 14, the difference in energy between metal coordination of the hydroxyl of 4CABP and a corresponding hydrogen bond for the hydroxyl of 2CABP. Consideration of the stereochemistry of the reaction, a likely cis ene-diol intermediate (III), coordination of oxygens two and three of RuBP and intermediates by metal throughout the reaction sequence (polarizing RuBP to facilitate abstraction of the C–3 proton, stabilization of the ene-diolate and facilitating proton transfer between oxygens 2 and 3, polarization of the carbonyl of CKABP to facilitate attack by water, lowering the pK of the hydroxyl of the gem-diol to facilitate cleavage of the alkoxide, and stabilization of the aci–carboxylate carbanion of 3PGA), and a cis, stepwise addition of CO_2 and water to the same face of the metal-chelate "ring" (oriented toward solvent), would suggest that the hydroxyl of the gem–diol corresponding to the C–3 hydroxyl of 4CABP is metal coordinated. The difference in affinity between 2CABP and 4CABP falls within the range expected for a difference between a magnesium-alcohol and hydrogen bond. Although Mn^{2+} is known to bind the carboxyl and C-2 hydroxyl of 2CABP in the complex of this inhibitor with the manganese-activated enzyme (Miziorko and Sealy, 1984), there is as yet no direct evidence for the coordination of metal by the C-3 hydroxyl of 4CABP, as illustrated in Figure 14. 2CABP and 4CABP do have quite different effects on the ESR spectrum of Mn^{2+}–activated carboxylase (Gutteridge et al., 1985), however, suggesting a likely coordination of the C-3 hydroxyl of one of these two inhibitors. Although the carboxyl of CKABP is not shown to coordinate to metal in Figure 14, this can be accomplished, with two "fused", five membered chelate rings. For simplicity of illustration (three dimensional structures are difficult to portray in two dimensions, and except where bonds are clearly different (VIa) Fischer projections are intended) this known interaction has been omitted.

Direct evidence for the first intermediate of the carboxylase (or oxygenase) reaction was initially provided by Jaworowski et al. (1984). By use of chemical quench, it was shown that an intermediate was obtained upon acid denaturation of enzymic reaction mixtures, which lost the C–1 phosphate of RuBP, even in the presence of sodium borohydride (Jaworowski et al., 1984; Jaworowski and Rose, 1985). This intermediate differs from CKABP, which, although labile, and will lose its C–1 phosphate after acid quench, is stabilized after reduction by sodium borohydride with retention of the C–1 phosphate. The ene-diol of RuBP decomposes to inorganic phosphate and, presumably, a 1–deoxy-2,3–dioxo-pentitol monophosphate, with a half-time of less than 6 ms. But, decomposition of the ene-diol to inorganic phosphate can be prevented by reacting it under acidic conditions with iodine (Jaworowski et al., 1984). The monophosphate product obtained from the ene-diol upon acid quench of enzymic reaction mixtures of the *Rhodospirillum rubrum* carboxylase has also been observed (Butcher et al., 1989). Label was derived from [5-^{3}H]RuBP, and, after acid quench and sodium borohydride reduction, eluted in the same chromatographic position as the monophosphate obtained by base degradation of RuBP, which had been similarly treated.

Evidence for the ene–diol has also been provided by the observed exchange of the C–3 proton of RuBP with solvent under turnover conditions (Saver and Knowles, 1982; Sue and Knowles, 1982a). The extent of labelling of RuBP by ^{3}H$_2$O or loss of label from [3-^{3}H]RuBP, as a function of the extent of the overall reaction to 3PGA, suggested that the step(s) that follows the proton abstraction step is energetically less favorable (slower) (Saver and Knowles, 1982; Sue and Knowles, 1982a). The steps that immediately follow proton abstraction (Figure 14) would be tautomerization of the C–3 carbanion (II) to the ene-diolate (III and IV) and its subsequent tautomerization to the C–2 carbanion (V).

A more detailed examination of the rates of interconversion of trappable intermediates has been carried out by use of [5-^{3}H]RuBP and chromatographic resolution of the products obtained at each reaction time (Butcher et al., 1989). The concentrations of RuBP and *Rhodospirillum rubrum* carboxylase employed in these studies were quite high (0.05 mM and 0.43 mM protomer, respectively), so as to minimize the time required to form the initial enzyme-substrate complex. The rate of RuBP consumption at several bicarbonate concentrations, 2, 10, and 100 mM (pH 8), was first order, and the rate constant for its disappearance was used in simulations of the rate of interconversion of RuBP, ene-diol, CKABP, and 3PGA. The maximum rates for these steps were 17, 550, and 240 s^{-1}, as summarized in Figure 15. Simulated profiles for the levels of substrate, product, and intermediates between 0 to 200 ms are illustrated in Figure 16. The levels of ene-diol and CKABP intermediates are quite low, as RuBP loss can be accounted for almost entirely by the rate of 3PGA appearance. This indicates that the internal steps of the enzymic reaction are much more rapid than the rate of formation of the ene-diol. The levels of the ene-diol and CKABP intermediates at saturating, non-inhibitory concentrations of bicarbonate are approximately 3 and 6 % of total radioactivity at 10 and 20 ms, respectively (Figure 17). These values are similar to those observed by other investigators for the same enzyme (Jaworowski and Rose, 1985; Jaworowski et al., 1984). However, the maximum rates of interconversion, which are somewhat higher than those seen previously, are likely to be closer to rates that are not limited by association of

Rhodospirillum rubrum enzyme

$$E + RuBP \longrightarrow E \cdot ene\text{-}diol \longrightarrow E \cdot CKABP \longrightarrow E + 2\ 3PGA$$

$$\frac{(17\ s^{-1})(CO_2)}{4.3\ mM + CO_2} \qquad \frac{(550\ s^{-1})(CO_2)}{5\ mM + CO_2} \qquad 240\ s^{-1}$$

Figure 15. Rates of interconversion of RuBP, ene–diol, CKABP, and 3PGA determined by chemical quench.

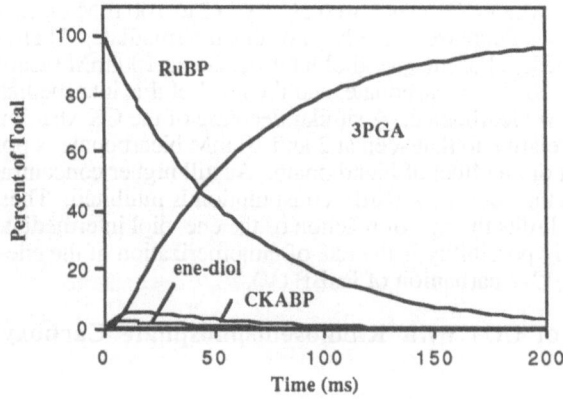

Figure 16. Simulated time course for RuBP, ene–diol, CKABP, and 3PGA under single turnover conditions.

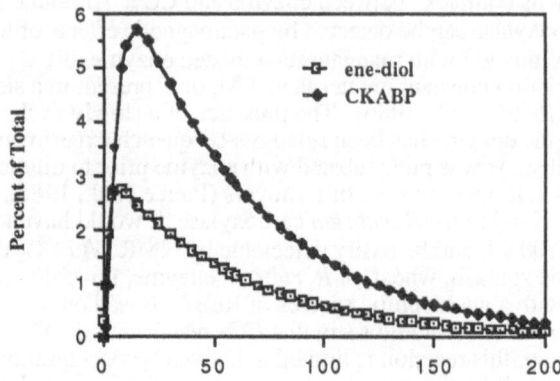

Figure 17. An expanded view of the simulated levels of the ene–diol of RuBP and CKABP obtained under saturating conditions of CO_2.

RuBP and enzyme due to the much higher levels of both used in these studies. The limiting rate, conversion of RuBP to the ene-diol (17 s^{-1}), is only twice the published value for k_{cat} (8.5 s^{-1} at saturating CO_2, Schloss et al., 1979). Assuming that the initial association of RuBP with enzyme has to be at least as fast as $k_{cat}/K_{m\ (RuBP)}$ (2.5×10^5 M^{-1} s^{-1}) the minimum association rate of this substrate at its concentration in these experiments would be 12.5 s^{-1}. Since the calculated value for the minimum association rate is *slower* than the observed first step of the reaction, it is possible that binding limits the observed rate, and the true maximum rate of conversion of RuBP to the ene-diol is somewhat greater than 17 s^{-1}. The most important feature of these experiments, however, is the observed saturation in the rate of interconversion of the ene-diol to CKABP. If the same intermediate that eliminates phosphate on acid quench were to react directly with CO_2, then when CO_2 was raised 50–fold (2 to 100 mM bicarbonate) there should be a corresponding decrease in the level of this intermediate. There is at most a two–fold decrease in the level of the ene–diol intermediate at 100 mM bicarbonate relative to the level observed at 10 mM bicarbonate, and the level of this intermediate is almost identical at 2 and 10 mM bicarbonate. A similar decrease of the CKABP intermediate at 100 mM bicarbonate, relative to that seen at 2 and 10 mM bicarbonate is possibly due to an ionic strength effect or an effect of bicarbonate. At still higher concentrations of bicarbonate (200 mM) the net rate of RuBP consumption is inhibited. These results suggest that some step limits the rate of reaction of the ene–diol intermediate at high CO_2 concentrations. A likely possibility is the rate of tautomerization of the ene-diol (III and IV in Figure 14) to the C–2 carbanion of RuBP (V).

6. The Interaction of CO_2 with Ribulosebisphosphate Carboxylase as Studied by NMR

The exchange of the C–3 proton of RuBP with ^2H_2O is known to compete with CO_2, being suppressed at high concentrations of this substrate with both plant and *R. rubrum* carboxylases (Gutteridge et al., 1984; Pierce et al., 1986). This is likely to be due to consumption of the ene-diol at high CO_2, rather than suppression of exchange due to the formation of a "Michaelis complex" between enzyme and CO_2. Although binding of bicarbonate to the carboxylase can be detected by paramagnetic effects of Mn^{2+} on the ^{13}C–NMR spectrum of this ion with manganese-activated enzyme, if CO_2 binds to the enzyme it has a dissociation constant greater than 5 M, or is present in a slowly exchanging complex (Pierce et al., 1986). The presence of a slowly exchanging complex of substrate CO_2 with the enzyme has been ruled out by quench experiments, where isotopically labeled substrate was preincubated with enzyme prior to dilution into unlabeled substrate for a limited number of turnovers (Pierce et al., 1986). By this criteria, if CO_2 were to bind to the *R. rubrum* carboxylase, it would have to have a dissociation rate of > 100 s^{-1}, and be easily detectable by NMR. As CO_2 does not form a complex with either the spinach, wheat, or *R. rubrum* enzyme, it would seem to react, in a bimolecular manner, with a nucleophilic species of RuBP. Based on the results of chemical quench experiments, it would seem that CO_2 reacts with the C–2 carbanion of RuBP, and that the rate of this reaction is limited at high CO_2 concentrations (where the rate of this step is saturated) by the rate of tautomerization of the ene-diolate. Whether exchange of the C–3 proton of RuBP with solvent proceeds via the C–3 carbanion (II), either of the protonated forms of the ene–diolate (III and IV), or at the stage of the C–2 carbanion (V), remains to be seen.

7. Isotope Effects on the Carboxylase Reaction

Isotope effects on an enzymic reaction can be quite useful in determining the significance of internal steps of the mechanism. Several types of isotope effects have been determined for the carboxylase: solvent deuterium isotope effects with ^2H_2O (Christeller, 1982), carbon isotope effects for $^{13}CO_2$ (discrimination, i. e. effects on V_{max}/Km) (Roeske and O'Leary, 1984; 1985), primary deuterium isotope effects with [3–^{2}H]RuBP (Van Dyk and Schloss, 1986; Sue and Knowles, 1982a), tritium discrimination against incorporation of label from ^3H_2O into C–3 of RuBP (Saver and Knowles, 1982) or into C–2 of the "upper" 3PGA (Hurwitz et al., 1956; Saver and Knowles, 1982), and tritium isotope effects with [3–^{3}H]RuBP (Sue and Knowles, 1982b; Van Dyk and Schloss, 1986; Fiedler et al., 1967). Although the secondary isotope effects have not been determined, ^{18}O–labeled RuBP (labeled at the C-2 carbonyl oxygen or at the C-3 hydroxyl) has been used to demonstrate the retention of the oxygens at C–2 and C–3 in the C-2 hydroxyl and C-1 carboxyl of the "upper" and "lower" 3PGA products, respectively (Lorimer, 1978; Sue and Knowles, 1978).

Isotope effects can be evaluated for any possible reaction mechanism by calculating the dependence of the expression of the full isotope effect for the isotope-sensitive step (^Dk, ^{13}k, or ^{18}k, for deuterium, ^{13}C, or ^{18}O isotope effects, respectively), which is called the intrinsic effect, on the kinetic parameters of the enzymic reaction such as V_{max} and V_{max}/Km (^DV and $^D(V/K)$ for the deuterium effects, respectively; ^{13}V and $^{13}(V/K)$ for the ^{13}C isotope effects, respectively) in terms of the various steps of the mechanism. The constants that define the extent to which the observed isotope effect will be reduced in size relative to the intrinsic isotope effect (the one on the isotope sensitive step of the reaction) are called commitment factors (Northrop, 1977). There are three types of commitment factors, C_f, C_{Vf}, and C_r, that are the forward commitment factor for an isotope effect on V_{max}/Km (or an isotope effect determined by discrimination, e. g. by trace radiolabel discrimination or change in the isotopic composition of a trace label determined by mass spectroscopy), the forward commitment factor for an isotope effect on V_{max}, and the reverse commitment factor (the same for isotope effects on either V_{max} or V_{max}/Km), respectively. The relationship between the observed isotope effect ($^D(V/K)$ and ^DV for a deuterium isotope effect), the intrinsic isotope effect (^Dk for a deuterium isotope effect), the commitment factors, and the isotope effect on the equilibrium of the reaction ($^DK_{eq}$ for a deuterium isotope effect) would be:

$$^D(V/K) = \frac{^Dk + C_f + C_r \ ^DK_{eq}}{1 + C_f + C_r \ ^DK_{eq}}$$

$$^DV = \frac{^Dk + C_{Vf} + C_r \ ^DK_{eq}}{1 + C_{Vf} + C_r \ ^DK_{eq}}$$

The forward commitment factor for an effect on V_{max}/Km (C_f) is the ratio of the rate of the isotope sensitive step to the net rate of release of the intermediate back to substrate. It is a measure of the partitioning between completing the step and aborting the catalytic sequence. If, for a given mechanism, there is a large commitment to catalysis, there will be no isotope effect on V_{max}/K_m, as nearly every molecule that binds to the enzyme will complete the catalytic sequence, with no possibility for discrimination between substrates that differ in isotopic composition. The forward commitment factor for an effect on V_{max} (C_{Vf}), also referred to as a ratio to catalysis (Northrop, 1977), is the ratio of the rate of the isotope sensitive step to the net rate of the overall reaction. For this commitment factor, if the isotope sensitive step is slow (rate limiting) C_{Vf} will be small, and a larger portion of the isotope effect on the isotope sensitive step (^Dk) will be observed on the maximum rate of the overall reaction (^DV). If, by contrast, the isotope sensitive step is very fast relative to the rate of the overall reaction, slowing the rate of that step due to an isotope effect will not be observed on the maximum rate of the reaction $(C_{Vf}$ will be large). The reverse commitment factor (C_r) is the ratio of the rate of the reverse reaction of the isotope sensitive step to the net rate at which the intermediate produced in this step is released from the enzyme as product. If the reversal rate of the isotope sensitive step is very rapid relative to the rate at which the reaction is completed from that point forward, then the step will come to isotopic equilibrium and the isotope effect on that step will not be observed on the overall reaction $(V_{max}$ or $V_{max}/K_m)$. A slow rate for the reverse of the isotope sensitive step, by contrast, ensures that the isotope effect that can be expressed (based on ^Dk and C_f or C_{Vf}) will be maintained for the completion of the reaction (and gives a small value for C_r). Even where the overall enzymic reaction is irreversible, as is the case for the reaction catalyzed by ribulosebisphosphate carboxylase, there is the possibility for a significant reverse commitment factor (C_r), if the isotope sensitive step in the mechanism is reversible.

The carbon isotope effects $(^{13}CO_2)$ for the spinach and *Rhodospirillum rubrum* carboxylases are 1.029 and 1.018, respectively (Roeske and O'Leary, 1984; 1985). When [3–^{2}H]RuBP is used to determine the carbon isotope effect with the spinach enzyme, the somewhat smaller value of 1.021 is obtained (Roeske and O'Leary, 1984). The spinach effect is not sensitive to the concentration of RuBP (0.2 or 1 mM), and varies only slightly with pH, values of 1.026 and 1.030 being obtained at pH 9 and 7, respectively. A somewhat smaller effect has been determined for a ^{13}C-isotope effect for discrimination at the C-3 position of RuBP, 1.008 (Roeske and O'Leary, 1984). These effects are actually quite large, as the intrinsic effect for carboxylation is only expected to be between 1.04–1.07.

By contrast with the carbon isotope effects, the primary deuterium isotope effects with [3–^{2}H]RuBP are strongly dependent on pH for both the spinach and *Rhodospirillum rubrum* carboxylases. For the spinach enzyme, $^D(V/K)$ is pH independent with a value of 2.1, while ^DV varies from 2.5 at high pH (7.5–9) to about 9 at low pH (6) (Van Dyk and Schloss, 1986). For the *R. rubrum* enzyme, both $^D(V/K)$ and ^DV vary with pH, from 1.0 and 1.5 at high pH, to 7 and 5 at low pH, respectively. These changes in the isotope effects correlate with the lower pK (essential base) of two essential ionizations observed in both spinach and *R. rubrum* pH profiles. The observed pK values from the V_{max}/K_m pH profile for the spinach enzyme are both 7.5 (essential base and essential acid); those from the V_{max} profile are 7.1 (essential base) and 8.3 (essential acid); and those for the pK_i $(-log K_i)$ profile of the substrate analog xylulose bisphosphate are 7.2 and 7.8. The observed pK values from the V_{max}/K_m pH profile for the *R. rubrum* enzyme are 7.1 and 8.1; only the lower pK, 7.3, was observed in the V_{max} pH profile (Van Dyk and Schloss, 1986). At pH 8, the isotope effects on V_{max} and V_{max}/K_m of the spinach enzyme were not sensitive to the concentration of CO_2 (1.2 to 100 mM bicarbonate). By use of a mixture of [3–^{3}H]RuBP and [^{14}C]RuBP, a tritium isotope effect of 6.5 was determined for the spinach carboxylase (Van Dyk and Schloss, 1986), similar to a value obtained by use of [3–^{3}H]RuBP and ^{14}CO$_2$ (Fiedler et al., 1967).

The observed change in the V_{max} isotope effect with the spinach carboxylase without a corresponding change in the V_{max}/K_m isotope effect, has some interesting consequences on the mechanism of protonation of the essential enzymic base and the binding of RuBP. In Figure 18, a simple protonation scheme is presented for an enzyme, where protonation of the enzyme renders it inactive, presumably by titrating an essential base. If the enzyme-

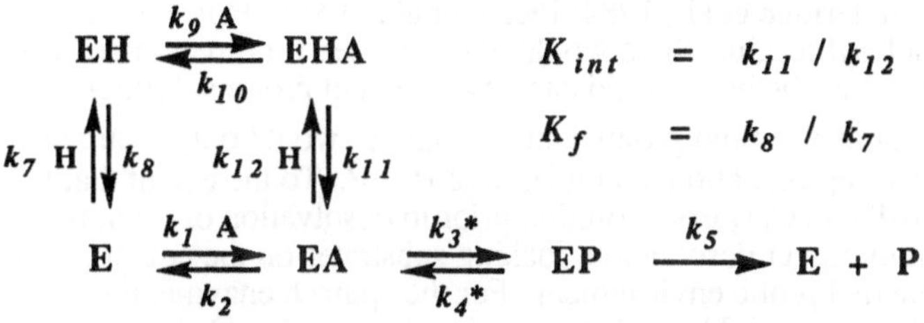

$$EH \underset{k_{10}}{\overset{k_9\,A}{\rightleftharpoons}} EHA \qquad K_{int} = k_{11}\,/\,k_{12}$$

$$k_7\,H \updownarrow k_8 \qquad k_{12}\,H \updownarrow k_{11} \qquad K_f = k_8\,/\,k_7$$

$$E \underset{k_2}{\overset{k_1\,A}{\rightleftharpoons}} EA \underset{k_4{}^*}{\overset{k_3{}^*}{\rightleftharpoons}} EP \overset{k_5}{\longrightarrow} E + P$$

$$C_f = \frac{k_3\,(k_{10}\,/\,k_{11} + 1)}{k_{10}\,H\,/\,K_{int} + k_2\,(k_{10}\,/\,k_{11} + 1)}$$

$$= 0 \qquad H \longrightarrow \infty \qquad \text{unless } k_{10} = 0$$

$$C_{vf} = \frac{k_3\,/\,k_5}{1 + H\,/\,K_{int}} = 0 \qquad H \longrightarrow \infty$$

$$C_r = k_4\,/\,k_5$$

Figure 18. A simple protonation scheme and the commitment factors for this mechanism.

substrate complex cannot be protonated directly (k_{12} is slow, and the "internal pK ", pK_{int}, is very small), then the isotope effects on both V_{max} and V_{max}/K_m would be pH independent. This is certainly not the case for either the spinach or $R.$ $rubrum$ enzymes, so the essential enzymic group in the enzyme-RuBP complexes (at least the initial complex) must be capable of interacting with solvent. If, however, substrate cannot be released from the protonated enzyme-substrate complex (k_{10} is slow), then the isotope effect on V_{max} will remain pH dependent, but the effect on V_{max}/K_m will no longer be pH dependent (C_f becomes equal to k_3/k_2). This is the rather unusual situation observed for the spinach (but not the $R.$ $rubrum$) carboxylase, where the isotope effect on V_{max} gets quite large at low pH, while that on V_{max}/K_m is pH independent. With both the spinach and $R.$ $rubrum$ enzymes the ionization constant with the lower pK observed in the pH profiles, that is a group essential to the enzyme in its free base form, is the one that is responsible for increasing the V_{max} isotope effect for the former and both V_{max} and

V_{max}/K_m effects for the latter enzyme. This behavior would identify the enzymic residue with a pK of about 7.5 in either enzyme as the essential enzymic base that abstracts the C–3 proton of RuBP (Cook and Cleland, 1981a; 1981b). The rather curious behavior of the spinach enzyme may make some sense in this context. After abstraction of the C-3 proton of RuBP, the carboxylase should desolvate the carbanions (C–2 and C–3) and ene–diolates to facilitate the reaction. It should be pointed out, that shielding carbons 2 and 3 from interaction with solvent is not necessarily the same thing as preventing the essential enzymic base from interacting with solvent. This is an important point since the enzyme is known to catalyze exchange of the C–3 proton with solvent (Saver and Knowles, 1982; Gutteridge et al., 1984; Pierce et al., 1986). However, if the essential enzymic base is a lysyl residue, there would not be a need for direct interaction of the residue with solvent, as the base would carry two solvent protons through each catalytic

cycle in the form of its ε–amino group, and "exchange" would only require rotation of the ammonium moiety, reprotonation, and release of RuBP. To the extent that the enzyme needs to bind RuBP and CO_2 from solution prior to desolvation of both, there needs to be some mechanism to trigger desolvation, locking substrates on the enzyme and shielding them from the external protic environment. For the spinach enzyme, this would seem to be protonation of the essential base that normally abstracts the C–3 proton of RuBP. A change in the charge of this group could easily be envisioned to initiate protein conformational changes that would shield RuBP. For the spinach enzyme, protonation of this group would be sufficient to initiate this change, whereas for the *R. rubrum* enzyme protonation of this group by solvent alone would not.

A kinetic mechanism that would account for the known deuterium and ^{13}C isotope effects is illustrated in Figure 19. In this scheme E is enzyme, ER is the enzyme-RuBP complex (I in Figure 14), ER' is the enzyme-ene–diol complex (II, III, IV, and V), EK is the enzyme-CKABP complex (VI), EK' is the enzyme-gem–diol complex (VII), EP'P is the complex containing the 3PGA carbanion and 3PGA (VIII), and EP_2 is the complex with two molecules of 3PGA (IX). The rate constants marked with asterisks would be deuterium sensitive steps, while only k_5 and k_6 would be sensitive to $^{13}CO_2$. Although the deuterium and carbon isotope effects would be consistent with a random kinetic mechanism (but not with ordered binding of substrates followed by proton abstraction), the inability to detect CO_2 binding by the carboxylase suggests a mechanism with direct combination of this substrate with a nucleophilic intermediate (Pierce et al., 1986). However, if CO_2 addition comes after abstraction of the C–3 proton of RuBP, then is not possible to explain the smaller ^{13}C isotope obtained with [3–^2H]RuBP (Roeske and O'Leary, 1984). To accommodate this observation in a "Theorell-Chance" type of mechanism, the deuterium abstracted from RuBP must be conserved and be involved in a later step of the mechanism. The only step where this would be likely would be the hydration and cleavage of CKABP (see Figure 14). As the proton abstracted from RuBP does not get transferred to the "upper" 3PGA (Fiedler et al., 1967; Sue and Knowles, 1982a; Pierce et al., 1986), the group responsible for initial proton abstraction is unlikely to protonate the carbanion of 3PGA. A reversible step after the addition of CO_2 is required to accurately predict the level of the six carbon reaction intermediate seen in chemical quench experiments (Van Dyk and Schloss, 1986). As the steps after cleavage of the gem–diol of CKABP should be essentially irreversible (Figure 14), the reversible step is probably hydration/dehydration of CKABP. Synthesis of the gem–diol of CKABP from two molecules of 3PGA is not only unlikely based on the extremely unfavorable chemistry, but formation of the needed carbanion of 3PGA is not likely to happen, as the carboxylase will not catalyze exchange of the C–2 proton of 3PGA with solvent (Fiedler et al., 1967; Hurwitz et al., 1956). This would increase the likelihood that carboxylation and hydration are discrete steps, and make a concerted addition of water to the ene–diol of RuBP, with concomitant reaction of the delocalized electrons with CO_2, less probable.

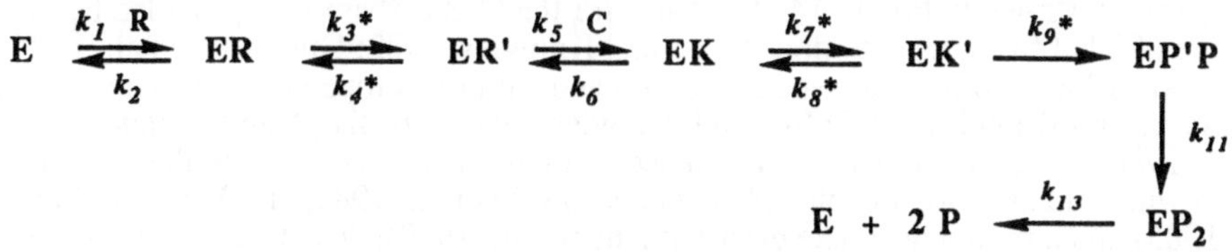

$$E \underset{k_2}{\overset{k_1 \quad R}{\rightleftharpoons}} ER \underset{k_4{}^*}{\overset{k_3{}^*}{\rightleftharpoons}} ER' \underset{k_6}{\overset{k_5 \quad C}{\rightleftharpoons}} EK \underset{k_8{}^*}{\overset{k_7{}^*}{\rightleftharpoons}} EK' \overset{k_9{}^*}{\longrightarrow} EP'P$$

$$\downarrow k_{11}$$

$$E + 2P \overset{k_{13}}{\longleftarrow} EP_2$$

^2H Isotope Effect

$$C_f = k_3 / k_2$$

$$C_{Vf} = \frac{k_3}{k_{13}} + \frac{k_3}{k_{11}} + \frac{k_3}{k_5 C} + {}^D k_9 \left(\frac{k_3}{k_9}\right) \left[1 + \frac{k_8}{k_7} {}^D K_{eq7} \left(1 + \frac{k_6}{k_5 C}\right)\right]$$

$$+ {}^D k_7 \left(\frac{k_3}{k_7}\right) \left(1 + \frac{k_6}{k_5 C}\right)$$

$$C_r = \frac{k_4}{k_5 C} \left[1 + \left(\frac{k_6}{k_7}\right) \left({}^D k_7 + \frac{k_8}{k_9} {}^D K_{eq7} {}^D k_9\right)\right]$$

^{13}C Isotope Effect

$$C_f = 0$$

$$C_{Vf} = \text{no isotope effect on } V_{max}$$

$$C_r = \frac{k_6 (k_8 + k_9)}{k_7 k_9}$$

Figure 19. A kinetic mechanism for ribulose bisphosphate carboxylase and the commitment factors for its ^2H and ^{13}C isotope effects.

8. A Summary of the Mechanism of Ribulosebisphosphate Carboxylase

There is evidence for each of the steps of the detailed mechanism of ribulosebisphosphate carboxylase illustrated in Figure 14. Evidence for the C–3 carbanion is provided by the exchange of label from that position (Saver and Knowles, 1982; Gutteridge et al., 1984; Pierce et al., 1986) and the observed deuterium and tritium isotope effects (Fiedler et al., 1967; Saver and Knowles, 1982; Sue and Knowles, 1982a; 1982b; Van Dyk and Schloss, 1986). The ene–diolates can be detected after chemical quench of the enzymic reaction with acid as inorganic phosphate (Jaworowski et al., 1984; and Jaworowski and Rose, 1985) and a monophosphate ester (Butcher et al., 1989). The C–2 carbanion can be inferred from the reaction of carbon 2 with carbon dioxide (Müllhofer and Rose, 1965) and the expression of a sizable carbon isotope effect for this step (Roeske and O'Leary, 1984; 1985). Also, a rate limiting step is required between the intermediate that eliminates phosphate on acid quench (the ene–diol of RuBP) and the formation of CKABP, to account for the maximum rate observed in the interconversion of these two intermediates at high CO_2 (Butcher et al., 1989). This step is likely to be tautomerization of the ene–diolate to the C–2 carbanion, as it is known that tautomerization of the ene–diolate in solution is a slow process relative to its rate of decomposition (half–time of less than 6 ms, Jaworowski and Rose, 1984). Trapping of the gem–diol of CKABP can be accomplished by chemical quench (Schloss and Lorimer, 1982; Jaworowski et al., 1984; Jaworowski and Rose, 1985) and demonstrated to be predominantly in its hydrated form at the time of acid quench by ^{18}O–labeling with $[3-^{18}O]RuBP$ or $H_2^{18}O$ (but not by $H_2^{18}O$ in the quench solution) (Lorimer et al., 1984). Although the gem–diol predominates in steady–state (or in solution, see Lorimer et al., 1984), the unhydrated form of CKABP is required to reconcile the results of the levels of the gem–diol obtained by chemical quench with a Theorell–Chance mechanism and the deuterium isotope results, and the interconversion of these two intermediates must be a reversible process (Van Dyk and Schloss, 1986). The carbanion of the "upper" 3PGA is consistent with half of the 3PGA produced in the enzymic reaction containing a solvent-derived hydrogen at C–2 (Fiedler et al., 1967; Sue and Knowles, 1982a; Pierce et al, 1986; Gutteridge et al., 1984). Although the carbanion of 3PGA cannot be detected by chemical quench, it must have a significant lifetime in the reaction mechanism, as evidenced by the discrimination

against incorporation of heavier hydrogen isotopes into the α–position (Hurwitz et al., 1956; Saver and Knowles, 1982) and the stereospecificity of the protonation [only D–3PGA is obtained from the "upper" 3PGA (Jaworowski et al., 1984)].

References

Andrews, T. J., M. R. Badger, and G. H. Lorimer (1975) 'Factors affecting interconversion between kinetic forms of ribulose diphosphate carboxylase-oxygenase from spinach', Arch. Biochem. Biophys. 171, 93–103.

Andrews, T. J., G. H. Lorimer, and N. E. Tolbert (1973) 'Ribulose diphosphate oxygenase. I. Synthesis of phosphoglycolate by fraction–1 protein of leaves', Biochemistry 12, 11–18.

Ayres, D. C. (1966) Carbanions in Synthesis, Oldbourne Press, London.

Badger, M. R., and G. H. Lorimer (1976) 'Activation of ribulose-1,5–bisphosphate oxygenase', Arch. Biochem. Biophys. 175, 723–729.

Bates, R. B., and C. A. Ogle (1983) Carbanion Chemistry, Springer-Verlag, Berlin.

Bowes, G., and W. L. Ogren (1972) 'Oxygen inhibition and other properties of soybean ribulose 1,5–diphosphate carboxylase', J. Biol. Chem. 247, 2171–2176.

Bowes, G., W. L. Ogren, and R. H. Hageman (1971) 'Phosphoglycolate production catalyzed by ribulose diphosphate carboxylase', Biochem. Biophys. Res. Commun. 45, 716–722.

Butcher, K. A., F. K. Fong, and J. V. Schloss (1989) 'A single turnover kinetic analysis of ribulosebisphosphate carboxylase', the present reference.

Christeller, J. T. (1982) 'Effects of divalent cations on the activity of ribulosebisphosphate carboxylase: interactions with pH and with D_2O as solvent', Arch. Biochem. Biophys. 217, 485–490.

Cook, P. F., and W. W. Cleland (1981a) 'pH variation of isotope effects in enzyme-catalyzed reactions. 1. Isotope- and pH-dependent steps the same', Biochemistry 20, 1797–1805.

Cook, P. F., and W. W. Cleland (1981b) 'pH variation of isotope effects in enzyme-catalyzed reactions. 2. Isotope-dependent step not pH dependent. Kinetic mechanism of alcohol dehydrogenase', Biochemistry 20, 1805–1816.

Cooper, T. G., D. Filmer, M. Wishnick, and M. D. Lane (1969) 'The active species of "$CO2$" Utilized by ribulose diphosphate carboxylase', J. Biol. Chem. 244, 1081-1083.

Donnelly, M. I., C. D. Stringer, and F. C. Hartman (1983) 'Characterization of the activator site of *Rhodospirillum rubrum* ribulosebisphosphate carboxylase/oxygenase', Biochemistry 22, 4346–4352.

Fiedler, F., G. Müllhofer, A. Trebst, and I. A. Rose (1967) 'Mechanism of ribulose–diphosphate carboxydismutase reaction', Eur. J. Biochem. 1, 395–399.

Gutteridge, S., M. A. J. Parry, C. N. G. Schmidt, and J. Feeney (1984) 'An investigation of ribulosebisphosphate carboxylase activity by high resolution ^{1}H NMR', FEBS Lett. 170, 355–359.

Gutteridge, S., M. Parry, G. Schmidt, and G. Lorimer (1985) 'Nature of the activation and active site of ribulose bisphosphate carboxylase from the electron paramagnetic resonance transition state enzyme-manganese complexes', Biochem. Soc. Trans. 13, 629–631.

Hurwitz, J., W. B. Jakoby, and B. L. Horecker (1956) 'On the mechanism of CO_2 fixation leading to phosphoglyceric acid', Biochim. Biophys. Acta 22, 194–195.

Jakoby, W. B., D. O. Brummond, and S. Ochoa (1956) 'Formation of 3–phospho-glyceric acid by carbon dioxide fixation with spinach leaf enzymes', J. Biol. Chem. 218, 811-822.

Jaworowski, A., F. C. Hartman, and I. A. Rose (1984) 'Intermediates in the ribulose-1,5–bisphosphate carboxylase reaction', J. Biol. Chem. 259, 6783–6789.

Jaworowski, A., and I. A. Rose (1985) 'Partition kinetics of ribulose-1,5–bisphosphate carboxylase from *Rhodospirillum rubrum*', J. Biol. Chem. 260, 944–948.

Jencks, W. P. (1969) Catalysis in Chemistry and Enzymology, McGraw-Hill, New York.

Laing, W. A., and J. T. Christeller (1976) 'A model for the kinetics of activation and catalysis of ribulose 1,5–bisphosphate carboxylase', Biochem. J. 159, 563–570.

Lorimer, G. H. (1978) 'Retention of the oxygen atoms at carbon–2 and carbon–3 during the carboxylation of ribulose 1,5–bisphosphate', Eur. J. Biochem. 89, 43–50.

Lorimer, G. H. (1979) 'Evidence for the existence of discrete activator and substrate sites for CO_2 on ribulose-1,5–bisphosphate carboxylase', J. Biol. Chem. 254, 5599–5601.

Lorimer, G. H. (1981) 'Ribulose bisphosphate carboxylase: amino-acid sequence of a peptide bearing the activator carbon dioxide', Biochemistry 20, 1236–1240.

344

Lorimer, G. H., T. J. Andrews, and N. E. Tolbert (1973) 'Ribulose diphosphate oxygenase. II. Further proof of reaction products and mechanism of action', Biochemistry 12, 18–23.

Lorimer, G. H., M. R. Badger, and T. J. Andrews (1976) 'The activation of ribulose-1,5–bisphosphate carboxylase by carbon dioxide and magnesium ions. Equilibria, kinetics, a suggested mechanism, and physiological implications', Biochemistry 15, 529–536.

Lorimer, G. H., J. Pierce, S. Gutteridge, and J. V. Schloss (1984) 'Some mechanistic aspects of ribulose bisphosphate carboxylase', in C. Sybesma (ed.), Advances in Photosynthesis Research, Vol. III, Nijhoff/Junk Publishers, The Hague, pp. 725–734.

Miziorko, H. M., and R. C. Sealy (1984) 'ESR of ribulosebisphosphate carboxylase: identification of activator cation ligands', Biochemistry 23, 479–485.

Müllhofer, G., and I. A. Rose (1965) 'The position of carbon-carbon bond cleavage in the ribulose diphosphate carboxydismutase reaction', J. Biol. Chem. 240, 1341–1346.

Northrop (1977) 'Determining the absolute magnitude of hydrogen isotope effects', in W. W. Cleland, M. H. O'Leary, and D. B. Northrop (eds.), Isotope Effects on Enzyme–Catalyzed Reactions, University Park Press, Baltimore, pp. 122–152.

Ogren, W. L., and G. Bowes (1971) 'Ribulose diphosphate carboxylase regulates soybean photorespiration', Nature (London), New Biol. 230, 159–160.

Perrin, D. D., and B. Dempsey (1974) Buffers for pH and Metal Ion Control, Chapman and Hall Ltd., London.

Pierce, J., G. H. Lorimer, and G. S. Reddy (1986) 'Kinetic mechanism of ribulosebisphosphate carboxylase: evidence for an ordered, sequential reaction', Biochemistry 25, 1636–1644.

Pierce, J., N. E. Tolbert, and R. Barker (1980) 'Interaction of ribulosebisphosphate carboxylase/oxygenase with transition–state analogues', Biochemistry 19, 934–942.

Roeske, C. A., and M. H. O'Leary (1984) 'Carbon isotope effects on the enzyme-catalyzed carboxylation of ribulose bisphosphate', Biochemistry 23, 6275–6284.

Roeske, C. A., and M. H. O'Leary (1985) 'Carbon isotope effect on carboxylation of ribulose bisphosphate catalyzed by ribulosebisphosphate carboxylase from *Rhodospirillum rubrum*', Biochemistry 24, 1603–1607.

Saver, B. G., and J. R. Knowles (1982) 'Ribulose-1,5–bisphosphate carboxylase: enzyme catalyzed appearance of solvent tritium at carbon 3 of ribulose 1,5–bisphosphate reisolated after partial reaction', Biochemistry 21, 5398–5403.

Schloss, J. V. (1988) 'Comparative affinities of the epimeric reaction-intermediate analogs 2– and 4–carboxy-D–arabinitol 1,5–bisphosphate for spinach ribulose 1,5–bisphosphate carboxylase', J. Biol. Chem. 263, 4145–4150.

Schloss, J. V., and G. H. Lorimer (1982) 'The stereochemical course of ribulosebisphosphate carboxylase', J. Biol. Chem. 257, 4691–4694.

Schloss, J. V., E. F. Phares, M. V. Long, I. L. Norton, C. D. Stringer, and F. C. Hartman (1979) 'Isolation, characterization, and crystallization of ribulosebisphosphate carboxylase from autotrophically grown *Rhodospirillum rubrum*', J. Bacteriol. 137, 490–501.

Siegel, M. I., and M. D. Lane (1972) 'Interaction of ribulose diphosphate carboxylase with 2–carboxyribitol diphosphate, an analogue of the proposed carboxylated intermediate in the CO_2 fixation reaction', Biochem. Biophys. Res. Commun. 48, 508–516.

Siegel, M. I., and M. D. Lane (1973) 'Chemical and enzymatic evidence for the participation of a 2–carboxy-3–ketoribitol-1,5–diphosphate intermediate in the carboxylation of ribulose 1,5–diphosphate', J. Biol. Chem. 248, 5486–5498.

Stowell, J. C. (1979) Carbanions in Organic Synthesis, John Wiley & Sons, New York.

Sue, J. M., and J. R. Knowles (1978) 'Retention of the oxygens at C–2 and C–3 of D–ribulose 1,5–bisphosphate in the reaction catalyzed by ribulose–1,5–bisphosphate carboxylase', Biochemistry 17, 4041–4044.

Sue, J. M., and J. R. Knowles (1982a) 'Ribulose-1,5–bisphosphate carboxylase: fate of the tritium label in [3–^{3}H]ribulose 1,5–bisphosphate during the enzyme-catalyzed reaction', Biochemistry 21, 5404–5410.

Sue, J. M., and J. R. Knowles (1982b) 'Ribulose-1,5–bisphosphate carboxylase: primary deuterium kinetic isotope effect using [3–^{2}H]ribulose 1,5–bisphosphate', Biochemistry 21, 5410–5414.

Tolbert, N. E. (1973) 'Glycolate biosynthesis', Curr. Top. Cell. Regul. 7, 21–50.

Umbreit, W. W., R. H. Burris, J. F. Stauffer (1964) Manometric Techniques, Burgess Publishing Company, Minneapolis.

Van Dyk, D. E., and J. V. Schloss (1986) 'Deuterium isotope effects in the carboxylase reaction of ribulose-1,5–bisphosphate carboxylase/oxygenase', Biochemistry 25, 5145–5156.

Weast, R. C. (ed.) (1971) CRC Handbook of Chemistry and Physics, 52nd Edition, The Chemical Rubber Co., Cleveland.

Weissbach, A., B. L. Horecker, and J. Hurwitz (1956) 'The enzymatic formation of phosphoglyceric acid from ribulose diphosphate and carbon dioxide', J. Biol. Chem. 218, 795–810.

RUBISCO:[1]   ACTIVE-SITE   CHARACTERIZATION   AND   MECHANISTIC IMPLICATIONS

FRED C. HARTMAN
Protein Engineering Program
Biology Division
Oak Ridge National Laboratory
P. O. Box 2009
Oak Ridge, Tennessee USA 37831-8077

ABSTRACT.   Diverse chemical methods and comparative sequence analyses have identified Lys-166, Lys-329, and Glu-48 of Rubisco as active-site residues and have suggested that these residues participate in catalysis.   These conclusions have been validated by site-directed mutagenesis; furthermore, hybridization of distinct site-directed mutant proteins demonstrates an intersubunit location of the active site in which side chains from each subunit are required.   The ability of mutant proteins, devoid of overall carboxylase activity, to catalyze certain partial reactions suggests that Lys-166 is the base that enolizes ribulose-P_2 and that Lys-329 facilitates addition of gaseous substrate to the enediol.

Introduction

Active-site characterization represents one important facet of elucidating the mechanism of Rubisco (EC 4.1.1.39) and of ultimately evaluating the feasibility of altering the carboxylase/oxygenase ratio.   Literature surveys reveal that virtually every type of functional group found in proteins has been assigned, through chemical modification, to the active site of the enzyme and described as essential [see citations in ref. 1].   This unsettling situation is not so much a testimony to the fallibility of chemical modification *per se* but rather a reflection of some conclusions that are not totally justified by experimental observations.   In many cases, reagent specificity has not been documented, observed lack of reagent specificity has been ignored, direct effects of modification of essential residues have not been distinguished from indirect effects via conformational changes, and inactivations have not been correlated with the modification of particular residues.   Thus, despite the volume of chemical modification studies of the carboxylase, few have conclusively revealed structural and functional features of the active site.

A notable exception to the enumerated weaknesses of some chemical modification studies is provided by the use of diazomethane to identify the lysyl ε-amino group of the

---

[1]Abbreviations: Rubisco, D-ribulose-1,5-bisphosphate carboxylase/oxygenase, ribulose-P_2, D-ribulose-1,5-bisphosphate; carboxyarabinitol-P_2, 2-carboxyarabinitol-1,5-bisphosphate.

347

M. Aresta and J. V. Schloss (eds.),
Enzymatic and Model Carboxylation and Reduction Reactions for Carbon Dioxide Utilization, 347–365.
© 1990 *Kluwer Academic Publishers.*

carboxylase that undergoes carbamylation (*i.e.* condensation with CO_2) in an obligatory activation process [2]. The carbamate (prepared with $^{14}CO_2$) can be trapped with the transition-state analogue 2-carboxyarabinitol-P_2, thereby permitting esterification of the carbamate to form a stable derivative [3]. Although the esterification is not specific, only one site is labeled with a radioactive marker. The carbamylated lysyl residue, which occupies position 201 in the spinach enzyme [3,4] and position 191 in the *Rhodospirillum rubrum* enzyme [5,6], provides one ligand for the catalytically essential divalent metal ion.

Among the numerous studies in which general protein reagents have been applied to active-site characterization of ribulose-P_2 carboxylase, the use of diethyl pyrocarbonate is instructive in that the inactivated enzyme was sufficiently well-characterized to implicate His-298 of the spinach enzyme as an active-site residue [7]. Although assignment of His-298 to the active-site domain is confirmed by crystallography [8], site-directed mutagenesis, in which the corresponding histidine (His-291) of the *R. rubrum* enzyme was replaced by alanine with retention of considerable carboxylase activity, had shown that the histidyl residue is not required for catalysis [9].

An emphasis has been placed on affinity labels to probe the active site of Rubisco because of their inherently high degree of specificity, which simplifies correlation of inactivation with modification. Collective data provided by several different affinity labels (see Fig. 1) have revealed the presence of two distinct lysyl residues at the active site [for a review, see ref. 10]. In the hetero-hexadecameric [11] enzyme from spinach, these lysines occupy positions 175 and 334 of the large subunit; and in the homo-dimeric [6]

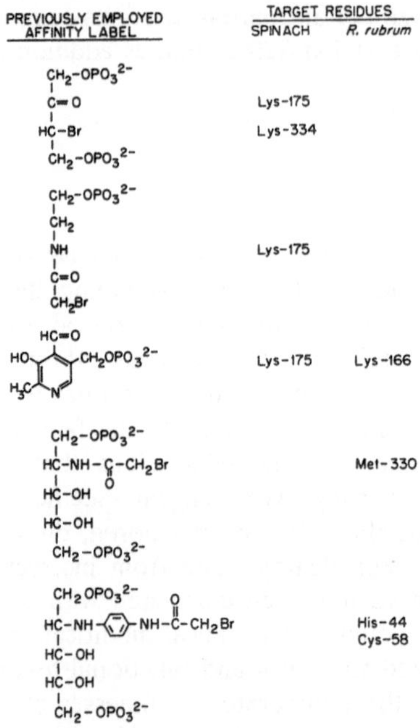

Figure 1. Affinity labels for Rubisco and the sites of reaction

enzyme from *R. rubrum*, they occupy positions 166 and 329. The absolute species invariance of both lysines and the sequence homologies of adjacent regions among the carboxylases from evolutionarily diverse organisms (*e.g.* spinach and *R. rubrum*) suggest that the two active-site lysines are directly or indirectly required for proper enzyme functions:

175
(spinach)   T-I-K-P-K-L-G-L-S
166
(*R. rubrum*)   I-I-K-P-K-L-G-L-R

334
(spinach)   G-K-L-E-G-E
329
(*R. rubrum*)   G-K-M-E-G-E

These homologies are especially striking in view of the total sequence homology of only 31% [6].

Another affinity label (see Fig. 1) suggests that the active-site domain also includes a region of the polypeptide near the NH$_2$-terminus. This reagent labels both His-44 and Cys-58 of the *R. rubrum* enzyme in a mutually exclusive fashion [12]; neither residue can be essential since they are not species invariant:

(spinach)   G-A-A-V-A-A-E-S-S-T-G-T-W-T-T-V-W-T
44   58
(*R. rubrum*)   A-A-H-F-A-A-E-S-S-T-G-T-N-V-E-V-C-T

However, the high degree of homology within this region indicates a prominent role for one or more of the conserved residues.

Judicious application of chemical modification, particularly with affinity labels, followed by cautious interpretation of results can provide compelling evidence of the identity of active-site residues but rarely can reveal their function. This present report describes three approaches being used to bridge the gap between assigning residues to the active-site and defining their function: (a) a determination of their pK_a values based on the pH-dependencies of chemical modifications; (b) a determination of their proximity based on covalent cross-linking with bifunctional reagents; and (c) a determination of the consequences of their selective replacement with other amino acids by site-directed mutagenesis. Each of these approaches was conceptualized and executed prior to publication of any high-resolution crystallographic data. The last section of this report considers the compatibility of chemical, mutagenesis, and structural studies now that the 3D structure of Rubisco has been elucidated [8,13-17].

pK_a Values of Active-Site Lysines.

During the screening of lysine-selective reagents for their ability to preferentially modify the active-site lysines of Rubisco, 2,4,6-trinitrobenzenesulfonate was observed to arylate Lys-166 of the *R. rubrum* enzyme and Lys-334 of the spinach enzyme with a very high degree of selectivity [18]. Hence, a single reagent, applied to the carboxylase from two different organisms, can provide the pK_a values for both active-site lysines. Based on the pH-dependencies of arylation, the following pK_a values and intrinsic reactivities (k_o) have been obtained: acetyllysine, pK_a = 10.8 and k_o =1250 $M^{-1}min^{-1}$; Lys-166 of the activated *R. rubrum* enzyme, pK_a = 7.9 and k_o = 670 $M^{-1}min^{-1}$; and Lys-334 of the activated spinach enzyme, pK_a = 9.0 and k_o = 4500 $M^{-1}min^{-1}$. Particularly noteworthy are the enhanced nucleophilicities of the two protein ϵ-amino groups compared to acetyllysine, despite their stronger acidities.

In addressing the question of a role in substrate binding versus a role in catalysis for the two active-site lysine residues, several observations argue against the former. The enhanced nucleophilicities of both lysine residues are more consistent with a catalytic involvement than with a function in binding. The incomplete protection of both enzymes against inactivation by saturating levels of carboxyribitol bisphosphate (a simple competitive inhibitor) is inconsistent with those lysine residues, which are targets for arylation, forming salt linkages with phosphate groups of ribulose-P_2. The inactivated spinach carboxylase, in which Lys-334 has been arylated, is still able to form the quaternary complex with CO_2, Mg^{2+}, and carboxyarabinitol-P_2. With respect to Lys-166 in the *R. rubrum* carboxylase, its extreme acidity (pK_a = 7.9) appears incompatible with effective utilization as a phosphate-binding site.

The initial step in both the carboxylation and oxygenation pathways as catalyzed by Rubisco is abstraction of the C-3 proton of ribulose-P_2 to form an enediol intermediate (Fig. 2) [for a review, see ref. 19]. The conjugate acid of the base that abstracts this proton has a pK_a of 7.5 based on the pH-dependencies of both V_{max}/K_m and the deuterium isotope effect with [3-^{2}H]ribulose-P_2 as substrate [1]. This pK_a is insensitive to the dielectric constant of the solvent, suggestive of a cationic acid, *i.e.* a protonated lysyl or histidyl residue. Given its enhanced nucleophilicity and unusually low pK_a, Lys-166 is an attractive candidate for the base that facilitates enolization of ribulose-P_2.

Although the unusual acidity and nucleophilicity can account for the selectivity of the arylating reagent for Lys-166 and Lys-334, their similar reactivities at pH 8.0 raise the question as to why only one of them is modified in each species of carboxylase examined. A likely explanation is that the two lysines are juxtaposed within the catalytic site so that derivatization of both is precluded on steric grounds.

Interresidue Distances Based on Chemical Cross-Links

To challenge the postulate of proximity of the two active-site lysines, the reactions of the *R. rubrum* enzyme with several chemical cross-linking agents have been explored [20]. One reagent that inactivates the carboxylase with a high degree of specificity is 4,4'-diisothiocyano-2,2'-disulfonate stilbene, which spans 12 Å.

R. rubrum Rubisco is a dimer so that any cross-links introduced could be intrasubunit, intersubunit, or intermolecular. Thus, the inactivated enzyme, subsequent to

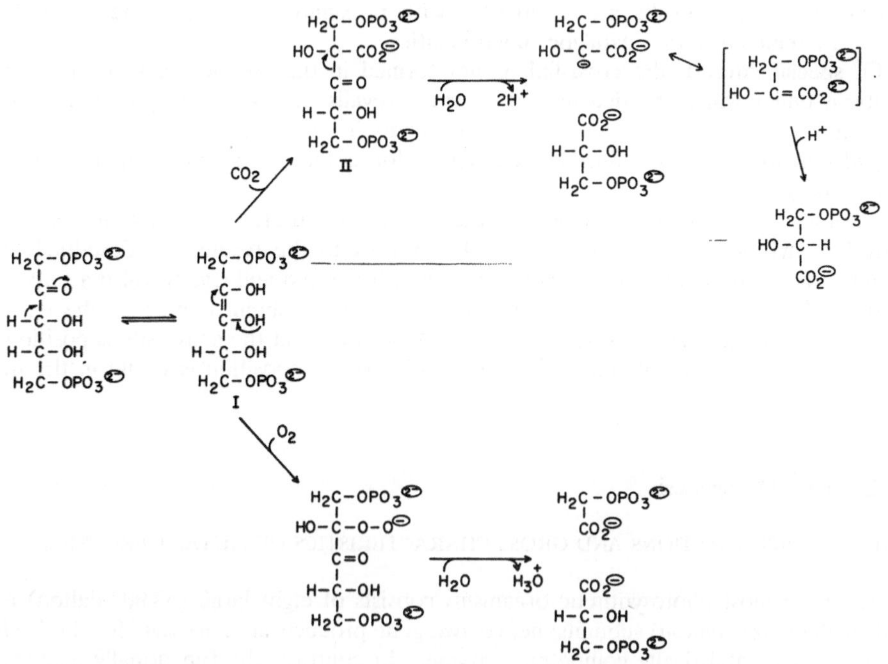

Figure 2. Carboxylation and oxygenation pathways as catalyzed by Rubisco.

carboxymethylation of sulfhydryls, was subjected to gel filtration in the presence of urea, thereby permitting the isolation of the monomeric polypeptide. Samples of the isolated monomeric fraction were digested with trypsin; these digests were inspected by HPLC to ascertain the degree of reagent specificity. Based on the A_{342nm} (the major visible absorbancy band of the reagent), over one-half of the incorporated reagent was associated with a single peptide. This peptide was purified by preparative ion-exchange chromatography on DEAE-cellulose followed by gel filtration on Sephadex G-25.

The amino acid composition and sequence of the purified peptide demonstrated that it is comprised of two chains encompassing position 149-168 and 314-337 of the original protein subunit [see ref. 6 for complete primary structure]:

166

NH₂-V-L-G-R-P-E-V-D-G-G-L-V-V-G-T-I-I-K-P-K-COOH

NH₂-L-Q-G-A-S-G-I-H-T-G-T-M-G-F-G-K-M-E-G-E-S-S-D-R-COOH

329

The location of the covalent cross-link between Lys-166 and Lys-329 was revealed by the complete absence of phenylthiohydantoin-lysine at their respective positions during Edman degradation. These results are compatible with proximity of Lys-166 and Lys-329 as dictated by their purported catalytic functionalities.

Of special interest, the cross-link is not formed in the absence of CO_2 and Mg^{2+}, conditions under which the deactivated enzyme prevails. A likely interpretation of this observation is that the two lysines are farther apart in the deactivated enzyme and that the conformational change induced by carbamate formation is crucial to proper alignment of catalytic groups.

Another cross-linking reagent that proved useful is 4,4'-difluoro-3,3'-dinitrodiphenylsulfone, which spans 9 Å. In contrast to the results just described, this reagent forms a cross-link between subunits in which Lys-166 and Cys-58 (also a target of an affinity label) participate [21]. This finding of an intersubunit contact in the vicinity of a well-documented essential residue (Lys-166) indicates that the active site is positioned at an interface between subunits. Confirmation of this supposition is found in the next section.

Site-Directed Mutagenesis

GENERAL CONSIDERATIONS AND GROSS CHARACTERISTICS OF MUTANT PROTEINS

Rubisco from most photosynthetic organisms consists of eight large (53,000-dalton) and eight small (14,000-dalton) subunits; hence, two gene products are necessary for the *in vivo* formation of a catalytically competent enzyme. In contrast, the functionally analogous enzyme from the photosynthetic bacterium *R. rubrum* is a dimer of 50,500-dalton subunits, altogether lacking small subunits, and thus an attractive candidate for site-directed mutagenesis. Its originally cloned gene was expressed as a fusion protein [22], but a reconstruction [23] gave rise to the wild-type enzyme and has thus been used to generate the mutant proteins described in this article. Mutant proteins were constructed by either the single-primer extension method utilizing an appropriate single-stranded M13 vector (Fig. 3) [9,24,25] or by a technique applied directly to an expression plasmid [26,27]. In most cases, *E. coli* JM107 or *E. coli* DH1 was used as the expression host; mutant proteins were purified to apparent homogeneity from whole-cell extracts by various chromatographic procedures.

Although numerous residues of Rubisco have been substituted by site-directed mutagenesis, an emphasis has been placed on Lys-166, Lys-329, and Glu-48 with the goal of illuminating their functions. These two lysines were selected for substitution because of the extensive data already presented that had implicated catalytic roles. The reasons for inspecting Glu-48 were less compelling. Glu-48 represents the only acid/base group, and hence a candidate for participation in catalysis, present within the conserved region flanked by His-44 and Cys-58, which are targets of an affinity label as noted earlier.

Because active-site residues were singled out for substitution, drastic effects on catalysis were anticipated. The general strategy then was to distinguish among the major likely causes of catalytic deficiency: (a) improper folding of the polypeptide, (b) failure of subunits to associate to form a dimer, (c) inability to undergo carbamylation as required for activation, (d) failure to bind substrates, or (e) loss of a group that participates directly in catalysis. Simple gel filtration reveals whether the mutant is a monomer or dimer. Because the transition-state analogue carboxyarabinitol-P_2 binds tightly only to the

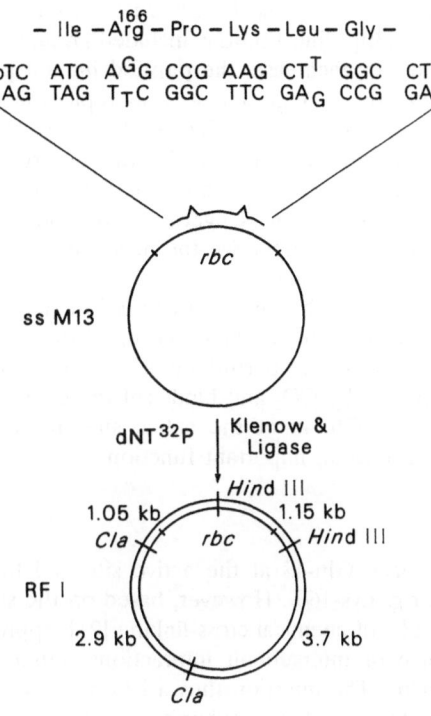

Figure 3. Oligonucleotide-directed mutagenesis of the carboxylase gene. The synthetic oligonucleotide shown introduces two mutations: one which alters the codon for Lys-166 to a codon for arginine and the other (a silent mutation) which creates a new restriction site and thereby facilitates subsequent plaque screening. The newly-introduced restriction site is the HindIII site within the carboxylase gene; the other HindIII and Cla sites are present in the original M13 vector.

activated (*i.e.* carbamylated) form of the carboxylase [28], demonstration of quaternary complex formation (protein•CO_2•Mg^{2+}•analogue) verifies carbamylation and ligand (and by inference ribulose-P_2) binding. Decreased k_{cat} as a consequence of a small conformational change cannot be excluded rigorously without crystallographic analysis.

Lys-166 has been replaced with a number of other amino acids including arginine (retention of a cationic side-chain), cysteine (retention of an acid/base group), or glycine (removal of side-chain) [29]. Each of the three mutant proteins, whose intact dimeric structures have been verified, lack carboxylase activity. However, the Gly-166 mutant does form a quaternary complex with CO_2, Mg^{2+}, and carboxyarabinitol-P_2, thereby demonstrating that Lys-166 is not required for the CO_2/Mg^{2+}-dependent activation or for substrate binding. These observations reinforce the postulate that Lys-166 is intimately involved in catalysis, possibly as the general base that enolizes ribulose-P_2.

The role of active-site Lys-329 of the *R. rubrum* carboxylase has also been explored by site-directed mutagenesis [30]. Substitutions included glycine, alanine, serine, cysteine, arginine, and glutamic acid. In each case, the purified mutant protein was shown to be a dimer, demonstrating that these single amino acid replacements are compatible with proper folding and association of subunits. The purified mutant proteins do not exhibit detectable enzyme activity nor do they form a stable quaternary complex with CO_2, Mg^{2+}, and carboxyarabinitol-P_2. However, based on ligand-selective elution of the mutant proteins from an affinity matrix (green A agarose), they do retain the ability to bind substrate analogues. These results emphasize the necessity of a lysyl residue at position 329 but do not define its precise function.

Greater than 99.9% of the carboxylase activity is lost upon replacement of Glu-48 with glutamine [31], which removes the acid/base group of the glutamyl side-chain without introducing major steric or electronic perturbation. As the mutant protein is a dimer which undergoes carbamylation by CO_2 and binds substrate, the drastic reduction in k_{cat} must reflect the presence of Glu-48 at the active site in the wild-type enzyme and indicates that this residue serves an important function.

ACTIVE-SITE LOCATION

The mutagenesis studies place Glu-48 at the active site and by deduction proximal to other active-site residues, *e.g.* Lys-166. However, based on the structure of the α-carbon backbone [13] and the results of chemical cross-linking [21], approximation of Glu-48 and Lys-166 are a consequence of intersubunit interactions rather than the folding of an individual polypeptide chain. The question thus arises as to whether a functional active site is wholly contained within a catalytic subunit or is located at an interface between catalytic subunits. Wente and Schachman [32] have distinguished between "independent" and "shared" active sites in the case of multimeric aspartate transcarbamoylase by *in vitro* formation of hybrids of site-directed mutants. Because of difficulties in establishing reversible dissociation conditions for the *R. rubrum* Rubisco, an *in vivo* procedure has been developed in which genes for the appropriate mutant subunits are coexpressed from separate, but compatible, plasmids [33].

If each subunit of *R. rubrum* Rubisco contains a complete, functional active site, a heterodimeric hybrid comprised of one $K166G^2$ subunit and one E48Q subunit will be devoid of activity just like each of the mutant homodimers. However, if a functional active site requires interacting domains of adjacent subunits, the potential exists for restoration of carboxylase activity by hybridization of two different inactive mutant proteins, *e.g.* K166G and E48Q (Fig. 4). Excepting the possibility of a bias in association, the carboxylase population should be 25% of the K166G homodimer, 25% of the E48Q homodimer, and 50% of the K166G/E48Q heterodimer. Since the latter would contain one functional active site per dimer, the specific activity of the total carboxylase population should be 25% that of wild-type.

---

[2]The single-letter code for amino acids is used to designate mutants. The first letter denotes the amino acid present in the wild-type enzyme at the numbered position. The final letter denotes the amino acid present at the corresponding position in the mutant. Designation of hybrids of two different mutant subunits are separated by a slash.

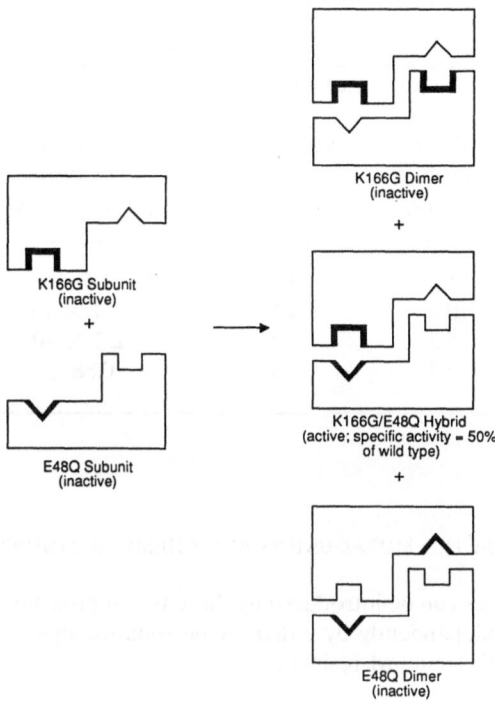

Figure 4. Schematic diagram of the different species that can be formed in a cell coexpressing the genes for the K166G and E48Q mutant subunits. One domain of the active site is illustrated by the rectangular indentations, and the other domain of the active site is illustrated by the triangular indentations. Indentations outlined in black represent nonfunctional domains due to specific amino acid substitutions. Heterodimer formation from the two mutant subunits generates a species with one wild-type active site per dimeric molecule.

These expectations are fully realized by coexpression in *E. coli* of the gene for K166G and E48Q (Table 1). Analysis of the carboxylase purified from these cells confirms that the activity reflects the presence of a heterodimer (one subunit with the Gly-166 substitution and one subunit with the Gln-48 substitution) with one active site per molecule (as compared to two in wild-type enzyme). This interallelic complementation conclusively demonstrates that domains from each subunit constitute the active site and that side chains from both are required, directly or indirectly, for catalysis [33].

TABLE 1. Carboxylase activity from *E. coli* cells expressing various forms of the gene for Rubisco.

Protein encoded	Crude extracts		Purified	
	Total protein (mg/ml)	Relative activity (% wild-type)	Specific activity (μmol/min/mg/protein)	Relative activity (% wild-type)
wt/wt	21.5	100	4.4	100
K166G	30.0	<1	$<4 \times 10^{-5}$	<0.001
E48Q	25.5	<1	2.2×10^{-3}	0.05
K166G/E48Q	17.2	26	0.88	20

CONCERTED SITE-DIRECTED MUTAGENESIS AND CHEMICAL MODIFICATION

Subtle structural changes can be introduced by these two approaches in combination which cannot be achieved independently by either. One such example is conversion of a lysyl residue to an aminoethylcysteinyl residue:

$$C_\alpha HCH_2CH_2CH_2CH_2NH_2 \longrightarrow C_\alpha HCH_2SH \xrightarrow{BrCH_2CH_2NH_2} C_\alpha HCH_2SCH_2CH_2NH_2$$

mutagenesis   modification

The overall structural change is the mere replacement of the lysyl γ-methylene by the sulfur atom.

This strategy has been applied to both Lys-166 and Lys-329 of *R. rubrum* Rubisco for several reasons [34]. Indications of catalytic functionality of these two active-site lysines included their strong acidities and enhanced nucleophilicities, presumably reflective of an unique microenvironment. If substitutions for Lys-166 and Lys-329 did not alter the active-site conformation, other reactive side chains at these positions should also exhibit unusual properties. The cysteinyl mutant carboxylases, devoid of enzyme activity, provide a direct chemical approach to inspecting the conformational integrity of the active site by aminoethylation as shown above. A second consideration prompting these experiments was to gain additional insights into the stringency of the requirements for lysines at positions 166 and 329. Another impetus for investigating the aminoethylation of the cysteinyl mutant proteins concerns the feasibility of altering the carboxylase/oxygenase activity ratio of the wild-type enzyme. The aminoethylations offer the opportunity to ascertain the effects on enzymatic properties of an extremely modest structural change at the active site.

Treatment of the K166C and K329C mutant proteins with 2-bromoethylamine partially restores enzyme activity (Fig. 5), presumably as a consequence of selective aminoethylation of the thiol group unique to each protein. Amino acid analyses, isoelectric focusing under denaturing conditions, slow inactivation of the wild-type carboxylase by bromoethylamine, and the failure of bromoethylamine to restore activity to the corresponding glycyl mutant proteins support this interpretation. The observed facile, selective aminoethylations are consistent with active-site microenvironments not dissimilar to that of the native enzyme. Catalytic constants of these novel carboxylases, which contain a sulfur atom in place of a specific lysyl γ-methylene group, are significantly lower than that of the wild-type enzyme. The k_{cat} for K329C is 40% that of wild-type, while k_{cat} for K166C is reduced 5-fold relative to wild-type. Furthermore, the aminoethylated mutant proteins form isolable complexes with carboxyarabinitol-P$_2$, but with compromised stabilities. These detrimental effects by such a modest structural change underscore the stringent requirement for lysyl side chains at positions 166 and 329. In contrast, the aminoethylated mutant proteins exhibit K_m values that are unperturbed relative to those for the native enzyme. Clearly, major reductions in k_{cat} with unaltered K_m values argue for direct roles of Lys-166 and Lys-329 in catalysis.

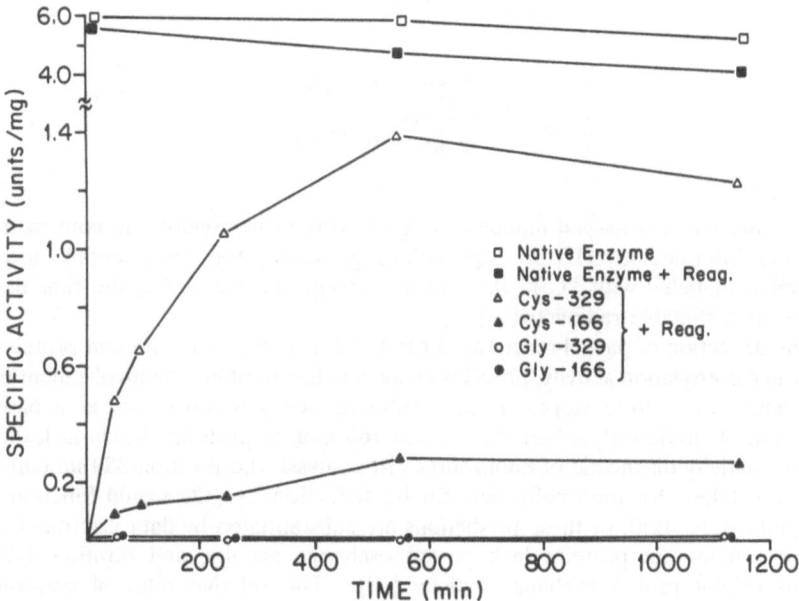

Figure 5. Carboxylase activity during treatment of wild-type Rubisco and mutant proteins (all at 0.4 mg/ml) with 2-bromoethylamine (100 mM).

Rather insensitive assays using an oxygen electrode do not reveal significant change in the carboxylase/oxygenase ratio, relative to that of wild-type enzyme, displayed by the aminoethylated mutant proteins.

An analogous strategy has been used to replace Lys-191, the residue that undergoes carbamylation during enzyme activation, with aminoethylcysteine [35]. In this case, the more reactive ethylene imine is necessary for the introduction of the aminoethyl side chain, because a sulfhydryl group at position 191 is difficult to alkylate. The k_{cat} for the aminoethylated protein is only 4-6% of the wild-type value. Some properties of the K191C mutant (without aminoethylation) are surprising. Despite absence of the amino group that condenses with CO_2 to form a carbamate, the K191C mutant binds stoichiometric levels of CO_2 and forms a quaternary complex with carboxyarabinitol-P_2 that is even more stable than the wild-type counterpart. The total lack of carboxylase activity of the mutant protein, even though it binds "activator CO_2" and Mg^{2+}, raises the possibility that the nitrogen of the carbamate of the wild-type enzyme is crucial to catalysis.

MUTANT PROTEINS AS CATALYSTS IN PARTIAL REACTIONS

As shown in Fig. 2, the two reaction pathways as catalyzed by Rubisco entail several discrete chemical steps. The initial step in both carboxylation and oxygenation pathways is abstraction of the C-3 proton of ribulose-P_2 to form an enediol intermediate (I); this reaction can be monitored by transfer of tritium from solvent (T_2O) into ribulose-P_2 or by loss of tritium from [3-^{3}H]ribulose-P_2 [36], as seen in the equation. Another partial

$$\text{Enz}-\text{B:}\overset{\displaystyle R}{\underset{\displaystyle R'}{\overset{\displaystyle |}{\underset{\displaystyle |}{\overset{\displaystyle C=O}{\underset{\displaystyle H-C-OH}{}}}}}} \;\rightleftharpoons\; T_2O \underset{E-BT}{\overset{E-BH}{\big\Updownarrow}} H_2O \;\;\; \overset{R}{\underset{R'}{\overset{C}{\overset{\displaystyle \diagdown}{\diagup}}\overset{O^-}{\underset{\displaystyle \diagup}{\overset{\|}{C}}}}}\text{OH} \;\rightleftharpoons\; \text{Enz}-\text{B:}\overset{\displaystyle R}{\underset{\displaystyle R'}{\overset{\displaystyle |}{\underset{\displaystyle |}{\overset{\displaystyle C=O}{\underset{\displaystyle T-C-OH}{}}}}}}$$

reaction that may be assayed independently of overall carboxylation in conversion of the six-carbon intermediate (II) to 3-phospho-D-glycerate; the assay entails use of the intermediate labeled with ^{14}C in the carboxyl group and measuring the time-dependent increase in acid-stable radioactivity [37].

The dissection of partial reactions, if catalyzed by site-directed mutant proteins devoid of overall carboxylation activity, provides an avenue for ascribing the involvement of active-site residues to discrete steps. If the postulate that Lys-166 serves as a base in the enolization of ribulose-P_2 is correct, position 166 mutant proteins should lack the proton exchange activity diagnostic of enolization. In contrast, the position 329 mutant proteins could be catalysts for the enolization, for by deduction Lys-329 would function at some subsequent step. Both of these predictions are substantiated by data illustrated in Fig. 6: position 166 mutant proteins lack proton exchange activity, and position 329 mutant proteins exhibit proton exchange activity [38]. The relative rates of enolization are provided in Table 2. Even though K166G is essentially devoid of proton exchange activity, it is a catalyst for conversion of the six-carbon reaction intermediate (compound II, Fig. 2) to phosphoglycerate [39], a property expected of a protein deficient only in the initial step in the overall pathway. The demonstration that mutant proteins lacking either

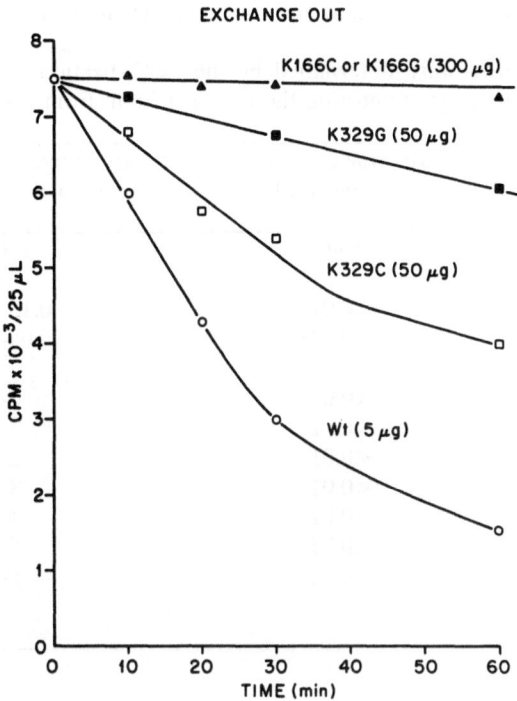

Figure 6. Enolization of [3-³H]ribulose-P₂ by wild-type Rubisco and mutant proteins as measured by the decrease in nonvolatile radioactivity. The μg-values denoted on the curves represent the amount of protein present in each 200-μl reaction mixture.

Lys-166 or Lys-329 are indeed enzymes, albeit not carboxylases, that act on a reaction intermediate or ribulose-P₂, respectively, provides strong evidence of active-site topologies very similar to that of wild-type enzyme. Thus, the catalytic deficiency of these mutant proteins in overall carboxylation appears to reflect the absence of a catalytic ε-amino group rather than conformational perturbations. Assignment of Lys-166 to the initial proton abstraction step is clearly consistent with a wide variety of experimental observations.

As the enolization activity of the position 329 mutant proteins requires both CO_2 and Mg^{2+}, Lys-329 cannot be required for the activation process. Reversible enolization of ribulose-P₂ is not accompanied by any net destruction nor decreased concentration of ribulose-P₂. The absence of dephosphorylation, isomerization, and/or epimerization shows that the enediol never dissociates from the enzyme but yet fails to react with either gaseous substrate. Perhaps the role of Lys-329 is to enhance the reactivity of the enediol through polarization and development of the nucleophilic center at C-2.

TABLE 2. Relative overall carboxylase and exchange activities of mutant carboxylases

Carboxylase activities were determined by the $^{14}CO_2$-fixation assay. The exchange activities were determined by monitoring the loss of tritium from [3-^{3}H]ribulose-P$_2$.

Protein	Carboxylase Activity (% wild-type)	Exchange Activity (% wild-type)
Wild-type	100	100
K166C	<0.01	0.15-0.25
K166G	<0.001	0.05-0.08
K329C	<0.01	4-5
K329G	<0.01	2-3
K329E	<0.01	2-3
K329R	<0.01	3-4
K329Q	<0.01	4-5
K329A	<0.01	5-6
K329S	<0.01	5-6

CONCLUSIONS BASED ON CHEMICAL VS. CRYSTALLOGRAPHIC STUDIES

Of the 3D structures of Rubisco published thus far, that of the carbamylated spinach enzyme with carboxyarabinitol-P$_2$ bound is the most revealing of active-site features [8]. In agreement with chemical and mutagenesis studies, Lys-175, Lys-334, and Glu-60 (analogous to Lys-166, Lys-329, and Glu-48 in the R. rubrum enzyme) are observed at the active site within contact distances to the bound carboxyarabinitol-P$_2$. An intersubunit ionic bond is identified between Glu-60 and Lys-177, verifying the assertion based on hybridization of mutant proteins that the active site requires interacting domains of adjacent subunits [33]. The ϵ-amino group of Lys-334 is within hydrogen bonding distance of the carboxyl group of carboxyarabinitol-P$_2$, a location consistent with a role in influencing the reactivity of the enediol of ribulose-P$_2$ as suggested by properties of position 329 mutants of the R. rubrum enzyme [38]. The ϵ-amino group of Lys-175 is, however, closer to the bridge oxygen of the C-1 phosphate group than to C-3, the location required for it to function as the base that enolizes ribulose-P$_2$. However, "model building outside the electron density could bring the nitrogen atom to within 3 Å from C-3" [8]. Furthermore, these crystallographers point out that the observed structure simulates a stage in the catalytic pathway that is well beyond the initial enolization; in other words, the precise positioning of the ϵ-amino group of Lys-175 in an enzyme•ribulose-P$_2$ complex could differ from that in the quaternary complex. Thus, the 3D structure of the quaternary complex does not exclude Lys-175 (Lys-166 in the R. rubrum enzyme) as the base which abstracts the C-3 proton from ribulose-P$_2$.

The non-carbamylated (deactivated) forms of Rubisco from both *R. rubrum* and tobacco have also been subjected to crystallographic analyses [13-17]. Conformational differences between non-carbamylated and carbamylated enzymes are revealed, which signal the advisability of exercising caution in extrapolating from structural features to mechanistic inferences. For example, the structure of the non-carbamylated *R. rubrum* enzyme did not permit the deductions that the active site is generated by subunit-subunit interactions and that side chains from both subunits comprise the active site. Furthermore, this structure, as well as that of non-carbamylated tobacco enzyme, did not show Lys-329 as an active-site residue. This lysyl residue is contained within a disordered, flexible loop and its active-site location is only visualized in protein complexed with carboxyarabinitol-P_2 [8,17]. The activation-state dependence of the interresidue distance between Lys-166 and Lys-329 as revealed by crystallography was predicted by earlier cross-linking studies described herein.

Based on a 3D structure of the non-carbamylated *R. rubrum* enzyme complexed with carboxyarabinitol-P_2, both Lys-166 and Lys-329 are assigned roles as ligands for one of the phosphate groups [17]. Such assignments discount considerable evidence to the contrary as presented in this paper, including the failure of arginine substitutions in mutant proteins to replace lysine functionally. While this 3D structure is relevant to understanding an ill-defined catalytic activity of unknown biological significance possessed by non-carbamylated protein (decarboxylation of the six-carbon reaction intermediate) [37], the structure may not be fully informative concerning the normal reaction pathway. The crystals for this study were grown at pH 5.6, where the enzyme displays little catalytic activity, and the carboxyarabinitol-P_2 is bound in an upside-down orientation relative to its binding by carbamylated spinach enzyme, perhaps reflective of slight differences in active-site topologies of the two forms of the enzyme. At a pH so far below the optimum for catalysis, the ionization states of amino acid side chains and bound ligands will differ from those that predominate during catalysis, a situation that could complicate assignment of function of side chains. For example, irrespective of function, Lys-166 has a pK_a of ~7.9 and will exist substantially in the uncharged state during normal catalysis.

In summary, X-ray crystallography of Rubisco lends credence to many conclusions derived from chemical and mutagenesis studies, but precise functions of active-site residues remain to be established unequivocally.

Acknowledgments

My research was sponsored by the Office of Health and Environmental Research, U.S. Department of Energy under Contract DE-AC05-84OR21400 with Martin Marietta Energy Systems, Inc. The intellectual and technical contributions of local colleagues are gratefully acknowledged: Dr. Eva H. Lee (postdoctoral fellow) and Mr. Harry B. Smith (graduate student) of the University of Tennessee-Oak Ridge Graduate School of Biomedical Sciences; Dr. Frank W. Larimer, Dr. Richard J. Mural, and Dr. Thomas S. Soper of the Biology Division, Oak Ridge National Laboratory.

References

1. Van Dyk, D. E. and Schloss, J. V. (1986) "Deuterium isotope effects in the carboxylase reaction of ribulose-1,5-bisphosphate carboxylase/oxygenase", Biochemistry 25, 5145-5156.

2. Lorimer, G. H., Badger, M. R., and Andrews, T. J. (1976) "The activation of ribulose-1,5-bisphosphate carboxylase by carbon dioxide and magnesium ions. Equilibria, kinetics, a suggested mechanism, and physiological implication", Biochemistry 15, 529-536.

3. Lorimer, G. H. (1981) "Ribulosebisphosphate carboxylase - amino acid sequence of a peptide bearing the activator carbon dioxide", Biochemistry 20, 1236-1240.

4. Zurawski, G., Perrot, B., Bottomley, W., and Whitfeld, P.R. (1981) "The structure of the gene for the large subunit of ribulose 1,5-bisphosphate carboxylase from spinach chloroplast DNA", Nucl. Acids Res. 9, 3251-3270.

5. Donnelly, M. I., Stringer, C. D., and Hartman, F. C. (1983) "Characterization of the activator site of *Rhodospirillum rubrum* ribulosebisphosphate carboxylase/oxygenase", Biochemistry 22, 4346-4352.

6. Hartman, F. C., Stringer, C. D., and Lee, E. H. (1984) "Complete primary structure of ribulosebisphosphate carboxylase/oxygenase from *Rhodospirillum rubrum*", Arch. Biochem. Biophys. 232, 280-295

7. Igarashi, Y., McFadden, B. A., El-Gul, T. (1985) "Active site histidine in spinach ribulosebisphosphate carboxylase/oxygenase modified by diethyl pyrocarbonate", Biochemistry 24, 3957-3962.

8. Andersson, I., Knight, S., Schneider, G., Lindquist, Y., Lundqvist, T., Brändén, C.-I., and Lorimer, G. H. (1989) "Crystal structure of the active site of ribulose-bisphosphate carboxylase", Nature (Lond.) 337, 229-234.

9. Niyogi, S. K., Foote, R. S., Mural, R. J., Larimer, F. W., Mitra, S., Soper, T. S., Machanoff, R., and Hartman, F. C. (1986) "Nonessentiality of histidine 291 of *Rhodospirilllum rubrum* ribulose-bisphosphate carboxylase/oxygenase as determined by site-directed mutagenesis", J. Biol. Chem. 261, 10087-10092.

10. Hartman, F. C., Stringer, C. D., Milanez, S., and Lee, E. H. (1986) "The active site of Rubisco", Philos. Trans. R. Soc. Lond. (Biol.) 313, 379-395.

11. Rutner, A. C. (1970) "Estimation of the molecular weight of ribulose diphosphate carboxylase subunits", Biochem. Biophys. Res. Commun. 39, 923-929.

12. Herndon, C. S. and Hartman, F. C. (1984) "2-(4-Bromoacetamido)anilino-2-deoxypentitol 1,5-bisphosphate, a new affinity label for ribulose bisphosphate carboxylase/oxygenase from *Rhodospirillum rubrum*", J. Biol. Chem. 259, 3102-3110.

13. Schneider, G., Lindquist, Y., Bränden, C.-I., and Lorimer, G. (1986) "Three-dimensional structure of ribulose-1,5-bisphosphate carboxylase/oxygenase from *Rhodospirillum rubrum* at 2.9 Å resolution", EMBO Journal 5, 3409-3415.

14. Chapman, M. S., Suh, S. W., Cascio, D., Smith, W. W., and Eisenberg, D. (1987) "Sliding-layer conformational change limited by the quaternary structure of plant Rubisco", Nature (Lond.) 329, 354-356.

15. Chapman, M. S., Suh, S. W., Curmi, P. M. G., Cascio, D., Smith, W. W., and Eisenberg, D. S. (1988) "Tertiary structure of plant Rubisco: domains and their contacts", Science 241, 71-74.

16. Lundqvist, T. and Schneider, G. (1988) "Crystal structure of the binary complex of ribulose-1,5-bisphosphate carboxylase and its product, 3-phospho-D-glycerate", J. Biol. Chem. 263, 3643-3646.

17. Lundqvist, T. and Schneider, G. (1989) "Crystal structure of the complex of ribulose-1,5-bisphosphate carboxylase and a transition state analogue, 2-carboxy-D-arabinitol 1,5-bisphosphate", J. Biol. Chem. 264, 7078-7083.

18. Hartman, F. C., Milanez, S., and Lee, E. H. (1985) "Ionization constants of two active-site lysyl ϵ-amino groups of ribulosebisphosphate carboxylase/oxygenase", J. Biol. Chem. 260, 13968-13975.

19. Andrews, T. J. and Lorimer, G. H. (1987) "Rubisco: structure, mechanisms, and prospects for improvement", in M. D. Hatch and N. K. Boardman (eds.), The Biochemistry of Plants, A Comprehensive Treatise, Vol. 10: Photosynthesis, Academic Press, New York, pp. 131-218.

20. Lee, E. H., Stringer, C. D., and Hartman, F. C. (1986) "Distance between two active-site lysines of ribulose bisphosphate carboxylase from *Rhodospirillum rubrum*", Proc. Natl. Acad. Sci. USA 83, 9383-9387.

21. Lee, E. H., Soper, T. S., Mural, R. J., Stringer, C. D., and Hartman, F. C. (1987) "An intersubunit interaction at the active site of D-ribulose-1,5-bisphosphate carboxylase/oxygenase as revealed by cross-linking and site-directed mutagenesis", Biochemistry 26, 4599-4604.

22. Somerville, C. R. and Somerville, S. C. (1984) "Cloning and expression of the *R. rubrum* ribulosebisphosphate carboxylase gene in *E. coli*", Mol. Gen. Genet. 193, 214-219.

23. Larimer, F. W., Machanoff, R., and Hartman, F. C. (1986) "A reconstruction of the gene for ribulose bisphosphate carboxylase from *Rhodospirillum rubrum* that expresses the authentic enzyme in *Escherichia coli*", Gene 41, 113-120.

24. Zoller, M. J. and Smith, M. (1983) "Oligonucleotide-directed mutagenesis of DNA fragments cloned into M13 vectors", Methods Enzymol. 100, 468-500.

25. Nisbet, J. T. and Beilharz, M. W. (1985) "Simplified DNA manipulations based on *in vitro* mutagenesis", Gene Anal. Techn. 2, 23-40.

26. Mural, R. J. and Foote, R. S. (1986) "'Bandaid' mutagenesis: A novel technique for oligonucleotide-directed site-specific mutagenesis", DNA 5, 84.

27. Childs, J., Villanueba, K., Barrick, D., Schneider, T. D., Stormo, G. D., Gold, L., Leitner, M., and Caruthers, M. (1985) "Ribosome binding site sequences and function", in R. Calendar and L. Gold (eds.), Sequence Specificity in Transcription and Translation, Alan R. Liss, New York, pp. 341-350.

28. Miziorko, H. M. and Sealy, R. C. (1980) "Characterization of the ribulosebisphosphate carboxylase-carbon dioxide-divalent cation-carboxypentitol bisphosphate complex", Biochemistry 19, 1167-1171.

29. Hartman, F. C., Soper, T. S., Niyogi, S. K., Mural, R. J., Foote, R. S., Mitra, S., Lee, E. H., Machanoff, R., and Larimer, F. W. (1987) "Function of Lys-166 of *Rhodospirillum rubrum* ribulosebisphosphate carboxylase/oxygenase as examined by site-directed mutagenesis", J. Biol. Chem. 262, 3496-3501.

30. Soper, T. S., Mural, R. J., Larimer, F. W., Lee, E. H., Machanoff, R., and Hartman, F. C. (1988) "Essentiality of Lys-329 of ribulose-1,5-bisphosphate carboxylase/oxygenase from *Rhodospirillum rubrum* as demonstrated by site-directed mutagenesis", Protein Eng. 2, 39-44.

31. Hartman, F. C., Larimer, F. W., Mural, R. J., Machanoff, R., and Soper, T. S. (1987) "Essentiality of Glu-48 of ribulose bisphosphate carboxylase/oxygenase as demonstrated by site-directed mutagenesis", Biochem. Biophys. Res. Commun. 145, 1158-1163.

32. Wente, S. R. and Schachman, H. K. (1987) "Shared active sites in oligomeric enzymes: model studies with defective mutants of aspartate transcarbamoylase produced by site-directed mutagenesis", Proc. Natl. Acad. Sci. USA 84, 31-35.

33. Larimer, F. W., Lee, E. H., Mural, R. J., Soper, T. S., and Hartman, F. C. (1987) "Intersubunit location of the active site of ribulosebisphosphate carboxylase/oxygenase as determined by *in vivo* hybridization of site-directed mutants", J. Biol. Chem. 262, 15327-15329.

34. Smith, H. B. and Hartman, F. C. (1988) "Restoration of activity to catalytically deficient mutants of ribulosebisphosphate carboxylase/oxygenase by aminoethylation", J. Biol. Chem. 263, 4921-4925.

35. Smith, H. B., Larimer, F. W., and Hartman, F. C. (1988) "Subtle alteration of the active site of ribulosebisphosphate carboxylase/oxygenase by concerted site-directed mutagenesis and chemical modification", Biochem. Biophys. Res. Commun. 152, 579-584.

36. Sue, J. M. and Knowles, J. R. (1982) "Ribulose-1,5-bisphosphate carboxylase: fate of tritium label in [3-^{3}H]ribulose 1,5-bisphosphate during the enzyme-catalyzed reaction", Biochemistry 21, 5404-5410.

37. Pierce, J., Andrews, T. J., and Lorimer, G. H. (1986) "Reaction intermediate partitioning by ribulose-bisphosphate carboxylases with differing substrate specificities", J. Biol. Chem. 261, 10248-10256.

38. Hartman, F. C. and Lee, E. H. (1989) "Examination of the function of active site lysine 329 of ribulosebisphosphate carboxylase/oxygenase as revealed by the proton exchange reaction", J. Biol. Chem. (in press).

39. Lorimer, G. H. and Hartman, F. C. (1988) "Evidence supporting lysine 166 of *Rhodospirillum rubrum* ribulosebisphosphate carboxylase as the essential base which initiates catalysis", J. Biol. Chem. 263, 6468-6471.

STRUCTURAL AND FUNCTIONAL ASPECTS OF THE PHOTOSYNTHETIC FIXATION OF CARBON DIOXIDE

G. Schneider, I. Andersson, C.-I. Brändén, S. Knight
Y. Lindqvist and T. Lundqvist
Swedish University of Agricultural Sciences
Uppsala Biomedical Center
Department of Molecular Biology
P.O. Box 590
S-751 24 Uppsala, Sweden

INTRODUCTION

Ribulose-1,5-bisphosphate carboxylase/oxygenase (Rubisco) has attracted a lot of interest due to its central role in the carbon metabolism of plants and photosynthetic microorganisms (for a review see (1)). The dual function of this enzyme, catalyzing the primary steps in both photosynthetic carbon dioxide fixation and photorespiration (Figure 1), makes it a challenging target for attempts to improve the efficiency of photosynthesis. Recombinant DNA-techniques provide a promising tool to modify the carboxylase/oxygenase ratio by genetic engineering. However, the application of these techniques requires a detailed knowledge of the catalytic mechanism of the enzyme and the structure of its active site.

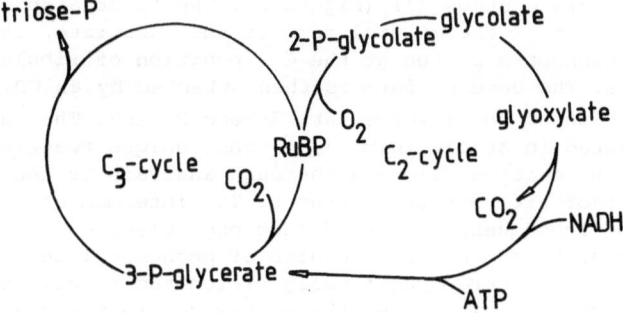

Figure 1:   Rubisco at the interface between the C_3 carbon reduction cycle and the C_2 carbon oxidation cycle.

The initial step during the dark reactions of photosynthesis consists of the addition of CO_2 to ribulose-1,5-bisphosphate. The

M. Aresta and J. V. Schloss (eds.),
Enzymatic and Model Carboxylation and Reduction Reactions for Carbon Dioxide Utilization, 367–376.

intermediate six-carbon compound is then cleaved into two molecules of phospho-glycerate. The latter is then partly recycled in the Calvin cycle, where it is used to regenerate ribulose bisphosphate and partly converted to starch.

Rubisco also catalyses the initial oxygenation step in photorespiration, during which a considerable amount of the stored energy is converted to heat thereby limiting crop yield. In this step, oxygen instead of carbon dioxide is added to ribulose bisphosphate, thus yielding one molecule of phospho-glycerate and one molecule of phospho-glycolate. The latter is metabolized in the glycolate pathway, where it is ultimately converted to CO_2. The reaction mechanism of the oxygenation reaction is much less understood than the mechanism of the carboxylation reaction. In the following, we will only discuss aspects of the carboxylation reaction.

The catalytic activities of Rubisco require an activation process, during which a lysine residue reacts with an activator CO_2 molecule (2), which is not the substrate CO_2. The labile carbamate formed in this reaction is stabilized by binding of a magnesium ion. The substrate ribulose-1,5-bisphosphate binds to the activated ternary complex and is subsequently either carboxylated by CO_2 or oxygenated by O_2. All biochemical evidence indicates that these two reactions occur at the same site in the protein (1).

In the following, we will describe the chemical series of events during the carboxylation and the structure of the enzyme and correlate the structural information to biochemical and genetic data.

The carboxylation reaction

Chemically, the carboxylation reaction consists of the addition of CO_2 to ribulose-1,5-bisphosphate, yielding two molecules of phospho-glycerate. The overall reaction can be dissected into a series of discrete steps (1) (Figure 2). The first step in catalysis is the formation of the enediol form of the substrate, initiated by the abstraction of a proton at the C-3 position of ribulose bisphosphate. The enediol form is then attacked by an CO_2 molecule under formation of the intermediate 3-keto-2-CABP. The intermediate can be isolated in stable form (3,4). This unique feature of the carboxylation reaction allows a thorough analysis of the catalytic deficiencies of site-directed mutants. The intermediate is then hydrated and subsequently cleaved into one molecule of phospho-glycerate and a C-2 carbanion of phospho-glycerate. The carbanion is then stereospecifically protonated to give the second molecule of phospho-glycerate. Stereochemical restraints require that the base which initiates catalysis by abstraction of a proton from the C-3 position of the substrate is different from the one which protonates the carbanion (1). The carboxylation reaction can thus be

desribed in five steps: enolization, carboxylation, hydration, C-C
cleavage and protonation.

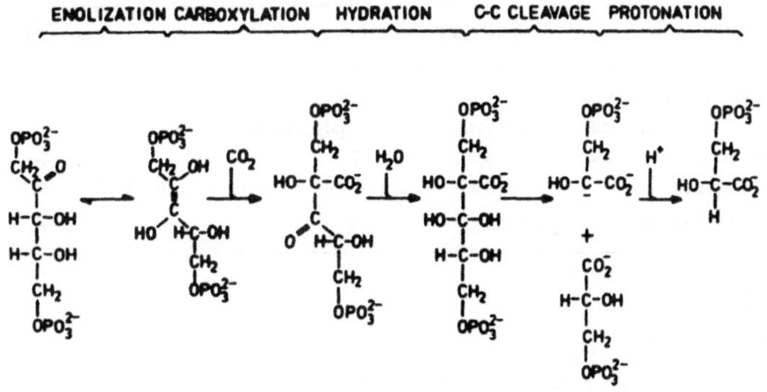

Figure 2:   The mechanism of the carboxylation reaction.

The enzyme, ribulose-1,5-bisphosphate carboxylase/oxygenase

Rubisco from higher plants, algae and most photosynthetic
microorganisms is a multisubunit complex built up of eight large (mol
wt. 56 kd) and eight small (mol wt. 14 kd) subunits. The catalytic
activities for both the carboxylation and oxygenation reaction reside
on the large subunit. The primary structure of the large subunits of
higher plants and algal carboxylases studied so far exhibit a high
degree of amino acid homology, in the range of 70-90 % (1).

In contrast to these L_8S_8 type carboxylases, the enzyme from
the photosynthetic bacterium *Rhodosprillium rubrum* differs
considerably in primary and quaternary structure. This carboxylase is
only a dimer of large subunits and lacks the small subunits (5). The
overall amino acid homology to the large subunit of higher plant
carboxylases is 25% (6,7). Despite this low overall amino acid
homology, some peptide regions are highly conserved in all the
carboxylases. Three of these conserved peptide regions have been
identified as active site peptides (2,8,9), indicating a common
active site and thus a similar three-dimensional structure for all
the carboxylase large subunits.

A number of crystallographic studies have focussed on Rubisco
and structural information is now available for both the L_2 and the
L_8S_8 type of the enzyme and for diffent complexes. Table 1 summarizes
the Rubisco structures, which have been solved at present. From these
studies, the following picture of the overall structure and assembly
of the enzyme has emerged.

TABLE 1. Crystal structures of Rubisco

Source	Species	Resolution	Reference
Rh. rubrum	native	1.7 Å	10,13
Rh. rubrum	enzyme – phosphoglycerate	2.9 Å	11
Rh. rubrum	enzyme – CABP	2.6 Å	12
spinach	enzyme – Mg(II) – CO_2 – CABP	2.8 Å	13
tobacco	native	2.8 Å	14,15

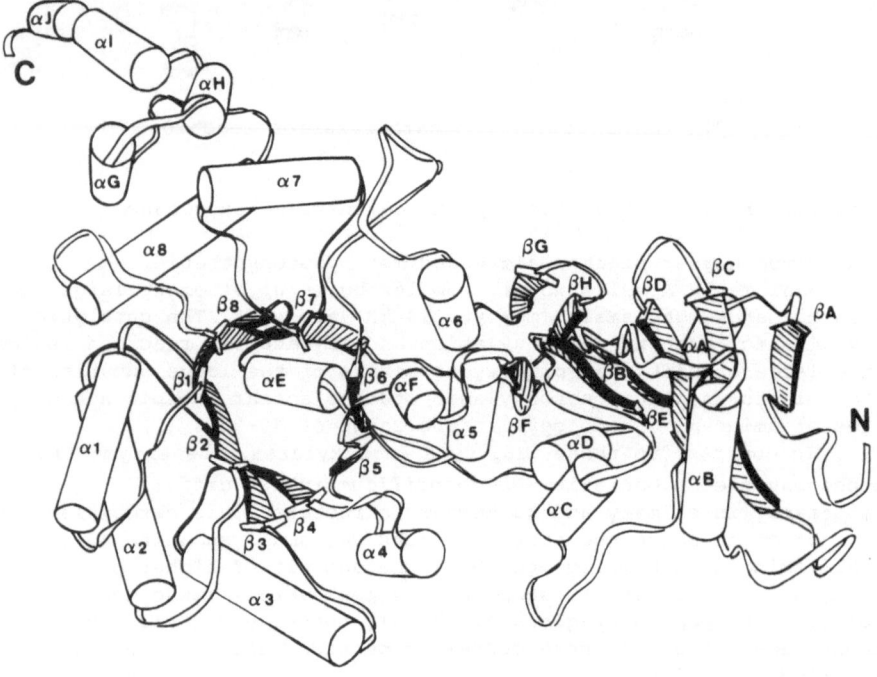

Figure 3:   Schematic view of the large subunit of Rubisco. The
              secondary structural elements are indicated
              (cylinders represent α–helices and arrows
              represent β–strands).

The large subunit is divided in two domains, one smaller
N-terminal domain linked to a C-terminal domain which has an
eight-stranded barrel type structure (10). The domain arrangement and
the secondary structure of these domains is shown in Figure 3. The
two subunits interact tightly to form the functional L_2-dimeric

Rubisco molecule of *Rh. rubrum* (Figure 4). The core of this binding area consists of interactions between the C-terminal domains around a local twofold axis. In addition, two regions from the N-terminal domain of one subunit interact with regions from the C-terminal domain from the second subunit. These subunit interactions are of functional importance as some of the residues involved occur in or close to the active site region. Each active site of the L_2 dimer is thus built up from residues of both subunits.

Figure 4: A schematic view of the subunit arrangement in the L_2-dimer of Rubisco from *Rh. rubrum*. The location of the active sites are indicated by the position of the active site Mg(II) ion. (Drawing by U. Uhlin)

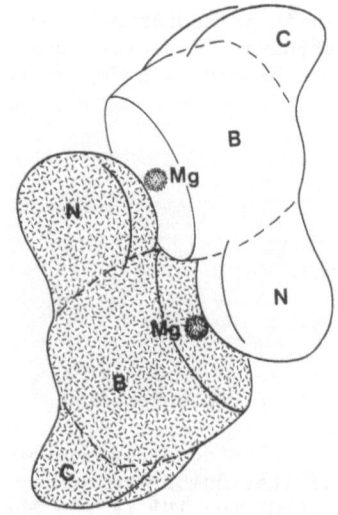

The corresponding functional L_2 dimer occcurs as part of the $L_8 S_8$ Rubisco molecule from spinach (13) and from tobacco (14,15). In the spinach enzyme, four such dimers are arranged around a fourfold axis, building up the L_8 core of the molecule (Figure 5). There are large crevices between the ends of the L_2 dimers at the top and the bottom of the L8 core. The small subunits occupy these crevices. The overall shape of the $L_8 S_8$ type Rubisco resembles a cylinder with a diameter of approximately 110 A and height 100 Å. The eight active sites are on the outside of the molecule facing the solution. They are spaced widely apart; the shortest distance between two active-site metals is 36 Å. There is no contribution of residues from the small subunits to the active site.

The active site of Rubisco is located at the carboxy-ends of the eight β-strands in the barrel. The site is shaped like a funnel and is mainly formed by the eight loop regions that connect the eight b-strands with the corresponding helices in the barrel domain. The

N-terminal domain from the second subunit in the L_2 dimer covers part of the top of the active site. In particular, two loop regions of this domain provide residues to the active site.

Figure 5: Schematic illustration of the assembly of L_8S_8 type Rubisco from L_2 type dimers. The position of one β/α barrel domain is indicated. The small subunits are represented as spheres. (Drawing by B. Furugren).

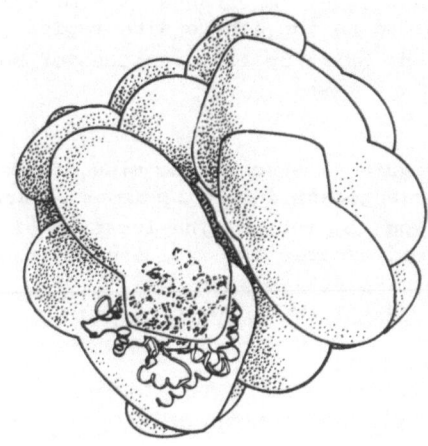

The activation process

The amino acid residue of central importance for activation is Lys 191 (the number refers to the *Rh. rubrum*, sequence, unless otherwise stated. Lys 191 is the last residue in β-strand 2 of the β/α-barrel and is located at the bottom of the active site. The immediate surroundings of the sidechain of Lys 191 are shown in Figure 6. The

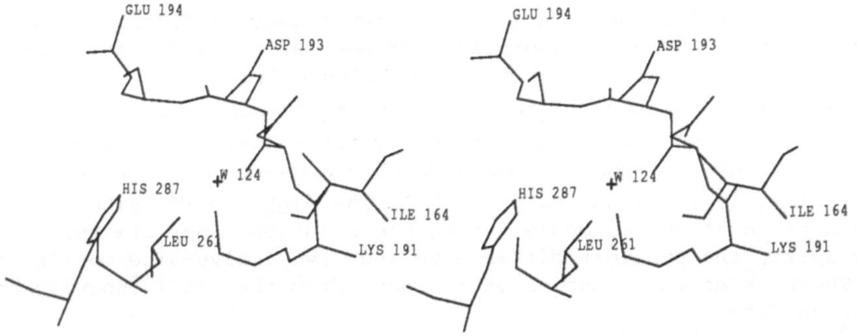

GLU 194   GLU 194

ASP 193   ASP 193

HIS 287 +W 124   HIS 287 +W 124

ILE 164   ILE 164

LEU 261   LYS 191   LEU 261   LYS 191

Figure 6: Stereo view of the surroundings of the activator lysine 191 in the non-activated Rubisco from *Rh. rubrum*.

sidechain forms hydrogen bonds to the main chain oxygen of Asn 192 and a water molecule. The sidechains of residues Asp 193, His 287, Leu 261 and Ile 164 are in van der Waals distance to the lys sidechain.
    The addition of a CO_2 molecule to the ε-amino group of lys 191 results in the formation of a carbamate. The activation process thus changes a positive charge, located at a central position in the active site to a negative charge, which can now accomodate a positivly charged metal ion. By binding Mg2+, the active site becomes poised to properly bind and orient the substrate.

Product and inhibitor binding

The binding of product and the inhibitor CABP has been studied in the non-activated and the activated form of the enzyme (Table 1). These

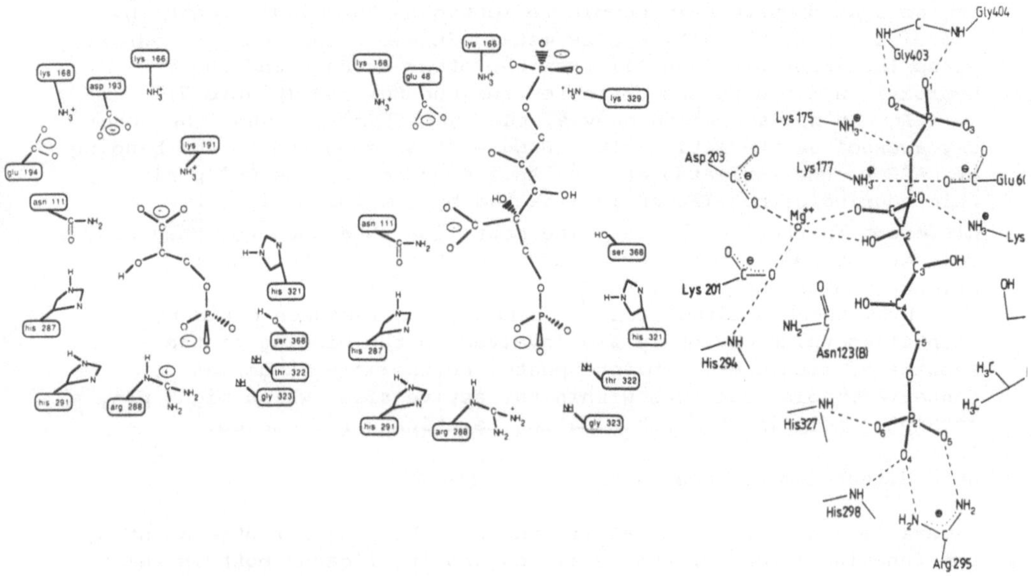

Figure 7:Binding of product and inhibitor to the active site of non-
    activated and activated Rubisco. left: binding of
    phospho-glycerate to non-activated Rubisco from *Rh. rubrum*,
    middle: binding of the inhibitor. 2-carboxy-arabinitol -1,5-
    bisphosphate to non-activated Rubisco from *Rh. rubrum*,
    right:binding of the inhibitor 2-carboxy-arabinitol-
    1,5-bisphosphate to activated Rubisco from spinach.

studies have revealed the location of the active site and the mode
of interaction of product and inhibitor respectively with groups on
the enzyme. The product, 3-phospho-glycerate binds at the active site
with the phosphate group interacting with residues Arg 288, His 321
and Ser 368 (Figure 7). The carboxyl group interacts with the side
chains of His 287, Lys 191 and Asn 111. The inhibitor CABP binds in a
rather extended conformation across the barrel. There are two
distinct phosphate binding sites at opposite sides of the funnel,
separating the two phosphate atoms of CABP by 9.7 Å. One phosphate
site is identical to the one, observed in the binary complex of the
enzyme with phospho-glycerate. The second phosphate binding site is
formed by residues from loop 8, which form a one and a half turn
helix. In addition, the sidechains of residues Lys 166 and Lys 329
interact with the phosphate group at this site (Figure 7).

The sugar bisphosphate molecule is thus anchored at its two ends
on opposite sides of the funnel by the phosphate binding sites and in
the activated quaternary complex oriented in the middle region by
coordination to the active site metal. In addition, the C-4 hydroxyl
group interacts with Asn 111 from the other subunit and the C-3
hydroxyl is close to the conserved residue Ser 368 (Figure 7).

In the non-activated enzyme, the inhibitor is bound 'the other
way around' at the active site (Figure 7) as compared to the binding
of CABP to the activated enzyme. As a conseqence, the C-2 position,
which during carboxylation is attacked by a CO_2 molecule, is
different in the two species. The metal ion thus does not only play a
role in catalysis, but seems to be important for the proper
orientation of the substrate.

From these binding studies, a number of residues could be
identified which obviously are involved in the binding of the
substrate, ribulose-1,5-bisphosphate. Furthermore, a series of
conserved, polar residues within the active site, which might play a
catalytic role in the carboxylation reaction were located.

Site-directed mutagenesis

Several mutants of the metal ligands have been constructed to study
the functional role of the metal ion and its ligands both in the
overall reaction and in partial reactions (16). The structural
information allows a correlation of the results of these studies to
the structure of the active site.

To determine whether the length or the charge of the lysine side
chain is the important feature of the activation process Lys 191 was
changed to Glu (17). The mutant was completely inactive and could not
form a stable complex with CABP. Model building experiments show that
the glu side chain is too short to form a proper metal binding site
instead of carbamylated Lys.

Mutation of the metal ligand Asp 193 to Asn abolishes the
carboxylation reaction. At high concentrations of magnesium however,
the mutant catalyses the formation of the product when presented with

the six-carbon intermediate. The mutant is thus deficient in the ability to catalyze the enolization of the substrate, ribulose-1,5-bisphosphate, an early step during catalysis. This step requires an intact metal binding site. Enzymatic studies of Rubisco where magnesium has been replaced with different metals show that the nature of the metal influences partitioning between carboxylation and oxygenation (18).

Mutants where the adjacent conserved residue Glu 194 has been changed to Gln or Val did not catalyze overall carboxylation, nor the partial enolization nor hydrolysis of the six-carbon intermediate (16). Clearly this residue has an important function, possibly as a metal ligand.

Many more site-directed mutagensis experiments have been done, probing the function of certain amino acid residues in the catalytic cycle (19-26). The structural information will allow both the further design of mutants and aid the interpretation of the results obtained. This work will lead to a structural and functional mapping of the active site of Rubisco, which in turn is a prerequisite for the rational design of mutants with a changed carboxylation/oxygenation ratio.

Acknowledgements: This work was supported by grants from the Swedish research councils NFR and SJFR and the E.I. duPont de Nemours Company.

REFERENCES

(1) Andrews,T.J. and Lorimer,G.H. (1987) in The Biochemistry of Plants (Hatch,M.D.,ed.), Vol.10,pp.131-218, Academic Press,Orlando,FL
(2) Lorimer,G. (1981), Biochemistry 20, 1236 – 1240
(3) Pierce,J., Andrews,T.J. and Lorimer,G. (1986), J. Biol. Chem. 261, 10248 – 10256
(4) Lorimer,G., Andrews,T.J., Pierce,J. and Schloss,J.V. (1986), Phil. Trans. Roy. Soc. London, Ser.B 313, 397 – 407
(5) Schloss,J.V., Phores,E.F., Long,M.W., Norton,I.L., Stringer,C.D. and Hartman,F.C. (1979), J. Bacteriol. 137, 490 – 501
(6) Hartman,F.C., Stringer,C.D., Omnaas,J., Donnelly,M.I. and Fraij,B. (1982) Arch. Biochem. Biophys. 219, 422 – 437
(7) Nargang,F., McIntosh, L. and Somerville,C., Molec. gen. Genet. 193,(1984) 220 – 224
(8) Herndon,C.S., Norton,I.C. and Hartman,F.C. (1982), Biochemistry 21, 1380 – 1385
(9) Fraij, B. and Hartman,F.C. (1982), J. Biol. Chem. 257, 3501 – 3505
(10) Schneider,G., Lindqvist,Y., Brändén, C.-I. and Lorimer,G.,(1986a) EMBO J. 5, 3409 – 3415
(11) Lundqvist,T. and Schneider,G. (1989a), J.Biol.Chem. 264, 3643 – 3646

(12) Lundqvist,T. and Schneider,G. (1989b), J.Biol.Chem., 264, 7078 – 7083

(13) Andersson,I., Knight,S., Schneider,G., Lindqvist,Y., Lundqvist,T., Brändén, C.-I. and Lorimer,G. (1989), Nature 337, 229-234

(14) Chapman,M.S., Se Won Suh, Curmi,P.M.G., Cascio,D., Smith ,W.W. and Eisenberg,D. (1987), Nature 329, 354 – 356

(15) Chapman,M.S., Se Won Suh, Curmi,P.M.G., Cascio,D., Smith ,W.W. and Eisenberg,D. (1988), Science 241, 71 – 74

(16) Lorimer,G., Gutteridge,S. and Madden,M. (1987), in Plant Molecular Biology (eds. D.v.Wettstein and N.-H. Chua) 31 -31 (Nato ASI Series A, 140)

(17) Estelle,M., Hanks,J., McIntosh,L. and Somerville, C. , (1985), J. Biol. Chem. 260, 9523 – 9526

(18) Robison, P.D., Martin,N.N. and Tabita, F.R. (1979), Biochemistry 18, 4453 – 4458

(19) Gutteridge, S., Sigal, I., Thomas, B., Arentzen, R., Cordova,A. and Lorimer, G., (1984), EMBO J. 3, 2737-2742

(20) Hartman, F.C., Soper,T.S., Niyogi, S.K., Mural, R.J., Foote, R.S., Mitra, S., Lee, E.H., Machanoff,R. and Larimer,W.F., (1987a), J. Biol. Chem., 262, 3496-3501

(21) Hartman, F.C., Foote, R.S., Larimer, F.W., Lee, E.H., Machanoff,R., Milanez, S., Mitra, S., Mural R.J., Niyogi, S.K., Smith, H.B., Soper, T.S., and Stringer, C.D. (1987b), In Plant Molecular Biology, (D.V.Wettstein and N.-H. Chua, eds.), pp. 9 – 20, Plenum Press, New York

(22) Hartman, F.C., Larimer, F.W., Mural, R.J., Machanoff, R. and Soper, T.S. (1987c) , Biocem. Biophys. Res. Comm. 145, 1158-1163

(23) Larimer, F.W., Lee, E.H., Mural, R.J., Soper, T.S. and Hartman, F.C. (1987), J.Biol. Chem. 262, 15327-15329

(24) Lorimer, G. and Hartman, F.C. (1988), J.Biol.Chem. 263, 6468-6471

(25) Niyogi, S.K., Foote, R.S., Mural, R.J., Larimer, F.W., Mitra, S., Soper, T.S., Machanoff, R. and Hartman, F.C. (1986), J. Biol. Chem. 261, 10087-10092

(26) Terzaghi, B.E., Laing, W.A., Christeller, J.T., Petersen, G.B. and Hill, D.F. (1986), Biochem. J. 235, 839-846.

COUPLING BETWEEN THE LIGHT AND DARK REACTIONS OF OXYGEN
EVOLUTION AND CO_2 FIXATION IN PHOTOSYNTHESIS: EARLY
EXPERIMENTS IN PHOTOSYNTHESIS REVISITED

F. K. Fong,[a] K. A. Butcher,[b] A. Agostiano,[c] M. della Monica,[c] M. S.
Showell,[d] and J. V. Schloss[e]
[a]Research Division, Christiana Research Corp, The Ferguson Building,
Suite 300, 2134 West Washington Street, IN 46222 USA, [b]Department of
Chemistry, Purdue University, West Lafayette, IN, 47907 USA,
[c]Department of Chemistry, Bari University, Via Amendola 173, 70126
Bari, Italy, [d]Ivorydale Technical Center, Procter and Gamble Co.,
Cincinnati, OH 45217, USA., [e]Central Research & Development
Department, E. I. du Pont de Nemours & Co., Wilmington, DE USA

ABSTRACT. The early experiments that led to our current understanding of the pathway
of carbon assimilation during photosynthesis are reviewed. Some of the observations
made during the course of this work that remain unexplained to date are highlighted. In
particular, the details of how captured luminant energy is used to drive photosynthesis
remains to be fully dissected. Joining the disparate areas of light capture and the
enzymology of CO_2 assimilation is a challenge for modern research in photosynthesis.

1. Introduction

For more than four decades it has been believed that CO_2 fixation, or carboxylation in
photosynthesis *in vivo*, occurs in the dark as a cyclic reaction that is coupled only
indirectly to the chlorophyll light reaction. Known as the photosynthetic carbon cycle, or
the Calvin cycle, this mechanism for carbon reduction has been taught in standard
textbooks to generations of biochemistry and plant physiology students. The Calvin cycle
has been regarded as a complete solution to that aspect of the photosynthesis question
dealing with the reduction of CO_2 to carbohydrate.

However, close examination of the work that led to our current picture of the path of
CO_2 assimilation in the light[1-14] suggests that many intriguing questions were left
unresolved and beg reinvestigation. In particular, whether experimental results obtained
under conditions of *in vivo* illumination are completely explained by known *in vitro*, light-
independent reactions. That perhaps carbon fixation is a non-cyclic, light-driven
process[4,12] and that it is instantaneously terminated in the dark,[7] a view that was held by
some of the early investigators in photosynthetic research. That an elusive and essential
part of the light-driven process *in vivo* remains to be defined. There is little question that
in vitro ribulosebisphosphate carboxylase, the first enzyme of the Calvin cycle, catalyzes

M. Aresta and J. V. Schloss (eds.),
Enzymatic and Model Carboxylation and Reduction Reactions for Carbon Dioxide Utilization, 377–396.
© 1990 *Kluwer Academic Publishers.*

the formation of two molecules of D-glycerate-3-phosphate (PGA) from D-ribulose-1,5-bisphosphate (RuBP) and CO_2:

$$\text{RuBP} + \text{CO}_2 \xrightarrow{\text{dark}} \text{PGA} + \text{PGA} \qquad (1)$$

This reaction seems to contradict some of the observations made by Calvin and his associates[1,4,6-13] however, with regard to the path of CO_2 under conditions of *in vivo* illumination. A detailed analysis of these experiments concerning the *in vivo* path of carbon during photosynthesis is presented in section 2. The possibility that under conditions of *in vivo* illumination the two products of carboxylation might have different fates, and the consequences of such metabolic discrimination (as suggested by Wilson and Calvin,[4] Calvin and Massini,[12] and Bassham and Kirk[13]), are discussed in section 3.

The Chl *a* photoreactions are reviewed in section 4. This review includes a discussion of the P_{700} and P_{680} Chl *a* light reactions in photosynthesis. The P_{680} reaction center, which is directly responsible for the water splitting reaction via the manganese-containing proteins,[15-19] can be modeled by the simpler in vitro water splitting and carbon reduction reactions photocatalyzed by chlorophyll.[14,20-22] The oxygen evolution in water splitting is a four electron process, and occurs in a reaction sequence of four stages, as demonstrated[22,23] by a series of flashes given to a suspension of dark-incubated chloroplasts.

In section 5 it is shown that CO_2 reduction, which is also a four-electron process, may be stoichiometrically[25] and kinetically coupled to the oxygen evolution from water oxidation.

2. Calvin's Discovery of the Photochemical Origin of CO_2 Fixation in Photosynthesis

a. Carbon fixation as a dark reaction. Carbon fixation as a dark reaction was proposed by Ruben and Kamen,[27-29] who pioneered the use of $^{11}CO_2$ as tracer in the study of the photosynthetic process. This proposal was initially met with complete disbelief.[30] It was unclear how plants would obtain the free energy needed to drive the carboxylation reaction. All known model dark reactions *in vitro* at that time were strongly endergonic, whereas the "dark" carboxylation in photosynthesis was exergonic.[30] A decade later, the role of ATP in photophosphorylation was discovered,[31] and the dark reductions in photosynthesis became widely accepted.

b. The dark enzymatic RuBP carboxylase reaction. By then, Benson, Calvin and their co-workers had applied ^{14}C as a tracer to the study of photosynthesis. These authors discovered PGA and RuBP among the early products of photosynthesis.[32-34] The details of a photosynthetic carbon cycle were described by Bassham *et al.* (1954).[6]

Weissbach *et al.* (1956),[2] Jakoby *et al.* (1956),[3] and later Racker (1957)[35] documented the *in vitro* dark enzymatic ribulose bisphosphate carboxylation reaction and the presence in photosynthetic tissues of two new enzymes, phosphoribulokinase and ribulose bisphosphate carboxylase (carboxydismutase). Although some authors believed[35] that the Benson-Calvin cycle "became greatly strengthened as a result," the experimental delineation of the *in vitro* carboxylation reaction enabled Calvin and his associates to reach the conclusion[1,4,12,13] that photosynthetic carboxylation *in vivo* was possibly a photochemical process, and occurred via a mechanism other than that given by the *in vitro* carboxylation.[2,3,35]

    c. <u>Photosynthetic carbon fixation as a photoreductive carboxylation reaction</u>. Shortly after the Calvin cycle was proposed,[6] Calvin and his associates discovered that several crucial aspects of that proposal could not be reconciled with the experimental behavior:[1,4,5,7,12,13]

      (1)   The uptake of CO_2 by RuBP in accordance with the dark enzymatic reaction in vitro,[2,3,35] proposed by Bassham *et al.*[6] (1954) as the initiating step in the Benson-Calvin cycle, was instead found by Bassham *et al.*[7] to be the terminating step in photosynthetic carbon fixation.

      (2)   Carbon fixation occurred via an asymmetric split of the reductive carboxylation intermediate.[1,4,5,12]

      (3)   The principal path of reductive carboxylation in sucrose synthesis was non-cyclic.[1,4,12]

      (4)   No measurable amounts of radiolabeled sucrose were obtained within the first fifty seconds from the upper three-carbon fragment of the six-carbon carboxylation intermediate.[5] By contrast, sucrose obtained from the lower three-carbon fragment of the carboxylation intermediate was observed simultaneously with PGA as the two earliest products of photosynthesis.[4]

The demonstration[2,3,35] of reaction 1 in vitro makes possible a comparison of the processes observed *in vivo*. Noting that reaction 1 was not obtained under photosynthetic conditions,[4] but was instead observed *in vivo* only when photosynthesis is stopped upon cessation of illumination, Bassham and Calvin wrote on pp. 52-55 of their monograph:[1]

    "An important question regarding the carboxylation reaction is whether the addition product splits to give two molecules of PGA, or whether only one molecule of PGA is formed, the other half of the addition product being reduced directly to triose phosphate ... A mechanism for the latter possibility is shown in the second reaction (L) ... Reaction (D) is the reaction indicated by the light-dark transient studies and by the studies of carboxydismutase, *in vitro*. Reaction (L) is a reaction which might occur only in the light, predominating at higher concentrations of reducing agent (photochemically produced) in which half of the product should appear at the oxidation level of carbohydrate."

The photochemical origin of the carboxylation reaction *in vivo* was thus recognized.

    d. <u>Calvin's observations of non-cyclic photoreductive carboxylation</u>. Wilson and Calvin reported their finding of a photoreductive path of carbon fixation in photosynthesis:[4]

$$\text{RuBP} + \text{CO}_2 \xrightarrow{\text{Light}} \underset{\underset{\text{PGA}}{+}}{\text{Triose}} \longrightarrow \text{Sucrose} \qquad (2)$$

whose product distribution differs from reaction 1. Instead of two PGA molecules, one molecule of PGA and another at the level of carbohydrate, observed as the new carbon incorporated in sucrose, were obtained. Reaction 2 was first proposed in 1952 by Calvin and Massini.[12] In confirming reaction 2, Wilson and Calvin used greatly improved techniques of sampling and chromatography.[4] In 1960, Bassham and Kirk presented[13] an argument based on kinetic data in further corroboration of reaction 2.

3. Experimental Basis for the Light-Driven Non-Cyclic Path of Carbon in Photosynthesis

a. Path of the upper three-carbon fragment. The experiments used by Calvin *et al.* to develop[1,6] the reductive pentose phosphate (RPP) cycle involved samples that were exposed to $^{14}CO_2$ for short times. There only the upper three-carbon fragment of the six-carbon intermediate, obtained from the addition of $^{14}CO_2$ to carbon 2 of RuBP, would be radiolabeled (indicated by *). The experimental condition is schematically shown as follows:

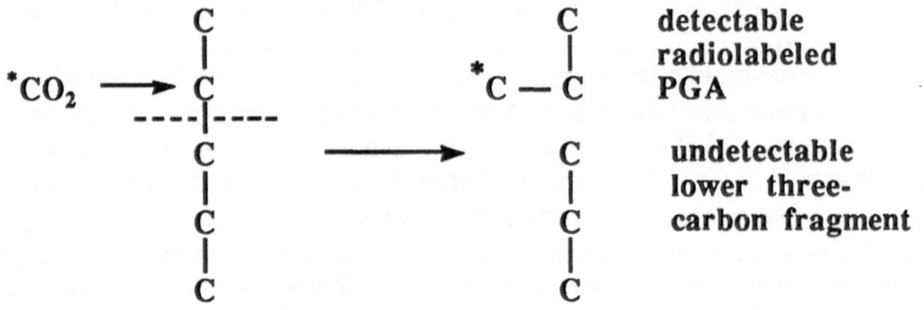

Scheme "A"

b. Path of the lower three-carbon fragment. A second group of experiments, which corroborated[4,10-12] the non-cyclic path of sucrose formation, were conducted on samples pre-saturated with ^{14}C obtained from long exposures to $^{14}CO_2$. In these experiments both the upper and lower three-carbon fragments from the six-carbon intermediate, obtained upon a rapid increase in the pressure of labeled CO_2, were detectable. Under these conditions, and not those of scheme "A", sucrose was observed as an early product. Calvin concluded that the lower three-carbon fragment, unlike the upper PGA, was directly, in a non-cyclic path, converted to the oxidation level of carbohydrate:[4]

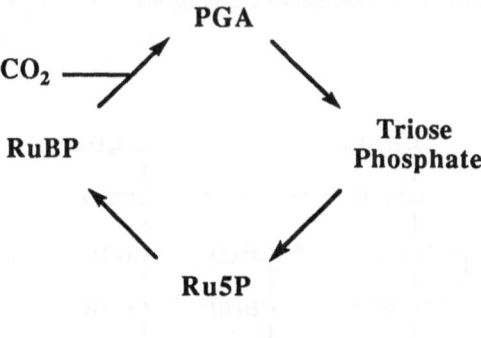

$^*CO_2 \longrightarrow$

$$\begin{array}{c}
\overset{*}{C} \\
| \\
\overset{*}{C} \\
\text{-----} \\
\overset{*}{C} \\
| \\
\overset{*}{C} \\
| \\
C
\end{array}
\qquad
\begin{array}{c}
\overset{*}{C} \\
| \\
C - \overset{*}{C} \\
\end{array}
\quad
\begin{array}{l}
\text{labeled} \\
\text{upper} \\
\text{PGA}
\end{array}$$

$$\begin{array}{c}
\overset{*}{C} \\
| \\
\overset{*}{C} \longrightarrow\!\!\longrightarrow \, ^*\text{Sucrose} \\
| \\
\overset{*}{C}
\end{array}$$

Scheme "B"

labeled
lower
PGL

c. <u>Development of the reductive pentose phosphate cycle: metabolic path of the upper PGA</u>. From the intramolecular distributions of radiocarbon found in the sugar phosphates and PGA, obtained under the experimental conditions of scheme "A", Bassham and Calvin developed the RPP cycle as follows:[1,6]

i. PGA as a first product. In experiments, in which short times of exposures of samples to $^{14}CO_2$ were required to limit the labeling of compounds to the first few steps of the carbon assimilation pathway, the predominant labeled product was PGA. Degradation of the PGA showed most of the radiocarbon to be in the carboxyl carbon and the formation of PGA via a carboxylation of the RuBP, as shown in scheme "A", was suggested as a first step in carbon dioxide assimilation during photosynthesis.

ii. Evidence for a cyclic process. The cyclic path for the pentose phosphate reduction was confirmed[4] by Wilson and Calvin. These authors postulated the essential constituents of the RPP cycle:

PGA

$CO_2 \longrightarrow$

RuBP

Triose
Phosphate

Ru5P

Scheme "C"

They argued that a sudden depletion in the CO_2 of an *in vivo* system in steady-state photosynthesis would result in sudden rises in the concentrations of the metabolites.

Based on the relative positions of these metabolites in the reaction cycle given by scheme "C", the predicted increases in the concentrations would follow a time sequence that is reverse to that of their respective appearances in the cyclic scheme: RuBP, Ru5P, triose phosphate and PGA. These predictions were entirely borne out by the observed experimental behavior.[4]

iii. Formation of fructose-6-P. The distribution of radiocarbon found in PGA and in hexose phosphate agreed with the postulated mechanism for the reduction and condensation of PGA via reactions which were presumed to be similar to a reversal of the Emden-Meyerhoff pathway for glycolysis.

$$
\begin{array}{lll}
& * \;\; CH_2OPO_3 = & * \;\; CH_2OH \\
2 & * \;\; CHOH & * \;\; C=O \\
& *** \;\; CO_2 - & *** \;\; CHOH \\
& & *** \;\; CHOH \\
& & * \;\; CHOH \\
& & * \;\; CH_2OPO_3 =
\end{array} \qquad (3)
$$

iv. Formation of sedoheptulose phosphate. In the transketolase reaction written below, reaction 4, which gives a satisfactory labeling pattern for the observed radiocarbon distribution in sedoheptulose monophosphate, the hexose is labeled to reflect the observation by Kandler and Gibbs[8] that degraded glucose obtained from short exposures of $Chlorella$ to $^{14}CO_2$ at lower light intensities contains less ^{14}C at C-3 than C-4:

$$
\begin{array}{lllll}
CH_2OP & CH_2OP & CH_2OH & & CH_2OH \\
C=O & C=O & C=O & \longrightarrow & C=O \\
**CHOH & **CH_2OH & -*CHOH & **CH_2O & **CHOH \\
*CHOH & & **CHOH & CHOH & CHOH \\
**CHOH & & CHOH & CH_2OP & CH_2OP \\
CHOH & & CH_2OP \\
CH_2OP
\end{array} \qquad (4)
$$

383

v. Formation of pentose phosphates. For the formation of pentose phosphate in agreement with the observed labeling pattern, an additional postulated transketolase reaction was given as follows:

$$
\begin{array}{ccccccc}
\text{CH}_2\text{OH} & & {}^{**}\text{CH}{=}\text{O} & & \text{CH}_2\text{OH} & & {}^{*}\text{CH}{=}\text{O} \\
| & & | & & | & & | \\
\text{C}{=}\text{O} & + & \text{CHOH} & \longrightarrow & \text{C}{=}\text{O} & + & {}^{*}\text{CHOH} \\
| & & | & & | & & | \\
{}^{*}\text{CHOH} & & \text{CH}_2\text{OP} & & {}^{**}\text{CHOH} & & {}^{*}\text{CHOH} \\
| & & & & | & & | \\
{}^{*}\text{CHOH} & & & & \text{CHOH} & & \text{CHOH} \\
| & & & & | & & | \\
{}^{*}\text{CHOH} & & & & \text{CH}_2\text{OP} & & \text{CH}_2\text{OP} \\
| \\
\text{CHOH} \\
| \\
\text{CH}_2\text{OP}
\end{array}
$$

$$
\left.
\begin{array}{c}
{}^{*}\text{C} \\
| \\
{}^{*}\text{C} \\
| \\
{}^{***}\text{C} \\
| \\
\text{C} \\
| \\
\text{C}
\end{array}
\right\} \quad (5)
$$

SMP PGL Xu5P R5P

b. <u>Development of the non-cyclic path of sucrose synthesis: metabolic path of the lower three-carbon fragment</u>. The results obtained by Calvin *et al.* in experiments involving pre-saturation of the intermediate compounds with ^{14}C (scheme "B") provided additional information, based on which Fong and Butcher suggested a non-cyclic path of sucrose synthesis as originally proposed by Wilson and Calvin:[4,10,11]

i. RuBP carboxylation. The molecular mechanism for RuBP carboxylation may be considered[10,11] as follows. Carbon dioxide fixation occurs with addition of the CO_2 molecule at C-2 of RuBP, followed by cleavage of the C_6 reaction intermediate at C-2 and C-3.[9,37] It was suggested that the *cis*-2,3-enediol form of RuBP is formed as a precursor to CO_2 uptake:[38]

$$
\begin{array}{ccc}
& \text{CH}_2\text{OP} & & \text{CH}_2\text{OP} \\
& | & & | \\
\text{-B:} \quad & \text{C}{=}\text{O} & \text{-BH} & \text{C}{-}\text{O}{-} \\
& | & & \| \\
\text{H}{-}\text{C}{-}\text{OH} & \longrightarrow & \text{C}{-}\text{OH} \quad (6) \\
& | & & | \\
\text{H}{-}\text{C}{-}\text{OH} & & \text{H}{-}\text{C}{-}\text{OH} \\
& | & & | \\
& \text{CH}_2\text{OP} & & \text{CH}_2\text{OP} \\
& \mathbf{1} & & \mathbf{2}
\end{array}
$$

The six-carbon intermediate is most likely 2-carboxy-3-keto-D-arabinitol-1,5-bisphosphate (CKABP).[39,40] Addition of CO_2 at the *si* face, at C-2 of the enediol, 2, followed by hydration of the reaction intermediate results in the formation of the *gem*-diol, 3,

$$
\begin{array}{ccc}
\text{CH}_2\text{OP} & \text{O} & \text{CH}_2\text{OP} \\
| & \| & | \\
\text{C}-\text{OH} & *\text{C} & \text{HO}-\text{C}-*\text{CO}_2- \\
\| & \| & | \\
\text{C}-\text{OH} & \text{O} & \text{HO}-\text{C}-\text{OH} \\
| & & | \\
\text{H}-\text{C}-\text{OH} & \text{O} & \text{H}-\text{C}-\text{OH} \\
| & \diagup\hspace{-0.3em}\diagdown & | \\
\text{CH}_2\text{OP} & \text{H}\quad\text{H} & \text{CH}_2\text{OP} \\
& & \mathbf{3}
\end{array}
\qquad (7)
$$

The inversion and retention of configuration respectively at C-2 and C-4 are required in the hydrolytic splitting of 3, as two molecules of D-glycerate-3-phosphate are produced in the enzymatic reaction.[2,3,34] Current views favor a carbanion intermediate, 4.[38,39] This pathway could reasonably be described by the SE1 mechanism proposed by Cram *et al*.[41,42] The inversion at C-2 subsequently results from protonation of the anion center from the opposite side of the solvation site, ultimately yielding the enzyme-bound adduct 5:

$$
\begin{array}{ccc}
 & & \text{CH}_2\text{OP} \\
 & & | \\
\text{CH}_2\text{OP} & -\text{O}_2*\text{C}-\text{C}-\text{OH} & \\
| & \text{Mg}^{2+}\ \ \text{H} & \\
\text{HO}-\text{C}-*\text{CO}_2- & & \\
\text{Mg}^{2+}\ |\quad \text{H} & \quad\quad\quad \text{H} & \\
\quad\quad | & \quad\quad | & \\
\text{O}=\text{C}-\text{C}-\text{CH}_2\text{OP} & \text{O}=\text{C}-\text{C}-\text{CH}_2\text{OP} & \\
\quad | \quad\ | & \quad | \quad\ | & \\
\quad \text{O}-\ \text{OH} & \quad \text{O}-\ \text{OH} & \\
\mathbf{4} & \mathbf{5} &
\end{array}
\qquad (8)
$$

In reaction 8, the presence of the Mg^{2+} ion, which presumably provides additional stabilization energy along the reaction coordinate for the carbon inversion, has been introduced as a possible role for the essential metal ion. Mg^{2+} ion is the normal activator, but Mn^{2+}, Co^{2+}, Ni^{2+}, Fe^{2+}, Ca^{2+} and Cu^{2+} can also support activity to various extents. The proposed use of the Mg^{2+} ion in reaction 8 would provide different environments for the two PGAs.

ii. Reaction pathway for the lower three-carbon fragment. Upon pre-saturation of the RuBP with radiocarbon, the uptake of a $^{14}CO_2$ molecule by the RuBP, followed by splitting of 3 at C-2 and C-3, would result in the formation of two radioactive three-carbon fragments, in accordance with scheme "B". Under these conditions, Wilson and Calvin reported[4] that the 5-s point of their experiment showed that the increase in total carbon in PGA was approximately 3/5 the decrease in the total carbon of RuBP. In other words, there was an increase of only one mole of PGA for a decrease of one mole of RuBP, suggesting that about half of the expected PGA was observed in some other form. Wilson and Calvin further concluded that the appearance of the sucrose in this 5-s point accounts for the remaining 2/5 decrease in the total carbon of RuBP plus the newly assimilated carbon from CO_2. The observed total carbon disappearing from the six-carbon RuBP carboxylase intermediate was thus equally distributed among the new carbon incorporated in the two earliest products of RuBP carboxylation, PGA and sucrose.[1,10]

The early appearance of sucrose was not observed[5] under the experimental conditions of scheme "A". Wilson and Calvin's observations of sucrose under the conditions of scheme "B" thus provide positive evidence for the non-cyclic reductive conversion of the lower three-carbon fragment from 3 to carbohydrate.

iii. Specificity of the biochemical fates of the upper and lower fragments of CKABP. The collective significance of the above observations may be further emphasized as

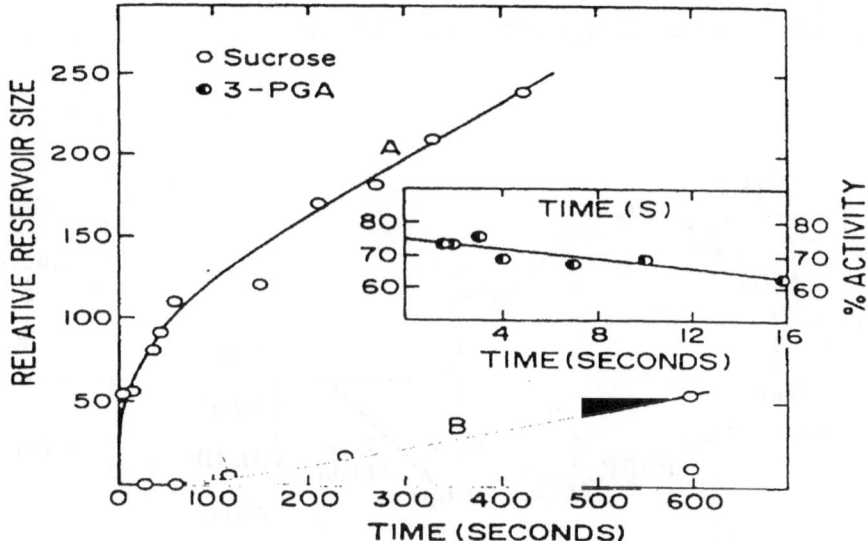

Figure 1. Changes in relative reservoir sizes of sucrose. Curve A: Data abstracted from Fig. 5 of Ref. 4 on the appearance of sucrose as the earliest and principal product of photosynthesis in ^{14}C pre-saturated *Scenedesmus* (scheme "B"). Curve B: Appearance of sucrose in short [14]CO_2 exposure experiments according to scheme "A" (data abstracted from Ref. 5). The data points for the relative reservoir sizes in curves A and B are calibrated against the 5-s point for PGA in Ref. 4. Inset: Extrapolation to zero time of the percent of ^{14}C radioactivity found in PGA (Ref. 6).

follows.[11] The kinetic behavior for the conversion of the labeled PGA in scheme "A" to sucrose is given in Figure 1 by curve B according to the measurements by Benson *et al.*[5] The sigmoidal shape of this curve clearly suggests that sucrose synthesis from the upper PGA involves a number of intermediate steps. The yield of sucrose is negligibly small within the first fifty seconds.

By contrast, the conversion of sucrose from the lower three-carbon fragment in scheme "B" is given in Figure 1 by curve A according to the measurements by Wilson and Calvin.[4] In this case, the yield of sucrose within the first few seconds is several orders of magnitude greater than that observed in curve B for the upper PGA. The non-sigmoidal behavior of curve A in Figure 2 provides evidence for the direct conversion of the lower three-carbon fragment in scheme "B" to PGL, then sucrose.[4]

The dramatic difference in the experimental behavior of curves A and B in Figure 2 provides strong evidence for the differentiable metabolic paths of the upper and lower three-carbon fragments obtained upon splitting of the CKABP. A schematic representation of this splitting may be given as follows:

$$
\begin{array}{l}
\text{[H]} \quad *\text{PGA} \\
*\text{PGL} \\
\text{[C}_2\text{]} \\
\text{Ru5P} \\
\text{ATP} \\
\text{RuBP} \quad *\text{CO}_2 \quad 2\,\text{H}_2\text{O} \\
*\text{C C} \\
\text{C C C} \quad \text{CHO} \\
\text{O}_2 \quad P_{680}\,\text{Light} \quad \text{CHOH} \quad \text{CH}_2\text{OP} \quad \text{PGL} \\
\text{Sucrose} \quad \text{hexose-P}
\end{array}
\tag{9}
$$

in which the non-cyclic pathway of sucrose synthesis from the lower three-carbon fragment is distinguished from the regenerative participation of the labeled upper three-carbon fragment in the RPP cycle, where the [C$_2$] denotes the C-1, C-2 fragment from the sedoheptulose-7-P or the hexose monophosphate in Reactions 4 and 5 respectively.

In Reaction 9, the two-electron reduction of CKABP to one equivalent of PGA and another of triose phosphate, glyceraldehyde-3-P (PGL), is shown to require two reducing equivalents, 2 [H] . The asymmetric arrangement of the upper and lower three-carbon fragments in Reaction 9, in which the Mg^{2+} ion is bound to the upper fragment, enables a branching of two different reductive PGA pathways. From this point one lower (unlabeled) PGL exits each single turnover of the RPP cycle. This PGL subsequently undergoes condensation to yield hexoses, then sucrose. The upper labeled *PGA is likewise reduced to PGL, using two additional reducing equivalents, 2[H]. A total of four reducing equivalents are thus used up per turn of the RPP cycle in the uptake of one CO$_2$ molecule, balancing the requirement of four oxidizing equivalents in the evolution of one molecule of oxygen.

In Figure 2, the details of the path of carbon given by Reaction 9 are shown, in which the CKABP is tentatively represented as having split into the two non-equivalent enzyme-bound PGA molecules according to 5.

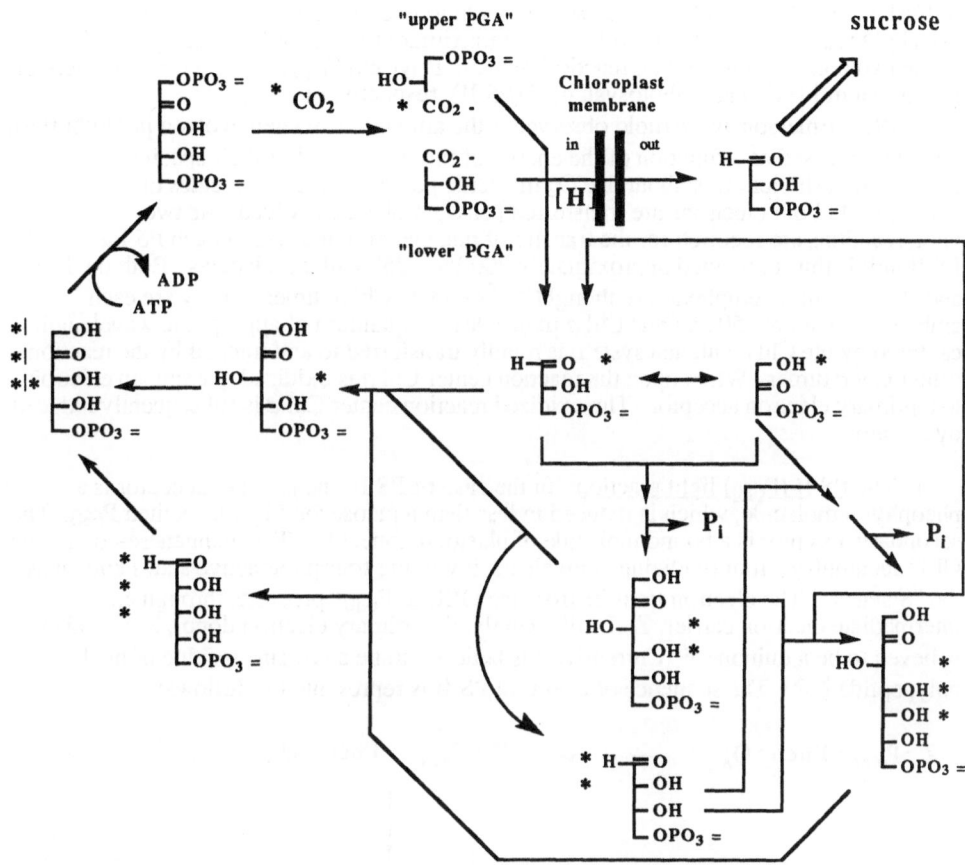

Figure 2. Schematic representation of the path of carbon in photosynthesis (see, Reaction 15 for alternative mechanism).

5. Oxygen Evolution, CO_2 Fixation, and the P_{680} and P_{700} Light Reactions

a. The Hill-Bendall scheme and the photosynthetic unit. According to the Hill-Bendall scheme,[45] two Chl a photoreactions are connected in series by a chain of electron carriers, whereby the electrons obtained from the oxidation of water are driven against an electrochemical gradient to reduce CO_2 to the carbohydrate level.[46]

The oxidizing end of the two photosystems is equipped with the oxygen evolving complex (OEC), an enzyme complex that catalyzes the oxidation of water and the evolution of molecular oxygen. At the reducing end, a flavoprotein mediates the reduction of NADP, and then the CO_2. The two Chl a photoreactions are known as the P_{700} and P_{680} light reactions after the wavelengths of maximum bleaching obtained upon oxidation of the two corresponding Chl a reaction centers. P_{700} and P_{680} are the reaction centers of photosystem I (PS I) and photosystem II (PS II), respectively.

In 1932, Emerson and Arnold observed[47] the amount of oxygen evolved per light flash (10^{-5} s) increased as a function of the energy of the flashes and of their spacing. The maximum flash yield was about one O_2 molecule per 2000-2500 molecules of chlorophyll. Four electrons are transferred per O_2 molecule evolved. As two photoreactions are required for the transfer of one electron, the size of each PS I and each PS II unit is thus computed approximately $2000/8 \approx 250$ Chl a molecules. Both the P_{700} and P_{680} reaction complexes are thought[14,48,49] to be Chl a dimers. They are each embedded in about 250 antenna Chl a molecules. A quantum of appropriate wavelength captured by the Chl a antenna system is rapidly transferred to and trapped by the reaction center Chl a dimer. Whereupon the reaction center Chl a is oxidized, passing an electron to a primary electron acceptor. The oxidized reaction center Chl a is subsequently reduced by a donor.

b. The PS II (P_{680}) light reaction. In the case of PS II, the primary acceptor is a pheophytin molecule, which is reduced in less than a picosecond by the excited P_{680}. The secondary acceptor is a bound molecule of plastoquinone, Q_A. The manganese-containing OEC accumulates four oxidizing equivalents, involving four photoactive states known as the "S states." The electron transfer from the OEC to P_{680}^+ proceeds through an intermediate electron carrier, Z. Until recently, the primary electron donor, Z, was also believed to be a quinone.[46] Currently, Z is believed to be a tyrosine residue of the D1 polypeptide.[50,51] The sequence of events at PS II is represented as follows:

$$Z \cdot P_{680} \cdot \text{Pheo} \cdot Q_A \longrightarrow Z \cdot P_{680}^* \cdot \text{Pheo} \cdot Q_A$$

$$\downarrow$$

$$Z^+ \cdot P_{680} \cdot \text{Pheo}^- \cdot Q_A \longleftarrow Z \cdot P_{680}^+ \cdot \text{Pheo}^- \cdot Q_A$$

$$\downarrow$$

$$Z^+ \cdot P_{680} \cdot \text{Pheo} \cdot Q_A^- \tag{10}$$

Reaction 10 proceeds on further with the reduction of Z^+ by the OEC. The electrons for this reduction are derived from the oxidation of water, which results in O_2 evolution.[52]

i. Periodic pattern of single-flash yields of oxygen evolution. The amounts of oxygen evolved by single flashes were first measured by Allen and Franck.[53] It was later demonstrated by the action spectrum of the activation process that only PS II was involved.[54,55]

The development of sensitive polarographic techniques[56] allowed precise measurement of the small amounts of O_2 evolved in very weak continuous illumination or by single brief (10^{-5} s) flashes. In a typical experiment, the yields of the first two flashes are close to zero, and the subsequent yields show a damped oscillation with a period of 4, involving the four stages of the so-called "S states."[57] The flash width is about 10^{-5} s.

ii. The charge-state model for proton release in oxygen evolution. The OEC consists of several membrane-bound proteins. Manganese is bound to some of these proteins and is responsible for the accumulation of the four positive charges in the transitions between the S states. Upon the accumulation of four positive charges, the oxidation of two water molecules gives rise to an O_2 molecule. Four equivalents of protons are liberated inside of the thylakoid lumen during the S states transitions.[58-60]

A 2-Mn model of the water-cleavage cycle was given by Saygin and Witt.[61] The reaction sequence at pH 7 accumulates the first positive charge in the $S_0 \rightarrow S_1$ transition by a $Mn^{2+} \rightarrow Mn^{3+}$ transition.[61] Subsequent accumulation of two additional positive charges in $S_1 \rightarrow S_2$ and $S_2 \rightarrow S_3$ involves[58,62] $Mn^{3+} \rightarrow Mn^{4+}$ transitions. The proton releases attending the $S_0 \rightarrow S_1$ and $S_2 \rightarrow S_3$ transitions are accompanied by the conversion of OH^- to O^{2-}. At pH 7 the release of the two H^+ equivalents in the $S_3 \rightarrow S_0$ transition concomitant with O_2 evolution may be represented as follows:

$$P_{680}^+ + Mn(IV)Mn(IV)(O^{2-})_2 \longrightarrow P_{680} + Mn(II)Mn(III)(OH^-)_2$$

$$S_3 \qquad\qquad\qquad 2\ H_2O \quad O_2 + 2\ H^+ \qquad\qquad S_0 \qquad\qquad (11)$$

In O_2-evolving PS II, P_{680}^+ is reduced[63-65] with half-lives in the nanosecond time domain. Z is oxidized in accordance with the reduction kinetics of P_{680}^+.[51] On the other hand, the half-lives of the reduction of Z^+ [66] and the oxidation of the S states[61,67,68] also coincide but occur in the micro-millisecond time-range.

iii. Dimer of Chl a dihydrate as model for P_{680} reaction center. The oligomer of Chl a dihydrate, when platinized, is capable of photocatalyzing the splitting of water to yield two parts of hydrogen and one part of oxygen.[21a] In the presence CO_2, the O_2 is evolved from the photolytic reaction, but the hydrogen from water splitting, instead of being evolved, is consumed in the reduction of the CO_2 to formaldehyde and formic acid:[21b]

$$2\ H_2O + 2\ CO_2 \longrightarrow H_2CO + HCOOH + 3/2\ O_2 \qquad (12)$$

The O_2 present in the water is converted to hydrogen peroxide, ylelding hydroxy and perhydroxy radicals.[21c] The formate ion in turn is re-oxidized to CO_2:[69]

$$\cdot OH + HCOO^- \longrightarrow H_2O + \cdot CO_2^- \qquad (13)$$

$$\cdot CO_2^- + O_2 \longrightarrow O_2^{\cdot -} + CO_2 \qquad (14)$$

resulting in a scrambling of the ^{18}O from an isotopically enriched water mixture into the O atoms of the CO_2.[22] Reactions 12-14 are *in vitro* analogs of the photocarboxylation and photorespiration reactions in vivo, respectively. The net yield of Reaction 12 thus depends on a regulated separation of O_2 from the photosynthetic products in Reaction 12. In photosynthesis *in vivo* the carboxylase is also the oxygenase that catalyzes photorespiration. The unraveling of the molecular mechanisms of the enzymatic reactions of RuBP carboxylase/oxygenase is thus important in our attempt to demonstrate reductive carboxylation *in vitro*.

The dihydrate oligomer, $(Chl\ a \cdot 2H_2O)_n$, is unusual among *in vitro* forms of the chlorophyll. It is the only known Chl *a* species *in vitro* to have a reduction potential (E_m = 1.1 V) sufficiently high to oxidize water (E_m = 0.82 V at pH 7).[14] Agostiano *et al.* demonstrated[20] that the dimer, $(Chl\ a \cdot 2H_2O)_2$, is stabilized in a 1:1 acetone-water mixture (50 volume % or 0.8 mole % water). From a measurement of the current intensities under the polarographic peaks attributable to monomeric Chl *a* and the dihydrate dimer, the Chl *a* monomer was found to have an area of 270 $Å^2$ compared to that, 526 $Å^2$, for $(Chl\ a \cdot 2H_2O)_2$.[20] The dihydrate dimer of Chl *a* obtained in 1:1 acetone-water has an absorption maximum at 680 nm.[70] A probable structure for this dimer, which has been proposed[14] as a model for the P_{680} reaction Chl *a* complex in PS II, is shown in Figure 3.

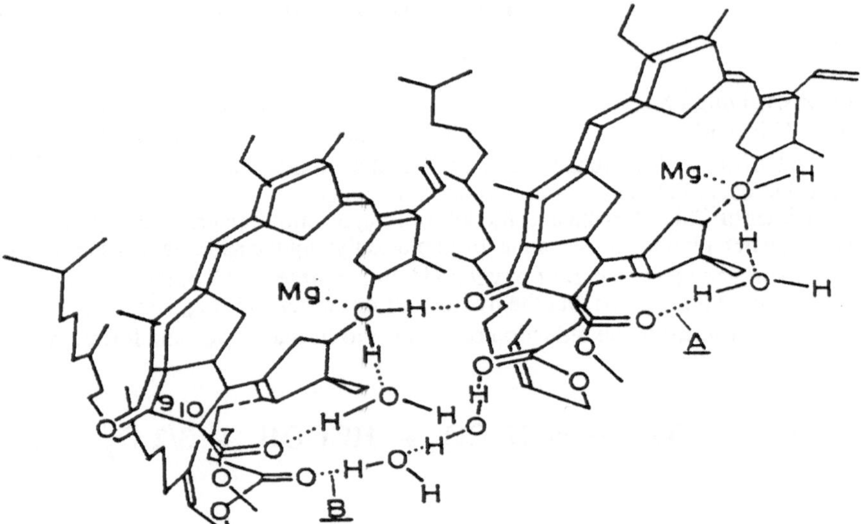

Figure 3. The Chl *a* dihydrate dimer as a model for the P_{680} reaction center in PS II.

c. The reducing side of PS II. On the reducing side of PS II, the protein-bound quinone, Q_A, a plastoquinone, accepts one electron from pheophytin. Q_A is reduced in 0.6-1 ns. Q_A reduction is temperature independent, and occurs at appreciable rates at liquid nitrogen temperature.[46,52,59,71] Q_A is re-oxidized by another molecule of plastoquinone, PQ_B, which is bound to a 32 KDa protein on the outer surface of the thylakoid membrane.

PQ_B, which is reduced in two steps while bound to the protein, is protonated by the external water phase.[59,72] The doubly reduced plastoquinone, PQ_BH_2, now dissociates from the protein and dissolves into a soluble pool of plastoquinone, and is substituted on the 32 KDa protein by a molecule of PQ. The reoxidation of PQH_2 is strongly dependent on the presence of HCO_3^- or CO_2, which is believed to accelerate the PQ-PQH_2 exchange on the 32 KDa protein. This step is the one-electron-two-electron gate of electron transport on the reducing side of PS II. CO_2, the final electron acceptor of the electrons from water, thus serves an important role in the "feed forward" regulation of the photosynthetic electron transport system.

The electron acceptor for PQH_2 reoxidation is the cytochrome f-b_6 complex, a five polypeptides complex containing the Fe-S center, one cyt.f and two molecules of cyt.b_6. Cyt.f is reduced by the Fe-S center, at the internal side of the membrane.[59,72] Two protons are released into the inner aqueous phase, contributing to the generation of a proton gradient across the membrane.

d. The PS I P_{700} light reaction. Cyt.f, which has a mid-point potential E_m = 360-365 mV, is re-oxidized, with a half-time of ca. 70 μs, by the copper protein plastocyanin (PC, E_m = 370 mV). PC is the donor to the PS I light reaction. The P_{700} reaction center complex (E_m = 450 mV) is oxidized in 30-50 ps in the photochemical reaction of PS I. The acceptor is believed to be a Chl a molecule, which is reduced to its anionic radical.[59] This radical anion reduces a series of three membrane-bound Fe-S centers. Then the electron is transferred to ferredoxin (Fd, E_m = -440 mV), a water-soluble 10 KDa iron sulfur protein dissolved in the external water phase present in the stroma of chloroplasts.[73]

Ferredoxin forms a 1:1 complex with the membrane-bound flavoprotein ferredoxin-NADP reductase (FNR). NADP also binds to the flavoprotein and is reduced and released into the external water phase, where it is utilized by the CO_2 reducing enzyme systems.

5. Consequences of the Coupling Between Photosynthetic Light and Dark Reactions

a. Unresolved questions. Reaction 9, which is based on the experimental conclusions of Wilson and Calvin[4] in corroboration of the earlier work by Calvin and Massini,[12] suggests a direct coupling between the oxygen evolution process and carbon fixation. The remainder of this review is devoted to a summary of the many unresolved questions arising from Calvin's discovery of the photoreductive carboxylation reaction.

b. Possible coupling between the oxygen evolution process and carbon fixation. The photosynthetic quotient, O_2/CO_2 = 1,[25] requires that the oxygen evolution and carbon fixation are stoichiometrically related. This relationship is reflected in Reaction 9, in which the reductive assimilation of one CO_2 molecule per turn of the RPP cycle requires four reducing equivalents, corresponding to each turn of the 5 states in the evolution of one O_2 molecule.

It is known that the maximal yield of oxygen evolution is obtained when each cycle of the four flashes takes up on the order of 200 ms.[25] Emerson and Arnold reported[47] that the O_2 yields were half maximal at a dark spacing between the flashes of about 40 ms. If the four-electron process involved in O_2 evolution is kinetically coupled to the four-electron process for CO_2 reduction, the time it takes for the RPP cycle turnover should be closely matched. Indeed, the turnover time for the *in vitro* carboxylase reaction has been observed to be on the order of 240 ms.[40] The time constants discussed above for the other electron transport processes are orders of magnitude shorter. It is conceivable that CO_2 fixation in photosynthesis may be kinetically coupled to the oxygen evolving process.

c. <u>Asymmetric splitting of CKABP in photoreductive carboxylation</u>. Reaction 9 requires an asymmetrical splitting of 3 at C-2 and C-3 in reductive carboxylation. A metabolic distinction between the two three-carbon fragments derived from this splitting involves some as yet unknown process.

In Figure 2, the unsymmetrical arrangement of the upper and lower PGA in 5 is assumed to be the precursor to the differentiable paths of the two three-carbon fragments. In this case, the upper and lower PGA molecules would be indistinguishable if they were released from the RuBP carboxylase to a common metabolic pool. Accordingly, it is necessary to postulate that either the lower PGA molecule is reduced to PGL while bound to the carboxylase active site, or it is released to a metabolic pool different from that to which the upper PGA is released. The former possibility would require an alternate means (RuBP carboxylase dependent) for converting the lower PGA to PGL, that would also be tightly coupled to sucrose synthesis (on the surface an unlikely possibility). The latter possibility would require some hitherto unknown channeling mechanism.

An alternative scheme that would explain the *in vivo* results, involves phosphohydroxy pyruvate (PHP) and PGL as products of the carboxylase reaction *in vivo* in the light. The six carbon reaction intermediate (if formed) would split to yield an upper molecule of PHP and a lower molecule of PGL. The PHP then undergoes a two-electron enzymatic reduction to yield PGA. Some evidence for this rather fanciful scheme may come from Benson *et al.*,[5] who observed PHP as a primary product of *in vivo* photosynthesis in very rapid equilibration with PGA. Finally, in 1959, Calvin and Pon presented the "best evidence" for photoreductive carboxylation *in vivo* in more-precise measurements of the rate of approach to ^{14}C saturation of the pools of PGA and RuBP in algae and in a more nearly true steady state of photosynthesis than had theretofore been achieved. They observed that the specific activity (^{14}C) of the RuBP remained higher than that of the α- and β-carbon atoms of PGA in photosynthetic carboxylation *in vivo*[76] and concluded that only a single PGA molecule was liberated for each molecule of CO_2 entering the algae, whereas the remaining three carbon atoms of CKABP went directly to the sugar level of oxidation when the light was on. These workers proposed an asymmetric splitting of the CKABP closely similar to that discussed, in which the lower and upper three-carbon fragments respectively yielded one molecule of triose phosphate and another of enzyme-bound PHP. The latter then underwent a two-electron reduction to give PGA.[76]

d. <u>Balance of carbon</u>. In higher plants, primary metabolism is believed to involve a compartmentation between the plastid and the cytosol.[77] While starch is synthesized and stored exclusively in the plastid, the synthesis of sucrose is restricted to the cytosol.[78] In

Figure 2, the sucrose synthesis is accordingly shown to occur via a rapid export of the triose phosphate to the cytosol outside of the thylakoid membrane. However, it must be cautioned that there exists no direct biochemical evidence that requires such a representation of sucrose synthesis from the lower three-carbon fragment of CKABP for the following reasons:

(1)   It has not been established that sucrose synthesis in *Scenedesmus*, as in higher plants, takes place outside of the chloroplast.

(2)   The recent radiolabel studies on sucrose synthesis in higher plants have mostly been conducted under the conditions given by scheme "A".[78,79] The corresponding interpretations relative to compartmentation of carbohydrate metabolism therefore strictly apply to the metabolic path of the upper three-carbon fragment of the CKABP.

(3)   Significantly, the data for sucrose synthesis given by the sigmoidal behavior in curve B of Figure 1, measured by Benson *et al.*[5] in 1952 using *Scenedesmus* whole cells under the experimental conditions of scheme "A", closely resemble the corresponding data obtained by Stitt *et al.*[78] more than three decades later using protoplasts prepared from wheat (*Tritcum aestivum var Timmo*) leaves.

If the transport of triose phosphate across the thylakoid membrane in sucrose synthesis is required as indicated in Figure 2, for each CO_2 molecule assimilated, one molecule of PGL is exported. In order to maintain the steady-state carbon content of the RPP pool, it will then be necessary to balance the requisite flow of P_i, PGA and triose phosphate using the phosphate translocator.[77,78]

e. Concluding remarks. Other light-activated enzymatic reactions in the RPP cycle are known. For example, the phosphoribulokinase is inactive in the oxidized form,[80] but is activated by light under the photoreductive conditions generated by the Chl *a* light reaction. The enzymatic synthesis of sucrose in higher plants is also known to be activated by light.[79,81] The regulatory mechanism for the activation of RuBP is mediated by light. This light-dependent activation of RuBP carboxylase/oxygenase is attributed to the increase in pH and Mg^{2+}, which has been shown to occur in the chloroplast stroma upon illumination.[82,83] Calvin and his associates thus presented[1,4-7,12,13,76] convincing evidence for photosynthetic carbon fixation as a photoreductive carboxylation reaction. This evidence, which is corroborated by the work of other investigators,[2,3,8-11,35,37-40,78] supports a scheme similar to that illustrated in Reaction 9 as the path of carbon in photosynthesis. Taken as a whole, Calvin's work requires a logical extension of currently accepted views to areas of inquiry long considered closed.

A demonstration of photosynthetic carboxylation *in vitro* could conceivably incorporate the characteristic properties of the Chl *a* light reaction as given by Reaction 12, coupled to enzymatic reactions using purified RuBP carboxylase/oxygenase and other appropriate enzymes and co-enzymes, in the reductive assimilation of CO_2 to yield carbohydrates outside of the living cell. After having succeeded in such a demonstration, we then will be in a position to state (as was done over thirty years ago)[74,75]: "We now know what the path of carbon is in photosynthesis, and it has been carried out in its entirety in the absence of living materials or formed parts of living cells."

Acknowledgments. One of us (FKF) acknowledges the extensive discussions with Professors Ralph Thauer and Georg Fuchs on the PHP-forming reaction; Professor Mark Stitt and Dr. Paul Quick on sucrose synthesis in higher plants; Professor Horst Senger and his group on *Scenedesmus* photosynthesis; and Drs. Fred Hartman and Gunther Schneider on the function and structure of the RuBP carboxylase.

References

1. Bassham, J.A. and Calvin, M. (1957) "The Path of Carbon in Photosynthesis," Prentice-Hall, Inc., Englewood Cl i f f s, N. J ., p . 5 5.
2 . Weissbach, A., Horecker, B.L. and Hurwitz, J. (1956) J. Biol . Chem. 218, 795-810.
3. Jakoby, W.B., Brummond, D.O., and Ochoa, S.J. (1956) J. Biol . Chem. 218, 811-822 .
4. Wilson, A.T. and Calvin, M. (1955) J. Am. Chem. Soc. 77, 5948-5957 .
5. Benson, A.A., Kawaguchi, S., Hayes, P., and Calvin, M. (1952) J. Am. Chem. Soc. 74, 4477-4482.
6. Bassham, J.A., Benson, A.A., Kay, L.D., Harris, A.Z., Wilson, A.T., and Calvin, M. (1954) J. Am. Chem. Soc. 76, 1760-1770 .
7. Bassham, J.A., Shibata, K., Steenberg, K., Bourdon, J., and Calvin, M. (1956) J. Am. Chem. Soc. 78, 4120-4124.
8. Kandler, O. and Gibbs, M. (1956) Plant Physiol. 31, 411-412 .
9. Fiedler, F., Mullhofer, G., Trebst, A. and Rose, T.A. (1967) Eur. J. Biochem. 1, 395-399.
10. Fong, F.K. and Butcher, K.A. (1988) Biochem. Biophys. Res. Commun. 150, 399-404.
11. Fong, F.K. and Butcher, K.A. (1987) Biochem. Biophys. Res. Commun. 142, 732-737.
12. Calvin, M. and Massini, P. (1952) Experientia 8, 445-484.
13. Bassham, J.A. and Kirk, M. (1960) Biochim Biophys. Acta 43, 447-464.
14. Fong, F.K. (1982) in: "Path of Light Reaction in Photosynthesis," ed. Fong, F.K. Springer Verlag, Heidelberg, Ch.8.
15. Barber, J. (1984) Trends Biochem. Scl. 99, 1-2.
16. Kuwabara, T., and Murata, N. (1982) Plant and Cell Physiol. 23, 533-539.
17. Kuwabara, T. and Murata, N. (1982) Biochim. Biophys. Acta 680, 210-215.
18. Hunziker, D., Abramowicz,m D.A., Damoder, R., and Dismukes, G.C. (1987) Biochim. Biophys. Acta 890, 614.
19. Abramowicz, D.A. and Dismukes, G.C. (1984) Biochim. Biophys. Acta 765, 318-328.
20. Agostiano, A., Caselli, M., Della Monica, M., Gotch, A.J., and Fong, F.K. (1988) Biochim. Biophys. Acta 936, 171-178.
21. (a) Fong, F.K. and Galloway, L. (1978) J. Am. Chem. Soc. 100, 3594-3596; (b) Fruge, D.R., Fong, D.G. and Fong, F.K. (1979) J. Am. Chem. Soc. 101, 3694-3697; (c) You, J.-L. and Fong, F.K. (1986) Biochem. Biophys. Res. Commun. 139, 1124-1129.
22. Showell, M.S., You, J-.L., Fong, G.D., Butcher, K.A., and Fong, F.K. (1987) in "Carbon Dioxide as a Source of Carbon," NATO ASI Series, ed. Aresta,M. and Forti,G., D. Reidel Publishing Co., Dordrecht, pp. 93-112.
23. Joliot, P. and Joliot, A. (1968) Biochim. Biophys. Acta 153 625.
24. Kok, B., Forbush, B., and McGloin, M. (1970) Photochem. Photobiol. 11, 457.

25. Allen, M.B., Arnon, D.I., Capinadale, J.B., Whatley, F.R., and Durham, L.J. (1955) J. Am. Chem. Soc. 77, 4149-4155.
26. Joliot, P. and Kok, B. (1975), in "Bioenergetics of Photosynthesis," ed. Govindjee, Academic Press, New York, pp. 368-413.
27. Ruben, S., Kamen, M.D. and Hassid, W.Z. (1940) J. Am. Chem. Soc. 62, 3443-3450.
28. Ruben, S., Kamen, M.D., and Perry, L.H. (1940) J. Am. Chem. Soc. 62, 3450-3451.
29. Ruben, S., and Kamen, M.D. (1940) J. Am. Chem. Soc. 62, 3451-3455.
30. Kamen, M.D. (1986) Ann. Rev. Biochem. 55, 1-34.
31. Arnon, D.I., Allen, M.B., Whatley, F.R. (1954) Nature 174, 394-396.
32. Benson, A.A. (1951) J. Am. Chem. Soc. 73, 2971.
33. Benson, A.A., Bassham, J.A., Calvin, M., Hall, A.G., Hirsch, H.E., and Kawaguchi, S., Lynch, V. and Tolbert,N.E. (1952) J. Biol. Chem. 196, 703-716.
34. Benson, A.A. and Calvin, M. (1947) Science 105, 648-649.
35. Racker, E. (1957) Arch. Biochem. Biophys. 69, 300-310.
36. Arnon, D.I. (1977) in: "Photosynthesis I. Photosynthetic Electron Transport and Photophosphorylation," ed. Trebst,A. and Avron,M., Springer-Verlag, Heidelberg, Germany, 7-56.
37. Mullhofer, G. and Rose, I. A. (1965) J. Biol. Chem. 240, 1341.
38. Jaworowski, A., Hartman, F. C. and Rose, I. A. (1984) J. Biol. Chem. 259, 6783.
39. Pierce, J., Tolbert, N. E., and Barker, R. (1980) Biochem. 19, 934.
40. Schloss, J. V. and Lorimer, G. H. (1982) J. Biol. Chem. 257, 4691.
41. Cram, D. J., Langemann, A., Allinger, J., and Kopecky, K. R. (1959) J. Am. Chem. Soc. 81, 5740.
42. (a) Cram, J. (1965) "Fundamentals of Carbanion Chemistry," Academic Press, New York, pp.138-158. (b) March, J. (1985) "Advanced Organic Chemistry. Reactions, Mechanisms, and Structure,n 3d ed., John Wiley and Sons, New York, pp. 517-518.
43. Lorimer, G. H. and Miziorko, H. M. (1980) Biochem. 19, 5321.
44. (a) Christeller, J. T. (1981) Biochem. J. 193, 839. (b) Lorimer, G. H. (1981) Ann. Rev. Plant Physiol. 32, 349.
45. Hill, R. and Bendall, F. (1960) Nature (London) 186, 136-137.
46. Forti,G., in "Carbon Dioxide as a Source of Carbon," ed. Aresta,M. and Forti,G., NATO ASI Series C: Mathematical and Physical Sciences Vol. 206, D. Reidel Publishing Company, Dordrecht, Holland, 1987.
47. Emerson, R. and Arnold, W. (1932) J. Gen. Physiol. 15, 391; 16, 191.
48. Den Blanken, H.J. and Hoff, A.J. (1983) Biochim. Biophys. Acta 724, 52-61.
49. Den Blanken, H.J., Hoff, A.J., Jongeselis, A.P.J.M., and Diner,B.A. (1983) FEBS Lett. 157, 21-27.
50. Debus, R.J., Barry, B.A., Babcock, G.T. and McIntosh, L. (1988) Proc. Natl. Acad. Sci. USA 85, 427-430.
51. Gerken, S., Brettel, K., Schlodder, E. and Witt, H.T. (1987) FEBS Lett. 223, 376-380.
52. Govindjee, Kambara, T., and Coleman, W. (1985) Photochem. Photobiol. 42, 187-210.
53. Allen, F. and Franck, J. (1955) Arch. Biochem. Biophys. 58, 510.
54. Joliot, P. (1961) J. Chim. Phys. 58, 584.
55. Kok, B. and Cheniae, G.M. (1966) Current Top. Bioenerg. 1, 1.
56. Joliot, P. and Joliot, A. (1968) Biochim. Biophys. Acta 153, 625.

57. Forbush, B., Kok, B., and McGloin, M. (1971) Photochem. Photobiol. 14, 307.
58. Junge, W. and Jackson, J.B. (1982) in: "Photosynthesis," ed. Govindjee.
59. Hahenel, W. (1984) Ann. Rev. Plant Physiol. 85, 659-693.
60. Meyer, B., Schlodder, E., Dekker, J.P., and Witt, H.T. (1989) Biochim. Biophys. Acta, 974, 36-43.
61. Saygin, O. and Witt, H.T. (1987) Biochim. Biophys. Acta 766, 403-415.
62. Dekker, J.P., Van Gorkom, H.J., Wensink, J. and Ouwehand,L. (1984) Biochim. Biophys. Acta 767, 1-9.
63. Van Best, J.A. and Mathis, P. (1978) Biochim. Biophys. Acta 503, 178-188.
64. Brettel, K. and Witt, H.T. (1983) Photobiochem. Photobiophys. 6, 253-260.
65. Schlodder, E., Brettel, K., Schatz, G.H. and Witt, H.T. (1984) Biochim. Biophys. Acta 765, 178-185.
66. Babcock, G.T., Blankenship, R.E. and Sauer, K. (1976) FEBS Lett. 61, 286-289.
67. Dekker, J.P., Plijter, J.J., Ouwehand, L. and Van Gorkom, H.J. (1984) Biochim. Biophys. Acta 767, 176-179.
68. Renger, G. and Weiss, W. (1985) Biochem. Soc. Trans. 14, 17-20.
69. Behar, D., Czapski, G., Rabani, J., Dorfman, L.M., and Schwarz, H.A. (1972) J. Phys. Chem. 74, 3209.
70. Agostiano, A., Butcher, K.A., Showell, M.S., Gotch, A.J. and Fong, F.K. (1987) Chem. Phys. Lett. 137, 37-41.
71. Witt, H.T. (1979) Biochim. Biophys. Acta 505, 355-370.
72. Cramer, W.A. and Crofts, A.R. (1982) in: "Photosynthesis," ed. Govindjee, Vol. I. pp.387-467, Academic Press, N.Y.
73. Forti, G. (1987) in: "New Comprehensive Biochemistry," Elsevier, North Holland, Amsterdam.
74. Calvin, M. (1956) J. Chem. Soc. 1956, 1895-1915.
75. News Report (July 4, 1955) Chem. Eng. News. 33, 2809.
76. Calvin, M. and Pon, N.G. (1959) J. Cellular Comp. Physiol . 54, Suppl . 1, 51-74.
77. Dennis, D.T. (1986) "The Biochemistry of Energy Utilization in Plants," Blackie, Glasgow.
78 . Stitt, M. (1987) Plant Physiol . 84, 201-204 .
79. Stitt, M., Wirtz ,W., and Heldt,H.W. (1983) Plant Physiol. 72, 767-774.
80. Buchanan, B.B. (1980) Ann. Rev. Plant Physiol. 31, 341-374 .
81. Battistelli, A. and Leegood, R.C. (1989) Proceedings of NATO-ASI Summer School on "Enzymatic and Model Carboxylation and Reduction Reactions for Carbon Dioxide Utilization," Ginosa, Italy.
82. Portis, A.R., Chon, C.J., Mosbach, A. and Heldt, H.W. (1977) Biochim. Biophys. Acta. 461, 313-325.
83. Purczeld, P., Chon, C. J., Po rtis , A . R., Heldt, H.W., and Heber, U. (1978) Biochim. Biophys. Acta. 501, 488-498.

Conclusions

by

M.Aresta

The carbon dioxide emission in the atmosphere is a problem of great concern today: the increase of the concentration of this gas in the atmosphere is a well demonstrated fact, but the relationship existing between carbon dioxide level and global warming is still a matter of discussion. Due to the complexity of the system and to the available data, mathematycal models do not offer the possibility of defining exactly the long term scenarios and, thus, the effects on the planet temperature and productivity of the carbon dioxide accumulation are still unknown.

The existing uncertainty has suggested to take an action for limiting the carbon dioxide emission: this can be realized using various technologies related to the energy production and utilization.

Science is offering a new strategy for limiting the carbon dioxide emission into the atmosphere.

This School has demonstrated, infact, that we have the possibility of developing "artificial systems" for either carbon dioxide fixation into useful chemicals or reduction to energy richer molecules.

This represents a powerful tool, because, if we succeed in developing selective reactions for carbon dioxide fixation into long-lasting materials, we can sequester it and limit its emission into the atmosphere, and if we develop fast processes for carbon dioxide reduction, we shall be able to use the natural fossil fuels in lower amounts.

The most precious suggestion coming from this School was that in order to make this strategy "economically efficient" and brought into practice, interdisciplinary research programmes must be developed by the international scientific community and funded at a level adequate to the complexity of the problem we wish to solve and to the results we expect and this research can afford.

Reaction mechanisms of biological and chemical systems must be further investigated and new catalysts for carbon dioxide conversion synthetized.

M. Aresta and J. V. Schloss (eds.),
Enzymatic and Model Carboxylation and Reduction Reactions for Carbon Dioxide Utilization, 397–398.
© 1990 *Kluwer Academic Publishers.*

398

New combined biological and chemical systems seem to be very promising and should be developed.

Topics of interest are:

- Chemical catalysts that promote carboxylation reactions;

- Coupled enzymes and semiconductors or active biological complexes supported on electrodes; (Little information is available on these topics.)

- Chemical catalysts that mimic photosynthetic enzymes and other carbon dioxide reducing enzymes;

- Biocatalysts relevant to carbon dioxide concentration and energy transfer.

These areas seem to be highly promising for the fixation of large amounts of carbon dioxide to be of practical application and assure, thus, the real possibility of recycling carbon dioxide from those industrial sources that actually produce it.

Once again, I would to thank NATO for having sponsored this School, that resulted very interesting thank to the great interest and active participa tion of the Students and to the high quality of the speeches delivered by the Lecturers and to their continuous presence and disponibility.
All of them contributed to generate a friendly atmosphere and, thus, to the success of the School.

To all of you I wish a pleasant journey home with the hope that you will keep alive the memory of Riva dei Tessali.

List of Short Communications Presented
at the NATO-ASI Summer School on :

ENZYMATIC AND MODEL CARBOXYLATION AND REDUCTION
REACTIONS FOR CARBON DIOXIDE UTILIZATION.

Riva dei Tessali, Ginosa (ITALY)

June 17-28, 1989.

CHEMICAL SECTION.

M. Aresta,[a] E. Quaranta,[a] I. Tommasi [a] and R. Gobetto [b],
^{13}C and ^{31}P n.m.r. Spectroscopy for Structural Characterization of Carbon Dioxide Metal Complexes : the Case of $Ni(PCy_3)_2(CO_2)$.
a : Dipartimento di Chimica, Università di Bari ,70126 Bari, Italy.,
b : Dipartimento di Chimica Inorganica, Università di Torino, 10125 Torino, Italy

C. Jegat and J. Mascetti,
Vibrational Studies of Some Transition Metal Carbon Dioxide Complexes.
Laboratoire de Spectroscopie Moléculaire et Cristalline (U.R.A.124-C.N.R.S.) ,Université de Bordeaux I, 351 Cours de la Libération 33405 Talence Cedex , Bordeaux, France.

E. Carmona, M.A. Munoz, M. Paneque, P.J. Pérez and M. Poveda,
Observation of $M-CO_2$ Rotational Isomers in $Bis-CO_2$ Adducts.
Departamento de Quìmica Inorgànica-Instituto de Ciencia de Materiales, Universidad de Sevilla-CSIC, Apartado 553. 41071-Sevilla Spain.

M. Aresta, E. Quaranta and I. Tommasi,
Electrophilic Attack at the Carbon Dioxide Moiety in Phosphacarbamates: an Easy Way to the Synthesis of Organic Carbamates.
Dipartimento di Chimica, Università di Bari ,70126 Bari, Italy.

D. Darensbourg and H. Pickner Wiegreffe,
Carbon Monoxide Ligand Substitutional Processes Involving Anionic Group 6 metal Carboxylates and their Relevance to Decarboxylation Mechanisms.
Department of Chemistry, Texas A&M University, College Station, TX 77843

M.Y. Darensbourg and C. Riordan,
Nickel(II) Hydrides and their Relevance to Processes of Metalloenzymes in Methanogenic Bacteria.
Department of Chemistry, Texas A&M University, College Station, TX 77843

M. Aresta,[a] E. Quaranta,[a] I. Tommasi [a] and R. Gobetto [b],
Reduction of Coordinated Carbon Dioxide via Electrophilic Attack at Oxygen.
a : Dipartimento di Chimica, Università di Bari ,70126 Bari, Italy,
b : Dipartimento di Chimica, Università di Torino, 10125 Torino, Italy.

J.C. Tsai, and K.M. Nicholas,
Reactivity of Coordinated Carbon Dioxide. Thermal and Photochemical Reactions of $(C_5H_5)_2Mo(^{2-} CO_2)$.
Department of Chemistry and Biochemistry,
University of Oklahoma, OK 73019, USA.

M. Venturi,[a,b] M. D'Angelantonio,[a] M. Ciano,[a] Q.G. Mulazzani[a] and M.Z. Hoffman,[c]
Radiolytic Generation and Characterization of One-Electron Reduced Ru(II)-Diimine Complexes in Aqueous Solution.
a : Istituto FRAE-CNR, Bologna, Italy.
b : Dipartimento di Chimica "G. Ciamician" Bologna, Italy.
c : Department of Chemistry, Boston University, Boston, USA.

E. Dunach Clinet,
Electrochemical Activation of CO_2 : Application to the Ni-Catalyzed Electrosynthesis of Carboxylic Acids from Alkynes.
Laboratoire d'Electrochimie, Catalyse et Synthèse Organique (U.M. n° 28), C.N.R.S. , 2 rue Henri Dunant, 94430 Thiais, France.

M. Ulmann and J. Augustynski,
Bicarbonate-mediated Electrochemical CO_2 Reduction at Composite Metallic Electrodes.
Department de Chimie Minérale, Analytique at Appliquée, Université de Geneve, Switzerland.

J. Fournier, C. Bruneau and P. Dixneuf,
Phosphine Catalyzed Synthesis of Unsaturated Cyclic Carbonates and Carbamates from Carbon Dioxide and Propargylic Alcohols.
Unité de Recherche CNRS DO415, Campus de Beaulieu, Université de Rennes, 35042 Rennes Cedex, France.

BIOCHEMISTRY SECTION.

E.H. Lee and F.C. Hartman,
Examination of the Function of Active-Site Lysine-329 of Ribulose-Bisphosphate Carboxylase/Oxygenase as Revealed by the Proton Exchange Reaction.
Protein Engineering Program, Biology Division, Oak Ridge National Laboratory and the University of Tennessee-Oak Ridge Graduate School of Biomedical Sciences, Oak Ridge Tennessee 37831-8077.

A. Battistelli and R.C. Leegood,
Sucrose Phosphate Synthase Activation State and the Control of the Sucrose Synthesis in Spinach Leaves.
Robert Hill Institute, University of Sheffield,
26, Taptonville Road, Sheffield, S10 2TN, U.K.

H.B. Smith and F.C. Hartman,
Subtle Structural Perturbations at the Active Site of Rubisco by Concerted Site-Directed Mutagenesis and Chemical Modification.
Protein Engineering Program, Biology Division, Oak Ridge National Laboratory and the University of Tennessee-Oak Ridge Graduate School of Biomedical Sciences, Oak Ridge Tennessee 37831-8077, USA.

K.A. Butcher,[1] F.K.Fong [2] and J.V. Schloss,[3]
A Single Turnover Kinetic Analysis of Ribulose-bisphosphate Carboxylase.
1 : Chemistry Department, Purdue University, West Lafayette, I Indiana.
2 : Carbon Reduction Laboratory, West Lafayette, Indiana
3 : Central Research & Development Department, Du Pont Co., Wilmington, Delaware, USA.

E. Alonso, V. Rubio and J. Cervera,
Use of Oxidation to Try to Characterize the ATP Binding Sites and Conformational Changes in Carbamoyl Phosphate Synthetase (CPS).
Instituto de Investogaciones Citologicas de la Caja de Ahorros de Valencia, Amedeo de Saboja 4, Valencia, Spain.

L. Abell and M.H. O'Leary,
Isotope Effects on Pyruvoyl and Pyridoxal 5'-Phosphate
Dependent Histidine Decarboxylase.
Central Research & Development Department, Du Pont Co.,
Wilmington, Delaware, USA.

R. Fischer and R.K. Thauer,
Methyltetrahydromethanopterin as an Intermediate in
Methanogenesis from Acetate in Methanosarcina barkeri.
Laboratorium für Mikrobiologie, FB Biologie, Karl-von-Frisch-Str.,
3550 Marburg , FRG.

N. Bresciani-Pahor, L. Randaccio and E. Zangrando,
Relationship between Structure and Solution Properties in Vitamin
B_{12} Models.
Dipartimento di Scienze Chimiche, Università di Trieste,
34127 Trieste, Italy.

T. Scott,
Models for Phosphorylation of Biotin-bicarbonate Adducts.
Department of Chemistry, University of Toronto, 80 St. George
Street, Toronto, Canada M5S 1A1.

List of Participants in the Summer School on:

ENZYMATIC AND MODEL CARBOXYLATION AND REDUCTION REACTIONS
FOR CARBON DIOXIDE UTILIZATION.

Riva dei Tessali, Ginoșa (ITALY)

June 17-28, 1989.

List of Participants.

COUNTRY : CANADA

A.BENNET	Department of Chemistry, E3-43 Chemistry Building East, University of Alberta, T6G 2G.
R.S.BROWN	Department of Chemistry, E3-43 Chemistry Building East, University of Alberta, T6G 2G.
J.KEILLOR	Department of Chemistry, E3-43 Chemistry Building East, University of Alberta, T6G 2G.
R.KLUGER	Department of Chemistry, University of Toronto, Toronto, Ontario, M5S A1.
S.TAYLOR	Department of Chemistry, University of Toronto, Toronto, Ontario, M5S A1.
G.THATCHER	Department of Chemistry, Queen's University, Kingston, Ontario, M5S IA1.
B.TSAO	Department of Chemistry, University of Toronto, Toronto, Ontario, M5S A1.

COUNTRY: FRANCE

A.DEDIEU	Laboratoire de Chimie Quantique, Université Louis Pasteur, 4 Rue B. Pascal, 67000 Strasbourg.
P.DIXNEUF	Laboratoire de Chimie de Coordination Organique, Campus de Beaulieu, 35042 Rennes Cédex.
E.DUNACH	Laboratoire d'Electrochimie, Catalyse et Synthése Organique (L.E.C.S.O.), 2AB Rue Henry Dunant, B.P. 28-94320 Thiais.
J.FOURNIER	Laboratoire de Chimie de Coordination Organique, Campus de Beaulieu, 35042 Rennes Cédex.
C.JEGAT	Laboratoire de Spectroscopie Moléculaire et Cristalline, Université de Bordeaux I, 351 Cours de la Libération, 33405 Talence Cédex.
J.MASCETTI	Laboratoire de Spectroscopie Moléculaire et Cristalline, Université de Bordeaux I, 351 Cours de la Libération, 33405 Talence Cédex.
M.PETRIGNANI	L'Air Liquide, CRCD/SGPC, B.P. 126 Les Loges en Josas, 78350 Joui en Josas.

| R.ZIESSEL | Institut le Bel, Université Louis Pasteur, 4 Rue B. Pascal, 67000 Strasbourg. |

COUNTRY: **GERMANY**

A.BALLESTEROS	Max-Planck-Institut fur Kohlenforschung, Kaiser-Wilhelm-Platz 1, D-4330 Mulheim a.d. Ruhr.
A.BREUNIG	Angewandte Mikrobiologie, Oberer Eselsberg M 23, D 7900 Ulm.
R.FISCHER	Philipps-Universitat- Marburg, Fachbeireich Biologie, Laboratorium fur Mikrobiologie, Karl-von-Frisch-Strasse, 3550 Marburg.
G.FUCHS	Angewandte Mikrobiologie, Oberer Eselsberg M 23, D 7900 Ulm.
H.HOBERG	Max-Planck-Institut fur Kohlenforschung, Kaiser-Wilhelm-Platz 1, D-4330 Mulheim a.d. Ruhr.
E.STUPPERICH	Angewandte Mikrobiologie, Oberer Eselsberg M 23, D 7900 Ulm.

COUNTRY: **ITALY**

C.G.ARENA	Dipartimento di Chimica Inorganica e Struttura Molecolare, Università di Messina, Salita Sperone 31, 98010 Vill. S. Agata, Messina.
M.ARESTA	Dipartimento di Chimica, Università degli Studi di Bari, Campus Universitario, 70126 Bari.
L. BANCI	Dipartimento di Chimica, Università degli Studi di Firenze, Via G. Capponi 7, 50121 Firenze.
I.BERTINI	Dipartimento di Chimica, Università degli Studi di Firenze, Via G. Capponi 7, 50121 Firenze.
F.CALDERAZZO	Dipartimento di Chimica e Chimica Industriale, Università degli Studi, v.le Risorgimento 35, 56100 Pisa.
C.FRAGALE	Dipartimento di Chimica, Università degli Studi di Bari, Campus Universitario, 70126 Bari.
P.GRAZIANO	Industrie Pergine, v.le Lavagnini 42, 50129 Firenze.
M.MAESTRI	Dipartimento di Chimica "G. Ciamician", Università di Bologna, v. Selmi 2, 49126 Bologna.

E.QUARANTA	Dipartimento di Chimica, Università degli Studi di Bari, Campus Universitario, 70126 Bari.
G.SILVESTRI	Istituto di Ingegneria Chimica, Facoltà di Ingegneria, v.le delle Scienze, 90128 Palermo.
I.TOMMASI	Dipartimento di Chimica, Università degli Studi di Bari, Campus Universitario, 70126 Bari.
M.VENTURI	Istituto di Fotochimica e Radiazioni di Alta Energia del C.N.R., via dei Castagnoli 1, 40126 Bologna.
E.ZANGRANDO	Dipartimento di Scienze Chimiche, Università di Trieste, p.le Europa 1, 34127 Trieste.

COUNTRY: **PORTUGAL**

A.B. DA SILVA	Departamento de Biologia Vegetal, Faculdade de Ciencias da Universidade de Lisboa, C2 E. de Vasconcelos, 1700 Lisboa.
J.C.DUARTE	LNETI, Biotechnology, Estrada das Palmeiras, Queluz de Baixo, 2745 Queluz.
A.EUSEBIO	LNETI, DTIQ, Estrada das Palmeiras, Queluz de Baixo, 2745 Queluz.
P.GOMES	Centro de Quimica Estrutural Complexo I, Instituto Superior Técnico, Av. Rovisco Pais, 1096 Lisboa Codex.
C.ROMAO	Centro de Quimica Estrutural Complexo I, Instituto Superior Técnico, Av. Rovisco Pais, 1096 Lisboa Codex.

COUNTRY: **SPAIN**

E.ALONSO	Instituto de Investigaciones Citologicas de la Caja de Ahorros de Valencia, Amedeo de Saboya 4, 46010 Valencia.
I.CLIMENT	Instituto de Investigaciones Citologicas de la Caja de Ahorros de Valencia, Amedeo de Saboya 4, 46010 Valencia.
A.LUQUE	Vicerrector de Investigaciones, Departamento de Biologia, Universidad Politecnica de Canarias, c/Alfonso XIII 2, 35003 Las Palmas de G.C..
M.PANEQUE	Departamento de Quimica Inorganica, Universidad de Sevilla, C/Prof. Garcia Gonzales S/n. 41012 Sevilla.

P.PEREZ-ROMERO Departamento de Quimica Inorganica, Universidad de Sevilla, C/Prof. Garcia Gonzales S/n. 41012 Sevilla.

M.POVEDA Departamento de Quimica Inorganica, Universidad de Sevilla, C/Prof. Garcia Gonzales S/n. 41012 Sevilla.

V. RUBIO Instituto de Investigaciones Citologicas de la Caja de Ahorros de Valencia, Amedeo de Saboya 4, 46010 Valencia.

COUNTRY: **TURKEY**

S.CELEBI Dokuz Eylul Universitesi, Muh-Mim Fakultesi, Department of Chemistry, Bornova, Izmir.

Z.GAYRETLI Department of Chemistry, METU, 06531 Ankara, Turkey.

C.TANYELI Department of Chemistry, METU, 06531 Ankara.

S.YIGIT Middle East Tech. University of Chemistry, Inonu Bulvari, 06531 Ankara.

COUNTRY: **UNITED KINGDOM**

A.BATTISTELLI Institut of Photosynthesis, University, 26 Taptonville Rd., Sheffield 5105 BR.

S.FUNTOWICZ Department of Mathematics, Sheffield 5105 BR.

K.K.RAO Department of Biology, King's College University of London, Campden Hill Road, London W 8 7AH.

M.N.SIVAK Research Institute of Phosynthesis, Department of Botany, Sheffield 5105 BR.

COUNTRY: **UNITED STATES OF AMERICA**

L.M.ABELL Department of Chemistry, Penn. State University 152 Davey Laboratory, University Park, PA 16802.

K.BUTCHER Purdue University, Chemistry Department, West Lafayette, IN 41906.

D.DARENSBOURG Department of Chemistry, Texas A & M University, College Station, Texas 77843.

K.K.FONG	Purdue University, Chemistry Department, West Lafayette, IN 41906.
F.C.HARTMAN	Biology Division, Oak Ridge National Laboratory, BOX 2009, Oak Ridge, Tennessee 37831.
W.B.KNIGHT	Merck Sharpe & Dohme, Research Labs., Dept. of Enzymology, BLDG. 80Y-150, P.O. Box 2000 Rashway, NJ 07065.
E.LEE	Biology Division, Oak Ridge National Laboratory, BOX 2009, Oak Ridge, Tennessee 37831.
K.MERZ JR.	Department of Chemistry, Penn. State University, University Park, Penn 16802.
A.MILDVAN	Department of Biological Chemistry, John Hopkins Univerity, School of Medicine, 725 N Wolfe St., Baltimore, MD 21205.
D.K.MILLS	Department of Chemistry, Texas A & M University, College Station, Texas 77843.
K.M.NICHOLAS	Department of Chemistry,University of Oklahoma, Norman, Oklahoma 73019.
H.PICKNER	Department of Chemistry, Texas A & M University, College Station, Texas 77843.
Y.POCKER	Department of Chemistry, University of Washington, Seattle, Washington.
B.RAMAGE	Department of Biochemistry, BSW 528, University of Arizona, Tucson AZ85721.
C.G.RIORDAN	Department of Chemistry, Texas A & M University, College Station, Texas 77843.
J.V.SCHLOSS	Central Research & Development Dept. Experimental Station, E328/246, E.I. du Pont de Nemours & Co., Wilminfgton, DE 19898.
H.SMITH	Department of Chemistry, Texas A & M University, College Station, Texas 77843.
L.VASKA	Department of Chemistry, Clarkson University, Postdam, N.Y. 13676.

NON NATO COUNTRIES.

SWEDEN

G.SCHNEIDER	Biomedical Centre, Agriculture Sciences University, Department of Molecular Biology, Box 590, S-75124 Uppsala.

SWITZERLAND

B.KRAUTLER   Institut fur Organische Chemie, ETH, Zurich.
M.ULMANN   Départment de Chimie Minérale Analytique at
                   Appliquée, Quai Ernest Anserment 30, 1211
                   Genève 4.

416

APPENDIX   A

Carboxylation Reactions Using Carbon Dioxide —
Simple and Double Functionalization of Olefins

Heinz Hoberg,

Max-Planck-Institut für Kohlenforschung, Kaiser-Wilhelm-Platz 1,

D-4330 Mülheim/Ruhr (FRG).

To most chemists, CO_2 is an unsaturated compound with electrophilic properties. However, to organometallic chemists, especially those working with transition metals, the outstanding feature of this molecule is the cumulated π-system.

One of the fundamental concepts of organometallic chemistry is that the bonding between the transition metal atom and the unsaturated substrate involves σ and π interactions between the HOMO orbitals of the one and the LUMO orbitals of the other. As a result, when the π-system is complexed to the metal fragment, the double bond system is no longer in the ground state but has become "activated". As is well known, this type of complexation can be influenced considerably by varying the electronic and steric properties of the other ligands. These generalizations also hold true for the complexation of CO_2. The most convincing demonstration of this is the bent structure taken up by the CO_2 ligand instead of its usual linear geometry in the free molecule.

The variety of ways that CO_2 can be bound to various transition metals will be described in detail. Above all, the importance of the other ligands will be shown. One consequence is that, under certain circumstances, several CO_2 molecules can be bound to one metal atom and under suitable conditions further reactions can be induced such that two CO_2 molecules undergo a C–O coupling reaction to form a metalla–cyclopentane ring.

This homocoupling reaction is largely suppressed in the presence of suitable co–substrates such as aldehydes, imines or unsaturated CC systems. Instead a 1/1 reaction with C–O, C–N or C–C coupling takes place between the CO_2 and the co–substrate to afford the corresponding metalla–complexes. In the formation of the ring, which requires the creation of three new σ-bonds, the metal acts as a reducing agent. Its oxidation state changes by two units, for example, $M^0 \rightarrow M^{2+}$ or $M^{+1} \rightarrow M^{3+}$.

It will be shown how the ligands influence the reversibility of the C–C coupling reaction. The consequences of this on the regioselectivity of the reaction of mono–substituted alkenes will be analysed in detail.

Protolysis of the carboxylate formed by the C–C coupling leads to the liberation of the corresponding carboxylic acids. The consequences of these novel series of reactions will be discussed.

PART 2

It has been known for a long time that the electrophilic CO_2 molecule undergoes insertion reactions into metal–C σ–bonds to form C–C bonds. This reaction occurs with both main group metals and transition metals. These "classical" Grignard reactions will not be discussed further here.

As I showed in the first part of my lecture, it has been discovered in the last few years how a new type of C–C coupling reaction between CO_2 and unsaturated systems with CC multiple bonds can be achieved on transition metals. When the oxametalla–ring systems which are formed are hydrolysed then the corresponding carboxylic acids are produced.

The oxametalla–ring complexes formed by 1/1 coupling can undergo further reactions typical of organometallic compounds. The most important of these are β–H elimination and insertion of unsaturated co–substrates in the remaining metal–C σ–bond. Which of these reactions occurs depends both upon the transition metal and to a great extent upon the electronic and steric properties of the other ligands on the metal. In this way the range of carboxylic acids that are obtained after hydrolysis can be expanded considerably.

When alkenes are used as the co–substrate, then a β–H elimination opens up a route to α–β–unsaturated carboxylic acids. This constitutes a model reaction for a potential catalytic process, but unfortunately no such catalysis has yet been achieved in practice. Other unsaturated substrates can be inserted into the metal–C σ–bond of these 1/1 coupling products. The insertion of alkenes produces systems with larger rings. Here too, depending upon the other ligands on the metal, a β–H elimination can be induced, allowing unsaturated carboxylic acids with longer chains to be prepared. On the other hand, if the β–elimination is suppressed, then the protonolysis liberates only saturated carboxylic acids.

The significance of the electronic and/or steric properties of the ligands will be discussed in detail. The geometry of the metalla–5–ring will be examined and the importance of the 18e rule explained.

By selecting suitable ligands it is also possible to insert a second molecule of CO_2. The metalla–dicarboxylates thus formed are inherently stable compounds, so that they do not undergo β–H elimination. Their hydrolysis yields the corresponding dicarboxylic acids.

It will be shown in detail how with the right choice of ligands it is possible to convert the reaction of conjugated dienes and CO_2 on transition metal complexes, which is otherwise stoichiometric, into a catalytic process.

Special features of the reaction will be discussed and consequences considered.

NATO-ASI Summer School - Ginosa (Italy) - June 17 - 28, 1989

The Communication of Scientific Uncertainty

Silvio O. Funtowicz
Visiting Scientist, Joint Research Centre, Ispra Establisment
S.E.R. - Ed. 32A, 21020 Ispra (VA), Italy

Abstract

When research results are used to provide input for decision-making, as in environmental science and risk analysis, the scientist's problems of managing uncertainty are severe. Firstly, the original data are very rarely as well-controlled as in the laboratory. Well-structured theories, normally expected in basic or applied science, are conspicuous by their absence. Furthermore, such research is interdisciplinary, involving fields of varying states of maturity and with very different practices in their theoretical, experimental and social dimensions. Scientists must use results from fields with which they are unfamiliar, so they cannot make the same quality judgements as in their own subject. The outcome is the dilution of the quality control on the research process and a weaker quality assurance of results.

Such problems are compounded by the traditional training of scientists, in which the learning of quality assessment is a largely tacit skill. This fosters a healthy prudence about passing judgement on the results of others and any tendency to meddle in others' fields is discouraged. Unfortunately, in policy-related research, such tact can be very counterproductive, because criticism does not occur in sufficient strength. We therefore need a new methodology for evaluating scientific work which should provide clear, explicit and public guidelines for communication, while retaining the skills and judgements of the best traditional scientific practice.

The problems of uncertainty in policy-related research are further increased in its public dimension. Science is judged by decision-makers and the general public on its performance in such politically sensitive areas as greenhouse effect, ozone depletion, genetic manipulation, radioactive fall-out and hazardous wastes. All these involve much uncertainty, as well as inescapable social and ethical issues. Simplicity and precision in scientific explanations and predictions are not feasible, yet policy-makers tend to expect straightforward information as input to their own decision-making process. In such circumstances, the maintenance of confidence in Science becomes increaingly difficult.

We cannot simply banish uncertainty and the dangers it presents for Science, but we can make it manageable. Practical tools are needed whereby researchers and those they advise (including the mass media) can communicate better on crucial aspects of scientific information. The task is coping with, rather than conquering, uncertainty. Recognition of this principle represents the first step towards solving scientific problems in the context of policy.

422

LNETI- Department of Chemical Industries

Estrada das Palmeiras - Queluz de Baixo
2745 QUELUZ - PORTUGAL

Tel.: (01)435 2186/7/8

Growth and Physiology of Methanogenic Bacteria. Bioconversions.

DUARTE
José

BIOTECHNOLOGICAL ASPECTS OF CARBON DIOXIDE CONVERSION INTO METHANE AND OTHER USEFUL COMPOUNDS

José C. Duarte, A. Eusébio

Methanogenic bacteria can grow autotrophically, using carbon dioxide as a source of carbon. In addition carbon dioxide may be used as an energy source through a pathway which causes its complete reduction to methane gas a process with a free energy change $G° = -131$ KJ/mol. In this way methanogenic bacteria fix CO_2 into biomass at the expense of its reduction to CH_4.

This process is of significant importance in natural ecosystems and it is most probably related with the biogenesis of our fuel fossil reserves, oil and natural gas. Therefore it is evident that bio-methanogenesis offers an interesting route to CO_2 fixation with economical potential either for energy production either for the production of chemicals.

Information on the mechanisms of CO_2 fixation and on the metabolism of CO_2 for biomass formation is relatively scarce on some aspects is almost absent.

The techniques of genetic engineering have been used on the study of these bacteria by the cloning of its genes into bacteria amenable to the use of the r-DNA technology. Therefore a significant amount of molecular information about methanogenic archaebacteria has been accumulated, but a good reproducible method for introducing the genes back into the original methanogen strain has not yet been developed.

When growing on CO_2 methanogenic bacteria use hydrogen as a source of electrons for the reductive process. The Y_{CH4} of methanogenic bacteria seems to depend on the supply of H_2, increasing when the growth rate is H_2-limited suggesting that methanogenesis and phosphorilation of ADP are uncoupled at high concentrations of H_2.

The incorporation of CO_2 into biomass depends on the "activation" of CO_2 into acetate or its metabolic equivalent acetyl-CoA. From acetyl-CoA all other cell compounds are generated by reductive carboxylation to pyruvate and the Tricarboxylic Acid Cycle. A key enzyme in this process may be acetate kinase which catalyses the interconversion between acetate and acetyl-phosphate, which in turn is converted to acetyl-CoA by a phosphotransacetilase.

An important class of proteins on the process of CO_2 fixation into biomass and energy production are the corrinoid-bound proteins. This is reflected by the high content of corrinoids present on the methanogenic bacteria, which may become an economic source of valuable compounds of the family of cobalamines, such as vitamin B-12.

424

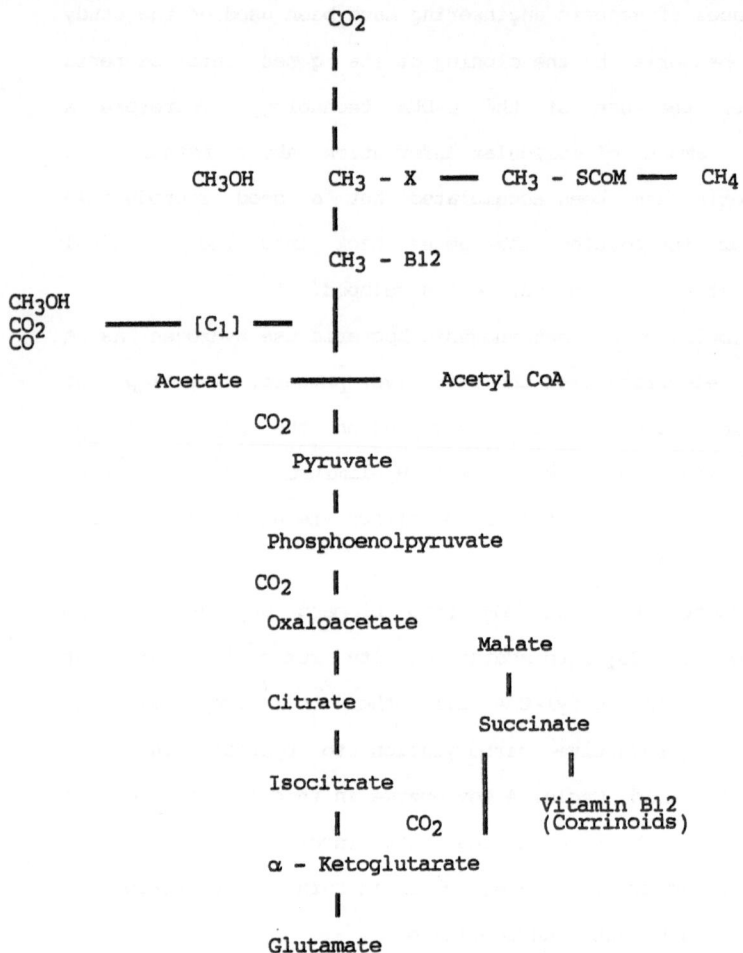

Figure 1: Proposed scheme for the CO_2 assimilation and its coupling with methanogenesis by Methanosarcina barkeri. Autotrophic CO_2 fixation by Methanobacterium Thermoautotrophicum proceeds by a different pathways in wich malate is generated from oxaloacetate to produce aminoacids and corrinoids.

Bibliography:
• Kenealy, W.R. and Zeikus, J.G. (1982). J. Bacteriol.; 151.(2): 932-941.
• Fuchs, G.; Stupperich, E. (1984). In: Microbial Growth on C_1-Compounds, pp.199-202. ASM (Washington).
• Worrell, V.E.; Nagle, JR., D.P.; McCarthy, D. and Eisenbraun, A. (1988) J. Bacteriol., 170.(2): 653-656.
• Duarte, J. and Eusébio, A. (1989). Enzymology of the CO_2 fixation on Methanosarcina barkeri. FEBS Meeting. In press.

Department of Chemistry   VASKA, Lauri
Clarkson inversity
Potsdam, New York 13676, USA
Tel. 315-268-2393/2389

Research interests: inorganic chemistry, catalysis, organometallic and
                      bioinorganic chemistry.

FROM CO_2, NH_3 AND H_2 TO CH_4, HCN, $Ar-C_3H_3N_3$ AND OTHER SPECIES VIA
CATALYTIC SYNTHESIS, HYDROGENATION AND DECOMPOSITION OF FORMAMIDE

Lauri Vaska

The work described in this paper consists of three interrelated
parts and is summarized by the equations given below. Abbreviations:
[Ir] = trans-[Ir(Cℓ)(CO)(Ph$_3$P)$_2$] = catalyst precursor; Me = CH_3;
ROH = MeOH or MeCH$_2$OH; Ar = arene; $Ar-C_3H_3N_3$ = 1,3,5-triazine; reaction
temperatures and pressures represent average or typical values.

   A.   Homogeneous catalytic synthesis of formamide.[1]

$$CO_2 + H_2 + NH_3 \xrightarrow[\text{125°C; 100 atm}]{\text{[Ir]; toluene or ROH}} HC(O)NH_2 + H_2O \qquad (1)$$

   B.   Catalytic hydrogenation (eq 2) and thermal decomposition
(eq 3) of formamide at elevated pressures.[1]

$$HC(O)NH_2 + H_2 \xrightarrow[\text{150°C, 150 atm}]{\text{[Ir]}} CH_4 + HC(O)NHMe + HC(O)NMe_2 +$$

$$NH_3 + CO + CO_2 + (H_2) + H_2O + C + ... \qquad (2)$$

$$HC(O)NH_2 + H_2 \xrightarrow[\text{150°C, 100 atm}]{} NH_3 + CO + CO_2 + (H_2) + H_2O + C + ... \qquad (3)$$

Subtracting the thermal decomposition products of eq (3) from those of
eq (2), we obtain the net catalytic hydrogenation of formamide:

$$HC(O)NH_2 + H_2 \xrightarrow{\text{[Ir]}} CH_4 + HC(O)NHMe + HC(O)NMe_2 \qquad (4)$$

And the formation methane alone can be expressed by the balanced eq (5):

$$HC(O)NH_2 + 3H_2 \xrightarrow{[Ir]} CH_4 + H_2O + NH_3 \qquad (5)$$

The overall stoichiometry of the catalytic methanation of carbon dioxide may thus be formulated by adding eq (1) and (5):

$$CO_2 + 4H_2 \xrightarrow{[Ir], NH_3} CH_4 + 2H_2O \qquad (6)$$

It is seen that ammonia emerges as a co-catalyst for this particular process.

C. Thermal and catalytic decomposition of formamide at low pressures.[2]

$$HC(O)NH_2 \xrightarrow[100-190°C, \ 0-700 \ torr]{N_2, \ He, \ H_2, \ O_2 \ or \ vac/[Ir]} NH_3 + CO + HCN + Ar-C_3H_3N_3 + H_2O$$

$$+ \ C \ (solid \ residue) + (CO_2 + H_2) + ([NH_4]^+[H_2NC(O)O]^-) \qquad (7)$$

The above reactions (eq 7) take place thermally but are accelerated in the presence of the iridium complex. The most noteworthy result is that the products include $Ar-C_3H_3N_3$. Thus, by combining eq (1) and the applicable part of eq (7), and adding appropriate coefficients for balance, we obtain eq (8) which summarizes the overall route to triazine from its inorganic precursors:

$$3CO_2 + 3H_2 + 3NH_3 \xrightarrow{[Ir]} Ar-C_3H_3N_3 + 6H_2O \qquad (8)$$

The significance and possible mechanisms of these very complex catalytic and thermal reaction systems (eq 1,2,3,4,7) will be discussed.

References:

(1) L. Vaska, S. Schreiner, R.A. Felty, J.Y. Yu, J. Mol. Catal., in press.

(2) L. Vaska, S. Schreiner, J.Y. Yu, unpublished work.

APPENDIX B

^{13}C and ^{31}P n.m.r. Spectroscopy for Structural Characterization of Carbon Dioxide Metal Complexes: the Case of Ni(PCy$_3$)$_2$(CO$_2$).

M.Aresta,E.Quaranta,I.Tommasi
Dipartimento di Chimica,Università,70126 Bari,Italy
and
R.Gobetto,
Dipartimento di Chimica Inorganica,Chimica Fisica e Chimica dei Materiali,
Università,10125 Torino,Italy.

Solid state magnetic resonance spectroscopy is particularly useful for the determination of structural features of those species which either have a short lifetime or show low energy dynamic processes that average solution n.m.r. spectra. The use of cross polarization techniques combined with magic angle spinning provide an important contribution to the knowledge of the change of conformation of (PCy$_3$)$_2$Ni(CO$_2$) (I)[1] from solid to solution state.

In solution the ^{31}P n.m.r. spectra of (I) show a single sharp resonance at 36.1 ppm in the temperature range 197 - 303 K.[2] The same feature is observed in liquid CO$_2$.[3]

Three different explanations can be suggested :

i) a σ M-CO$_2$ co-ordination

ii) a fast dinamic process that averages the two possibilities of n^2-co-ordination of the carbon dioxide moiety to the metal:

iii)a fast rotation around the Ni-^{2}n(CO$_2$) bond.

In the solid state CPMAS ^{31}P n.m.r. of (I) two different isotropic peaks at 45.2 and 22.0 ppm are observed with a large variation in their chemical shift anisotropy as shown in the spinning sideband pattern, in accordance with the n^2-co-ordination of carbon dioxide. Unfortunally, the intensities of the spinning sidebands are insufficient to obtain values of the principal elements of the shielding tensors calculated with the graphical method proposed by herzfeld and Berger. The phosphorous signal in solution is close to the average value of the two solid state resonances.

430

^{13}C CPMAS n.m.r. of a sample of (I) prepared with ^{13}C enriched CO_2
affords a resonance at 158.4 ppm that represents,to our knowledge,
the first solid state n.m.r. observation of a CO_2 molecule co-ordinated
to a metal atom. Interestingly,this value is close to that found in
solution at 158.61 ppm (t,J(C-P)=14.8 Hz). This value is unique for
the CO_2-transition metal complexes that usually show ^{13}C resonances
around 200 ppm. Dioxygen or water caused the disappearance of the signal.
  Moreover,to our knowledge,this represents the first solid state
n.m.r. observation of carbon dioxide co-ordinated to a metal centre.

  The experimental conditions used for solid state measurements were:
69.8 and 109.2 MHz respectively for ^{13}C and ^{31}P , JEOL GX 270 apparatus.
The rotors were made of Delrin and the spinning rate was 3.5 - 4.0 kHz.

References :

1) M.Aresta,C.F.Nobile,V.G.Albano,E.Forni,M.Manassero,JCS Chem Comm,
   1975 , 636.
2) M.Aresta,E.Quaranta and I.Tommasi, JCS Chem Comm, 1988, 450.
3) M.G.Mason and J.A.Ibers, J Am Chem Soc , 104 , 1982 , 5153.

Acknowledgements : NATO (Grant 45/88) is gratefully acknowledged.

VIBRATIONAL STUDIES OF SOME TRANSITION
METAL CARBON DIOXIDE COMPLEXES.

Corine JEGAT and Joëlle MASCETTI

Laboratoire de Spectroscopie Moléculaire et Cristalline (U.R.A.124 - C.N.R.S.)
Université de Bordeaux I, 351, Cours de la Libération 33405 TALENCE Cedex.

Matrix isolation spectroscopy studies [1] have shown that structural conclusions about the coordination mode and the value of the OCO angle in transition metal CO_2 complexes can be drawn from isotopic shifts observed in their infrared spectra.

To improve our analysis, we have synthesized the isotopically labelled derivatives (^{13}C, ^{18}O) of some compounds, i.e. $(PCy_3)_2Ni(CO_2)$ [2], $(PMe_3)_4Fe(CO_2)$ [3], $Cp_2Ti(PMe_3)(CO_2)$ [4] and $(PM_3)_4Mo(CO_2)_2$ [5].

F.T.I.R. and Raman spectra have been obtained and normal coordinate analysis performed.

From this study, some structural conclusions can be proposed for such complexes for which no X-ray structure was known. Valence-force field constants of $M(CO_2)$ moieties have been calculated and conclusions about chemical bonding and reactivity in these complexes will be presented.

1   J.MASCETTI and M.TRANQUILLE, J.Phys.Chem., 1988, 92, 2177.

2   In collaboration with Professor M.ARESTA, University of Bari, Italy.

3   H.H.KARSCH, Chem.Ber., 1977, 110, 2213.

4   H.G.ALT and coll., J.Organometal Chem., 1987, 321, C9.

5   E.CARMONA and coll., J.Am.Chem.Soc., 1986, 108, 2286.

432

OBSERVATION OF M-CO$_2$ ROTATIONAL ISOMERS IN BIS-CO$_2$ ADDUCTS.
trans -[Mo(CO$_2$)$_2$L$_4$]

Ernesto Carmona, Miguel A. Muñoz, Margarita Paneque,
Pedro J. Pérez and Manuel L. Poveda

Departamento de Química Inorgánica-Instituto de Ciencia de Materiales.
Universidad de Sevilla-CSIC. Apartado 553.
41071-Sevilla. SPAIN.

we have recently reported the formation of the bis - CO$_2$ adducts trans-[Mo(CO$_2$)$_2$(PMe$_3$)$_4$] and trans,mer-[Mo(CO$_2$)$_2$(CNR)(PMe$_3$)$_3$] and noticed their fluxional behaviour in solution[1]. In this communication we wish to present a detailed variable temperature multinuclear (^{1}H, ^{31}P and ^{13}C) NMR study, carried out with these and other new members of the series including species of the type trans-[Mo(CO$_2$)$_2$(PMe$_3$)$_2$(P-P)] and trans -[Mo(CO$_2$)$_2$-(P-P)$_2$], where P-P = R$_2$P(CH$_2$)$_n$PR$_2$ (R = Me, n = 1, 2; R = Et, n = 2). This study has allowed the observation of M-CO$_2$ rotational isomers. For example, the compound trans-[Mo(CO$_2$)$_2$(CNCMe$_3$)(PMe$_3$)$_3$] has been found to exist in CD$_3$OD, at -90°C, as a mixture of two rotamers (A and B) in a 1:4 ratio.

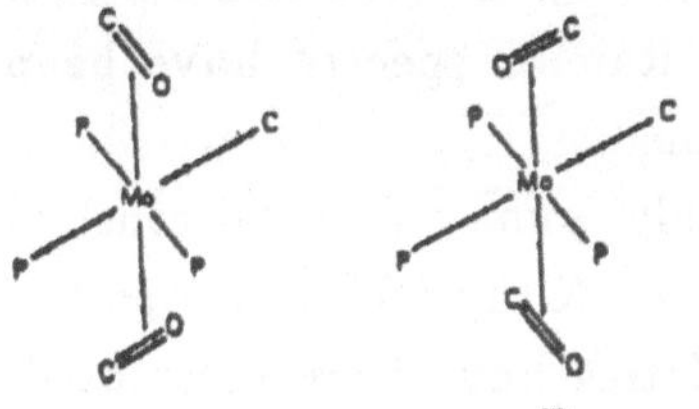

A   B

Above this temperature and with rates that are strongly dependent on the solvent, these species readily interchange giving average fast limiting spectra above room temperature. A detailed analysis of the NMR data allows to reach the conclusion that in all cases a fast concerted conrotation of the two CO$_2$ ligands is responsible for the observed interchange process. Interestingly, this type of fluxionality had been predicted theoretically[2].

1) R. Alvarez, E. Carmona, J.M. Marín, M.L. Poveda, E. Gutierrez-Puebla, A. Monge. J. Am. Chem. Soc., 108, 1986, 1326.

2) E. Sánchez Marcos, R. Caballol, G. Trinquier, J.C. Bartherlat. J. Chem. Soc., Dalton Trans. 1987, 2373.

Electrophilic Attack at the Carbon Dioxide Moiety in Phosphacarbamates: an Easy Way to the Synthesis of Organic Carbamates.

M. Aresta, E. Quaranta, I. Tommasi

Dipartimento di Chimica
Campus Universitario
70126 Bari, Italy

The insertion of carbon dioxide into the P-N bond of aminophosphines is a reaction which is known since a long time: the reaction can occur in absence[1] or in presence of metal ions [2] and phosphacarbamates of formula $Z_xP(OOCNR_2)_{3-x}$ (Z= halogen, alkyl, alkylamino) can be prepared which contain one or two carbamate groups covalently bound to a phosphorous center.

The presence of carbamate groups in this class of compounds makes phosphacarbamates very attractive from a synthetic point of view for the preparation of organic carbamates.

However, it is worth noting that the direct reaction between phosphacarbamates and an organic halide does not afford the desired products: N- alkylation is always observed which gives amines and CO_2. A similar behaviour has been observed by other authors with transition metal carbamates.[3]

We have observed O-alkylation, leading to organic carbamates, only when phosphacarbamates are reacted with organic halides in presence of a metal halide (MY) and an appropriate crown-ether (CE). [4]

$$MY+CE$$
$$(R_2N)_{3-x}P(OOCNR_2)_x + xR'X ======= xR_2NCOOR' + (R_2N)_{3-x}PY_x$$

The yield depends on several factors: the operative conditions, the nature of the metal salt and of the organic halide.

The reaction proceeds in two steps:

-(i) formation of a metal carbamate ($R_2NCOOM \cdot CE$) in which the alkaline metal is bound to the macrocyclic ether,

-(ii) electrophilic attack by R'X at one of two oxygen ends of the carbamate ion to produce the organic carbamate.

A decisive role is plaied by the macrocyclic ligand: if it is absent, both steps i) and (ii) are suppressed and only N-alkylation occurs.

Among Group I metal ions investigated (Li^+, Na^+, K^+), K^+ seems to be the best transfer agent. Fluoride is the most active and selective of the halide ions.

This results suggested that alkaline metal carbamates prepared in whatever condition could be used in the synthesis of organic carbamates in presence of the proper crown-ether and, indeed, this is the case.[5]

References.
1) R.W. Light, L.D.Hutchins, R.T. Paine, C.F.Campana, Inorg. Chem., 1980 , 19. 3597
2) M. Aresta, M. De Fazio, P. Bruno, Inorg.Chem. 1982, 21.441
3) D. Belli dell'Amico, F. Calderazzo, B. Giovanitti, G.Pelizzi, JCS Dalton Trans 1984, 647.
4) M. Aresta, E.Quaranta, J.Org.Chem., 1988,53.4153
5) M. Aresta,E.Quaranta, Domanda di brevetto 22507 A/86; M.Aresta, E. Quaranta

434

CARBON MONOXIDE LIGAND SUBSTITUTIONAL PROCESSES INVOLVING ANIONIC GROUP 6 METAL CARBOXYLATES AND THEIR RELEVANCE TO DECARBOXYLATION MECHANISMS. Donald J. Darensbourg and Holly Pickner Wiegreffe, Dept. of Chemistry, College Station, TX 77843

Oxygen donor ligands are known to labilize cis carbonyl ligands toward substitution. Upon examination of the following system:

$$[PPN][W(CO)_5O_2CCH_3] + P(OCH_3)_3 \rightleftharpoons [PPN][W(CO)_4(P(OCH_3)_3)O_2CCH_3] + CO$$

the rate of ligand substitution was found to be only slightly dependent on the phosphite concentration in tetrahydrofuran. The kinetic data is consistent with a dissociative interchange mechanism through the intermediate formation of a chelate, i.e., $W(CO)_4(\eta^2-O_2CCH_3)^-$, with activation parameters ΔH^* and ΔS^* being 23.3 ± 0.8 kcal $\cdot$ mole^{-1} and 5.4 ± 2.6 e.u. respectively. Upon photolysis of $W(CO)_5O_2CCH_3^-$, the chelating species was identified by IR and ^{13}C NMR spectroscopy. Additional studies involving the analogous metal carboxylates, where M = Cr, Mo will be reported. These substitution reactions are being investigated to examine the relevance of the chelating intermediate in the decarboxylation reaction of $Cr(CO)_5O_2CH^-$.

NICKEL(II) HYDRIDES AND THEIR RELEVANCE TO PROCESSES OF METALLOENZYMES IN METHANOGENIC BACTERIA
Marcetta Y. Darensbourg and Charles G. Riordan, Department of Chemistry, Texas A&M University, College Station, TX 77843

Solutions of $[Ni(PCy_3)_2]_2N_2$ liberate N_2 under a stream of Ar gas to yield the reactive 16 e$^-$ species "Ni(PCy_3)_2(solv)" which oxidatively adds a variety of sulfur and selenium acids, REH (E = S, R = H, Ph, p-tol; E = Se, R = Ph) to yield $trans$-HNi(ER)(PCy_3)_2. The Ni-SR functionality reacts with electrophiles as well as other thiols to yield thioethers and thiols, eqs 1 and 2.

$$HNi(SPh)(PCy_3)_2 + MeI \longrightarrow HNi(I)(PCy_3)_2 + MeSPh \quad (1)$$

$$HNi(SH)(PCy_3)_2 + HSPh \longrightarrow HNi(SPh)(PCy_3)_2 + H_2S \quad (2)$$

These complexes do not react with CO_2. Preliminary studies indicate $[Ni(PCy_3)_2]_2N_2$ reacts with 2 molar equivalents of $MeS(CH_2)_3SH$ to liberate methane. When the reaction is carried out at -78 °C, the low temperature ^1H NMR spectrum exhibits a high field triplet indicative of S-H addition to the Ni(0) complex yielding a Ni(II) hydride species. Upon warming this resonance disappears. Possible mechanisms for methane production will be explored.

REDUCTION OF CO-ORDINATED CARBON DIOXIDE via ELECTROPHILIC ATTACK AT OXYGEN

M.Aresta,E.Quaranta,I.Tommasi
Dipartimento di Chimica,Campus Univesritario,70126 Bari , Italy
and
R.Gobetto
Dipartimento di Chimica

The reduction of carbon dioxide to bound-carbon monoxide has been proposed as a fundamental step in the utilization of carbon dioxide in the synthesis of methane by some anaerobic bacteria (Methanogenic Bacteria). The enzymatic system involved seems to contain Ni,Fe,Zn and free -SH groups. The Ni center has been proposed as the active site,being still under investigation the exact nature of the red-ox couple. In the inactive form a Ni(II) seems to be present in the enzyme that is e.p.r. silent.[1]

Ni-complexes are able to co-ordinate carbon dioxide and the mode of bonding of the co-ordinated cumulene has been ascertained via cristallographic analysis.[2] The ^{13}C and ^{31}P n.m.r. spectra both in the solid state and in solution have shown the "non-rigid" character of the bonded carbon dioxide in solution.[2b]

It is still open the question if the cumulene in solution is "n^1-bonded" to Ni via the C-atom or if an olefin-like rotation or a rapid interconversion involving the "external" oxygen atom take place.

The co-ordinated carbon dioxide is easily protonated by Broensted acids and carbon monoxide is obtained as the only product. Thiols are particularly active agents and Ni(II) sulphides are isolated. Intermediate formation of disulphides is postulated that easily add to Ni(0) species present in the reaction medium and originated from the Ni-CO_2 complex.[3]

Ni-sulphides are inactive towards carbon dioxide co-ordination and reduction,also in the presence of light.

Ni(I)-species also react with carbon dioxide to afford carbonyl- and carbonate-complexes.

The reaction with protonating agents of the Ni(I) complexes in the presence of carbon dioxide gives carbonyl complexes,but the mechanism is not clear.

References :

1) G.Fuchs, FEMS Mikrobiol.Rev., 39 (1986) 181
2a)M.Aresta,C.F.Nobile,V.G.Albano,E.Forni,M.Manassero, JCS Chem Comm , 1975,636.
 b)M.Aresta,E.Quaranta,I.Tommasi,R.Gobetto,manuscript in preparation.
3) M.Aresta,E.Quaranta,I.Tommasi, JCS Chem.comm.,1988,450.

REACTIVITY OF COORDINATED CARBON DIOXIDE. THERMAL AND PHOTOCHEMICAL REACTIONS OF $(C_5H_5)_2Mo(\eta^2\text{-}CO_2)$.

Jing-Cherng Tsai and Kenneth M. Nicholas*, Department of Chemistry and Biochemistry, University of Oklahoma, Norman, OK 73019.

Recently, we reported the first example of a photoinduced reaction of coordinated carbon dioxide (*J. Am. Chem. Soc.*, **1988**, *110*, 2004) which occurs upon near ultraviolet irradiation of $(C_5H_5)_2Mo(\eta^2\text{-}CO_2)$ (1). Following a discussion of recent results relating to the mechanism of this novel photoinduced disproportionation of carbon dioxide, we will present our findings on the *dark* reactions of 1 with various electrophilic reagents. Complex 1 has been found to react rapidly with a variety of Lewis acids to afford products derived from O-centered electrophilic attack on coordinated CO_2. In the case of Me_3SiCl, $^tBuMe_2SiCl$ and $Me_3SiOSO_2CF_3$, complete oxygen transfer occurs according to the following equation.

$$(C_5H_5)_2Mo(\eta^2\text{-}CO_2) + 2\ R_3SiX \text{ -------> } [(C_5H_5)_2Mo(CO)X]X + R_3SiOSiR_3$$

A similar process results when 1 is treated with HCl producing water as the coproduct, a stoichiometric reverse water gas shift reaction. The reaction of 1 with $BF_3\cdot Et_2O$ and BCl_3, on the other hand, produces adducts which apparently retain the Lewis acid in the metal's coordination sphere. Reactive carbon electrophiles such as $MeOSO_2CF_3$ and carboxylic acid chlorides also react with 1 to induce CO_2 cleavage. The scope and mechanistic aspects of these first documented reactions of an $\eta^2\text{-}CO_2$ complex towards external reagents will be discussed.

RADIOLYTIC GENERATION AND CHARACTERIZATION OF ONE-ELECTRON REDUCED Ru(II)-DIIMINE COMPLEXES IN AQUEOUS SOLUTION.

M. Venturi,[a,b] M. D'Angelantonio,[a] M. Ciano,[a] Q.G. Mulazzani,[a] M.Z. Hoffman.[c]

[a] Istituto FRAE-CNR, Bologna, Italy. [b] Dipartimento di Chimica "G. Ciamician", Bologna, Italy. [c] Department of Chemistry, Boston University, Boston, USA.

Ruthenium(II) diimine complexes have been used for a long time as photosensitizers in model systems for the conversion of solar energy into chemical energy via the cleavage of H_2O to H_2 and O_2. Recent efforts were also directed toward the photoreduction of CO_2 to organic fuels.

In one of the model systems examined by Willner et al. (J.A.C.S., 1987, 109, 6080), the reduction of CO_2 to CH_4 is obtained in a cycle in which the excited state of $Ru(bpz)_3^{2+}$ (bpz = 2,2'-bipyrazine) is reduced by TEOA (triethanolamine) acting as sacrificial quencher; the subsequent step can be viewed as electron transfer from the reduced complex to a CO_2 molecule, mediated by a heterogeneous catalyst (colloidal Ru or Os).

With this model system and its analogues as a basis, we have examined the photosensitized reduction of CO_2 utilizing the techniques of radiation chemistry in the continuous and fast kinetics mode.

The majority of the results reported here deals with the generation and characterization of the one-electron reduced forms of ten Ru(II) complexes of the family $Ru(bpy)_{3-m-z}(bpm)_m(bpz)_z^{2+}$, where bpy = 2,2'-bipyridine, bpm = 2,2'-bipyrimidine, m and z = 0,1,2,3, and m + z $\leq$ 3. The reduction of the complexes was achieved by means of the carboxyl radical, CO_2^{-}; the rate constants for these reactions were obtained, and were correlated with the standard free energy changes of the electron transfer reactions. For each monoreduced complex, the following information was also obtained: absorption spectrum, redox potentials, acid-base properties, stability, and reactivity toward MV^{2+} (1,1'-dimethyl-4,4'-bipyridinium dication), the most commonly used electron relay in model photocatalytic systems.

ELECTROCHEMICAL ACTIVATION OF CO_2 : APPLICATION TO THE NICKEL-CATALYZED ELECTROSYNTHESIS OF CARBOXYLIC ACIDS FROM ALKYNES

Elisabet DUNACH-CLINET

Laboratoire d'Electrochimie, Catalyse et Synthèse Organique (U.M. n° 28), C.N.R.S., 2, rue Henri Dunant,   94320 THIAIS (France)

Electrogenerated low-valent nickel complexes are active catalysts for the carbon dioxide incorporation into unsaturated molecules. In situ electroreduction of a stable Ni(II) precursor such as Ni(bipy)$_3$ (BF$_4$)$_2$ (bipy = 2,2'-bipyridine) enables the double activation of CO_2 and alkynes. The reactions are performed under mild conditions (P_{CO_2} = 1 atm, T = 20-80° C, DMF) in an undivided cell fitted with a sacrificial magnesium anode. Terminal alkynes are regioselectively carboxylated to yield α-substituted acrylic acids of type 1.

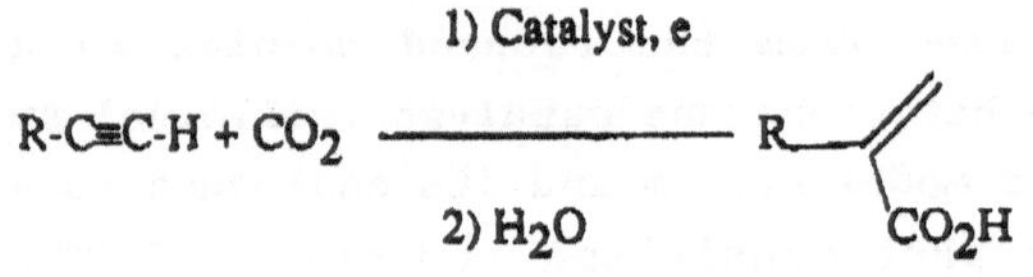

$$R\text{-}C{\equiv}C\text{-}H + CO_2 \xrightarrow[\quad 2)\ H_2O \quad]{\quad 1)\ Catalyst,\ e \quad} R\diagup\!\!\diagdown CO_2H$$

1

Disubstituted alkynes afford mono-and di- carboxylated derivatives. A strong influence of the ligand's nature on the selectivity of the carboxylation process has been found, affording an example of high ligand-directed product specificity.

The synthesis of functionalized molecules will be presented. Mechanistic considerations concerning CO_2 fixation and electroreduction will be discussed.

Bicarbonate–mediated electrochemical CO_2 reduction at composite metallic electrodes

M. Ulmann and J. Augustynski

Département de chimie minérale, analytique et appliquée, Université de Genève, Switzerland

For a long time the electrochemical reduction of carbon dioxide in aqueous solutions was considered as a strongly irreversible reaction, taking effectively place only at high–hydrogen–overvoltage cathodes. These views are contradicted by several recent works indicating that such electrode materials as copper or gold enable the reduction of CO_2 to occur with moderate overvoltages, yielding rather unexpected products including methane, ethylene or carbon monoxide. The present authors have demonstrated that bicarbonate (HCO_3^-) ions act, in fact, as true electroactive species at a palladium cathode, leading to the formate formation at potentials close to the equilibrium potential of the hydrogen electrode. Recently, this work has been extended to the cathodes consisting of palladium deposits on different metallic and non–metallic supports. Among the new results obtained, those regarding bicarbonate reduction at palladium/zirconium and palladium/titanium cathodes will be discussed in some detail. Of special interest is the case of the Pd/Zr electrode, both of these metals forming hydrides under conditions of the cathodic polarization.

Phosphine catalysed synthesis of unsaturated cyclic carbonates and carbamates from carbon dioxide and propargylic alcohols.

Jean Fournier, Christian Bruneau and Pierre H. Dixneuf.

Unité de recherche CNRS D0415, Campus de Beaulieu, Université de Rennes, 35042 Rennes Cedex France.

Cyclic carbonates have been shown to be useful intermediates for the synthesis of functional carbamates or esters. Their unsaturated analogs have industrial potential for the access to transparent polymers.

We report a new one-step synthesis of α–methylene cyclic carbonates directly from CO_2 and α-ethynyl alcohols.

The α-methylene cyclic carbonates are reactive towards nucleophiles such as amines or alcohols to give functional carbamates or linear carbonates.

Under similar conditions α-methylene cyclic carbamates can be produced from α-ethynyl alcohols, CO_2 and primary amines.

Examination of the Function of Active-Site Lysine-329 of Ribulose-Bisphosphate
Carboxylase/Oxygenase as Revealed by the Proton Exchange Reaction

Eva H. Lee and Fred C. Hartman
Protein Engineering Program, Biology Division, Oak Ridge National Laboratory,
and the University of Tennessee-Oak Ridge Graduate School of Biomedical Sciences,
Oak Ridge, Tennessee 37831-8077

Diverse approaches that include site-directed mutagenesis have indicated a catalytic role of Lys-329 of ribulose-bisphosphate carboxylase/oxygenase from *Rhodospirillum rubrum*. To determine whether Lys-329 is required for the initial enolization of ribulose bisphosphate or for some subsequent step in the overall reaction pathway, the competence of position-329 mutant proteins (devoid of carboxylase activity) in catalyzing exchange of solvent protons with the C-3 proton of substrate has now been examined. Irrespective of the amino acid substitution for Lys-329, the mutant protein retains 2-6% of the wild-type activity in the proton exchange reaction. The complete stability of ribulose bisphosphate during the enolization catalyzed by mutant protein suggests that the major effect of Lys-329 is to facilitate the addition of gaseous substrates (CO_2 or O_2) to the enediol intermediate. The exchange reaction requires Mg^{2+}, is CO_2-dependent, and is inhibited by the transition-state analogue 2-carboxyarabinitol-1,5-bisphosphate. A mutant protein in which Lys-191, the site for carbamylation by CO_2 in an obligatory activation step, is replaced by a cysteinyl residue totally lacks proton exchange activity. Barely detectable exchange activity (~0.2% of wild-type) is displayed by the Lys-166→Cys mutant protein, consistent with the previously implicated role of Lys-166 in the deprotonation of ribulose bisphosphate. Retention of exchange activity by the Glu-48→Gln mutant protein, which is slightly active in overall carboxylation, demonstrates that active-site Glu-48, like Lys-329, exerts its major effect at some step subsequent to the initial enolization. (Research sponsored by Martin Marietta Energy Systems, Inc. with the U.S. Department of Energy.)

SUCROSE PHOSPHATE SYNTHASE ACTIVATION STATE AND THE CONTROL OF THE SUCROSE SYNTHESIS IN SPINACH LEAVES.

Alberto Battistelli* and Richard C. Leegood

Robert Hill Institute, University of Sheffield,
26, Taptonville Road, Sheffield, S10 2TN, U.K.

ABSTRACT

Carbon dioxide fixation of higher plants can be limited by a decreased utilisation of the product of the carboxylation process. All the reactions of the Reductive Pentose Phosphate Pathway (RPPP) are localised inside the chloroplast and, as the exported product of the RPPP is the triosephosphate,(TP) it is clear that not only CO_2 but also inorganic phosphate (Pi) must enter the chloroplastic envelope to allow the process to continue at high rates. Pi must enter the chloroplast by the action of the TP–translocator that exchanges it with TP in a 1:1 ratio. Pi must be available in the cytoplasm to re–enter the chloroplast envelope in exchange with TP. Pi is recycled in the cytoplasm, mainly by the sucrose synthesis. Such a process is regulated by the action of the cytoplasmic fructose bisphosphatase (EC 3.1.3.11) and sucrose–phosphate synthase (SPS) (EC 2.4.1.14).

If the release of Pi in the cytoplasm is impaired, then the chloroplastic level of Pi will drop leading to a limitation of the ATP synthesis. Conversely if the sucrose synthesis runs too fast, the chloroplastic level of TP will drop and a depletion of the RPPP intermediates level will cause a reduction in the regeneration of ribulose–bis phosphate. Both the situations would lead to a decrease in CO_2 fixation.

The aim of our work was to elucidate in which extent the activation state of SPS changes in relation to short term (hours) acclimation of spinach leaves to environmental factors (temperature and light) or to metabolic conditions, in order to evaluate the role of SPS activation state in the control of sucrose synthesis and its feedback effect on the CO_2 assimilation rate.

(*) Permanent address: Istituto per l'Agroselvicoltura. (C.N.R.). Viale Marconi 2. 05010 Porano (TR),Italy.

Subtle Structural Perturbations at the Active Site of Rubisco by Concerted Site-Directed Mutagenesis and Chemical Modification

Harry B. Smith and Fred C. Hartman
Protein Engineering Program, Biology Division, Oak Ridge National Laboratory,
and the University of Tennessee-Oak Ridge Graduate School of Biomedical Sciences,
Oak Ridge, Tennessee 37831-8077

Four amino acid residues of ribulosebisphosphate carboxylase/oxygenase from *Rhodospirillum rubrum* (viz. Glu-48, Lys-166, Lys-191, and Lys-329) have been well-established as catalytically essential through chemical, mutagenesis, and sequence studies. In order to define their catalytic functions and to ascertain their possible roles in influencing substrate specificity, we have introduced subtle structural perturbations into the pertinent side chains through a combination of site-directed mutagenesis -- specifically, replacement of the pertinent residue with cysteine -- and subsequent chemical modification. In this way, carboxymethylation of the E48C protein effectively lengthens an acidic side chain, while aminoethylation of the K166C, K191C, or K329C protein entails the mere replacement of a given γ-methylene group by a sulfur atom. While substitution of any of the four native residues with cysteine obliterates catalytic activity, the subsequent, appropriate chemical modification partially restores enzyme activity: Treatment of the E48C protein with iodoacetate restores 2-3% of wild-type carboxylase activity. Incubation of the K166C and K329C proteins with bromoethylamine results in novel enzymes characterized by k_{cat} values which are, respectively, 20% and 60% of the wild-type value; ethylene imine similarly restores to the K191C protein a k_{cat} that is 4-7% of the wild-type value. The significant impairment of k_{cat} that accompanies each of the slight structural alterations of the side chains underscores the stringency of their catalytic essentiality. Substrate specificities of the novel carboxylases are being evaluated. (Research sponsored by Martin Marietta Energy Systems, Inc. with the U. S. Department of Energy.)

A SINGLE TURNOVER KINETIC ANALYSIS OF RIBULOSEBISPHOSPHATE CARBOXYLASE

Karen A. Butcher,[1] Francis K. Fong,[2] & John V. Schloss[3]

[1]Chemistry Department, Purdue University, West Lafayette, Indiana, [2]Carbon Reduction Laboratory, West Lafayette, Indiana, and [3]Central Research & Development Department, Du Pont Co., Wilmington, Delaware, USA.

The rates of conversion of 0.05 mM [5-^{3}H]ribulose 1,5-bisphosphate (RuBP) to its enediol to 2-carboxy-3-keto-D-arabinitol 1,5-bisphosphate (CKABP) to 3-phosphoglycerate (PGA) were determined at several bicarbonate concentrations (2, 10, and 100 mM), in the presence of 24 mg/ml (0.43 mM protomer) of the *Rhodospirillum rubrum* RuBP carboxylase, 5 mM $MgCl_2$, 50 mM bicine-NaOH, pH 8, 0.05 mM EDTA, 2.5 mM glutathione, and 0.025 mg/ml carbonic anhydrase. The enediol of RuBP was detected as the monophosphate elimination product obtained upon acid quenching the enzymic reaction mixture and reduction of the products with sodium borohydride. Similarly, CKABP was detected after acid quench and reduction as a mixture of 2- and 4-carboxyarabinitol 1,5-bisphosphate, and remaining RuBP was obtained as a mixture of ribitol and arabinitol bisphosphate. Kinetic results were modelled by adjusting the rates of successive irreversible steps (RuBP —> enediol —> CKABP —> PGA) until the predicted levels of substrate, intermediates, and product matched those obtained. The rates of RuBP consumption (5.5, 12, and 16 s^{-1}, at 2, 10, and 100 mM NaHCO$_3$, respectively) and conversion of enediol to CKABP (200, 310, and 540 s^{-1}, at 2, 10, and 100 mM NaHCO$_3$, respectively) exhibited saturation kinetics, with maximal rates of 17 and 550 s^{-1} and saturation constants for NaHCO$_3$ of 4.3 and 5 mM, respectively. These results are inconsistent with a simple Theorell-Chance type of kinetic mechanism, in which CO$_2$ adds directly in a bimolecular fashion with the enzyme-bound enediol. A slow step would be required after the formation of the enediol, prior to its reaction with CO$_2$, to adequately describe these data. This slow step need not involve a Michaelis complex with CO$_2$, and as such an extended Theorell-Chance mechanism (with rate limiting "enzyme isomerization" after the formation of the enediol) is still a plausible mechanism for the carboxylase.

Isotope Effects on Pyruvoyl and Pyridoxal 5'-Phosphate Dependent Histidine Decarboxylase.

Lynn M. Abell and Marion H. O'Leary

Pyridoxal 5'-phosphate (PLP) is the coenzyme utilized by a majority of amino acid decarboxylases. A smaller and relatively new class of amino acid decarboxylases lack PLP but contain instead a covalently bound pyruvate residue which serves as a prosthetic group. Although, the PLP cofactor is bound to the enzyme through a imine linkage and the carbonyl oxygen of the pyruvate is not, the two cofactors appear to operate by similar mechanisms: Both involve an imine linkage between the substrate and cofactor and the cofactor in each case serves as the electron sink in the decarboxylation step. However, the chemical details of these two mechanisms are somewhat different. The pyruvate dependent enzyme forms the imine linkage with the substrate directly from the carbonyl group while the PLP enzyme catalyzes a kinetically more favorable Schiff base interchange. Also the PLP enzyme utilizes the aromatic pyridine ring to neutralize the negative charge generated in the decarboxylation step whereas the pyruvoyl enzyme must use the amide carbonyl for this purpose, presumably forming a charged intermediate. We wished to study the relative catalytic advantages of these two cofactors. Histidine decarboxylase is ideally suited for such a study as it is the only decarboxylase known to date that exists in both a PLP and pyruvoyl form. These enzymes were studied using carbon, nitrogen, deuterium and solvent isotope effects. In both cases the energetics of the chemical steps differed by less than 1 KJ/mole with the decarboxylation step being slightly more rate limiting. This result indicates that despite the seemingly less sophisticated electron sink and the kinetic disadvantage of direct Schiff base formation the pyruvoyl moiety appears to be just as effective as PLP in catalyzing the decarboxylation of histidine. The nitrogen isotope effects also showed that in both cases the nitrogen of the Schiff base intermediate is protonated on the enzyme. The choice of PLP or pyruvate for catalysis may be based on more subtle considerations.

USE OF OXIDATION TO TRY TO CHARACTERIZE THE ATP BINDING SITES AND CONFORMATIONAL CHANGES IN CARBAMOYL PHOSPHATE SYNTHETASE (CPS).

Eulalia Alonso, Vicente Rubio and Javier Cervera. Instituto de Investigaciones Citologicas de la Caja de Ahorros de Valencia.

We showed [1] earlier that CPS from rat liver mitochondria is oxidatively inactivated in a chemical system (Fe-Oxygen-ascorbate) which mimics enzymatic mixed-function oxidation. Inactivation is markedly accelerated by the allosteric activator of the enzyme, acetylglutamate, and, in the presence of acetylglutamate, it is prevented by ATP. We proposed [1] that acetylglutamate exposes one of the two ATP sites in the enzyme and that ATP protects by binding to this site.

To try to locate this ATP-binding site we labelled with [^{3}H]borohydride the carbonyl groups generated in the enzyme by the oxidation. About one ^{3}H atom was incorporated per enzyme molecule after oxidative inactivation in the presence of acetylglutamate. However, essentially the same amount of labelling was observed in the absence of acetylglutamate, and the labelling was increased nearly 3-fold when ATP was present. Thus, the overall extent of labelling was unrelated to the degree of inactivation. Nevertheless, different amino acid residues may be oxidized in the presence or absence of acetylglutamate and/or ATP. This was supported partly by experiments in which the labelling of the different enzyme domains was analyzed by limited proteolysis and SDS-PAGE. Four domains of 40, 40, 60 and 20 kDa, in this order from the N-terminus, are demonstrated in this way. The N-terminal domain of 40 kDa was faintly labelled in all cases. In contrast, the 2nd and particularly the 3rd domains (40 and 60 kDa) were labelled more intensely in the presence than in the absence of ATP. Putative ATP-binding sequences have been proposed in these two domains, but our results are not consistent with the oxidation to carbonyl groups of residues in these domains that are involved in ATP binding. Our results support the triggering by ATP binding of conformational changes in these two domains.

Formation of carbonyl groups from basic amino acid residues is not reverted by SH-reagents, but about 50% of the oxidative inactivation in the presence of acetylglutamate was reverted by dithioerythritol. This suggests that oxidation of essential thiol groups is involved in the inactivation that is prevented by ATP, and that thiol group(s) are present at the ATP binding site(s) in the enzyme. (Supported by grant PB85-0198 of the CAICYT).

1. Alonso, E. and Rubio, V. (1987) Arch. Biochem. Biophys. 258, 342-350.

Methyltetrahydromethanopterin as an intermediate in methanogenesis from acetate in *Methanosarcina barkeri*

R. Fischer, R.K. Thauer, Laboratorium für Mikrobiologie, FB Biologie, Karl-von-Frisch-Str., 3550 Marburg.

Methanosarcina barkeri is a strict anaerobic archaebacterium able to ferment acetate to methane and carbondioxide ($\Delta G^{o'} = -36$ kJ/mol). Cell extracts (100 000 x g) of acetate grown cells catalyzed this reaction with specific activities of 50 $nmol \cdot min^{-1} \cdot mg$ $protein^{-1}$. In Sephadex G-25 treated extracts of *M. barkeri* methane formation from acetyl phosphate was dependent on coenzyme A, tetrahydromethanopterin (app. $K_M = 4$ µM), coenzyme M and 7-mercaptoheptanoylthreonine phosphate and independent of methanofuran and coenzyme F_{420}. When coenzyme M was omitted methyltetrahydromethanopterin accumulated. These results indicate that acetate is converted to methane via acetyl phosphate, acetyl-coenzyme A, methyltetrahydromethanopterin and methyl-coenzyme M as intermediates. The disproportionation of form- aldehyde to methane and carbondioxide was dependent on tetrahydro- methanopterin, methanofuran, coenzyme F_{420}, coenzyme M and 7-mer- captoheptanoylthreonine phosphate.

Lit.: R. Fischer, R.K. Thauer (1989) Methyltetrahydromethanopterin as an intermediate in methanogenesis from acetate in *Methanosarcina barkeri*. Arch Mikrobiol In press

RELATIONSHIP BETWEEN STRUCTURE AND SOLUTION PROPERTIES IN VITAMIN
B_{12} MODELS.

N.Bresciani-Pahor, L.Randaccio and E.Zangrando

Dipartimento di Scienze Chimiche, Universita' di Trieste, 34127
Trieste, Italy

A knowledge of the factors which affect the strength of the
Co-C bond is important to clarify the mechanism of the Co-C bond
homolysis which is an essential step in the enzymic processes
involving the B_{12} coenzyme. The corrin ring distortion, possibly
originated by protein conformational variations, are thought (1-
3) to be important in weakening the Co-C bond, by inducing variations in the cobalt coordination sphere. Due to the intrinsic
difficulty in obtaining suitable single crystals for X-ray analyses, the structural characterizations of cobalamins are few and
of low accurancy. Nevertheless, available data suggest that the
electronic nature of the R alkyl group should influence the Co-
N(axial) distance, while the bulk of R seems to influence the Co-
C bond length and determine unusually large values of the ideally
tetrahedral Co-C-X angle. Organocobalt complexes RCo(chel)L
(R=alkyl group and L=neutral ligand), with small equatorial
macrocycles, have been shown (3) to be very useful in providing a
foundation for understanding the structural aspects of the more
complex cobalamins.
    The study of models, containing the Co(III)(chel) equatorial
moiety (Figure 1), has furnished a significant amount of data,
both in solution and in the solid state. Available data show that
the Co-C bond lengths increase with the increasing bulk of R,
regardless the nature of the chel and L ligands. This trend is
mainly determinad by the steric interaction between the R and
chel ligands (steric cis-influence (3)), and follows the same
trend reported for the Co-C Bond Dissociation Energy (BDE), in
the series pyCo(saloph)R with R of different bulk (1). The large
value of the $Co-CH_2-X$ angle (125°) found in the coenzyme
structure (4), is another consequence of this steric interaction,
but it should not destabilize the Co-C bond, since angles up to
130° have been found in many stable $LCo(chel)CH_2X$ derivatives
(3,10), where X is a bulky group. Furthermore, the Co-C distances
in complexes with the same L-Co-R axial fragment are not significantly influenced by the nature of the (chel) ligand, while in
complexes, with the same (chel), the nature of the L ligand
appears to influence the Co-C bond lengths (Table 2). Halpern et
al. (1) have shown that the Co-C BDE in the series
$LCo(DH)_2CH(Me)Ph$ linearly decreases with the increasing bulk of L
(L= phosphine with different bulk), as well as with the
decreasing basicity of L when L= 4-substituted pyridines.
    Comparison of structural results in simple models and those
of cobalamins, suggests that values of length (and energy) for
Co-C bonds are similar for both and could respond to distortions
in a similar manner.
    It appears likely that the enzyme induced coenzyme Co-C
bond weakeninig is due to steric influences, involving an upward
distortion of the corrin ring which provokes the displacement of

the adenosyl group (1,3,6,7).

On the other hand, recently it has been hypothesized that the benzimidazole residue may also play a role in weakening the Co-C bond (5). Structural data for models indicate that the Co-N distances are strongly influenced by the σ-donor power of the trans R group and at least when L contains N- or O-donors, by the nature of the equatorial ligand. (Table 1). The following order of increasing Co-N distances for the same axial fragment L-Co-R may be derived: (DH)2 < (DO)(DOH)pn < saloph < corrin

Recent results (8,9) suggest that a significant amount of the Co-N lengthening in (DO)(DOH)pn complexes having planar L ligands, when compared with (DH)₂ analogues, is due to a different orientation of L with respect to the equatorial moiety. A similar relationship between orientation and axial distances was also observed in saloph derivatives (5). This observation suggests that a lengthening of the Co-N(axial) bond due to corrin distortions may derive also from changes in the orientation of the benzimidazole moiety. This overall weakening decreases the electron donation to cobalt, destabilizing Co(III) with respect to the Co(II) and hence, decreases the Co-C bond energy, as found for the series 4X-pyCo(DH)₂CH(Me)Ph (1) with X= NH₂, Me, H, CN. Therefore, if we assume that the corrin upwards distortion, which displaces the adenosyl group, may also induce (trough its side chains), a change of the benzimidazole orientation, a further weakening of the trans Co-C bond occurs. In this hypothesis, the cleavage of the Co-C bond will occur with less severe corrin distortions than those required for only a direct weakening of the Co-C bond.

This work was supported by a grant (to L.R.) from MPI, Rome, Italy.

1. J.Halpern, Science 227, 869 (1985)
2. B.P.Hay and R.G.Finke, J.Amer.Chem.Soc. 108, 4820 (1986) and references therein.
3. N.Bresciani-Pahor, M.Forcolin, L.G.Marzilli, L.Randaccio, M.F.Summers and P.J.Toscano, Coord.Chem.Rev. 63, 1 (1985) and references therein.
   L.Randaccio, N. Bresciani-Pahor, E.Zangrando and L.G.Marzilli, submitted to Chem. Soc. Rev.
4. J.P..Glusker in D.Dolphin (Ed.), "B12", vol.1, p.23, Wiley, New York.
5. L.G.Marzilli, M.F.Sumers, N.Bresciani-Pahor, E.Zangrando, J.P.Charland and L.Randaccio, J.Amer.Chem.Soc. 107, 6880 (1985).
6. N.Bresciani-Pahor, M.Calligaris, G.Nardin and L.Randaccio, JCS Dalton Trans. 2549 (1982)
7. S.M.Chemaly and J.M.Pratt, J.Chem.Soc. Dalton Trans. 2274 (1980) and references therein.
8. W.O.Parker, E.Zangrando, N.Bresciani-Pahor, L.Randaccio and L.G.Marzilli, Inorg.Chem. 25, 3489 (1986)
9. W.O.Parker, E.Zangrando, N.Bresciani-Pahor, P.A.Marzilli, L.Randaccio and L.G.Marzilli, Inorg.Chem, 27,2170 (1988)
10. Q.Chen, L.G.Marzilli, N.Bresciani-Pahor, L.Randaccio and E.Zangrando, Inorg.Chim.Acta, 144, 241 (1988)

TABLE 1. Co-N(axial) distances (Å) and α bending angles (°) in some LCo(chel)R complexes.

chel	R= CH₂CF₃	Me	Et	CH₂CMe₃	adam
		L=py			
(DH)₂	2.041(4)	2.068(3)	2.081(3)	2.081(4)	
	1.0	3.2	9.1	-5.2	
(DO)(DOH)pn	2.106(3)		2.121(3)		
	6.9		14.3		
saloph	2.124(9)		2.215(4)		
	17.		-25.0		

chel	CH₂CF₃	Me	Et	CH₂CMe₃	adam
		L=Me₃Bzm			
(DH)₂		2.060(3)			2.137(4)
		4.7			-6.1
(DO)(DOH)pn	2.060(3)	2.100(3)	2.105(3)		
	16.6	13.8	16.6		

TABLE 2. Influence of the equatortial ligand on the axial bond lengths in RCo(chel)L complexes.

RCo(chel)L	Co-R(Å)	Co-L(Å)	α(°)
MeCo(acacen)py	1.99(1)	2.16(1)	7.0
Me[Co(DO)(DOH)pn]py	2.003(3)	2.106(3)	6.9
MeCo(DH)₂py	1.998(5)	2.068(3)	6.3
MeCo(GH)₂py	2.005(4)	2.064(3)	5.6
EtCo(saloph)py	2.042(6)	2.215(4)	-25.0
EtCo(DH)₂(4R-py)	2.035(5)	2.081(3)	9.0
EtCo(GH)₂py	2.020(7)	2.067(6)	1.1
Et[Co(DO)(DOH)pn]NH₂Ph	2.030(4)	2.174(3)	-7.1
EtCo(DH)₂NH₂Ph	2.030(3)	2.147(2)	2.8

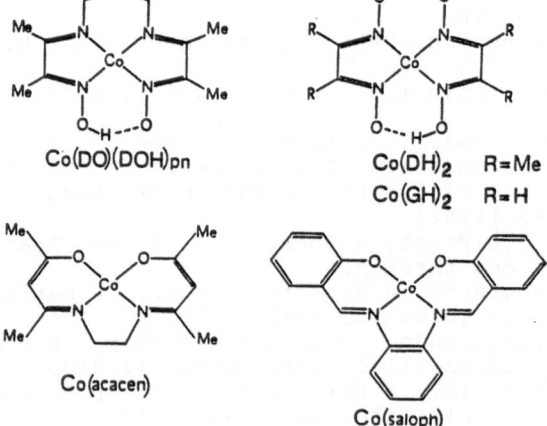

Co(DO)(DOH)pn

Co(DH)₂ R=Me
Co(GH)₂ R=H

Co(acacen)

Co(saloph)

Fig 1

Department of Chemistry

University of Toronto

80 St. George Street

Toronto, Canada M5S 1A1

416-978-8186

TAYLOR

SCOTT

MODELS FOR PHOSPHORYLATION OF BIOTIN-BICARBONATE ADDUCTS

Abstract

Carbon dioxide fixation through formation of an adduct between biotin and bicarbonate may

occur efficiently if that adduct can be trapped by reaction with ATP. The reactivity of such an

adduct to a phosphate center has been determined for a system in which the functional groups

are located within a single molecule. It is found that such a reaction occurs very readily

through a mechanism involving the addition of a carbonyl hydrate to a phosphate ester.

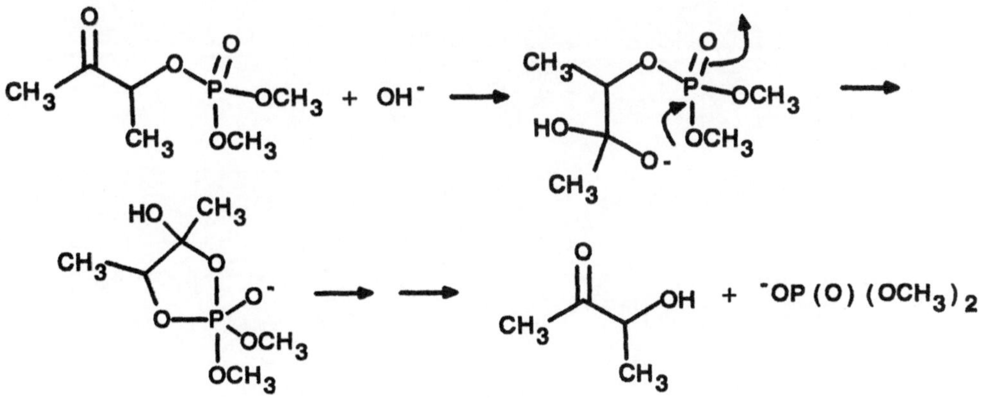

Department of Chemistry

University of Toronto

80 St. George Street

Toronto, Canada M5S 1A1

15 473 8185

MODEL FOR THE PHOSPHORYLATION OF BIOTIN BICARBONATE ADDUCTS

Permission to in situ in the poor formation of an adduct between biotin and bicarbonate may occur although it that adduct can be stopped by reaction with ATP. The reactivity of such an adduct to a phosphate center has been determined for a system in which the functional groups are located within a single molecule. It is found that such a reaction occurs very readily through a disfavoring involving the addition of a carbonyl hydrate to a phosphate ester.